T0189456

Lecture Notes in Computer Science 14332

The series Lecture Notes in Computer Science (LNCS), including its subseries Lecture Notes in Artificial Intelligence (LNAI) and Lecture Notes in Bioinformatics (LNBI), has established itself as a medium for the publication of new developments in computer science and information technology research, teaching, and education.

LNCS enjoys close cooperation with the computer science R & D community, the series counts many renowned academics among its volume editors and paper authors, and collaborates with prestigious societies. Its mission is to serve this international community by providing an invaluable service, mainly focused on the publication of conference and workshop proceedings and postproceedings. LNCS commenced publication in 1973.

Xiangyu Song · Ruyi Feng · Yunliang Chen ·
Jianxin Li · Geyong Min
Editors

Web and Big Data

7th International Joint Conference, APWeb-WAIM 2023
Wuhan, China, October 6–8, 2023
Proceedings, Part II

Editors
Xiangyu Song
Peng Cheng Laboratory
Shenzhen, China

Ruyi Feng
China University of Geosciences
Wuhan, China

Yunliang Chen
China University of Geosciences
Wuhan, China

Jianxin Li
Deakin University
Burwood, VIC, Australia

Geyong Min
University of Exeter
Exeter, UK

ISSN 0302-9743 ISSN 1611-3349 (electronic)
Lecture Notes in Computer Science
ISBN 978-981-97-2389-8 ISBN 978-981-97-2390-4 (eBook)
https://doi.org/10.1007/978-981-97-2390-4

This Springer imprint is published by the registered company Springer Nature Singapore Pte Ltd.
The registered company address is: 152 Beach Road, #21-01/04 Gateway East, Singapore 189721, Singapore

Preface

This volume (LNCS 14332) and its companion volumes (LNCS 14331, LNCS 14333, and LNCS 14334) contain the proceedings of the 7th Asia-Pacific Web (APWeb) and Web-Age Information Management (WAIM) Joint Conference on Web and Big Data, called APWeb-WAIM 2023. Researchers and practitioners from around the world came together at this leading international forum to share innovative ideas, original research findings, case study results, and experienced insights in the areas of the World Wide Web and big data, thus covering web technologies, database systems, information management, software engineering, knowledge graphs, recommend systems and big data.

The 7th APWeb-WAIM conference was held in Wuhan during 6–8 October 2023. As an Asia-Pacific flagship conference focusing on research, development, and applications in relation to Web information management, APWeb-WAIM builds on the successes of APWeb and WAIM. Previous APWeb conferences were held in Beijing (1998), Hong Kong (1999), Xi'an (2000), Changsha (2001), Xi'an (2003), Hangzhou (2004), Shanghai (2005), Harbin (2006), Huangshan (2007), Shenyang (2008), Suzhou (2009), Busan (2010), Beijing (2011), Kunming (2012), Sydney (2013), Changsha (2014), Guangzhou (2015), and Suzhou (2016); and WAIM was held in Shanghai (2000), Xi'an (2001), Beijing (2002), Chengdu (2003), Dalian (2004), Hangzhou (2005), Hong Kong (2006), Huangshan (2007), Zhangjiajie (2008), Suzhou (2009), Jiuzhaigou (2010), Wuhan (2011), Harbin (2012), Beidaihe (2013), Macau (2014), Qingdao (2015), and Nanchang (2016). The APWeb-WAIM conferences were held in Beijing (2017), Macau (2018), Chengdu (2019), Tianjin (2020), Guangzhou (2021), Nanjing (2022), and Wuhan (2023). With the ever-growing importance of appropriate methods in these data-rich times and the fast development of web-related technologies, APWeb-WAIM will become a flagship conference in this field.

The high-quality program documented in these proceedings would not have been possible without the authors who chose APWeb-WAIM for disseminating their findings. APWeb-WAIM 2023 received a total of 434 submissions and, after the double-blind review process (each paper received at least three review reports), the conference accepted 133 regular papers (including research and industry track) (acceptance rate 31.15%), and 6 demonstrations. The contributed papers address a wide range of topics, such as big data analytics, advanced database and web applications, data mining and applications, graph data and social networks, information extraction and retrieval, knowledge graphs, machine learning, recommender systems, security, privacy and trust, and spatial and multi-media data. The technical program also included keynotes by Jie Lu, Qing-Long Han, and Hai Jin. We are grateful to these distinguished scientists for their invaluable contributions to the conference program.

We would like to express our gratitude to all individuals, institutions, and sponsors that supported APWeb-WAIM2023. We are deeply thankful to the Program Committee members for lending their time and expertise to the conference. We also would like to acknowledge the support of the other members of the Organizing Committee. All of them helped to make APWeb-WAIM 2023 a success. We are grateful for the guidance of the honorary chair (Lizhe Wang), the steering committee representative (Yanchun Zhang) and the general co-chairs (Guoren Wang, Schahram Dustdar, Bruce Xuefeng Ling, and Hongyan Zhang) for their guidance and support. Thanks also go to the program committee chairs (Yunliang Chen, Jianxin Li, and Geyong Min), local co-chairs (Chengyu Hu, Tao Lu, and Jianga Shang), publicity co-chairs (Bohan Li, Chang Tang, and Xin Bi), proceedings co-chairs (David A. Yuen, Ruyi Feng, and Xiangyu Song), tutorial co-chairs (Ye Yuan and Rajiv Ranjan), CCF TCIS liaison (Xin Wang), CCF TCDB liaison (Yueguo Chen), Ph.D. consortium co-chairs (Pablo Casaseca, Xiaohui Huang, and Yanan Li), Web co-chairs (Wei Han, Huabing Zhou, and Wei Liu), and industry co-chairs (Jun Song, Wenjian Qin, and Tao Yu).

We hope you enjoyed the exciting program of APWeb-WAIM 2023 as documented in these proceedings.

October 2023

Yunliang Chen
Jianxin Li
Geyong Min
David A. Yuen
Ruyi Feng
Xiangyu Song

Organization

General Chairs

Guoren Wang	BIT, China
Schahram Dustdar	TU Wien, Austria
Bruce Xuefeng Ling	Stanford University, USA
Hongyan Zhang	China University of Geosciences, China

Program Committee Chairs

Yunliang Chen	China University of Geosciences, China
Jianxin Li	Deakin University, Australia
Geyong Min	University of Exeter, UK

Steering Committee Representative

Yanchun Zhang	Guangzhou University & Pengcheng Lab, China; Victoria University, Australia

Local Co-chairs

Chengyu Hu	China University of Geosciences, China
Tao Lu	Wuhan Institute of Technology, China
Jianga Shang	China University of Geosciences, China

Publicity Co-chairs

Bohan Li	Nanjing University of Aeronautics and Astronautics, China
Chang Tang	China University of Geosciences, China
Xin Bi	Northeastern University, China

Proceedings Co-chairs

David A. Yuen	Columbia University, USA
Ruyi Feng	China University of Geosciences, China
Xiangyu Song	Swinburne University of Technology, Australia

Tutorial Co-chairs

Ye Yuan	BIT, China
Rajiv Ranjan	Newcastle University, UK

CCF TCIS Liaison

Xin Wang	Tianjin University, China

CCF TCDB Liaison

Yueguo Chen	Renmin University of China, China

Ph.D. Consortium Co-chairs

Pablo Casaseca	University of Valladolid, Spain
Xiaohui Huang	China University of Geosciences, China
Yanan Li	Wuhan Institute of Technology, China

Web Co-chairs

Wei Han	China University of Geosciences, China
Huabing Zhou	Wuhan Institute of Technology, China
Wei Liu	Wuhan Institute of Technology, China

Industry Track Co-chairs

Jun Song	China University of Geosciences, China
Wenjian Qin	Shenzhen Institute of Advanced Technology CAS, China
Tao Yu	Tsinghua University, China

Program Committee Members

Alex Delis	University of Athens, Greece
Amr Ebaid	Google, USA
An Liu	Soochow University, China
Anko Fu	China University of Geosciences, China
Ao Long	China University of Geosciences, Wuhan, China
Aviv Segev	University of South Alabama, USA
Baoning Niu	Taiyuan University of Technology, China
Bin Zhao	Nanjing Normal University, China
Bo Tang	Southern University of Science and Technology, China
Bohan Li	Nanjing University of Aeronautics and Astronautics, China
Bolong Zheng	Huazhong University of Science and Technology, China
Cai Xu	Xidian University, China
Carson Leung	University of Manitoba, Canada
Chang Tang	China University of Geosciences, China
Chen Shaohao	China University of Geosciences, China
Cheqing Jin	East China Normal University, China
Chuanqi Tao	Nanjing University of Aeronautics and Astronautics, China
Dechang Pi	Nanjing University of Aeronautics and Astronautics, China
Dejun Teng	Shandong University, China
Derong Shen	Northeastern University, China
Dong Li	Liaoning University, China
Donghai Guan	Nanjing University of Aeronautics and Astronautics, China
Fang Wang	Hong Kong Polytechnic University, China
Feng Yaokai	Kyushu University, Japan
Giovanna Guerrini	University of Genoa, Italy
Guanfeng Liu	Macquarie University, Australia

Guoqiong Liao	Jiangxi University of Finance & Economics, China
Hailong Liu	Northwestern Polytechnical University, USA
Haipeng Dai	Nanjing University, China
Haiwei Pan	Harbin Engineering University, China
Haoran Xu	China University of Geosciences, Wuhan, China
Haozheng Ma	China University of Geosciences, Wuhan, China
Harry Kai-Ho Chan	University of Sheffield, UK
Hiroaki Ohshima	University of Hyogo, Japan
Hongzhi Wang	Harbin Institute of Technology, China
Hua Wang	Victoria University, Australia
Hui Li	Xidian University, China
Jiabao Li	China University of Geosciences, Wuhan, China
Jiajie Xu	Soochow University, China
Jiali Mao	East China Normal University, China
Jian Chen	South China University of Technology, China
Jian Yin	Sun Yat-sen University, China
Jianbin Qin	Shenzhen University, China
Jiannan Wang	Simon Fraser University, Canada
Jianqiu Xu	Nanjing University of Aeronautics and Astronautics, China
Jianxin Li	Deakin University, Australia
Jianzhong Qi	University of Melbourne, Australia
Jianzong Wang	Ping An Technology (Shenzhen) Co., Ltd., China
Jinguo You	Kunming University of Science and Technology, China
Jizhou Luo	Harbin Institute of Technology, China
Jun Gao	Peking University, China
Jun Wang	China University of Geosciences, Wuhan, China
Junhu Wang	Griffith University, Australia
K. Selçuk Candan	Arizona State University, USA
Krishna Reddy P.	IIIT Hyderabad, India
Ladjel Bellatreche	ISAE-ENSMA, France
Le Sun	Nanjing University of Information Science and Technology, China
Lei Duan	Sichuan University, China
Leong Hou U.	University of Macau, China
Li Jiajia	Shenyang Aerospace University, China
Liang Hong	Wuhan University, China
Lin Xiao	China University of Geosciences, Wuhan, China
Lin Yue	University of Newcastle, UK

Lisi Chen	University of Electronic Science and Technology of China, China
Lizhen Cui	Shandong University, China
Long Yuan	Nanjing University of Science and Technology, China
Lu Chen	Zhejiang University, China
Lu Qin	UTS, Australia
Luyi Bai	Northeastern University, China
Miaomiao Liu	Northeast Petroleum University, China
Min Jin	China University of Geosciences, Wuhan, China
Ming Zhong	Wuhan University, China
Mirco Nanni	CNR-ISTI Pisa, Italy
Mizuho Iwaihara	Waseda University, Japan
Nicolas Travers	Pôle Universitaire Léonard de Vinci, France
Peiquan Jin	University of Science and Technology of China, China
Peng Peng	Hunan University, China
Peng Wang	Fudan University, China
Philippe Fournier-Viger	Shenzhen University, China
Qiang Qu	SIAT, China
Qilong Han	Harbin Engineering University, China
Qing Xie	Wuhan University of Technology, China
Qiuyan Yan	China University of Mining and Technology, China
Qun Chen	Northwestern Polytechnical University, China
Rong-Hua Li	Beijing Institute of Technology, China
Rui Zhu	Shenyang Aerospace University, China
Runyu Fan	China University of Geosciences, China
Sanghyun Park	Yonsei University, South Korea
Sanjay Madria	Missouri University of Science & Technology, USA
Sara Comai	Politecnico di Milano, Italy
Shanshan Yao	Shanxi University, China
Shaofei Shen	University of Queensland, Australia
Shaoxu Song	Tsinghua University, China
Sheng Wang	China University of Geosciences, Wuhan, China
ShiJie Sun	Chang'an University, China
Shiyu Yang	Guangzhou University, China
Shuai Xu	Nanjing University of Aeronautics and Astronautics, China
Shuigeng Zhou	Fudan University, China
Tanzima Hashem	Bangladesh University of Engineering and Technology, Bangladesh

Tianrui Li	Southwest Jiaotong University, China
Tung Kieu	Aalborg University, Denmark
Vincent Oria	NJIT, USA
Wee Siong Ng	Institute for Infocomm Research, Singapore
Wei Chen	Hebei University of Environmental Engineering, China
Wei Han	China University of Geosciences, Wuhan, China
Wei Shen	Nankai University, China
Weiguo Zheng	Fudan University, China
Weiwei Sun	Fudan University, China
Wen Zhang	Wuhan University, China
Wolf-Tilo Balke	TU Braunschweig, Germany
Xiang Lian	Kent State University, USA
Xiang Zhao	National University of Defense Technology, China
Xiangfu Meng	Liaoning Technical University, China
Xiangguo Sun	Chinese University of Hong Kong, China
Xiangmin Zhou	RMIT University, Australia
Xiao Pan	Shijiazhuang Tiedao University, China
Xiao Zhang	Shandong University, China
Xiao Zheng	National University of Defense Technology, China
Xiaochun Yang	Northeastern University, China
Xiaofeng Ding	Huazhong University of Science and Technology, China
Xiaohan Zhang	China University of Geosciences, Wuhan, China
Xiaohui (Daniel) Tao	University of Southern Queensland, Australia
Xiaohui Huang	China University of Geosciences, Wuhan, China
Xiaowang Zhang	Tianjin University, China
Xie Xiaojun	Nanjing Agricultural University, China
Xin Bi	Northeastern University, China
Xin Cao	University of New South Wales, Australia
Xin Wang	Tianjin University, China
Xingquan Zhu	Florida Atlantic University, USA
Xinwei Jiang	China University of Geosciences, Wuhan, China
Xinya Lei	China University of Geosciences, Wuhan, China
Xinyu Zhang	China University of Geosciences, Wuhan, China
Xujian Zhao	Southwest University of Science and Technology, China
Xuyun Zhang	Macquarie University, Australia
Yajun Yang	Tianjin University, China
Yanfeng Zhang	Northeastern University, China

Yanghui Rao	Sun Yat-sen University, China
Yang-Sae Moon	Kangwon National University, South Korea
Yanhui Gu	Nanjing Normal University, China
Yanjun Zhang	University of Technology Sydney, Australia
Yaoshu Wang	Shenzhen University, China
Ye Yuan	China University of Geosciences, Wuhan, China
Yijie Wang	National University of Defense Technology, China
Yinghui Shao	China University of Geosciences, Wuhan, China
Yong Tang	South China Normal University, China
Yong Zhang	Tsinghua University, China
Yongpan Sheng	Southwest University, China
Yongqing Zhang	Chengdu University of Information Technology, China
Youwen Zhu	Nanjing University of Aeronautics and Astronautics, China
Yu Liu	Huazhong University of Science and Technology, China
Yuanbo Xu	Jilin University, China
Yue Lu	China University of Geosciences, Wuhan, China
Yuewei Wang	China University of Geosciences, Wuhan, China
Yunjun Gao	Zhejiang University, China
Yunliang Chen	China University of Geosciences, Wuhan, China
Yunpeng Chai	Renmin University of China, China
Yuwei Peng	Wuhan University, China
Yuxiang Zhang	Civil Aviation University of China, China
Zhaokang Wang	Nanjing University of Aeronautics and Astronautics, China
Zhaonian Zou	Harbin Institute of Technology, China
Zhenying He	Fudan University, China
Zhi Cai	Beijing University of Technology, China
Zhiwei Zhang	Beijing Institute of Technology, China
Zhixu Li	Soochow University, China
Ziqiang Yu	Yantai University, China
Zouhaier Brahmia	University of Sfax, Tunisia

Contents – Part II

Computing Maximal Likelihood Subset Repair for Inconsistent Data

Anzhen Zhang(✉), Shengji Hu, Chuanyu Zong, Jiajia Li, and Xiufeng Xia

Shengyang Aerospace University, Shenyang, China
{azzhang,zongcy,lijiajia,xiaxiufeng}@sau.edu.cn, hushengji@stu.sau.edu.cn

Abstract. In this paper, we study the problem of subset repair under integrity constraints. For an inconsistent data set, a subset repair removes a minimal set of tuples such that the integrity constraints are no longer violated in the remaining tuples. There usually exist multiple subset repairs and it is difficult to determine which one is optimal. Most previous work prefer the one with minimum number of deleted tuples to avoid excessive removal and information loss. However, it will delete clean tuples and retain dirty tuples when the majority of tuples are dirty in a local scope. We intuitively notice that under a proper model, the correctness probabilities of clean tuples are often larger than that of dirty tuples, and therefore we propose to determine the subset repair with maximum likelihood, which retain tuples with large correctness probability as many as possible. In this paper, we first formalize the maximum likelihood subset repair problem and analyze the hardness. Then we propose a correctness probability model, together with a scalable inference approach. Finally, an efficient approximate algorithm is proposed to compute the maximum likelihood subset repair. Extensive experiments on real-world datasets show that our proposal can achieve higher precision and recall compared with state-of-the-art methods.

Keywords: Data cleaning · Inconsistency error · Subset repair

1 Introduction

Data integrated from multiple sources may contain inconsistencies that violate integrity constraints. Data inconsistencies have a severe negative impact on the quality of downstream analytical results. To eliminate inconsistencies, subset repair has been proposed [1,2], which removes a minimal set of tuples such that the remaining tuples satisfy the integrity constraints. For an inconsistent database, there often exist multiple subset repair plans. Previous work usually assume the majority of input data to be clean and use the principle of *minimality* to perform repairs. As a result, the one with minimum change with respect to the original data is preferred. However, as we will show in the following example, the minimal repair do not necessarily correspond to the correct repair.

X. Song et al. (Eds.): APWeb-WAIM 2023, LNCS 14332, pp. 1–15, 2024.
https://doi.org/10.1007/978-981-97-2390-4_1

Example 1. Figure 1 shows a partial order table with the address information of customers (zip code ZIP, city CT, street STR, area code AC). The errors are marked in red and the ground-truth values are in parentheses. A functional dependency (FD) on this table is: ZIP $\rightarrow$ CT, which indicates that the zip code can uniquely determine the city. However, the ZIP values of t_1, t_2, t_3, t_4 and t_8 are the same (6037), but the CT values are different (Berlin for t_1 and t_2, Brighton for t_3, t_4 and t_8). To eliminate these violations, there exist two subset repair plans: one is to remove t_1 and t_2 and the other is to remove t_3, t_4 and t_8. According to the principle of minimality, the first plan will be chosen since it deletes less tuples. Unfortunately, the accuracy of this plan is far from satisfactory since the clean tuples (t_1 and t_2) are deleted and the dirty tuples (t_3, t_4 and t_8) are retained. In contrast, the second plan with more tuples deleted is the correct repair.

	ZIP	CT	STR	AC	p
t_1	6037	Berlin	Cindy	203	0.93
t_2	6037	Berlin	Cindy	203	0.93
t_3	6037	Brighton (Berlin)	Cindy	203	0.06
t_4	6037 (2135)	Brighton	Vinal	857	0.34
t_5	2135	Brighton	Vinal	857	1
t_6	2135	Brighton	Vinal	857	1
t_7	2135	Brighton	Vinal	857	1
t_8	6037 (2135)	Brighton	Vinal	857	0.34
t_9	2135	Brighton	Vinal	857	1

(a) Order information

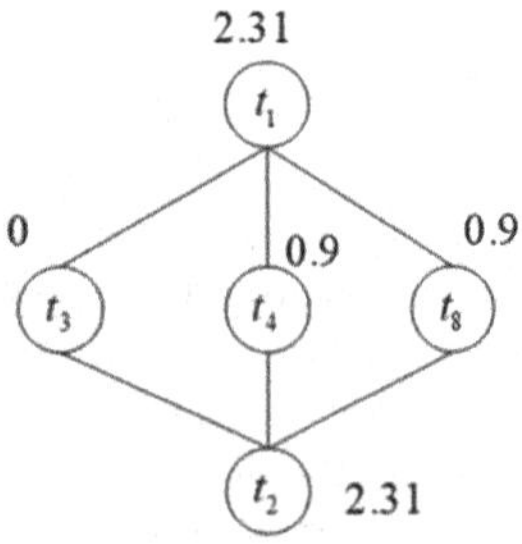

(b) Conflict graph

Fig. 1. A sampled order table and the corresponding conflict graph.

The reason for the failure of the minimal repair is that it assumes the majority of the tuples are clean and follows the principle of minority obeying majority to perform repair. While this assumption is true in a global scope since the error rate is usually small, it may fail in a local scope. For example, the majority of the tuples in the order table are clean but the majority of the tuples in the subset with ZIP = 6037 are dirty, which highlight the need for other signals, instead of the quantity of tuples, to perform subset repairs.

Motivated by this, we propose to determine the subset repair with maximum likelihood to take advantage of the quality of tuples. Informally, the more high quality tuples retained in a subset repair, the larger the likelihood is. The quality of a tuple is quantified by its correctness probability, which is the conditional probability of the noisy attributes values given the clean attributes values. For example, the correctness probabilities of tuples in the order table are shown in Fig. 1 (the detailed inference will be given in Sect. 3). Note that the correctness probabilities of clean tuples are generally larger than that of dirty tuples, and the subset repair with maximum likelihood is the one with all clean tuples retained. The contribution of this paper are as follows:

- We formalize the maximum likelihood subset repair problem under integrity constraints, which is to determine the subset repair with maximum likelihood, and explicitly analyze the hardness.
- We propose a correctness probability model to quantify the quality of tuples. Since the exact inference is shown to be NP-hard, we propose a scalable inference approach based on the Markov Blanket, which can significantly prune the search space without degrading the inference accuracy.
- We transform the maximum likelihood subset repair problem to the minimum weighted vertex cover problem, and then introduce an efficient approximate algorithm based on the linear programming.
- We present an extensive experimental evaluation to demonstrate the effectiveness and efficiency of our approach on real-world datasets.

The rest of the paper is as follows: Sect. 2 defines the problem of maximal likelihood subset repair. Section 3 presents the correctness probability model and the scalable inference approach. Section 4 presents the approximate algorithm to compute maximum likelihood subset repair. We demonstrate the validity of our approach and experimental results in terms of efficiency and scalability in Sect. 5. We conclude the paper in Sect. 6.

2 Problem Statement

In this section, we first introduce some basic terminologies and notations used in this paper and then formalize the maximum likelihood subset repair problem.

2.1 Function Dependency

Function dependency (FD) expresses relationships between attributes of a database relation and is commonly used in the subset repair problem. For convenience of discussion, we also take function dependency as the integrity constraint. It is worth mentioning that the approach proposed in this paper can be applied to other types of integrity constraints through some extensions.

Given a relation schema $\mathcal{R}$ with $\mathcal{A}$ denoting its set of attributes. The domain of an attribute $A \in \mathcal{A}$ is denoted by $dom(A)$. I is an instance over $\mathcal{R}$. A function dependency $\phi : X \rightarrow A$ is a statement over a set of attributes $X \subseteq \mathcal{A}$ and an attribute $A \in \mathcal{A}$ denoting that all tuples in X uniquely determine the values in A. More formally, let $t_i[A]$ be the value of tuple t_i in attribute A; the FD:$X \rightarrow A$ holds iff for all pairs of tuples $t_i, t_j \in I$ the following is true: if $t_i[B] = t_j[B]$ for all $B \in X$, then $t_i[A] = t_j[A]$. We call X the *left hand side* (LHS) of the FD, and A *right hand side* (RHS).

For a functional dependency $\phi : X \rightarrow A$ and a pair of tuples $t_i, t_j \in I$, we say t_i and t_j violates ϕ, denoted by $(t_i, t_j) \not\models \phi$, if $t_i[B] = t_j[B]$ for all $B \in X$, but $t_i[A] \neq t_j[A]$. Otherwise, we write $(t_i, t_j) \models \phi$ to denote t_i and t_j satisfy the constraint. For a relational instance I of R, I does not satisfy ϕ, denoted by $I \not\models \phi$, if there exists at least a pair of tuples $t_i, t_j \in I$ such that $(t_i, t_j) \not\models \phi$. Otherwise, I satisfy ϕ, denoted by $I \models \phi$.

Let Σ be a set of function dependencies. A tuple $t_i \in I$ is said to be *inconsistent* with respect to Σ, if there exists another tuple $t_j \in I$ such that t_i and t_j violate at least one FD in Σ. Otherwise, t_i is *consistent.* If all tuples in I are consistent with respect to Σ, then I is consistent with respect to Σ, denoted by $I \models \Sigma$. Otherwise, I is inconsistent, denoted by $I \not\models \Sigma$.

2.2 Subset Repair

Given an inconsistent database, we can remove some tuples such that the remaining tuples satisfy the constraints. The set of remaining tuples is called a consistent subset.

Definition 1 *(Consistent Subset). Given a set of function dependencies Σ over $\mathcal{R}$, and an instance I of $\mathcal{R}$, $J \subseteq I$ is a consistent subset w.r.t Σ if $J \models \Sigma$.*

Obviously, the subset of a consistent subset is also a consistent subset. In order to avoid excessive removal and information loss, the subset repair model (a.k.a, S-repair) only consider the maximal consistent subset that is not strictly contained in any other consistent subset.

Definition 2 *(Subset Repair). A consistent subset J of I is a subset repair of I w.r.t Σ if there does not exist any other consistent subset $J^{'}$ of I, such that $J \subset J^{'}$.*

Example 2. Continue Example 1. Since t_1, t_2 violate with t_3, t_4, t_8, We can remove either t_1, t_2 or t_3, t_4, t_8 to obtain a consistent subset. The corresponding subsets are $J_1 = \{t_3, t_4, t_5, t_6, t_7, t_8, t_9\}$ and $J_2 = \{t_1, t_2, t_5, t_6, t_7, t_9\}$. Note that both J_1 and J_2 are maximal consistent subsets, since putting any removed tuple back will introduce violations to the FD.

Note that there often exist multiple subset repairs for an inconsistent database, and we propose to determine the one with maximum correctness probability. Let $p(t_i)$ be the correctness probability of tuple $t_i \in I$, having $0 \leq p(t_i) \leq 1$. The correctness probability of a subset repair is defined as the product of the retained tuples' correctness probabilities times the product of the removed tuples' wrong probabilities, since the tuples in J are correct and only if the tuples not in J are wrong.

Definition 3 *(Subset Repair Likelihood). Given a subset repair J of I, the correctness probability of J is*

$$P(J) = \prod_{t_i \in J} p(t_i) \prod_{t_i \in I \setminus J} (1 - p(t_i)) \tag{1}$$

Then, the likelihood of a subset repair can be written (as log likelihood):

$$L(J) = \sum_{t_i \in J} \log p(t_i) + \sum_{t_i \in I \setminus J} (\log(1 - p(t_i))) \tag{2}$$

2.3 Problem Definition

The problem of computing maximum likelihood subset repair is formally defined as follows.

Definition 4 *(Maximum Likelihood Subset Repair). Given a relational instance I over schema R that violates a set of function dependencies Σ, the maximum likelihood subset repair problem is to find a subset repair J of I w.r.t. Σ such that $L(J)$ is maximized.*

Example 3. Continue Example 2. The correctness probabilities of tuples are shown in Fig. 1. Take J_1 as an example. The likelihood of J_1 is $L(J_1) = \log p(t_3) + \log p(t_4) + \log p(t_5) + \log p(t_6) + \log p(t_7) + \log p(t_8) + \log p(t_9) + \log(1 - p(t_1)) + \log(1 - p(t_2)) = -4.46$. Similarly, we can obtain $L(J_2) = -0.45$ and thus J_2 is the one with maximum likelihood.

According to the ground-truth shown in Fig. 1(a), the tuples retained in J_2 are all clean, and those deleted are all erroneous, which indicates J_2 is optimal in terms of repair accuracy. Nevertheless, computing maximum likelihood subset repairs is not easy. In fact, it is NP-complete, as shown in the following theorem.

Theorem 1. *The maximum likelihood subset repair problem is NP-complete.*

Proof. We show the hardness by considering the simple case where the probability of each tuple is equal, say 0.9. In this case, the maximum likelihood subset repair problem is equivalent to the subset repair with maximum size, which has been shown to be NP-complete [3,4].

Recognizing the hardness, we propose to first prune the search space and then compute the optimal subset repair approximately. Given an inconsistent instance I, let $I_e \subset I$ be the set of inconsistent tuples w.r.t. Σ, that is, $I_e = \{t_i \in I | \exists t_j \in I, i \neq j, (t_i, t_j) \not\models \Sigma\}$, and $I_c = I - I_e$ be the set of remaining consistent tuples. According the following proposition, we can safely retain all tuples in I_c and only need to search the optimal subset repair inside I_e.

Proposition 1. *Any optimal subset repair J^* with maximum likelihood always has $I_c \subset J^*$.*

Proof. The proof is straightforward. If there exists a tuple $t \in I_c$ and t is not in J^*, then J^* is not a subset repair, since we can add t in J^* without introducing any inconsistency.

Example 4. Continue Example 1. Since t_1, t_2, t_3, t_4, t_8 violate the FD, they are inconsistent tuples and the other tuples are consistent. That is, $I_e = \{t_1, t_2, t_3, t_4, t_8\}$ and $I_c = \{t_5, t_6, t_7, t_9\}$. According to Proposition 1, we only need to determine the maximum likelihood subset repair of I_e. There are two subset repairs in I_e, i.e., $J_3 = \{t_3, t_4, t_8\}$ and $J_4 = \{t_1, t_2\}$, and $L(J_3) = -4.46$, $L(J_4) = -0.45$. Therefore, the optimal subset repair $J^* = J_4 \cup I_c = \{t_1, t_2, t_5, t_6, t_7, t_9\}$.

3 Statistical Learning and Inference

Before introducing the approximate algorithm to compute the maximum likelihood subset repair, we first present the correctness probability model and the probabilistic inference approach.

3.1 Probability Modeling

For an inconsistent tuple $t \in I_e$ and a functional dependency $\phi : X \rightarrow A$ that t violates, there might be errors in the left hand side X or the right hand side A, and we cannot determine where the errors truly occur without further knowledge. Therefore, we add both the attributes in X and A in a *query* set Q, which represents the noisy attributes, on which errors may occur. If t violates multiple FDs, we add all the left hand side and right hand side attributes of the violated FDs in Q. We call the other attributes $E = \mathcal{A} - Q$ *evidence* attributes, which represents the clean attribute(Note that we only consider inconsistency errors in this paper). Therefore, an inconsistent tuple t can be divided into two parts: the noisy part $t\,[Q]$ and the clean part $t\,[E]$. Our objective is to use the clean attributes values to predict the probability for the noisy attribute for inconsistent tuples.

Suppose $t[E] = t[E_1, E_2, ..., E_L]$ and $t[Q] = t[Q_1, Q_2, ..., Q_K]$. Let $\mathcal{S}_E = dom(E_1) \times dom(E_2) ... \times dom(E_L)$ denotes the space of evidence attributes and $\mathcal{S}_Q = dom(Q_1) \times dom(Q_2) ... \times dom(Q_K)$ denotes the space of query attributes of t. Assuming that tuples generated randomly according to a probability distribution $P(Q, E)$ on $\mathcal{S}_Q \times \mathcal{S}_E$. We model the correctness probability of t as the conditional probability of the query attributes values given the evidence attributes values, i.e., $P(Q = t[Q] | E = t[E])$ (short for $P(t[Q]|t[E])$). According to the Bayesian theory, we have the following equation.

$$p(t) = P(t[Q]|t[E]) = \frac{P(t[Q], t[E])}{P(t[E])} \tag{3}$$

Example 5. Take the inconsistent tuple t_1 as an example. Since t_1 violate the FD:$\texttt{ZIP} \rightarrow \texttt{CT}$, we add both $\texttt{ZIP}$ and $\texttt{CT}$ in its query set, that is, $Q = \{\texttt{ZIP}, \texttt{CT}\}$, and the other attributes are put in the evidence set, i.e., $E = \{\texttt{AC}, \texttt{STR}\}$. Therefore, the correctness probability of t_1 is $P(\texttt{ZIP} = 6037, \texttt{CT} = \text{Berlin} | \texttt{STR} = \text{Cindy}, \texttt{AC} = 203)$, which can be computed by dividing $P(\texttt{STR} = \text{Cindy}, \texttt{AC} = 203)$ into $P(\texttt{ZIP} = 6037, \texttt{CT} = \text{Berlin}, \texttt{STR} = \text{Cindy}, \texttt{AC} = 203)$.

3.2 Scalable Inference

To infer the correctness probability, we employ a widely used probabilistic graphical model, Bayesian network, which provides a way to compactly describe the probability distribution of the attributes. According to the local Markov property in the Bayesian network, each variable is conditionally independent of its non-descendants given its parent variables. Therefore, the joint distribution of all variables in the network can be factorized as a product of the individual density functions, conditional on their parent variables. Given the Bayesian network on $\mathcal{A} = Q \cup E$, the numerator in Eq. 3, i.e., $P(t[Q], t[E])$, can be approximated by the following equation,

$$P(t[Q], t[E]) = \prod_{i=1}^{K} P(t[Q_i]|parent(Q_i)) \prod_{j=1}^{L} P(t[E_j]|parent(E_j)) \tag{4}$$

where $parent(Q_i)$ and $parent(E_j)$ denote the parent node set of Q_i and E_j respectively.Note that $P(t[Q_i]|parent(Q_i))$ and $P(t[E_i]|parent(E_i))$ are constants in the conditional probability table (CPT) in the Bayesian network since the values of all attributes are known.

When it comes to the denominator in Eq. 3, i.e., $P(t[E])$, the inference becomes a little complicated. Note that $P(t[E])$ is the marginal distribution of evidence attributes, which can be computed by marginalizing the joint distribution on query attributes, as shown in Eq. 5. Similar to our prior analysis, $P(q, t[E])$ can be determined efficiently according to the CPTs, for each q in the domain of $Q = \{Q_1, Q_2, ..., Q_K\}$. However, the number of q could be exponential if we allow the query attribute to obtain any value from the active domain. For example, suppose the domain size of each Q_i is d, then the space of all possible combination of Q is d^K.

$$P(t[E]) = \sum_{q \in \mathcal{S}_Q} P(q, t[E]) \tag{5}$$

We notice that the distribution of each query attribute Q_i is usually not independent of the other attributes. In fact, the distribution of Q_i depends on the joint distribution of the attributes in its Markov blanket. Therefore, we propose to prune the domain of Q_i based on the Markov blanket of Q_i. Specifically, let $MB(Q_i)$ denote the Markov blanket of Q_i and $MB(Q) = MB(Q_1) \cap MB(Q_2) ... \cap MB(Q_K)$ is the intersection of the Markov blankets of all query variables. Then we use the evidence attributes in $MB(Q)$ to prune the domain of Q, i.e., we only consider query attribute values that co-occur with the values of $E \cap MB(Q)$ at least once in the dataset. Since there are at most $|I|$ different combinations of query attributes in I, the inference cost reduce from d^K to $|I|$.

Example 6. Continue Example 5. The Bayesian network learned from the order table is shown in Fig. 2. $P(t_1[Q], t_1[E]) = P(\texttt{ZIP} = 6037|\texttt{CT} = \text{Berlin}) \cdot P(\texttt{CT} = \text{Berlin}|\texttt{AC} = 203) \cdot P(\texttt{STR} = \text{Cindy}|\texttt{CT} = \text{Berlin}, \texttt{ZIP} = 6037) \cdot P(\texttt{AC} = 203) = 1 \times 0.67 \times 1 \times 0.33 = 0.221$. Note that $MB(\texttt{ZIP}) = \{\texttt{CT}, \texttt{STR}\}$ and $MB(\texttt{CT}) = \{\texttt{AC}, \texttt{ZIP}, \texttt{STR}\}$, thus $MB(Q) = \{\texttt{STR}\}$. Given $\texttt{STR} = \text{Cindy}$, the combination of ZIP and CT could be {6037,Berlin} or {6037, Brighton}. Therefore, $P(t_1[E]) = P(\texttt{ZIP} = 6037, \texttt{CT} = \text{Berlin}, \texttt{STR} = \text{Cindy}, \texttt{AC} = 203) + P(\texttt{ZIP} = 6037, \texttt{CT} = \text{Brighton}, \texttt{STR} = \text{Cindy}, \texttt{AC} = 203) = 0.2211 + 0.0154 = 0.2365$. The correctness probability of t_1 is $P(t_1[Q], t_1[E])/P(t_1[E]) = 0.93$.

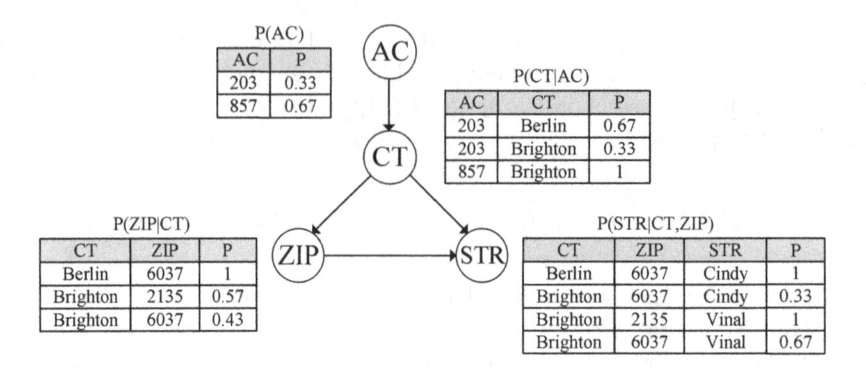

Fig. 2. The Bayesian network on the order table.

4 Subset Repair with Maximum Likelihood

In this section, we first transform the maximum likelihood subset repair problem to the minimum weighted vertex cover problem, and then introduce an efficient approximate algorithm based on the linear programming.

4.1 From Maximum Likelihood to Minimum Cost

For a subset repair J of I_e, the set of deleted tuples is called the removal set of J. Let $c(t_i)$ denote the cost of deleting $t_i \in I_e$, and the cost of J is defined as the cost of deleting all tuples in the removal set of J, that is,

$$C(J) = \sum_{t_i \in I_e \setminus J} c(t_i).$$

According to the following lemma, the subset repair with maximum likelihood is the one with minimum cost, where $c(t_i) = \log \frac{p(t_i)}{1-p(t_i)}$.

Lemma 1. *The likelihood of a subset repair J of I_e is maximum if*

$$J = argmin\{\sum_{t_i \in I_e \setminus J} \log \frac{p(t_i)}{1-p(t_i)} : J \in S_{I_e,\Sigma}\} \tag{6}$$

where $S_{I_e,\Sigma}$ is the set of all subset repairs of I.

Proof. Recall that the likelihood of J is defined in Eq. 2. First, we expand the second term in Eq. 2 and we have

$$L(J) = \sum_{t_i \in J} \log p(t_i) + \sum_{t_i \in I} (\log(1 - p(t_i))) - \sum_{t_i \in J} (\log(1 - p(t_i))) \tag{7}$$

Next, we combine the first term and third term, and $L(J)$ can be written as

$$L(J) = \sum_{t_i \in J} \log \frac{p(t_i)}{1-p(t_i)} + \sum_{t_i \in I} (\log(1 - p(t_i))) \tag{8}$$

Note that the second term $\sum_{t_i \in I}(\log(1 - p(t_i)))$ is identical for all J, and thus the subset repair with maximum likelihood is the one with maximum $\sum_{t_i \in J} \log \frac{p(t_i)}{1-p(t_i)}$, i.e., the one with minimum $\sum_{t_i \in I_e \setminus J} \log \frac{p(t_i)}{1-p(t_i)}$.

In the following, we will show that the subset repair with minimum cost can be obtained by computing the minimal vertex cover with minimum weight of the conflict graph of I_e. Let $G = (V, E, W)$ denote the conflict graph of I_e, where V is the vertex set, E is the edge set and W is the weight set. G can be constructed as follows. For each tuple t_i in I_e, we construct a vertex v_i in V. An edge between v_i and v_j will be constructed if t_i conflicts with t_j. Note that the conflict graph represents the violations in an intuitive manner.

Recall that a vertex cover is a subset of the vertexes such that every edge has at least one member in it as an endpoint. Therefore, if we remove the vertexes in a vertex cover of G, all edges will be removed (i.e., all violations are eliminated) and the tuples corresponding to the remaining vertexes will form a consistent subset of I_e. In addition, a minimal vertex cover is a vertex cover that is not a proper subset of any other vertex cover, which indicates that the corresponding consistent subset is maximal and it is thus a subset repair. Let the weight of v_i be $w_i = \log \frac{p(t_i)}{1-p(t_i)}$. Then the minimal vertex cover with minimum weight corresponds to the removal set of the subset repair with minimum cost.

4.2 Approximate Algorithm

Note that if the weight of each vertex is a non-negative number, then any existing solution for the minimum weighted vertex cover (MWVC) problem can be directly employed in our problem. However, when we dive into $w_i = \log \frac{p(t_i)}{1-p(t_i)}$, we find that w_i is negative if $p(t_i) < 0.5$. Therefore, we propose to scale the weight of each vertex to non-negative space. In particular, we add a constant $-\log \frac{\delta}{1-\delta}$ to w_i, $\delta = min\{p(t_i)|t_i \in I_e\}$, and the weights of all vertexes will be non-negative.

In the following, a classical approach based on linear programming is employed to determine the minimum weighted vertex cover. Specifically, given the conflict graph $G = (V, E, W)$ of I_e, define an identification variable x_i for each vertex $v_i \in V$ to indicate whether it is in a vertex cover $C \subseteq V$ or not, i.e., $x_i = 1$ if $v_i \in C$; otherwise $x_i = 0$. Then the minimum weighted vertex cover corresponds to an optimal solution of the following 0–1 integer linear programming:

$$\begin{aligned} & min \Sigma_{v_i \in C} w_i x_i \\ & s.t. \;\; x_i + x_j \geq 1, (v_i, v_j) \in E \\ & x_i = 0 \text{ or } 1, i = 1, 2, ..., n \end{aligned} \tag{9}$$

The optimization objective indicates that the weight of the optimal vertex cover is smallest, and the constraint means that at least one endpoint of each edge is in the vertex cover. Since integer linear programming is NP-hard, a typical approximation is to relax the integer constraints $x_i = 0$ or 1 to $0 \leq x_i \leq 1$, so that the optimal solution can be found in polynomial time. Let x_i^* be the optimal solution after relaxation for $i = 1, 2, ..., n$. Note that x_i^* may be a decimal number, thus we need to round it to 0 or 1 to get an approximate solution x_i^A. The rounding is as follows. $x_i^A = 1$ if $x_i^* \geq 1/2$; Otherwise, $x_i^A = 0$.

The maximum likelihood subset repair (MLSR) algorithm is shown in Algorithm 1. The input includes the inconsistent tuple set, the probability of each tuple and the probability threshold. The output is the approximate maximum likelihood subset repair J. The algorithm first constructs the conflict graph $G = (V, E, W)$ of I_e. The weight of each vertex is $w_i = \log \frac{p(t_i)}{1-p(t_i)} - \log \frac{\delta}{1-\delta}$ (line 1). An optimal solution x_i^* of Eq. 9 is computed by any existing linear programming solver (line 2). Then, we iterated over each optimal solution and rounded it to obtain the approximate solution x_i^A(lines 3–9). Finally, the tuples corresponding to $x_i^A = 0$ are added to J (line 10).

Algorithm 1: MLSR algorithm

Input : The inconsistent tuple set I_e; the probability of each tuple $p(t_i)$;the minimum probability $\delta = min\{p(t_i)|t_i \in I_e\}$

Output: The maximum likelihood subset repair J

1 Construct the conflict graph $G = (V, E, W)$ of I_e, where the weight of each vertex is $w_i = \log \frac{p(t_i)}{1-p(t_i)} - \log \frac{\delta}{1-\delta}$;
2 Compute the optimal solution of Eq. 9;
3 **for** $i \leftarrow 1$ **to** $|I_e|$ **do**
4   **if** $x_i^* \geq 1/2$ **then**
5         $x_i^A = 1$;
6   **else**
7         $x_i^A = 0$;
8   **end**
9 **end**
10 Add the tuple corresponding to $x_i^A = 0$ to J.

Proposition 2. *The MLSR algorithm returns an approximate maximum likelihood subset repair with time complexity $O(n^2)$, where $n = |I_e|$.*

Proof. The construction of the conflict graph takes $O(n^2)$ time when all tuples in I_e conflict with each other. Then, by calling the existing fast LP-solver, it takes $O(n)$ time to solve Eq. 9 and do the rounding. Therefore, the time complexity of MLSR is $O(n^2)$.

Example 7. Continue Example 1. We first compute the weight of each inconsistent tuple. Take t_1 for example. Note that $\delta = min\{p(t_i)|t_i \in I_e\} = 0.06$, thus the weigh of t_1 is $w_1 = \log \frac{0.93}{1-0.93} - \log \frac{0.06}{1-0.06} = 2.31$. The other weights can be computed in a similar way. The conflict graph is shown in Fig. 1(b). Obviously, the minimum weighted vertex cover is $\{t_3, t_4, t_8\}$ and thus $J = \{t_1, t_2\}$.

5 Experiments

In this section, we compare MLSR with the state-of-art methods in terms of data cleaning performance on a variety of real datasets with synthetic errors.

5.1 Experimental Steup

Datasets. We conduct experiments on three real-world datasets and one synthetic dataset. Table 1 shows statistics for these datasets. *Flights*[1] is a real-world dataset that contains information about flight departure and arrival times reported by different data sources on the web. *Hospital*[2] is a real-world dataset that has been used as a benchmark in several data cleaning papers.*Rayyan*[3] is a real-world dataset containing information about articles. *Tax*[4] is a synthetic dataset derived from the BART repository.

These datasets are assumed to be originally clean and we inject noises to the attributes related to functional dependencies. We introduce two types of noises: *typos* and *replacement errors*. Specifically, we randomly add a letter to an attribute value to construct a typo. For a replacement error, we replace a value with another value from the same domain.

Table 1. Information about evaluation datasets.

Dataset	Size	Error rate	Typo ratio	Error type	FD number
Flights	2376×6	0.05–0.3	0.2–1	R, T	4
Hospital	815×19	0.05–0.3	0.2–1	R, T	15
Rayyan	391×10	0.05–0.3	0.2–1	R, T	3
Tax	1021×15	0.05–0.3	0.2–1	R, T	4

Competing Methods. We compare MLSR with the following approaches. (1) **HoloClean** [5] is a state-of-the-art machine learning based data cleaning system. It combines existing qualitative data repair methods that rely on integrity constraints or external data with quantitative data repair methods that utilize statistical properties of input data. (2) **Raha** [6] is a configuration-free error detection system based on machine learning, which needs the existing error detection algorithms to generate sets of data errors and errors lebeled for training model. (3) **MS** [2] returns the minimum subset repair under the guidance of FD, following the minimum change principle.

[1] https://github.com/HoloClean/holo-clean/tree/master/testdata/flight.csv.
[2] https://github.com/HoloClean/holo-clean/tree/master/testdata/hospital.csv.
[3] https://github.com/minhptx/spade.
[4] https://db.unibas.it/projects/bart/.

Evaluation Methodology. We measure the effectiveness of the methods by using precision (P), recall (R), and F_1-score (F_1). F_1-score is defined as

$$F_1 = \frac{2 * Precision * Recall}{Precision + Recall} \quad (10)$$

where *precision* refers to correct deletion over the total deletions, and *recall* refers to correct deletion over the total dirty tuples. For data cleaning method which identify the specific error cells, e.g., Holoclean and Raha, we count the corresponding tuples as deleted tuples. All experiments were executed on a machine with Intel(R) Core(TM) i7-7700HQ CPU @ 2.80 GHz processor, 8 GB RAM, Windows 10 operating system.

5.2 Performance Evaluation

Error Rate Analysis. We study the performance of MLSR, Raha, HoloClean and MS by varying the error rate from 0.05 to 0.3, and report the F_1-score in Fig. 3. As shown in the figure, our proposed MLSR outperforms other data cleaning methods on all datasets except Rayyan. The reason is that there exist a lot of duplications in the Flight, Hospital and Tax datasets since they are integrated from multiple sources. Therefore, the injected errors are likely to incur inconsistency among duplicate tuples. However, this is not the case for the Rayyan dataset since it contains almost no duplication. As a result, both MLSR and MS fail to identify most of the errors in the Rayyan dataset, since they rely on the constrain rules. Nevertheless, MLSR still outperforms MS in the Rayyan dataset, which verifies our hypothesis that the quality of tuples leads to more accurate subset repairs than the quantity of tuples.

Error Type Analysis. We investigate the impact of different error types (replacement errors and typos) on the performance of MLSR, Raha, HoloClean and MS by varying the typos ratio from 0.2 to 1, and report the F_1-score. The corresponding experimental results are plotted in Fig. 4. The F_1-score of all methods perform stably under different proportions of error types. In addtion, our proposed MLSR outperforms the others in the Fights, Hospital and Tax datasets, which indicates that MLSR is robust to different types of errors. The reason for the low F1-Score on the Rayyan dataset is discussed above.

5.3 Runtime Evaluation

We now turn our attention to the runtime evaluation. Figure 5 shows the running time of MLSR with the other three methods on different datasets with different error rates and error type proportions. In general, MLSR is slower than the competing methods. Note that MS is most efficient in almost all cases, and it differs from MLSR in that MLSR needs to compute the correctness probability of inconsistent tuples. Therefore, the bottleneck of MLSR lies in the probabilistic inferences. Nevertheless, MLSR shows good repair performance and has good practicability to the cleaning of inconsistent data.

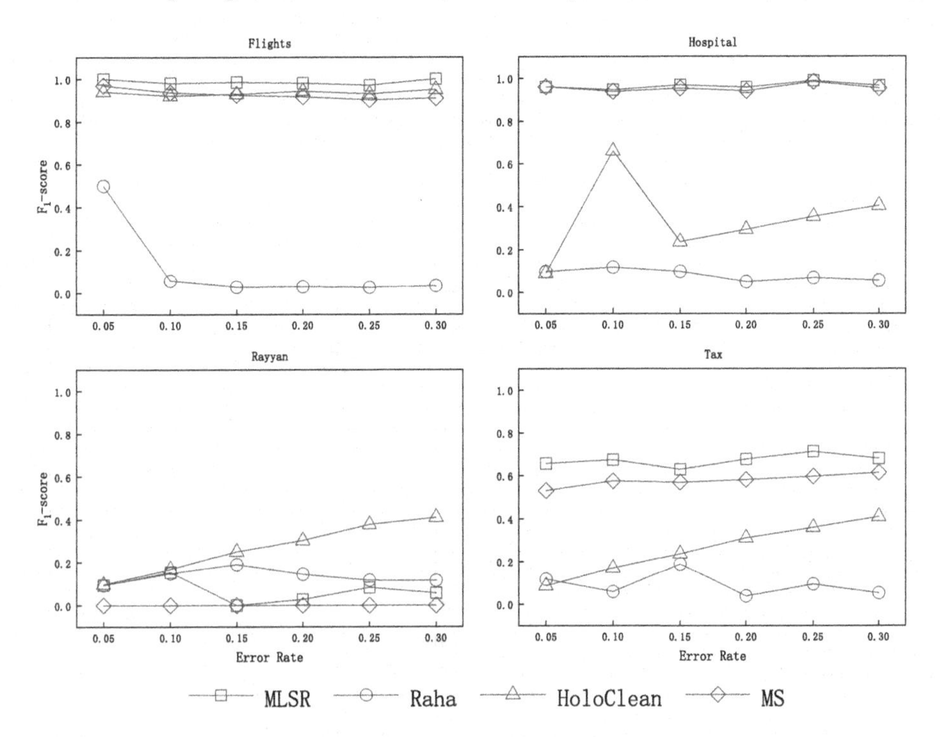

Fig. 3. F_1-score with different error rate.

6 Related Work

Existing inconsistency repair methods can be classified into two categories: subset repair and update repair. Subset repair makes the retained tuples consistent by deleting some tuples and update repair makes the modified tuples consistent by modifying the attribute values of some tuples.

Subset Repair. Existing approaches usually consider the subset repair with minimum number of deleted tuples as the optimal [1,2]. Arenas et al. [7] proved that computing the optimal subset repair is NP-complete even with only two FDs in the data. Livshits et al. [8] proved that the optimal subset repair problem is APX-complete in most cases. Miao et al. [3] show that it is usually NP-hard to approximate the optimal subset repairs within a factor better than 17/16, and an approximation algorithm with an approximation ration $(2 - 1/2^{|\Sigma|-1})$ is proposed.

Update Repair. Existing update repair works can be divided into two categories. One focuses on improving the approximation ratio of approximation algorithms to reduce the repair cost [4,8]. The other focuses on improving repair accuracy and proposes a variety of heuristic repair methods, for example, NADEEF [9], BigDansing [10], Temporal [11], Llunatic [12]. It's worth mentioning that in

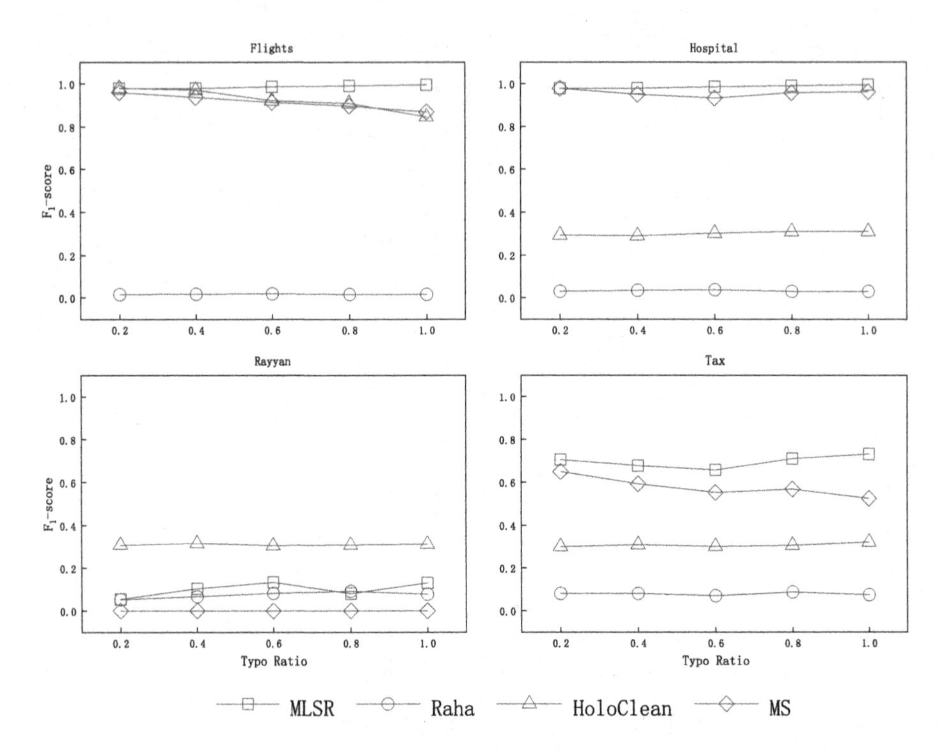

Fig. 4. F_1-score with different error type ratio.

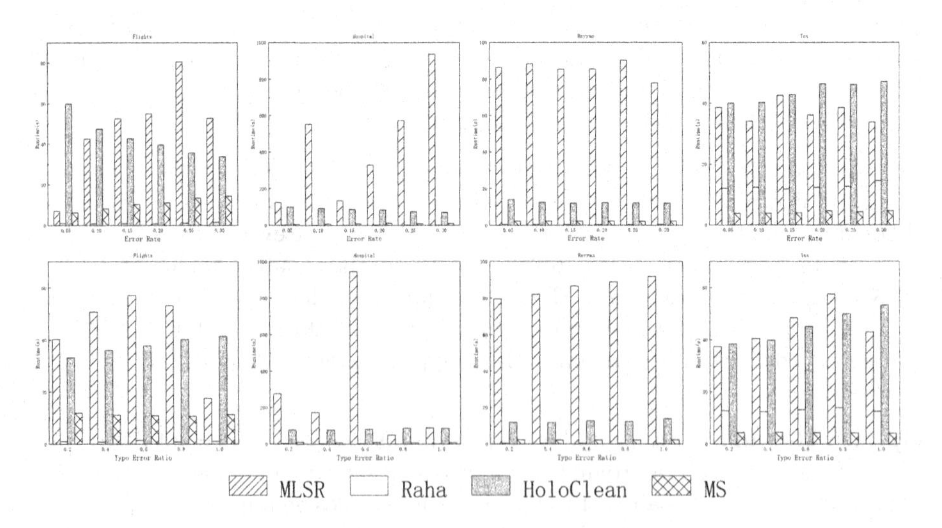

Fig. 5. Running time on datasets.

order to take advantage of integrity constraints and statistics, [5] proposed HoloClean and [6] proposed Raha; both showed good performance in repair accuracy.

7 Conclusions

In this paper, we propose to find the maximum likelihood subset repair for inconsistent data. We explicitly analyze the hardness of the problem and propose an approximate algorithm, MLSR, based on the correctness probability of inconsistent tuples. Extensive experiments over various datasets demonstrate that our proposal improves the data cleaning accuracy compared to the existing methods.

References

1. Afrati, F.N., Kolaitis, P.G.: Repair checking in inconsistent databases: algorithms and complexity. In: Proceedings of the 12th International Conference on Database Theory, pp. 31–41 (2009)
2. Chomicki, J., Marcinkowski, J.: Minimal-change integrity maintenance using tuple deletions. Inf. Comput. **197**(1–2), 90–121 (2005)
3. Miao, D., et al.: The computation of optimal subset repairs. Proc. VLDB Endow. **13**(12), 2061–2074 (2020)
4. Kolahi, S., Lakshmanan, L.V.S.: On approximating optimum repairs for functional dependency violations. In: Proceedings of the 12th International Conference on Database Theory, pp. 53–62 (2009)
5. Rekatsinas, T., et al.: HoloClean: holistic data repairs with probabilistic inference. arXiv preprint arXiv:1702.00820 (2017)
6. Mahdavi, M., et al.: Raha: A configuration-free error detection system. In: Proceedings of the 2019 International Conference on Management of Data, pp. 865–882 (2019)
7. Arenas, M., et al.: Scalar aggregation in inconsistent databases. Theor. Comput. Sci. **296**(3), 405–434 (2003)
8. Livshits, E., Kimelfeld, B., Roy, S.: Computing optimal repairs for functional dependencies. ACM Trans. Database Syst. (TODS) **45**(1), 1–46 (2020)
9. Dallachiesa, M., et al.: NADEEF: a commodity data cleaning system. In: Proceedings of the 2013 ACM SIGMOD International Conference on Management of Data, pp. 541–552 (2013)
10. Khayyat, Z., et al.: BigDansing: a system for big data cleansing. In: Proceedings of the 2015 ACM SIGMOD International Conference on Management of Data, pp. 1215–1230 (2015)
11. Abedjan, Z., et al.: Temporal rules discovery for web data cleaning. Proc. VLDB Endow. **9**(4), 336–347 (2015)
12. Geerts, F., et al.: The LLUNATIC data-cleaning framework. Proc. VLDB Endow. **6**(9), 625–636 (2013)

Design of Data Management System for Sustainable Development of Urban Agglomerations' Ecological Environment Based on Data Lake Architecture

Jiabao Li[1], Wei Han[1], Xiaohui Huang[1,2(✉)], Yuewei Wang[1], Ao Long[1], Rongrong Duan[1], Xiaohua Tian[1], and Yuqin Li[1]

[1] School of Computer Science, China University of Geosciences, Wuhan, China
xhhuang@cug.edu.cn

[2] The Hubei Key Laboratory of Intelligent Geo-Information Processing, Wuhan, China

Abstract. Research on the ecological environment of urban agglomerations plays a crucial role in enhancing environmental quality and ensuring sustainable development. In the research of sustainable development-oriented monitoring and assessing for ecological environments, the management and provision of data have gained significant prominence. However, the characteristics of vast data volumes, diverse data types, and inconsistent metadata descriptions hinder the comprehensive management of ecological environment data in urban agglomerations. This paper aims to investigate the data requirements that are necessary for sustainable development, with a particular focus on unified data management, online data product production and updates, and data service provision. To address these challenges, we have designed a metadata model that is capable of accommodating various types of datasets to facilitate their logical integration. Leveraging the data lake architecture, we have achieved semantic-level data governance driven by relational associations and proposed a data management solution for heterogeneous datasets for the unified management of terabyte-scale datasets. In addition, we have conducted the architectural design for the data management system. The prototype system is also developed to offer comprehensive data services, data product production, and other functionalities such as data visualization and analysis. This study provides extensive data services for monitoring and evaluating activities associated with the Sustainable Development Goals (SDGs) of SDG6, SDG11, and SDG15, while also supporting various application demonstrations and effectively facilitating the sustainable development of urban ecosystems.

Keywords: Data lake · Urban agglomerations · Ecological environment sustainable development · Data management · System design

X. Song et al. (Eds.): APWeb-WAIM 2023, LNCS 14332, pp. 16–27, 2024.
https://doi.org/10.1007/978-981-97-2390-4_2

1 Introduction

An urban agglomeration is an organizational form that emerges when cities reach a mature stage of development. It represents a vast and intricate cluster of cities that are characterized by multiple cores and hierarchical layers. Metropolises and large cities in an urban agglomeration are concentrated within a specific geographical area [8,26]. The rapid development of urban agglomerations poses numerous challenges, encompassing limited land availability, resource scarcity, congestion, and inadequate infrastructure. Moreover, it gives rise to a range of ecological and environmental issues, including the degradation of natural habitats, climate change, atmospheric pollution, and water contamination [19]. Undertaking research on the ecological environment of urban agglomerations contributes to the optimization of their functioning, enhancement of environmental quality, and facilitation of the formulation of measures for environmental protection [24,29]. Ultimately, the objective of this paper is to ensure the sustainable and harmonized development of the economy and the environment within urban agglomerations [28].

Currently, in the sphere of research concerning the sustainable development of ecological environments in urban agglomerations, there exists a broad and pressing need for data, particularly for monitoring, assessment, and associated studies [6,27]. As a result, the significance of managing ecological environment data for urban agglomerations is steadily growing. Therefore, it is imperative to expeditiously develop a data management system that satisfies diverse application requirements, aiming to enable the facilitation of various computational and evaluative applications to support the comprehensive provision of data services.

As the spatial realm encompasses intensive human activities and habitation, urban agglomerations' ecological environment encompasses diverse levels and scenarios [14,31]. It includes the vital water resource environment for human survival, the distribution and proportion of terrestrial elements within urban areas, the construction and planning of cities and human communities, as well as the state of protection and restoration of terrestrial ecosystems [9,15]. The intricate ecological environment of urban agglomerations involves data with distinctive characteristics, including varied sources, diverse types, inconsistent metadata descriptions, and substantial volumes, thereby presenting significant challenges to the management and sharing of urban agglomeration ecological environment data [32]. In addition, addressing the semantic-level data governance of such data and achieving the efficient production of data products are urgent issues that require immediate attention (Fig. 1).

This paper focuses on the data requirements within the realm of researching urban agglomeration ecological environments, exploring the domains of data aggregation and management, data product production, and data services. At first, in response to the diverse origins of urban agglomeration ecological environment data, a metadata model adaptable to various dataset types is designed. The proposed metadata model represents the attributes and characteristics of datasets, aiming to facilitate their integration at the logical level. Secondly, a solution is presented for the effective management of diverse ecological data per-

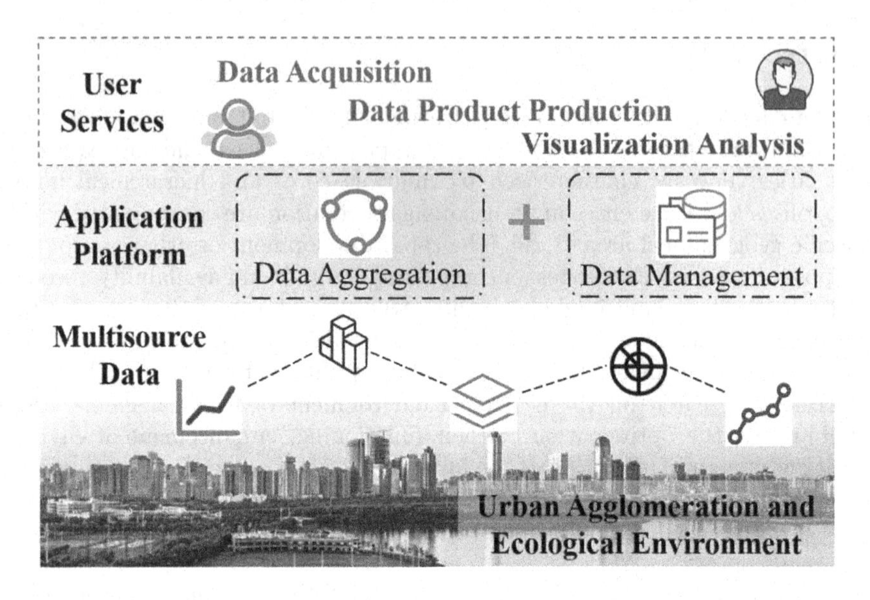

Fig. 1. Overview of the data service chain.

taining to urban agglomeration, originating from multiple sources. This solution leverages the data lake architecture to facilitate seamless integration and governance of relationship-driven datasets at the semantic level. As a result, a holistic approach is realized, enabling comprehensive management of datasets associated with urban agglomeration ecological environments across multiple dimensions, encompassing logical, semantic, and physical aspects. This approach effectively addresses the challenges posed by large-scale data volumes. Furthermore, we have developed a prototype system that offers data services, data production, and other functionalities such as data visualization analysis to support multiple research directions or domains. Apart from its data management and service capabilities, the system developed in this study also supports various application demonstration research, including ecological security risk identification, resilience assessment, and optimal allocation of resources. The proposed data management system effectively contributes to the sustainable development of urban agglomeration ecological environments.

The subsequent sections of this paper are organized as follows. Section 2 surveys the pertinent literature concerning the management of ecological environment data in urban agglomerations. Section 3 presents a comprehensive overview of the proposed system architecture design, which specifically caters to the management of ecological environment data in urban agglomerations. In Sect. 4, a meticulous exposition is provided, elucidating the implementation methods employed for the diverse principal modules integrated within the system. Section 5 delineates avenues for future research, while Sect. 6 concludes the paper by summarizing the key findings and implications.

2 Related Work

Urban agglomerations serve as crucial drivers for national economic development and play a vital role in fostering green growth and enhancing economic sustainability [25]. The effective coordination of ecological conditions, economic progress, and environmental preservation within urban agglomerations, alongside future planning and development strategies, holds paramount importance for achieving sustainable progress [7]. In the research work of monitoring and evaluating the ecological environment of urban agglomerations, relevant data plays an important role. In the big data era, the development of information technology enables people to obtain massive multi-source data with different formats [13,23]. The research on the ecological environment of urban agglomerations requires mining valuable knowledge from massive data, which can be used to evaluate the ecological environment status of urban agglomerations and provide a basis for planning and development or decision-making support [30]. The demand for data in various research fields has made the data management of the ecological environment in urban agglomerations more urgent and important [16].

Data management required for urban agglomerations' ecological environment research poses significant challenges in practice [5]. The types and sources of ecological environment data in urban agglomerations are diverse, including basic geographic data, remote sensing observation data, environmental monitoring and statistical data, social-economic data, biodiversity data, vegetation coverage data, water resource data, ecological environment status data, etc. [4]. Various data involved encompass raster, vector, statistical, and textual data, which exhibit variations in size and inconsistent attribute descriptions [17]. As a result, the endeavor to achieve unified data management entails intricate during the implementation process.

In terms of the ecological environment data management in urban agglomerations, there are currently two main data organization and storage methods: single node and distributed environment [1,3,22]. With the development of big data technology, data management solutions are gradually moving towards distributed architecture. Nevertheless, the existing distributed architecture-based solutions for organizing and storing urban agglomerations' ecological environment data still exhibit certain limitations. And the limitations of data management solutions are mainly reflected in the insufficient coverage of data types, which only target a certain or several types of data. Nevertheless, the theory and technology of a data lake, which is introduced in 2010, effectively tackles the challenges associated with storing and analyzing heterogeneous and multi-source big data [21]. A data lake is constructed upon a distributed infrastructure, which enables the archiving of data from diverse sources or of varying types in their original formats. Data lake facilitates data access, computation, and analysis for higher-level businesses or applications through metadata-driven data governance [12]. The current implementation solutions for data lakes predominantly rely on prominent big data ecosystems such as Hadoop and Spark [18]. Generally, a data lake solution comprises a hybrid storage architecture that integrates distributed file systems, relational databases, and NoSQL data management systems. Noteworthy open-source projects, such as Delta Lake, Apache Hudi, and Apache Iceberg, exhibit considerable activity in the implementation of data lakes [2,11,20]. These solutions facilitate the convergence of real-time and batch processing, adopt a table-based model for the logical organization of data within the lake, and employ Spark as the distributed processing engine. Besides, the above mentioned projects also provide users with diverse programming languages and SQL-like interfaces to facilitate interactive analysis [2,21]. However, the exploration of data lake technologies in the context of managing and processing data concerning urban agglomerations' ecological environments remains inadequate, necessitating further research and application efforts.

In the realm of research focused on ecological environments within urban agglomerations, there is a growing need for not just existing datasets, but also the production of data products [10]. The productive process involves real-time computational inference and the creation of data products using various raw data and algorithms. For multiple research scenarios on the sustainable development of urban agglomerations' ecological environment, it is imperative to establish a unified management for data and algorithms, while also providing a computational runtime environment. Completion of the above task facilitates efficient data processing and product production, ultimately establishing a data service platform for urban agglomerations' ecological environment, catering to the needs of users holistically. Nevertheless, existing system platforms suffer from significant deficiencies in supporting the data management of urban agglomerations' ecological environment, generating data products, and enabling data service sharing. These limitations predominantly stem from inadequate coverage of functional modules or incomplete incorporation of data types within centralized management systems. Consequently, this study endeavors to address the aforementioned challenges and explore the data requirements of business applications focused on urban agglomerations' ecological environment.

3 System Architecture Design

In response to the need for data management, production generation, and sharing services of urban agglomerations' ecological environment data, this research proposed a viable and transferable solution for managing such data. Specifically, we designed the data management system to support urban agglomerations' ecological environment sustainable development based on the data lake. The architecture of this system primarily consists of two components: a big data cloud platform built upon high-performance servers and other hardware infrastructure, and a data management system constructed on top of the cloud platform. Detailed descriptions of these components are provided below.

- Container Cloud Environment: The objective of the container cloud environment is to furnish infrastructure services for the successful implementation of the urban agglomerations ecological environment data management system.

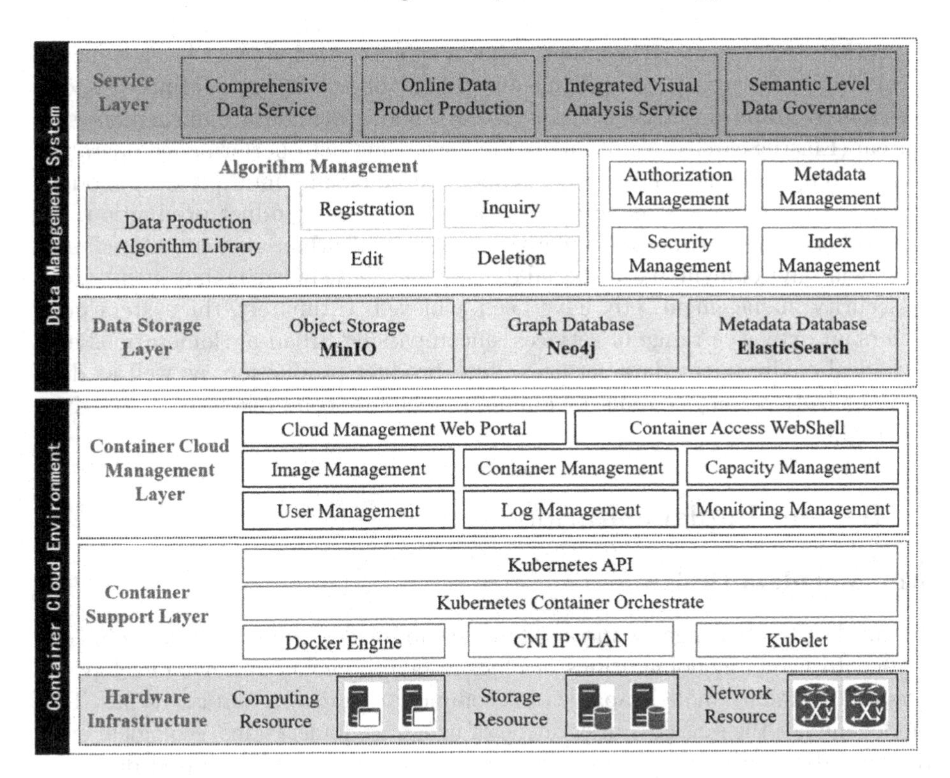

Fig. 2. The overview of the system architecture design.

The primary methodology entails the consolidation of computational, storage, and networking resources from server clusters and the dynamic allocation of virtualized resources in response to specific demand, thus enhancing overall resource utilization. This study employed container technology as a fundamental framework to construct the cloud environment, predominantly relying on the Docker engine and Kubernetes orchestration technology to virtualize diverse resources and deliver comprehensive management functionalities for the container cloud. Consequently, users and applications gain access to a highly adaptable and scalable cloud environment capable of accommodating evolving needs.

- Data Management System: The urban agglomerations' ecological environment data management system is meticulously developed on a robust container cloud platform, harnessing cutting-edge big data technologies to effectively attain continuous data acquisition capabilities, while ensuring the scalability, reliability, and efficiency of storage operations. And the technical proposal further boasts high-performance processing capabilities and embraces the versatility of intelligent data production and service functionalities, specifically tailored for urban agglomerations' ecological environment data. Primarily, taking into account the distinctive characteristics of urban agglomerations' ecological environment data, a sophisticated storage layer is meticulously constructed, comprising distributed object storage (MinIO), a distributed graph database system (Neo4j), and a metadata management system (ElasticSearch). Additionally, by harnessing high-performance operators and computational environments capable of managing multiple deep models, the system enables streamlined operations for data product production. On this basis, system functions like the ingestion of urban agglomerations' ecological environment data, metadata management, index management, data security management, etc. have been achieved. Ultimately, the system proffers an extensive range of services, encompassing urban agglomerations ecological environment data services, data product production, as well as data visualization analysis to users, aiming to facilitate comprehensive data-driven insights.

4 System Implementation

4.1 Metadata Design

To tackle the challenge of inconsistent attribute descriptions within datasets related to urban agglomerations' ecological environment, we have developed a versatile metadata model capable of accommodating diverse dataset types. This model enables the seamless integration of urban agglomerations' ecological environment data on a logical level. Moreover, it ensures a standardized depiction of the urban agglomerations' ecological environment data using the metadata model and facilitates index construction, data discovery, and access (Fig. 3).

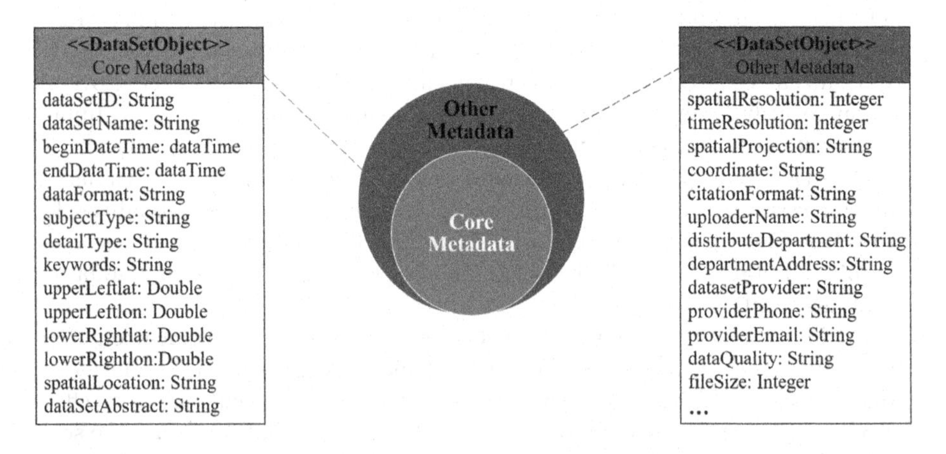

Fig. 3. Information of dataset's metadata model.

We proposed the design of a metadata model for urban agglomerations' ecological environment data set based on the ISO 19115 series of geospatial metadata standards. The metadata fields can be classified into core metadata and other metadata fields, with the core metadata representing key fields for data discovery. Figure 3 illustrates the details of the metadata model for urban agglomerations' ecological environment data.

4.2 Data Management

The data that underpin the sustainable development of the ecological environment in urban agglomerations exhibits characteristics such as multiple sources and diverse formats. To address the wide-ranging needs for data applications and sharing, we present a data management solution for urban agglomerations' ecological environment data based on the data lake architecture. Under the data lake architecture, data governance for urban agglomerations' ecological environment data can be achieved through metadata-based data governance using the proposed metadata model. Furthermore, building upon this foundation, graph databases can be employed to realize association-based data governance for urban agglomerations' ecological environment data.

Regarding the organization and storage of urban agglomerations' ecological environment data, the present study employed the ElasticSearch database system to effectively achieve metadata archiving while granting users access to data discovery based on the core metadata. For multi-source data with different volumes, the distributed object storage database system called MinIO is used to store the data, which can flexibly archive different mass data sets. Besides, we utilize the Neo4j graph database system to effectively manage the associations between data sets, as well as the connections between data sets and other entities. The management of data and its associated relationships can provide effective support for the evaluation and analysis of the sustainable development of the ecological environment in urban agglomerations.

4.3 Data Product Production

To meet the requirements of data production and real-time computing advocated by researchers across various domains, it is crucial for the system to seamlessly integrate multiple data production algorithms and offer computational processing support for urban agglomerations' ecological environment data products production. In this study, container technology was adopted to encapsulate various urban agglomeration ecological environment data processing and product production algorithms, which can achieve flexible integration and scheduling of multiple intelligent processing algorithms. Docker container virtualization offers the capability of allocating resources elastically while exhibiting relatively elevated deployment efficiency. Container encapsulation can effectively isolate algorithms, support flexible scheduling and execution, and have better portability. Based on the data support provided by underlying data management and encapsulation-driven algorithm management, algorithm scheduling, calculation processing, and production of data products can be flexibly carried out around various urban agglomerations' ecological environment data.

4.4 System Presentation

After designing a data management system architecture for the ecological environment of urban agglomerations, this study also developed a corresponding data management system. To validate the efficacy of the system in terms of data management and service sharing, data from the urban agglomerations located in the middle yangtze river was employed as a case study. The system exhibits comprehensive capabilities in managing ecological environment data within urban agglomerations, encompassing functions such as data uploading, querying, updating, and deletion. Moreover, the system can also provide functions such as user management, algorithm management, and visual analysis.

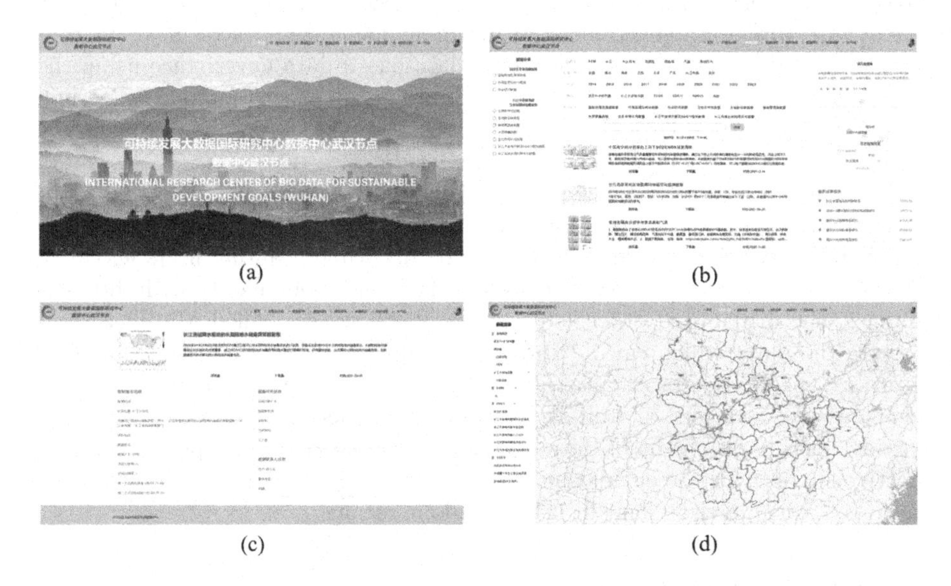

Fig. 4. Partial page display of the proposed system.

In the proposed system, we have gathered and managed data, including basic geographic and remote sensing data, water body data, land data, atmospheric data, and ecological monitoring data, with a volume of up to 6TB. By offering users comprehensive data discovery and data sharing services, the proposed system provides researchers investigating urban agglomerations' ecological environment with robust data support, contributing to the pursuit of sustainable development. The accompanying figure visually presents select interfaces from the developed system. This study presented a Web-based data management and application system developed using the B/S architecture. Figure 4 includes partial screenshots depicting key components of the system. Specifically, Fig. 4(a) serves as the homepage for the urban agglomerations' ecological environment data management system. Figure 4(b) showcases the data discovery page, where users can explore the complete set of managed urban agglomerations' ecological environment data and employ diverse methods to filter and retrieve specific datasets. Figure 4(c) provides a detailed presentation of particular datasets, which offers a comprehensive overview of the data's relevant information. At last, Fig. 4(d) presents the data visualization and analysis page, which allows users to obtain comprehensive visual representations of diverse data types.

5 Future Work

At the current stage, we have developed a comprehensive urban agglomerations ecological environment data management system, which effectively caters to users' needs for flexible and convenient data discovery and acquisition. Nevertheless, there are still shortcomings and further in-depth research is needed to improve the functionality of the system.

- According to the system architecture design based on data and algorithm management, we will continue to complete the data product production module and provide users with online data product production services. This functional module enriches the data sources and service forms of the system to meet the diverse data needs of users.
- Research on the ecological environment of urban agglomerations encompasses a wide range of areas or directions, involving diverse types of data. However, the current volume of integrated and managed data within the system is limited. Therefore, it is imperative to expand the number of urban agglomerations' ecological environment data within the system, aiming to facilitate more comprehensive data management and services. After the expansion of data volume, the system could provide better data services and support for relevant researchers.

6 Conclusion

This study is centered around the management and provision of services for ecological environment data in urban agglomerations. We present an architectural design for a data management system focused on the urban agglomerations' ecological environment data, aiming to address the diverse data requirements of researchers. This design scheme serves as a guiding framework at the system development level, facilitating the management and sharing of urban agglomeration and ecological environment data. Furthermore, a corresponding data management system has been developed, offering multiple services such as online data product production, data visualization analysis, and other functionalities, thereby providing essential data support for research on the sustainable development of urban agglomerations' ecological environment.

Acknowledgements. This work is supported by the International Research Center of Big Data for Sustainable Development Goals (CBAS2022GSP05) and the Open Research Project of the Hubei Key Laboratory of Intelligent Geo-Information Processing (KLIGIP-2022-B16).

References

1. Allam, Z., Dhunny, Z.A.: On big data, artificial intelligence and smart cities. Cities **89**, 80–91 (2019)
2. Armbrust, M., et al.: Delta lake: high-performance acid table storage over cloud object stores. Proc. VLDB Endow. **13**(12), 3411–3424 (2020)
3. Babar, M., Arif, F., Jan, M.A., Tan, Z., Khan, F.: Urban data management system: towards big data analytics for internet of things based smart urban environment using customized Hadoop. Futur. Gener. Comput. Syst. **96**, 398–409 (2019)
4. Barkham, R., Bokhari, S., Saiz, A.: Urban big data: city management and real estate markets. In: Artificial Intelligence, Machine Learning, and Optimization Tools for Smart Cities: Designing for Sustainability, pp. 177–209 (2022)
5. Chen, X., Li, F., Li, X., Hu, Y., Wang, Y.: Mapping ecological space quality changes for ecological management: a case study in the pearl river delta urban agglomeration, China. J. Environ. Manag. **267**, 110658 (2020)
6. Fang, C., et al.: Modeling regional sustainable development scenarios using the urbanization and eco-environment coupler: case study of Beijing-Tianjin-Hebei urban agglomeration, China. Sci. Total Environ. **689**, 820–830 (2019)
7. Fang, C., Liu, H., Wang, S.: The coupling curve between urbanization and the eco-environment: China's urban agglomeration as a case study. Ecol. Ind. **130**, 108107 (2021)
8. Fang, C., Yu, D.: Urban agglomeration: an evolving concept of an emerging phenomenon. Landsc. Urban Plan. **162**, 126–136 (2017)
9. Fang, C., Yu, X., Zhang, X., Fang, J., Liu, H.: Big data analysis on the spatial networks of urban agglomeration. Cities **102**, 102735 (2020)
10. Feng, R., Wang, F., Wang, K., Wang, H., Li, L.: Urban ecological land and natural-anthropogenic environment interactively drive surface urban heat island: an urban agglomeration-level study in china. Environ. Int. **157**, 106857 (2021)
11. Gebretsadkan Kidane, N.: Hudi on hops: incremental processing and fast data ingestion for hops (2019)
12. Giebler, C., Gröger, C., Hoos, E., Schwarz, H., Mitschang, B.: Leveraging the data lake: current state and challenges. In: Ordonez, C., Song, I.-Y., Anderst-Kotsis, G., Tjoa, A.M., Khalil, I. (eds.) DaWaK 2019. LNCS, vol. 11708, pp. 179–188. Springer, Cham (2019). https://doi.org/10.1007/978-3-030-27520-4_13
13. Glaeser, E.L., Kominers, S.D., Luca, M., Naik, N.: Big data and big cities: the promises and limitations of improved measures of urban life. Econ. Inq. **56**(1), 114–137 (2018)
14. He, J., Hu, S.: Ecological efficiency and its determining factors in an urban agglomeration in China: the Chengdu-Chongqing urban agglomeration. Urban Climate **41**, 101071 (2022)
15. He, X., Zhu, Y., Chang, P., Zhou, C.: Using tencent user location data to modify night-time light data for delineating urban agglomeration boundaries. Front. Environ. Sci. **10**, 860365 (2022)
16. Huang, H.J., Xia, T., Tian, Q., Liu, T.L., Wang, C., Li, D.: Transportation issues in developing China's urban agglomerations. Transp. Policy **85**, A1–A22 (2020)
17. Kharrazi, A., Qin, H., Zhang, Y.: Urban big data and sustainable development goals: challenges and opportunities. Sustainability **8**(12), 1293 (2016)
18. Khine, P.P., Wang, Z.S.: Data lake: a new ideology in big data era. In: ITM Web of Conferences, vol. 17, p. 03025. EDP Sciences (2018)

19. Liang, L., Wang, Z., Li, J.: The effect of urbanization on environmental pollution in rapidly developing urban agglomerations. J. Clean. Prod. **237**, 117649 (2019)
20. Mangold, N., et al.: Perseverance rover reveals an ancient delta-lake system and flood deposits at Jezero Crater, Mars. Science **374**(6568), 711–717 (2021)
21. Nargesian, F., Zhu, E., Miller, R.J., Pu, K.Q., Arocena, P.C.: Data lake management: challenges and opportunities. Proc. VLDB Endow. **12**(12), 1986–1989 (2019)
22. Nica, E., et al.: Urban big data analytics and sustainable governance networks in integrated smart city planning and management. Geopolit. Hist. Int. Relat. **13**(2), 93–106 (2021)
23. Priyashani, N., Kankanamge, N., Yigitcanlar, T.: Multisource open geospatial big data fusion: application of the method to demarcate urban agglomeration footprints. Land **12**(2), 407 (2023)
24. Surya, B., Salim, A., Hernita, H., Suriani, S., Menne, F., Rasyidi, E.S.: Land use change, urban agglomeration, and urban sprawl: a sustainable development perspective of Makassar city, Indonesia. Land **10**(6), 556 (2021)
25. Tang, P., Huang, J., Zhou, H., Fang, C., Zhan, Y., Huang, W.: Local and telecoupling coordination degree model of urbanization and the eco-environment based on RS and GIS: a case study in the Wuhan urban agglomeration. Sustain. Urban Areas **75**, 103405 (2021)
26. Tian, Y., Wang, R., Liu, L., Ren, Y.: A spatial effect study on financial agglomeration promoting the green development of urban agglomerations. Sustain. Urban Areas **70**, 102900 (2021)
27. Wang, Z., Liang, L., Sun, Z., Wang, X.: Spatiotemporal differentiation and the factors influencing urbanization and ecological environment synergistic effects within the Beijing-Tianjin-Hebei urban agglomeration. J. Environ. Manag. **243**, 227–239 (2019)
28. Yang, Y., Cheng, Y.: Evaluating the ability of transformed urban agglomerations to achieve sustainable development goal 6 from the perspective of the water planetary boundary: evidence from Guanzhong in China. J. Clean. Prod. **314**, 128038 (2021)
29. Yang, Y., Zhang, Y., Yang, H., Yang, F.: Horizontal ecological compensation as a tool for sustainable development of urban agglomerations: exploration of the realization mechanism of Guanzhong plain urban agglomeration in China. Environ. Sci. Policy **137**, 301–313 (2022)
30. Yu, D., Fang, C.: Urban remote sensing with spatial big data: a review and renewed perspective of urban studies in recent decades. Remote Sens. **15**(5), 1307 (2023)
31. Yuan, W., Li, J., Meng, L., Qin, X., Qi, X.: Measuring the area green efficiency and the influencing factors in urban agglomeration. J. Clean. Prod. **241**, 118092 (2019)
32. Zhang, S., Wei, H.: Identification of urban agglomeration spatial range based on social and remote-sensing data-for evaluating development level of urban agglomeration. ISPRS Int. J. Geo Inf. **11**(8), 456 (2022)

P-QALSH+: Exploiting Multiple Cores to Parallelize Query-Aware Locality-Sensitive Hashing on Big Data

Yikai Huang, Zezhao Hu, and Jianlin Feng(✉)

School of Computer Science and Engineering, Sun Yat-sen University, Guangzhou, China
fengjlin@mail.sysu.edu.cn

Abstract. Approximate nearest neighbor (ANN) search in high dimensional Euclidean space is a fundamental problem of big data processing. Locality-Sensitive Hashing (LSH) is a popular scheme to solve the ANN search problem. In the index phase, an LSH scheme needs to preprocess multiple hash tables, and in the query phase it exploits the preprocessed hash tables to speedup the ANN search. Query-Aware LSH (QALSH), a state-of-the-art LSH scheme, has rigorous theoretical guarantee on query accuracy, while suffering from high time overhead in the index and query phase. To improve the query efficiency, a multi-core parallel QALSH scheme called P-QALSH was proposed, which is mainly optimized for the query phase. In this paper, we further extend P-QALSH to P-QALSH+, which parallelizes QALSH in both the index and query phases based on multiple cores. Specifically, we first propose a Parallel Table Design to fully accelerate the index construction. Then, we follow P-QALSH to exploit a novel K-Counter Parallel Counting Technology and a novel Search Radius Estimation Strategy to improve the query performance. Using six real-world datasets and eight synthetic datasets, we have performed extensive experiments on a 16-core machine. Experimental results demonstrate the superiority of P-QALSH+ in terms of efficiency of parallel computing. Specifically, compared to QALSH, P-QALSH+ is 10-12X faster on index construction, and achieves 6-8X speedup on query search, and notably shows obvious improvement in query accuracy.

Keywords: Approximate Nearest Neighbor Search · Locality-Sensitive Hashing · Parallel Computing · Big Data Mining

1 Introduction

Nearest neighbor (NN) search in the high dimensional Euclidean space has wide applications in areas such as database, data mining and information retrieval [9]. Due to the curse of dimensionality [18], we usually spend higher search cost to find the exact NN in the high dimensional space. Hence, c-approximate NN search (c-ANN) is proposed, which improves the query speed by sacrificing some search accuracy. As one of the most important algorithms for c-ANN search problems, Locality-Sensitive Hashing (LSH) [11,13,20] displays effective search

X. Song et al. (Eds.): APWeb-WAIM 2023, LNCS 14332, pp. 28–43, 2024.
https://doi.org/10.1007/978-981-97-2390-4_3

performance and appealing success probability guarantee in the high dimensional Euclidean space. In 1998, Indyk first proposed the original LSH method [8] for the Hamming space. In 2004, the LSH scheme was extended for the Euclidean space, which was called E2LSH [2]. E2LSH requires a large number of hash tables to find c-ANN, leading to excessive storage space. Later, several LSH algorithms have been proposed, such as LSB-forest [17], C2LSH [5], SRS [16] and QALSH [6], which adopt more condensed data structures to reduce space consumption.

QALSH, a state-of-the-art LSH method, proposes a query-aware bucket partition strategy, which shows stable overall performances on various datasets and has rigorous theoretical guarantee. Recently, several new LSH variants such as VHP [15], R2LSH [14], PDA-LSH [19] and I-LSH [12] still follow the theoretical framework of QALSH for research. In the tutorials of SIGKDD'19 [1], ICDE'21 [4] and VLDB'21 [3], QALSH is consistently listed as the representative of LSH algorithms. Intuitively, the process of QALSH is divided into two parts: index phase and query phase. In the index phase, we construct multiple hash tables, which are called QALSH data structure and it is the basis of our entire algorithm. Then in the query phase, the QALSH data structure is used to find the c-ANN.

However, as a disk-based method, QALSH suffers from high latency of disk I/O, resulting in a slow query speed. Based on QALSH, P-QALSH [7] implements a multi-core parallel LSH, which is mainly optimized for the query tasks. However, P-QALSH mainly optimizes the query phase and does not sufficiently parallelize the process of construction of the QALSH data structure.

1.1 Our Contribution

In this article, we propose P-QALSH+, a more comprehensive parallel variant of QALSH algorithm. Specifically, we propose the Parallel Table Design which consists of the Inter-Table Parallel Design and Intra-Table Parallel Design to accelerate the index construction in the index phase. Finally, we follow P-QALSH to exploit a novel K-Counter Parallel Counting Technology and a novel Search Radius Estimation Strategy to improve the query efficiency in the query phase. On a 16-core machine, extensive experiments on six real-world datasets and eight synthetic datasets show that P-QALSH+ outperforms QALSH with 10–12X speedup on index construction and has 6–8X speedup over QALSH on query process, while achieving higher query accuracy.

The rest of this paper is organized as follows. Preliminaries are discussed in Sect. 2. Section 3 briefly presents the parallel designs on index construction. Section 4 introduces the detailed designs of parallel query process. Experimental studies are shown in Sect. 5. Finally, we conclude our work in Sect. 6.

2 Preliminaries

2.1 c-ANN Search Problem

Let D denote the dataset which contains n data objects in a d-dimensional Euclidean space R^d. $||o_1, o_2||$ is defined as the Euclidean distance between object

o_1 and o_2. Given the query object q in R^d and the approximation ratio $c > 1$, the c-ANN search problem is to construct a data structure that can find an object $o \in D$ such that $||o, q|| \leq c \cdot ||o^*, q||$, where o^* is the exact NN of q. Similarly, the c-k-ANN search is to find k objects $o_i \in D(1 \leq i \leq k)$ such that $||o_i, q|| \leq c \cdot ||o_i^*, q||$ holds, where o_i^* is the exact i^{th} nearest neighbor of q.

2.2 Framework of QALSH

QALSH [6] is proposed to answer c-ANN queries for Euclidean distance. QALSH adopts the following query-aware LSH function family based on the p-stable distribution [2]: $h_{\mathbf{a}}(o) = \mathbf{a} \cdot \mathbf{o}$. Here $\mathbf{o}$ is the vector representation of a data object, and $\mathbf{a}$ is a d-dimensional vector where each entry is drawn independently from the standard normal distribution $N(0, 1)$. Intuitively, the LSH function projects an object o onto a random line specified by $\mathbf{a}$ and the corresponding projected value (a real number) is regarded as the hash value.

To find a c-ANN, QALSH projects the whole dataset onto m random lines. When a query q arrives, QALSH first computes the projected values $(h_i(q), i \in \{1, ..., m\})$ of query, and then takes these projected values as the centers and performs a range search to traverse the objects falling in the range $[h_i(q) - \frac{w}{2}, h_i(q) + \frac{w}{2}]$. Each range is regarded as a hash bucket with a bucket width w. Then QALSH performs collision counting [5] for the objects in the m buckets, and an object with counting more than l is chosen as a candidate. After that, the search radius is enlarged by virtual rehashing technology [5] to search for more candidates. QALSH terminates the algorithm when more than β candidates are chosen or some candidate is close enough to the query in the original space. In practice, QALSH builds m hash tables in the index phase. For each hash table, QALSH projects all objects onto a random line firstly. The projected values of these objects and their id numbers form an initial hash table. Secondly, it sorts the hash table in ascending order. Then, QALSH exploits a B+-Tree to store the hash table and perform a query-centric range search to locate the corresponding query-centric bucket with width w in the query phase.

3 Parallel Table Design

3.1 Inter-Table Parallel Design

During the indexing phase, QALSH uses m query-aware LSH functions to create m hash tables. In serial execution, building the QALSH data structure requires m times of constructions on hash tables. To speed up this process, we propose an Inter-Table Parallel Design, which adopts k threads to execute the construction of k hash tables at the same time. In Fig. 1, suppose we have 4 threads, and 4 LSH functions are used to build the QALSH data structure. We let 4 threads perform the index construction on 4 random lines at the same time. For example, thread t_1 is responsible for the construction of the hash table on the random line defined by a_1. Similarly, other threads t_2, t_3, t_4 are also creating hash tables on their respective random lines.

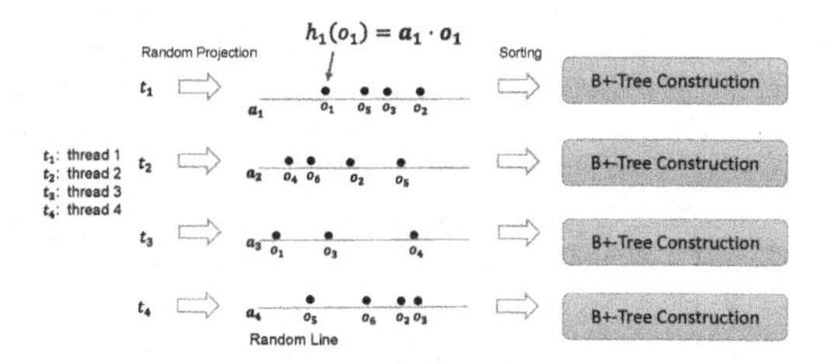

Fig. 1. Inter-Table Parallel Design of Index Construction

3.2 Intra-table Parallel Design

Parallelizing Random Projection. For each object, we perform the random projection operation on this object to get the projected value. Then we combine this projected value with its id number to get a hash pair. In QALSH, since we need to project all objects of the dataset onto a random line, we must perform n times of random projections. To accelerate this process, we propose parallel random projection technology by adopting k threads to randomly project k objects onto the random line simultaneously. As shown in Fig. 2, we have 4 objects o_1, o_2, o_3, o_4 in the dataset. We let 4 threads perform the random projections of the corresponding objects respectively. After projection, we get 4 hash pairs to form a hash table.

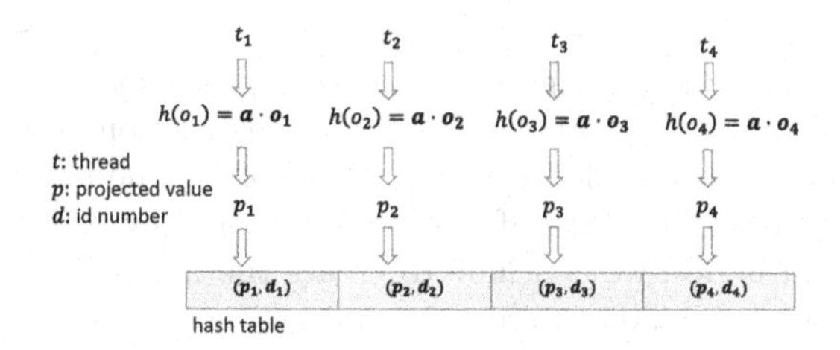

Fig. 2. Parallel Optimization For Random Projection

Parallelizing Sorting. After random projection, we create an origin hash table and sort it in ascending order. We propose a novel parallel technology to speedup the process of sorting. More precisely, we first divide the hash table into k equal parts, and each part is sorted independently by the corresponding thread. In Fig. 3, the hash table is created by random projection. Then it is divided into 4 parts, each part is locally sorted by the corresponding thread according to the projected values simultaneously.

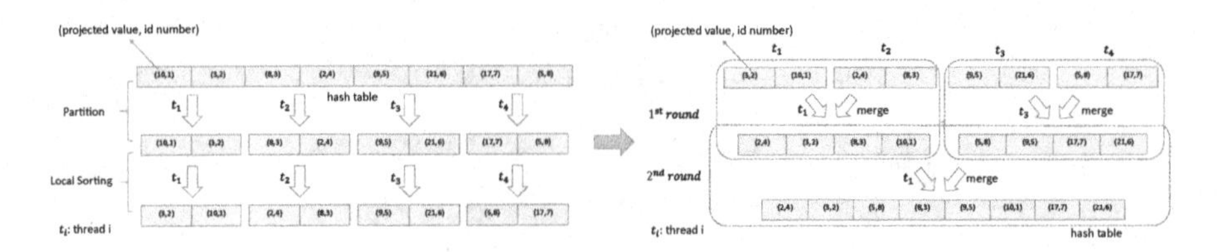

Fig. 3. Parallel Sorting of Hash Table

After the above steps, we get k locally ordered arrays. Next, we merge these k arrays into one sorted array. We propose a novel parallel merge-sort method. We divide the merge work into several rounds. In the first round, we divide the k arrays into $\frac{k}{2}$ groups, each of which is performed by a thread to execute the merge operation. When all threads finish the merge work, we get $\frac{k}{2}$ locally ordered arrays. Then we merge in a similar way until we finally merge into an ordered hash table. Our method subtly divides the merge tasks, so that the execution of multiple merge tasks can be parallelized, making full use of the multi-core resource.

Parallelizing B+-Tree Construction. After we finish the parallel sorting of the hash table, we store and organize this hash table with a B+-Tree. First, we make use of the sorted hash table to populate the leaf nodes of the B+-Tree. Then, a loop is used to create the non-leaf nodes in B+-Tree. Finally, we create a root node to index the non-leaf nodes. In practice, for QALSH, the load of serial B+-Tree construction is high on single core processor.

We propose a parallel design for B+-Tree construction. First, we propose a parallel technology to parallelize the creation of leaf nodes. More precisely, we exploit k threads to build k leaf nodes simultaneously. When we finish the construction of leaf nodes, we continue to create the non-leaf nodes in parallel. Similarly, we adopt k threads to create k non-leaf nodes simultaneously. After we construct these nodes, we write them to one B+-Tree file on disk. In order to parallel more effectively, we introduce two write strategies: buffer strategy and fusion strategy. For buffer strategy, we write all the nodes to the B+Tree file on disk after each thread has finished writing correspond nodes into memory. For fusion strategy, each thread will write their nodes into a local file on disk. When all the threads finish all write operations, we merge all of the local files into a whole B+-Tree file.

4 Parallel Query Design

4.1 Overview of Parallel Query

When a query q arrives, we execute the binary search to locate the projected positions of q on m random lines in parallel, which only executes $\lceil \frac{m}{k} \rceil$ search with a $O(\lceil \frac{m}{k} \rceil logn)$ time complexity, reducing the cost of locating process.

After we locate the m projected positions of q, we take these projected positions as the centers and perform m range queries in a round-robin manner, i.e., one query-centric range search on a random line. We propose the parallel range query technology. More precisely, we make k threads perform range search on the k random lines simultaneously. For each round, we only execute $\lceil \frac{m}{k} \rceil$ range search. In this process, we adopt parallel collision counting technology to select candidate objects. We choose the objects with counting more than l as candidates, and terminate the algorithm until enough candidates are found or one of the candidates is close enough to be a c-ANN.

4.2 Parallel Collision Counting Technology

Update Conflict Problem. QALSH adopts a global array of n counters for collision counting, i.e., one counter for recording the collision number of one data object. In parallel QALSH, if we still use a single array of n counters, and use a thread to check a random line at a time, then if multiple threads try to update the collision number of the same data object simultaneously, update conflict problem can happen, as illustrated in Fig. 4. Four threads $t_1 - t_4$ execute the range query on four random lines. When threads t_1, t_3 and t_4 search o_3 in their respective buckets, they will update the counting of o_3 at the same time, which may result in the update conflict. If the counting of objects is updated incorrectly, our algorithm can not find enough candidate objects, thus affecting the query accuracy. In serious cases, the program will not stop and even crash.

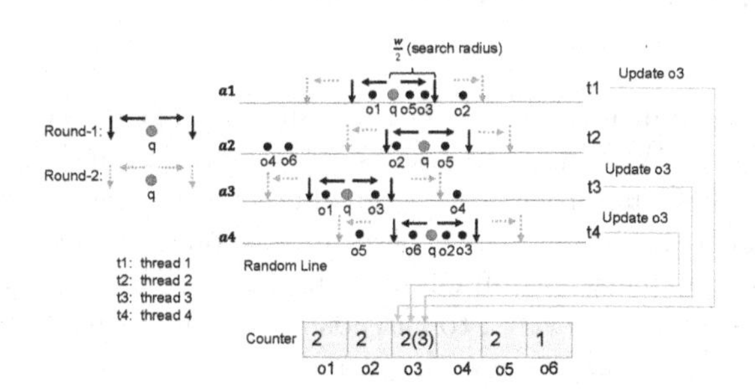

Fig. 4. The Conflict Problem of Parallel Collision Counting

K-Counter Parallel Counting. To avoid the update conflict problem, we allocate each thread a local array of n counters. If we use k threads for parallel collision counting, we distribute counting the whole collision number of each data object to k threads: each thread first collects the partial collision number of a object from $\lceil \frac{m}{k} \rceil$ random lines and updates the local counter array. After that, all the k local counter arrays are summarized to a global array. This method trades memory space ($k * n$ counters) for parallelism, and is called K-Counter Parallel Counting. Specifically, the collision counting process is divided into two parts: counting step and the merging step. In the counting step, we let k threads execute the range query on k random lines simultaneously and each thread t_i updates the collided object o_j to its corresponding counter (i, j) in the K-Counter Arrays (the jth counter in the ith local counter array). Besides, we introduce a collision recorder to record whether each object collides with the query, which is used to rationally allocate the thread workload and reduce the merge overhead during the merging step. In the merging step, we traverse the collision recorder once and then evenly distribute the id of collided objects to each thread. According to the allocated id, each thread merges all the countings of corresponding objects to the global counter.

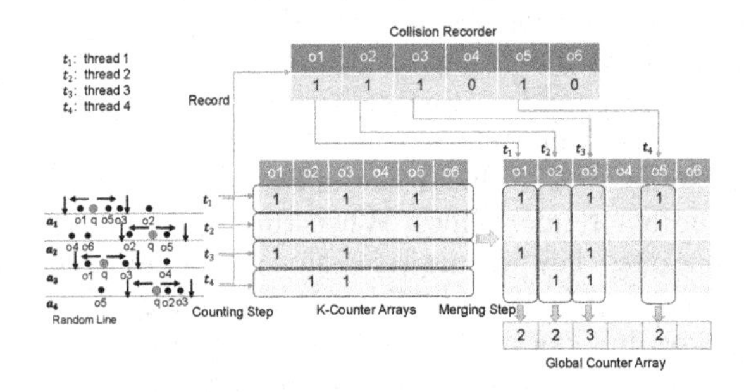

Fig. 5. The Process of K-Counter Parallel Counting

In Fig. 5, we let four threads $t_1 - t_4$ perform the range query on four random lines simultaneously. We can see that thread t_3 searches o_1 and updates the local counter $(3, 1)$ in K-Counter Arrays. Meanwhile, t_3 sets the 1st position of the collision recorder to 1. Other threads perform the collision counting in the same way. After the range query is completed, we find that o_1, o_2, o_3 and o_5 all collide with the query and allocate them to threads $t_1 - t_4$ in the merging step and let $t_1 - t_4$ execute their respective merging work.

4.3 Search Radius Estimation Strategy

To search the c-ANN, QALSH starts with small search radius ($R = 1$), and then expands the search radius in the following order: $R = \{1, c, c^2...\}$, which usually takes more rounds to find enough candidate objects. Since we traverse the collision recorder once and merge the records to the global counter in each round, multiple rounds of range query will cause high overhead for parallel query. Inspired by roLSH [10], we introduce a Search Radius Estimation Strategy to remedy this drawback. Notebly, QALSH enlarges the search radius in order $\{1, c, c^2, ..., T_{act}\}$, and finds enough candidate objects with terminated search radius T_{act}. Hence, if we use T_{act} as the starting search radius instead of 1, we will find enough candidates in just one round. To achieve this goal, we trained a model which predicts the terminated search radius T_{act} for a given query.

We exploit the Multilayer Perceptron (MLP) neural network to predict the T_{act}. Firstly, let $H(q) = [h_1(q) \dots h_m(q)]$ denote the feature representation of query q, where $h_i(q)$ is the projected value of the ith hash function. Besides, we define X_{tr} as the set of training queries that contains the feature representations of query objects. Meanwhile, we let Y_{act} denote the label set, each of which is the terminated search radius value T_{act} of the training query of X_{tr} to find enough c-ANNs. We take X_{tr} and Y_{act} as the input of the MLP for training. Finally, we use the trained neural network to predicted the terminated radius T_{pred} of a new query. Then, we take T_{pred} as the starting radius of range query. If in the first round we do not find enough candidates, we enlarge the T_{pred} linearly by $T_{pred} \cdot \epsilon$ (ϵ is user-speficied) to search enough candidates to terminate the algorithm. The other case is that the predicted terminated radius T_{pred} is larger than the actual one T_{act}, and we may search more objects and find more candidates compared with the case using T_{act} as search radius, which causes more disk I/Os. However, experiments results show that T_{pred} and T_{act} are close, and the overhead of extra disk I/O is still much less than multi-round query overhead.

Furthermore, we modify the termination condition of QALSH. When β candidate objects are found (case 1), or one of the candidate objects is close enough to be a c-ANN (case 2), QALSH will stop the query process. P-QALSH terminates the algorithm when one of the candidate results is close enough, which is the same as case 2. However, if β candidate objects are found, P-QALSH does not terminate the range query until all the subsequent random lines are searched in the current round. With this way, the candidate set of P-QALSH still includes the candidate set of QALSH. In fact, with Search Radius Estimation Strategy and the modification of termination condition, P-QALSH checks more nearby objects which are also likely to be the candidates than QALSH, resulting that the query accuracy can be further improved.

5 Experiments

5.1 Experiment Setup

All methods are implemented in C++ and are compiled with gcc 7.5. All experiments are carried on a machine with following configuration: Intel(R) Xeon(R) Gold 6130 sixteen-core CPU, 188 GB RAM, Ubuntu 18.04 operating system. Our parallel LSH methods are implemented by Pthread package. We use at most 16 CPU cores on our experiments and each thread is executed by a CPU core.

Datasets and Queries

- **Sun.** This 512-dimensional dataset has 79,106 objects. [1]
- **Nuswide.** This dataset contains 268,643 500-dimensional vectors.[2]
- **Glove.** This 100-dimensional dataset has 1,192,514 objects.[3]
- **Mnist5M.** This dataset contains 5,000,000 784-dimensional vectors.[4]
- **Deep10M.** We use 10,000,000 96-dimensional vectors of Deep10M. [5]
- **Sift10M.** This dataset has 11,164,766 128-dimensional vectors. [6]
- **Gauss.** We adopt eight synthetic datasets. By fixing dataset size to be 1,000,000, we set the dimensionality from 500 to 4,000. Each dimension value is drawn independently from the standard normal distribution $N(0,1)$.

Evaluation Metrics

- **Index Time:** consists of the time spent on index construction.
- **Index Speedup Ratio:** is the ratio of index time of QALSH to that of P-QALSH+ and P-QALSH.
- **Query Time:** measures the time cost for processing a query.
- **Query Speedup Ratio:** is the ratio of query time of QALSH to that of P-QALSH+.
- **Overall Ratio:** For a query q, it is defined as $\frac{1}{k}\sum_{i=1}^{k}\frac{dist(o_i,q)}{dist(o_i^*,q)}$. Here o_i is the ith object returned by the LSH methods and o_i^* is the exact i^{th} NN.
- **Recall:** is the ratio of number of returned true NN to the k NN of query q.
- **Mean Relative Error (MRE):** is defined as $\frac{1}{k}\sum_{i=1}^{k}\frac{|r_i-r_i^*|}{r_i^*}$. Where r_i is the predicted terminated radius of ith object and r_i^* is the exact terminated radius of ith object.

Benchmark Methods

- **QALSH** [6]. QALSH is a state-of-the-art LSH algorithm with theoretical guarantees on query quality.
- **P-QALSH** [7]. It only exploits the Inter-Table Parallel Design to accelerate the construction of QALSH data structure.
- **P-QALSH+.** P-QALSH+ is an improved version of P-QALSH, which proposes novel parallel designs for index construction and query tasks. This is the main method we proposed.

---

[1] http://groups.csail.mit.edu/vision/SUN.
[2] https://lms.comp.nus.edu.sg/wp-content/uploads/2019/research/nuswide/NUS-WIDE.html.
[3] https://nlp.stanford.edu/projects/glove/.
[4] http://yann.lecun.com/exdb/mnist.
[5] http://sites.skoltech.ru/compvision/noimi/.
[6] https://archive.ics.uci.edu/ml/datasets/SIFT10M/.

Parameters Setting. For c-k-ANN search, we set k to 100. The I/O page size is set to 16KB. The MLP neural network is implemented by Pytorch package. The hypermeters of MLP are set as follows: learning rate is 0.01, batch size is 256, epoch is 60, hidden feature dimension is [64, 32, 16, 8], Adaptive Moment Estimation (Adam) optimization and ReLU activation function. ϵ is set to 0.1. For each dataset, to balance the training cost and prediction error, we randomly select 10000 samples as training set and choose 1000 queries for evaluation.

5.2 Results and Analysis of the Index Phase

Analysis of Inter-table Parallel Design. In Fig. 6, our design achieves a 12–13X speedup over QALSH under the condition of 16 threads, indicating that our parallel optimization approaches an ideal speedup ratio. Besides, when processing larger datasets, the index speedup ratio remains relatively stable, showing that our design still has stable performance on large datasets. When we increase the thread number, the time cost for creating the QALSH data structure is significantly reduced while the index speedup ratio becomes higher.

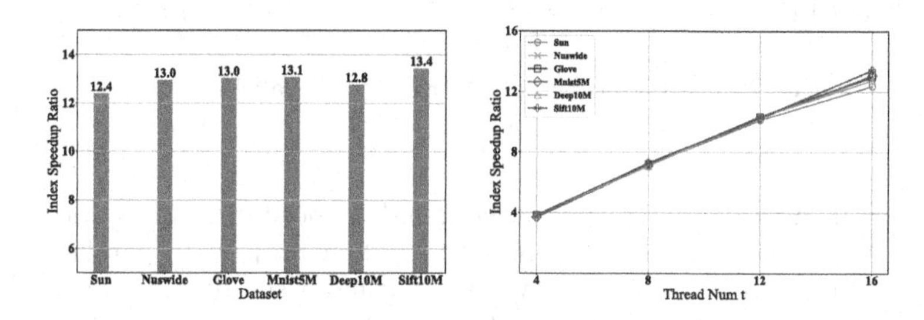

Fig. 6. The Performance of Inter-table Parallel Design

Analysis of Intra-Table Parallel Design. As shown in Fig. 7, we run several experiments to evaluate the performance of Intra-Table Parallel Design.

(a) Parallelizing Random Projection. Our design achieves the index speedup ratio of 8.0–11x over QALSH.As the dimensionality rises, the index speedup ratio increase. This is because the inner product operation in child threads need more time to complete in high dimensionality, while our design can better improve the CPU computing efficiency.

(b) Parallelizing Sorting. We can achieve speedup ratio of 1.6–6.7x in sorting of hash table. As the thread number increases, the index speedup ratio of sorting increases. Besides, when the size of dataset grows, the index speedup ratio increases because serial sorting on massive datasets consumes more time while parallel sorting can split massive datasets and sort them efficiently.

(c) Parallelizing B+-Tree Construction. Buffer strategy has a slightly higher speedup than fusion strategy while fusion strategy can save more DRAM occupancy. Since disk writing does not allow parallel writing of the same file, the performance of parallel construction is limited. However, the time of B+-Tree construction occupies only a small fraction of the total time of creation of whole hash table.

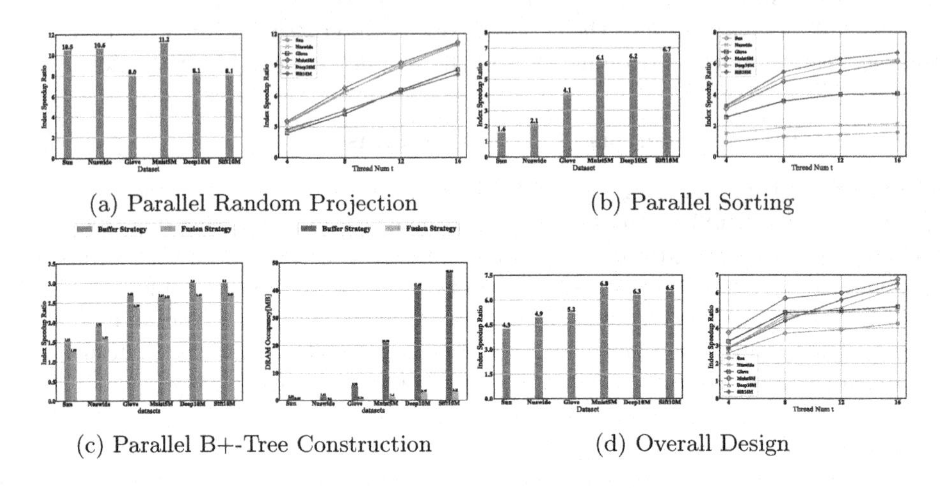

(a) Parallel Random Projection (b) Parallel Sorting

(c) Parallel B+-Tree Construction (d) Overall Design

Fig. 7. The Performance of Intra-Table Parallel Design

(d) Overall Analysis. We show overall index speedup ratio of Intra-Table Parallel Design by combining the parallel random projection, parallel sorting and parallel B+-tree construction. We find that index speedup ratio can achieve 4.3–6.8x with 16 threads. Besides, As the dataset becomes larger, we have better index efficiency, indicating that our design can perform more efficiently with larger datasets. Besides, the speedup ratio in Mnist5M dataset is higher than that in Deep10M and Sift10M datasets since parallel random projection has better performance under the condition of high dimension.

Analysis of Parallel Table Design. In this study, we compare the performance of P-QALSH+, which combines an Inter-Table Parallel Design with an Intra-Table Parallel Design, over P-QALSH which only exploits an Inter-Table Parallel Design. In P-QALSH+, we can make full use of all threads to build the QALSH data structure when there are more threads than the hash tables. In Fig. 8(a), we find that P-QALSH+ can achieve 8.4–10.7X speedup ratio while P-QALSH has only 3.7–3.8X speedup ratio. It shows that P-QLASH+ can make better use of multi-core computing resources and achieve better parallel performance.

In Fig. 8(b), as we increase the thread number, P-QALSH+ achieves better index speedup ratio, indicating that Parallel Table Design can fully make use of more CPU core resource to accelerate the index construction. Besides, Under the condition of same total thread number, when we use more threads in Inter-Table

Parallel Design, the index speedup ratio will be higher. This is because parallel construction of multiple hash tables is independent of each other, while parallel creation of a single hash table requires additional synchronization overhead.

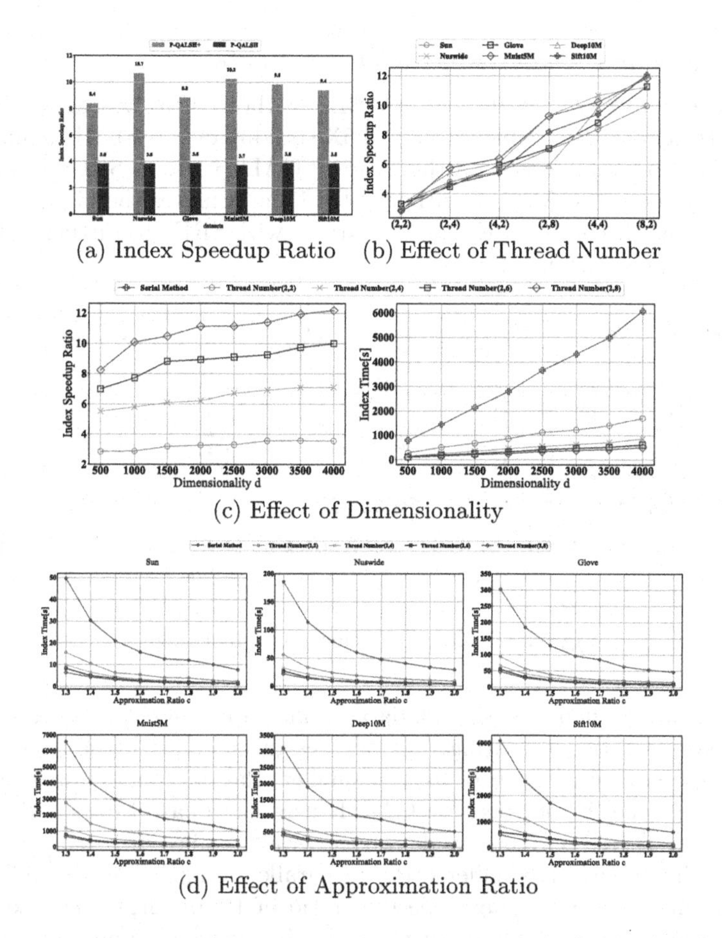

(a) Index Speedup Ratio (b) Effect of Thread Number

(c) Effect of Dimensionality

(d) Effect of Approximation Ratio

Fig. 8. The Performance of Parallel Table Design

We show the index performance over different dimensionality. In Fig. 8(c), as the dimensionality is higher, the index time of QALSH becomes larger because the time cost of random projection is positively related to the dimensionality. For comparison, P-QALSH+ still maintains a relatively smaller index time which indicating that it has better performance in creating the QALSH data structure in high-dimensional space. Besides, when we adopt more threads, the speedup ratio of P-QALSH+ in the index construction will be better.

In Fig. 8(d), As the approximation ratio c becomes smaller, the index time of QALSH becomes larger because QALSH needs to construct more hash tables serially. For comparison, P-QALSH+ always stays relatively less index time.

Besides, as the size of dataset is larger, the index time of QALSH also becomes larger, while P-QALSH+ still keeps the relatively smaller index time. Furthermore, by increasing the number of threads, we can further reduce the index time.

5.3 Results and Analysis of the Query Phase

Query Efficiency. In Fig. 9(a), P-QALSH+ without using Search Radius Estimation Strategy(SRES) achieves a 3.6–6.5X speedup over QALSH thanks to the design of K-Counter Arrays. Furthermore, P-QALSH+ with SRES has obvious improvement over P-QALSH+ without SRES and achieves the highest speedup of 6-8X because it combines K-Counter Arrays with SRES to further reduce the multi-round query overhead.

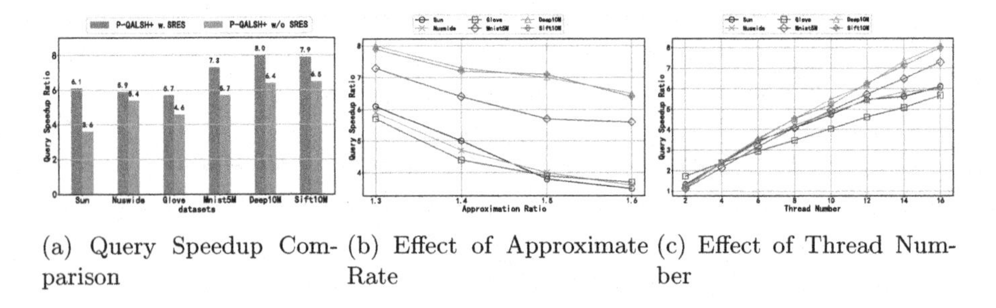

(a) Query Speedup Comparison (b) Effect of Approximate Rate (c) Effect of Thread Number

Fig. 9. The Query Efficiency of Our Design

Additionally, when the size of dataset increase, P-QALSH+ can achieve higher speedup ratio, indicating that our design has more obvious advantages in larger datasets. In Fig. 9(b), as c is smaller, the query speedup ratio of P-QALSH+ becomes higher because QALSH uses more hash tables and needs to execute range search on more random lines serially while our methods can keep the query time relatively smaller with our parallel designs. Thirdly, in Fig. 9(c), as we use more threads, query speedup ratio of P-QALSH+ increases significantly because more range query tasks can be executed simultaneously with more threads. Besides, the upward trend of query speedup on large datasets is obviously faster than that on smaller datasets, indicating that P-QALSH+ can achieve better query efficiency on larger datasets with less CPU core resource.

Query Accuracy. In Fig. 10, we present the query accuracy of P-QALSH+ and QALSH. QALSH achieves satisfactory overall ratio and P-QALSH+ has a further improvement in search accuracy. Besides, QALSH has slightly lower recall on Nuswide, Glove, and Mnist5M datasets, while P-QALSH+ achieves higher recall on all datasets. This is because our designs modify the termination condition of query process and enlarge the query range to search more nearby objects. In addition, when c is smaller, the overall ratio and recall of P-QALSH+ become better since we adopt more hash tables to filter out far objects.

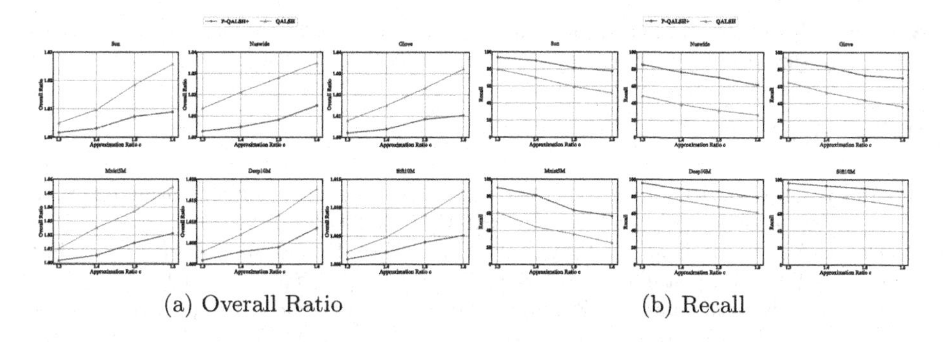

(a) Overall Ratio (b) Recall

Fig. 10. The Query Accuracy of Our Design

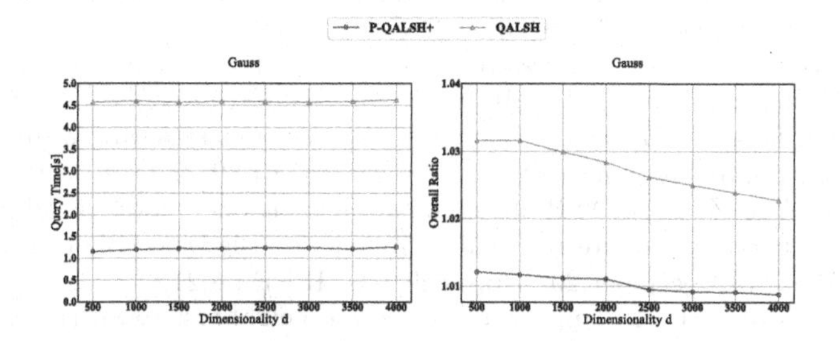

Fig. 11. Query Performance vs. Dimensionality d

Effect of Dimensionality. We evaluate the query performance of P-QALSH+ by varying the dimensionality from 500 to 4,000. In Fig. 11, both P-QALSH+ and QALSH remain stable query time under different dimensionality. Compared with QALSH, P-QALSH+ has significant improvement in query time and overall ratio. Additionally, P-QALSH+ can achieve better overall ratio in high-dimensional datasets, indicating that P-QALSH+ keeps better parallel search efficiency and query quality in high-dimensional Euclidean space.

6 Conclusion

In this paper, we mainly study the parallel optimization of the state-of-the-art LSH algorithm(QALSH). We propose a comprehensive and novel multi-core parallel LSH variant of QALSH, P-QALSH+, which proposes parallel optimizations in both the index construction and query process. Extensive experiments on six real-world datasets and eight synthetic datasets demonstrate the effectiveness and rationality of our methods.

Acknowledgments. The corresponding author of this work is Jianlin Feng. This work is partially supported by China NSFC under Grant No. 61772563.

References

1. Anastasiu, D.C., Rangwala, H., Tagarelli, A.: Are you my neighbor? Bringing order to neighbor computing problems. In: SIGKDD. KDD 2019 (2019)
2. Datar, M., Immorlica, N., Indyk, P., Mirrokni, V.S.: Locality-sensitive hashing scheme based on p-stable distributions. In: SoCG, pp. 253–262 (2004)
3. Echihabi, K., Palpanas, T., Zoumpatianos, K.: New trends in high-d vector similarity search: AI-driven, progressive and distributed. VLDB **14**(12), 3198–3201 (2021)
4. Echihabi, K., Zoumpatianos, K., Palpanas, T.: High-dimensional similarity search for scalable data science. In: ICDE. ICDE 2021 (2021)
5. Gan, J., Feng, J., Fang, Q., Ng, W.: Locality-sensitive hashing scheme based on dynamic collision counting. In: SIGMOD, pp. 541–552 (2012)
6. Huang, Q., Feng, J., Zhang, Y., Fang, Q., Ng, W.: Query-aware locality-sensitive hashing for approximate nearest neighbor search. VLDB **9**(1), 1–12 (2015)
7. Huang, Y., Yao, Z., Feng, J.: P-QALSH: parallelizing query aware locality-sensitive hashing for big data. In: International Conference on Big Data (Big Data), Orlando, FL, USA, 15–18 December 2021, pp. 629–634. IEEE (2021)
8. Indyk, P., Motwani, R.: Approximate nearest neighbors: towards removing the curse of dimensionality. In: STOC, pp. 604–613 (1998)
9. Jafari, O., Maurya, P., Nagarkar, P., Islam, K.M., Crushev, C.: A survey on locality sensitive hashing algorithms and their applications. arXiv:2102.08942 (2021)
10. Jafari, O., Nagarkar, P., Montaño, J.: Improving locality sensitive hashing by efficiently finding projected nearest neighbors. In: Satoh, S., et al. (eds.) SISAP 2020. LNCS, vol. 12440, pp. 323–337. Springer, Cham (2020). https://doi.org/10.1007/978-3-030-60936-8_25
11. Li, W., Zhang, Y., Sun, Y., Wang, W., Zhang, W., Lin, X.: Approximate nearest neighbor search on high dimensional data–experiments, analyses, and improvement. TKDE **32**, 1475–1488 (2016)
12. Liu, W., Wang, H., Zhang, Y., Wang, W., Qin, L.: I-LSH: I/O efficient C-approximate nearest neighbor search in high-dimensional space. In: ICDE, pp. 1670–1673 (2019)
13. Liu, W., Wang, H., Zhang, Y., Wang, W., Qin, L., Lin, X.: EI-LSH: an early-termination driven I/O efficient incremental c-approximate nearest neighbor search. VLDB J. **30**(2), 215–235 (2021)
14. Lu, K., Kudo, M.: R2LSH: a nearest neighbor search scheme based on two-dimensional projected spaces. In: ICDE, pp. 1045–1056 (2020)
15. Lu, K., Wang, H., Wang, W., Kudo, M.: VHP: approximate nearest neighbor search via virtual hypersphere partitioning. VLDB **13**(9), 1443–1455 (2020)
16. Sun, Y., Wang, W., Qin, J., Zhang, Y., Lin, X.: SRS: solving c-approximate nearest neighbor queries in high dimensional Euclidean space with a tiny index. VLDB **8**(1), 1–12 (2014)
17. Tao, Y., Yi, K., Sheng, C., Kalnis, P.: Efficient and accurate nearest neighbor and closest pair search in high-dimensional space. TODS **35**(3), 20:1–20:46 (2010)
18. Weber, R., Schek, H.J., Blott, S.: A quantitative analysis and performance study for similarity-search methods in high-dimensional spaces. In: VLDB, vol. 98, pp. 194–205 (1998)

19. Yang, C., Deng, D., Shang, S., Shao, L.: Efficient locality-sensitive hashing over high-dimensional data streams. In: ICDE, pp. 1986–1989 (2020)
20. Zhang, S., Huang, J., Xiao, R., Du, X., Gong, P., Lin, X.: Toward more efficient locality-sensitive hashing via constructing novel hash function cluster. Concurr. Comput. Pract. Exp. **33**(20) (2021)

Face Super-Resolution via Progressive-Scale Boosting Network

Yiyi Wang[1], Tao Lu[1](✉), Jiaming Wang[1], and Aibo Xu[2]

[1] Hubei Key Laboratory of Intelligent Robot, School of Computer Science and Engineering, Wuhan Institute of Technology, Wuhan 430073, China
lutxyl@gmail.com
[2] Wuhan Fiberhome Technical Services Co., Ltd., Wuhan, China
abxu@fiberhome.com

Abstract. Deep-learning-based face super-resolution (FSR) algorithms have performed more than traditional algorithms. However, existing methods need to pass multi-scale priors effectively constrained models. To alleviate this problem, we propose a progressive-scale boosting network framework, called PBN, which enables the progressive extraction of high-frequency information from low-resolution (LR) to reconstruct high-resolution (HR) face images. To ensure the accuracy of obtaining high-frequency signals, we introduce a constraint from HR to LR, which constructs supervised learning by progressively downsampling the reconstructed image to an LR space. Specifically, we propose a triple-attention fusion block to focus on different local features and prevent the secondary loss of facial structural information by removing the pooling layers. Experiments demonstrate the superior performance of the proposed method quantitatively and qualitatively on three widely used public face datasets (i.e., CelebA, FFHQ, and LFW) compared to existing state-of-the-art methods.

Keywords: Face super-resolution · Progressive-scale · Prior information · Attention

1 Introduction

Smart cities have become an essential trend in urban management, with recognizing personal identity information based on surveillance videos being a crucial foundation. However, captured facial images often suffer from low-resolution (LR) due to the limitations of the imaging environment and angle. Face super-resolution (FSR), also called face hallucination, can reconstruct a potential high-resolution (HR) image from an input LR image (or multiple frames). The FSR technology is used for effectively enhancing the resolution and quality of images, which can improve the performance of high-level visual tasks based, such as face detection and recognition.

Based on the rapid development of convolutional neural networks (CNNs), various deep-learning-based image super-resolution (SR) methods have been

X. Song et al. (Eds.): APWeb-WAIM 2023, LNCS 14332, pp. 44–57, 2024.
https://doi.org/10.1007/978-981-97-2390-4_4

proposed and obtained excellent subjective and objective reconstruction performance than traditional shallow learning methods. U-Net has been applied to computer vision tasks and application prospects, such as segmentation and reconstruction. Han *et al.* [1] proposed a multi-level U-Net residual structure composed of two different levels of U-Net structures to extract high-frequency and low-frequency information from LR images and reduce information loss during the convolution process. Guo *et al.* [2] proposed a dual-path version U-Net, which can improve the quality of reconstructed images by learning the mapping relationship between LR and HR images through an additional regression branch.

Although these above-mentioned U-Net-based SR networks have achieved good results, there are still some limitations: 1) These networks usually interpolate the LR image to HR size and restore the corresponding SR image by an encoding-decoding framework. However, this operation will cause artifacts and additional computational overheads, and it isn't easy to obtain accurate prior information from the interpolation image. 2) As shown in Fig. 1, the input low resolution contains limited facial structural information.

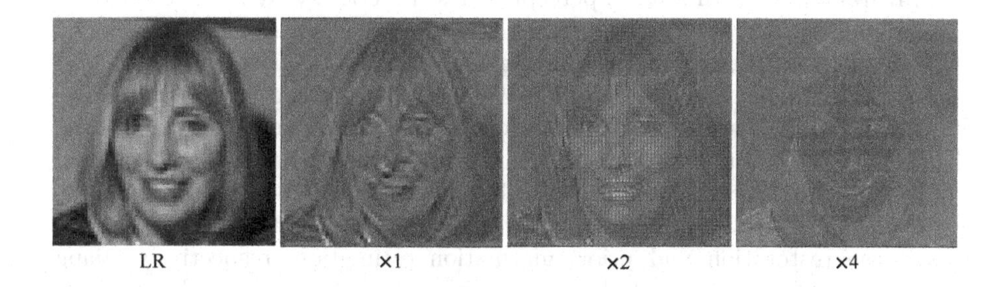

Fig. 1. Visualization of LR image and multi-scale feature maps.

To alleviate this situation and better learn the mapping relationship between LR and HR images, we propose an FSR network called PBN. Firstly, different from U-Net-based methods, we upsample LR images to the HR space progressively which can eliminate images' redundant encoding and decoding processes and reduce computational and parameter costs to some extent. Additionally, we gradually downsample HR images to the corresponding LR ones as an additional constraint for network improvement. Secondly, multiple attention feature fusion blocks (AFFB) are stacked as attention feature fusion groups (AFFG) to utilize LR image full features and capture long-distance spatial features for improved feature expression capability. Moreover, considering that pooling layers are essentially downsampling operations and most input face images have LR, to avoid secondary loss of facial structural information, we propose a new attention mechanism. In summary, the significant contributions are briefly highlighted as follows:

- We propose a progressive multi-scale FSR network, called PBN, to integrate features of different scales effectively and reduce network computation, which can better learn image prior knowledge and accelerate network convergence with these adding additional constraints.
- The proposed triple-attention fusion module focuses on local features and prevents secondary loss of facial structure information.
- We conduct extensive experiments on three popular face datasets (CelebA [3], FFHQ [4], and LFW [5]), which demonstrate the proposed method can recover more pleasant face images than other state-of-the-art methods.

2 Face Super-Resolution

The FSR task is a subfield of single image SR that has garnered significant attention and research in recent decades. With the advancement of CNNS and the availability of numerous public face datasets, FSR methods related have achieved remarkable progress.

Yang *et al.* [6] introduced a discriminative enhancement generation adversarial FSR framework with a novel perceptual loss function, which can reconstruct facial details while retaining the training gradients. Chen *et al.* [7] developed an end-to-end FSR method by utilizing facial geometry prior (such as facial landmark heatmaps and parsing maps) extracted from the coarse SR image, which can obtain a reconstructed face image with clearer texture. Considering that facial prior features, such as landmark and composition maps are estimated from rough SR images, leading to possible inaccuracies. To address this issue, Ma *et al.* [8] proposed two recurrent networks collaborating iteratively to achieve face image restoration and prior information estimation, respectively. Jiang *et al.* [9] proposed a dual-path deep fusion network that requires additional face priors. Specifically, it learns the global face shape and local face components through two separate branches. While these methods have shown satisfactory performance, accurately reconstructing clear identity information from tiny faces remains challenging without precise prior information.

Huang *et al.* [10] proposed a multi-scale wavelet CNN-based FSR method that predicts the HR wavelet coefficient sequence corresponding to LR for reconstruction. Grm *et al.* [11] utilized a cascaded architecture to incorporate multi-scale identity prior information into the network. Wang *et al.* [12] developed a progressive face prior estimation framework that integrates face prior knowledge and guides SR at each scale, resulting in higher-quality reconstructed face images.

The attention mechanism [13] has been widely used in high- and low-level visual tasks. Woo *et al.* [14] introduced a lightweight and versatile attention block to explore feature attention in space and channel dimensions, which can guide the model to focus on the target object regions of interest. In the SR task, Zhang *et al.* [15] employed the channel attention block to assign different weights to the channels in the feature map and readjust the difference of features. Liu *et al.* [16] proposed a lightweight enhanced spatial attention to obtain more focused residual features on key spatial content. Lu *et al.* [17] firstly proposed an FSR method based on the separated attention network, which can achieve the reconstruction of high-quality and HR facial images without using any prior facial information. Therefore, combining different attention mechanisms allows PBN to focus on other facial features effectively and extract more accurate prior knowledge.

3 Our Methods

In this section, we describe the design scheme for the proposed network and discuss its differences from other existing methods. As shown in Fig. 2, the overall architecture is divided into the primary and auxiliary paths. The primary path gradually enlarges the face image, which utilizes cascaded residual groups to learn the prior information distribution from LR to HR images. The auxiliary path provides an inverse constraint from HR to LR, which can explore specific prior information to reconstruct SR images.

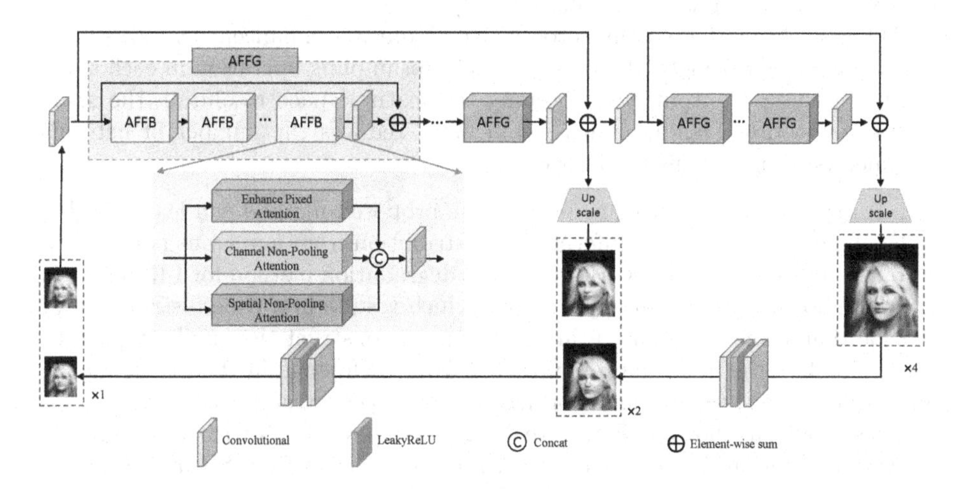

Fig. 2. The architecture of PBN with × 4. PBN contains a primary path from LR to SR and an auxiliary path from SR to LR. AFFG is the attention feature fusion group, and AFFB is the attention feature fusion block.

3.1 Network Architectures

Primary Path: As shown in Fig. 2, the primary path is composed of cascaded residual AFFGs. Within AFFG, a novel AFFB is introduced to extract high-frequency prior feature information from input face images effectively. Initially, an initial feature $\mathcal{F}_0$ is extracted from the LR image using a single convolutional layer ($\mathcal{C}$),

$$\mathcal{F}_0 = \mathcal{C}(I_{LR}), \tag{1}$$

where I_{LR} denotes the input LR image. Then, the primary path can be formulated as,

$$\begin{cases} \mathcal{F}_i = H_{AFFG}(\mathcal{F}_{i-1}) \\ I_i^{SR} = H_{UP}(\mathcal{F}_i) \end{cases} \quad (2)$$

where $\mathcal{F}_i$ denotes the i-th feature maps, $i \in \{1, 2, ..., n\}$ is the number of indexes. $H_{AFFG}(.)$ denotes the proposed attention feature fusion groups containing several attention feature fusion blocks. $H_{UP}(.)$ and I_i^{SR} represent the upscale operation and the face SR image of $\times 2^i$, respectively. Furthermore, a series of loss constraints are imposed on the intermediate SR results to ensure stability and effectiveness during training. The loss function of the primary path is defined as follows:

$$L_{PP} = \sum_{i=1}^{n} \left\| I_i^{HR} - I_i^{SR} \right\|, \quad (3)$$

where L_{pp} is the loss function of the primary path. n denotes the multiplication factor, I_i^{SR} represents the reconstructed face image of $\times 2^i$, and I_i^{HR} is a HR image corresponding to the resolution of I_i^{SR}.

Different from the existing progressive resolution enhancement strategy in multi-stage face SR methods, we employ an upsampling operation in each stage to obtain the SR output while keeping the feature space resolution the same as the input LR. This approach significantly reduces the number of network parameters. More details are in Sect. 4.3.

Auxiliary Path: The primary path of the proposed method can learn the LR to HR high-frequency prior information distribution from numerous paired data through progressive updates. However, the degradation method for LR images is often unknown in practical applications, which results in a wide distribution of the solution space. To address this problem, we construct the auxiliary path to constrain the prior information extracted from each stage in the primary path and explore the specific prior information of the reconstruction network.

Specifically, as shown in Fig. 2, we design a two-layer CNN with a LeakyReLU activation function $\iota(.)$ to generate downsampled images from SR images. Then we compute a loss of the same size for the LR images. The formula can be expressed as follows:

$$I_i^{PR} = \mathcal{C}(\iota(\mathcal{C}(I_i^{SR}))), \quad (4)$$

where I_i^{PR} is an LR face image generated by I_i^{SR} after downsampling.

We perform pixel-wise supervised learning in an LR space to reduce this solution space as much as possible and mine the network's latent prior information. The loss function of the auxiliary path is defined as follows:

$$L_{AP} = \sum_{i=1}^{n} \left\| I_i^{HR} - down(I_i^{SR}) \right\| = \sum_{i=1}^{n} \left\| I_i^{HR} - I_i^{PR} \right\|, \quad (5)$$

where L_{AP} is the loss functions of the auxiliary path.

Ultimately, the proposed PBN is optimized by minimizing the following overall objective function,

$$L_{Totall} = L_{PP} + \delta L_{AP}, \tag{6}$$

where δ represents the collective parameter of the loss.

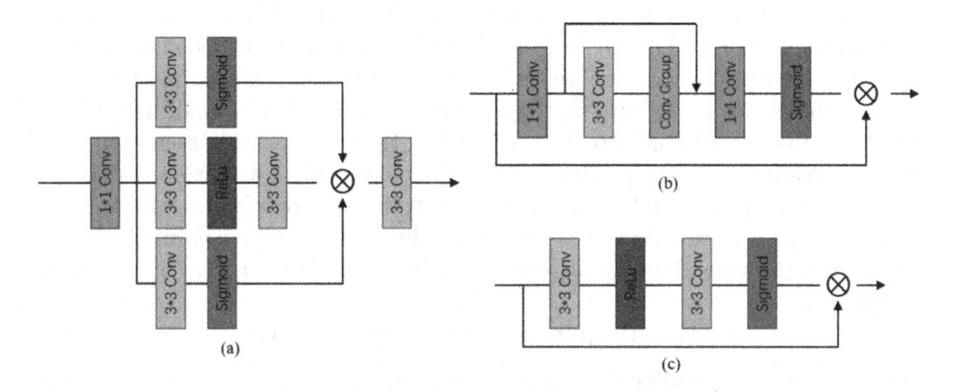

Fig. 3. Attention feature fusion block. Where (a), (b) and (c) are EPA, SNA, and CNA, respectively.

3.2 Attention Feature Fusion Block

Channel non-pooling attention (CNA). The channel attention (CA) block is a strategy of re-weighting features from different channels, considering each channel contains distinct and weighted information. However, the pooling layer leads to the secondary loss of high-frequency information, especially for LR face images with limited facial structure information, as it downsamples the input image. In contrast to previous versions of CA [14,15], we removed the pooling layer, as illustrated in Fig. 3:

$$m_{CNA} = \zeta(\mathcal{C}(\delta(\mathcal{C}(X^{h\times w\times c})))), \tag{7}$$

where ζ and δ represent sigmoid and ReLU activation functions, respectively. f_{CNA} is the output attention map. Finally, the input $X^{h\times w\times c}$ and the weights of the channel CNA are multiplied element-wise. The final attention feature is obtained as follows:

$$F_{CNA} = m_{CNA} \otimes X^{h\times w\times c}, \tag{8}$$

where $X^{h\times w\times c}$ represents input feature.

Spatial Non-Pooling Attention (SNA). We have introduced a novel type of SNA as a complement to CNA and EPA [18], as shown in Fig. 3. Different from previous approaches, such as [14,19], the proposed SNA does not incorporate a pooling layer, which aims to preserve high-frequency information and prevent its loss. The formulation for SNA is as follows:

$$F_{SNA} = X^{h \times w \times c} \otimes \zeta(\mathcal{C}_{1\times1}(f_{group}(\mathcal{C}_{3\times3}(\mathcal{C}_{1\times1})))), \quad (9)$$

Various attention mechanisms focus on different features, and combining multiple attention features can improve the expressive ability of features, leading to better performance in the FSR task. Multi-attention-based methods have gained popularity in computer vision recently. However, most of these methods use serial combinations for feature representation, resulting in the loss of transferred features between different attention modules. To overcome this issue and preserve the original features produced by each component of attention (as shown in Fig. 3), we propose EPA, CNA, and SNA to fuse output features from three attentions. Finally, we reduce dimensionality through a 1×1 convolutional layer to maintain the consistent input-output relationship. This process is expressed as follows:

$$O(F) = \mathcal{C}_{1\times1}(concat(F_{EPA}, F_{CNA}, F_{SNA})), \quad (10)$$

where $concat(\cdot)$ represents the fusion operation, F_{EPA}, F_{CNA}, and F_{SNA} represent the features of EPA, CNA, and SNA, respectively. We stack multiple AFFBs to create AFFG and use several AFFGs to form the backbone of our network. This backbone serves as the deep feature extractor. In the following experimental phase, we analyze the benefits of parallel fusion in more detail.

Table 1. Comparison results of average PSNR, SSIM, FSIM, and VIF for × 4 and × 8 face SR with the state-of-the-art approaches on FFHQ, CelebA, and LFW dataset and perform experiments on similar size. The best results are highlighted.

Algorithms	Scale	CelebA				LFW				FFHQ			
		PSNR	SSIM	FSIM	VIF	PSNR	SSIM	FSIM	VIF	PSNR	SSIM	FSIM	VIF
Bicubic	4	29.77	0.8597	0.8887	0.5041	28.90	0.8582	0.8787	0.4931	29.82	0.8459	0.8888	0.5245
EDGAN		31.64	0.8869	0.9272	0.5512	32.08	0.9003	0.9330	0.5792	30.84	0.8576	0.9229	0.5380
RCAN		32.99	0.9138	0.9356	0.6127	33.27	0.9249	0.9414	0.6362	32.65	0.8980	0.9335	0.6161
SRFBN		32.83	0.9117	0.9331	0.6088	33.10	0.9226	0.9389	0.6324	32.40	0.8947	0.9303	0.6077
HMRFN		32.98	0.9127	0.9347	0.6090	33.18	0.9225	0.9407	0.6290	32.56	0.8969	0.9325	0.6110
DIDnet		33.07	0.9146	0.9361	0.6128	33.43	0.9265	0.9428	0.6395	32.67	0.8981	0.9339	0.6129
LBNet		32.74	0.9117	0.9343	0.6064	32.98	0.9001	0.9387	0.6355	32.60	0.8979	0.9335	0.6143
Ours		**33.29**	**0.9175**	**0.9375**	**0.6220**	**33.67**	**0.9287**	**0.9449**	**0.6458**	**32.77**	**0.9001**	**0.9349**	**0.6191**
Bicubic	8	25.57	0.7343	0.7984	0.3049	24.24	0.7034	0.7724	0.2751	25.99	0.7312	0.7999	0.3404
EDGAN		25.71	0.7494	0.8558	0.3172	25.62	0.7471	0.8540	0.3216	25.89	0.7268	0.8550	0.3273
RCAN		28.12	0.8098	0.8654	0.3982	27.16	0.7902	0.8563	0.3802	28.09	0.7909	0.8582	0.4159
SRFBN		28.02	0.8055	0.8596	0.4055	27.24	0.7940	0.8512	0.3958	28.06	0.7906	0.8543	0.4205
HMRFN		28.06	0.8048	0.8668	0.3933	27.15	0.7887	0.8565	0.3775	28.08	0.7909	0.8599	0.4143
DIDnet		28.24	0.8129	0.8673	0.4170	27.02	0.7929	0.8573	0.3860	28.23	0.7956	0.8595	0.4213
LBNet		27.79	0.7999	0.8615	0.3876	26.37	0.7827	0.8513	0.3233	28.06	0.7908	0.8566	0.4155
Ours		**28.50**	**0.8214**	**0.8707**	**0.4251**	**27.83**	**0.8119**	**0.8644**	**0.4193**	**28.32**	**0.7989**	**0.8620**	**0.4300**

4 Experimental Results and Analysis

In this section, we initially introduce the dataset, evaluation indicators, and implementation details. Then, we compare the proposed method with the state-of-the-art approaches and conduct the analysis.

4.1 Datasets and Implementation Details

We conduct experiments on three public face datasets, such as LFW [5], CelebA [3], and FFHQ [4]. We randomly select 1,000 images for training and testing, of which 950 are used for training (850 for training, 100 for validation), and 50 are used for testing. The CelebA dataset is a large-scale facial attribute dataset with a total number of more than 200 thousand. Its face images include posing changes and backgrounds and 40 face attribute annotations. The original size of each image is 178×218 pixels. The LFW dataset is similar to CelebA. The original size of each image is 250×250 pixels. For the FFHQ dataset, the original size of each image is $1,024 \times 1,024$ pixels. In addition, for better performance of each method, we use four quality indicators to evaluate the quality of images: peak signal-to-noise ratio (PSNR), structural similarity (SSIM), feature similarity index measure (FSIM), and variance inflation factor (VIF).

4.2 Compared with State-of-the-Arts

We compare the proposed model with other state-of-the-art methods, including EDGAN [6], RCAN [15], SRFBN [20], HMRFN [21], DIDnet [22], and LBNet [23]. For a fair comparison, we retrain this model provided by the author in the same datasets with their best parameter settings.

Quantitative Comparison. As shown in Table 1, the PSNR values of the proposed method outperform the second-best method by 0.22/0.24/0.1 dB on the CelebA/LFW/FFHQ datasets with $\times$ 4. When the upsampling factor is 8, this difference can reach up to 0.26/0.59/0.09 dB, respectively. The reason for this improvement lies in our progressive structure that effectively mines and combines different prior knowledge at each stage to reconstruct a higher-quality image instead of directly amplifying LR to eight times lower SR like other methods. However, PSNR only focuses on pixel differences and does not consider perceptual quality, which may result in high PSNR values but poor perception results. Therefore, we also introduce the VIF to prove our method's effectiveness. Table 1 shows that PBN's performance improves on CelebA/LFW/FFHQ datasets with $\times$ 4 or $\times$ 8, indicating that our method produces better-quality images and validates the effectiveness of our proposed approach.

Qualitative Comparison. The reconstructed results of the proposed PBN and other methods on CelebA and FFHQ are illustrated in Figs. 4 and 5, respectively. The GAN-based method EDGAN can restore clearer contours and details but also introduces significant artifact effects, which do not fully meet our requirements. Compared to the other algorithms, our approach can recover sharper

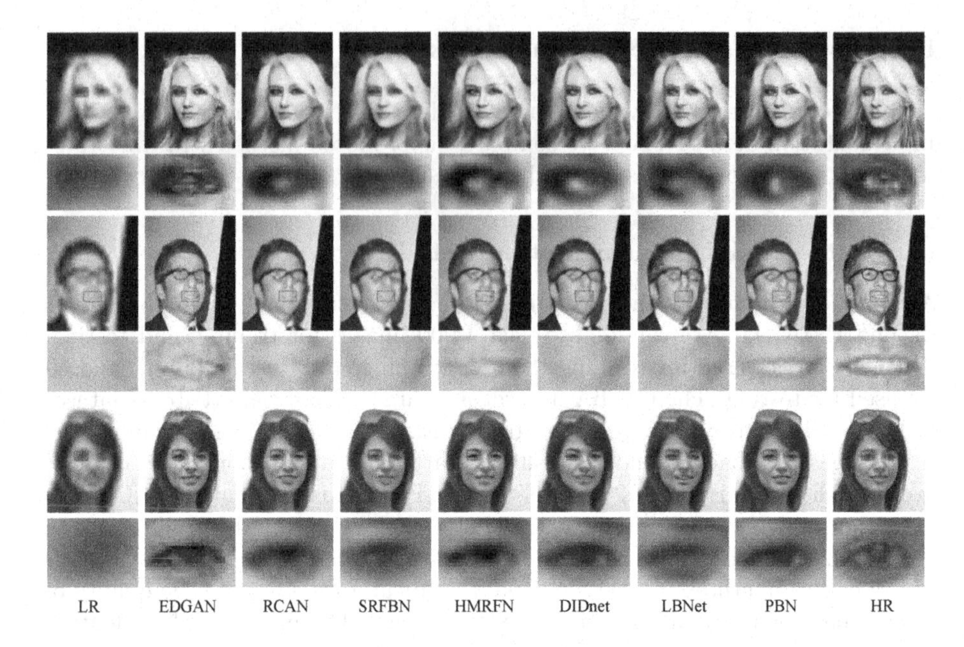

Fig. 4. Subjective comparison of our method with other algorithms on CelebA dataset with × 8.

facial contours and richer details, especially in critical facial structures such as the eyes, mouth, and others. This is because PBN adopts a progressive structure, effectively integrating features at different scales to improve feature expression ability. At the same time, it learns prior knowledge of images through additional constraints to accelerate network convergence. Especially for small faces, our proposed method can better guide the recovery of facial features.

4.3 Ablation Study

We conduct ablation studies on the CelebA dataset to verify the effectiveness of each component in the proposed method.

As presented in Table 2, we perform a series of ablation analyses involving the following four experiments. We remove multi-stage losses L_{PP} and L_{AP} while adding pooling layers in AFFB as Model 1. To evaluate the impact of the progressive structure, L_{PP} loss is introduced based on Model 1 as Model 2. The results demonstrate that Model 2 achieved significantly improved PSNR compared to Model 1. It shows that introducing L_{PP} is beneficial for the path to learn the distribution of high-frequency information gradually.Furthermore, in Model 3, we remove the pooling layer from Model 2 to verify the effect of pooling. We observe that removing the pooling layer further improved PSNR. This result can be attributed to the fact that small face images may experience secondary loss in facial structure due to the introduction of pooling layers, which hinders the accurate extraction of prior information. L_{PP} and L_{AP} loss functions

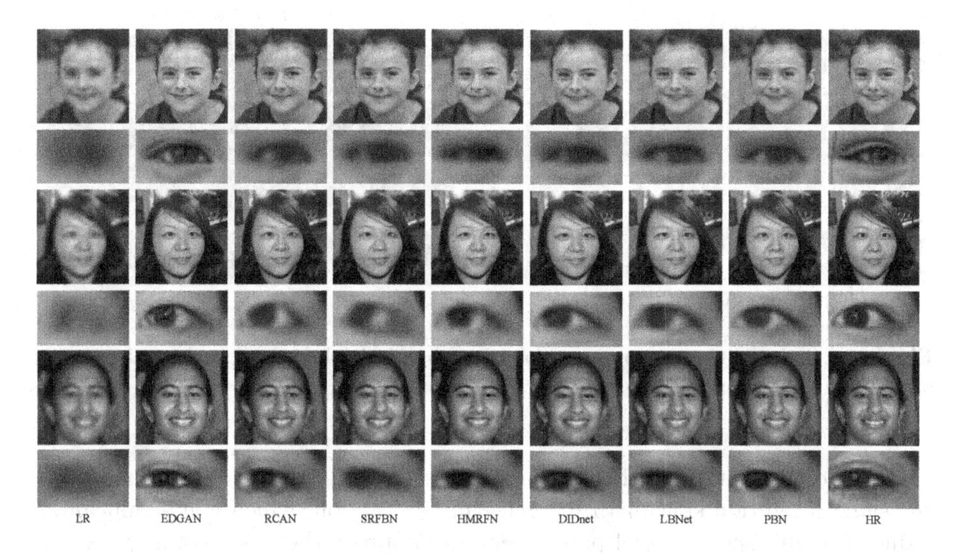

Fig. 5. Subjective comparison of our method with other algorithms on FFHQ dataset with × 8.

Table 2. Ablation study on CelebA dataset.

Models	Pooling	L_{PP}	L_{AP}	PSNR	SSIM
1	✓	×	×	28.00	0.8055
2	✓	✓	×	28.37	0.8155
3	×	✓	×	28.42	0.8169
4	✓	✓	✓	28.40	0.8161
5	×	✓	✓	28.50	0.8214

Table 3. Compare the objective results of floating point operations(FLOPs) and parameters(Params)

Methods	U-Net		Ours
n_feat	16	32	64
FLOPs(G)	262.83	1051.09	108.12
Params(M)	25.72	102.75	25.9

are introduced into the progressive network with pooling layers as Model 4. In contrast, the PSNR and SSIM values of Model 4 are higher than those of Model 2, indicating that the L_{AP} loss mines specific prior information to a certain extent. However, its evaluation metrics are insufficient compared to our proposed PBN. This is because adding the pooling layer amplifies some inaccurate factors in the reconstruction process, which hinders the use of multi-scale information for accurate face image reconstruction.

To better understand the impact of the pooling layer, we visualize the features of the proposed PBN and the network introduced into the pooling layer with × 8. Specifically, we separately visualized Models 4 and 5. Figure 6 (a) on the left is the original feature map of Model 4, while picture (b) on the right is that of Model 5. The red box in both pictures highlights a loss of facial structure in Fig. 6 (a), which becomes more pronounced with × 8. This makes it difficult to focus on clear facial contours using these feature maps. In contrast, richer facial structure information can be obtained from Fig. 6 (b).

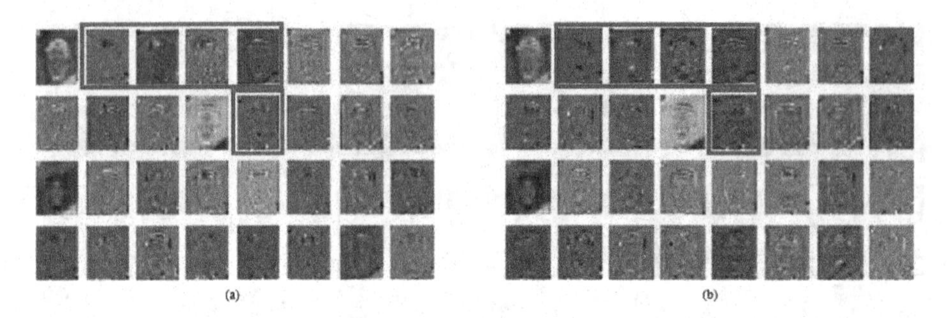

Fig. 6. (a) is the original feature map obtained by Model 4, and (b) is the feature map obtained by our proposed PBN. (Color figure online)

It is worth mentioning that the proposed multi-scale model always maintains the same resolution space as LR for feature extraction, which can significantly reduce the number of model parameters and improve the calculation speed. As shown in Table 3, keeping the rest of the conditions the same, when the initial channel of the U-net framework is 16, the amount of parameters is slightly lower than that of our PBN with 64 channels. Still, the amount of calculation is much higher than that of our framework. And when the initial channel of the U-net framework is set to 32, its calculation amount and parameter amount will increase exponentially.

4.4 Effectiveness of the Proposed Method

To verify the effectiveness of the proposed method, we employ the downstream task (feature matching) to evaluate reconstruction results. The scale-invariant feature transform (SIFT) results between HR and SR are shown in Fig. 7. It can be observed that even from the perspective of features, the proposed method still achieves optimal performance. To quantitatively evaluate the results, we list the number of matching points with different matching thresholds in Table 4. The number of similar feature points our method matches is better than all methods, including EDGAN.

Table 4. Number of feature point matches across different datasets and thresholds.

Dataset	Bicubic	EDGAN	RCAN	SRFBN	HMRFN	DIDNet	LBNet	Ours
FFHQ	7	13	16	14	11	16	18	19
CelebA	4	6	8	9	8	6	8	9

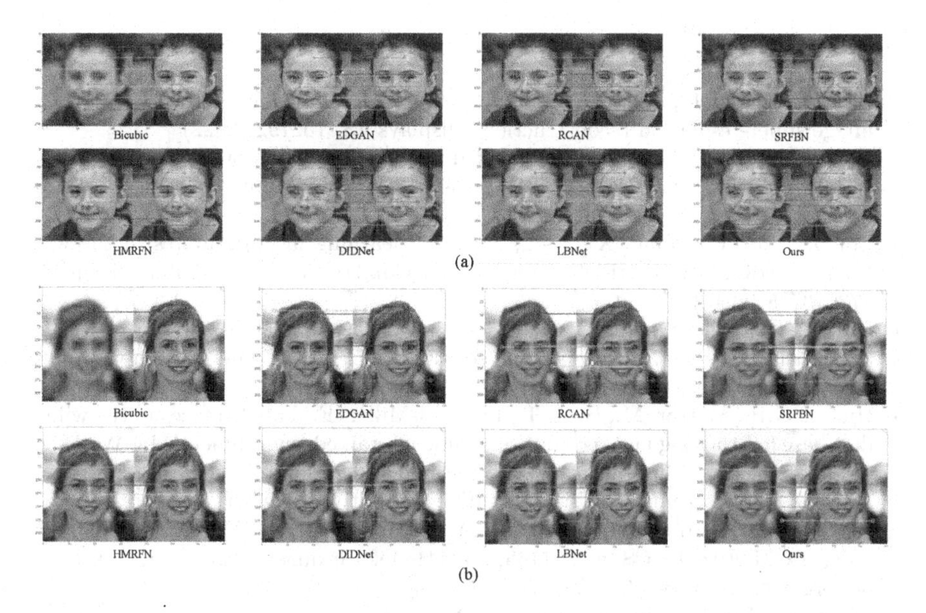

Fig. 7. (a) and (b) are feature-matching results on FFHQ and CelebA datasets.

5 Conclusion

In this paper, we propose a new progressive face super-resolution method. Specifically, we design a new advanced FSR framework that includes two paths. The primary path learns the mapping relationship between LR and HR images through cascaded residual groups to better extract prior information and reduce network computation and parameters. The auxiliary path gradually obtains corresponding LR space images through a progressive method, providing additional constraints for the network and assisting in exploring the prior information required for image reconstruction to guide face restoration. In addition, we propose a triple-attention fusion module that reduces facial structural information loss caused by secondary downsampling of images by removing pooling layers and reconstructing clearer textured face images. Extensive experiments on CelebA, FFHQ, and LFW datasets further demonstrate the superiority of our progressive face super-resolution method in enhancing network feature extraction and accelerating network convergence.

Acknowledgements. This work was supported in part by the National Natural Science Foundation of China under Grant 6207235 and Grant 62171328.

References

1. Han, N., Zhou, L., Xie, Z., Zheng, J., Zhang, L.: Multi-level u-net network for image super-resolution reconstruction. Displays **73**, 102192 (2022)
2. Guo, Y., et al.: Closed-loop matters: dual regression networks for single image super-resolution. In: Proceedings of the IEEE/CVF Conference on Computer Vision and Pattern Recognition, pp. 5407–5416 (2020)
3. Liu, Z., Luo, P., Wang, X., Tang, X.: Deep learning face attributes in the wild. In: Proceedings of the IEEE International Conference on Computer Vision, pp. 3730–3738 (2015)
4. Karras, T., Laine, S., Aila, T.: A style-based generator architecture for generative adversarial networks. In: Proceedings of the IEEE/CVF Conference on Computer Vision and Pattern Recognition, pp. 4401–4410 (2019)
5. Huang, G.B., Mattar, M., Berg, T., Learned-Miller, E.: Labeled faces in the wild: a database forstudying face recognition in unconstrained environments. In: Workshop on Faces in 'Real-Life' Images: Detection, Alignment, and Recognition (2008)
6. Yang, X., et al.: Enhanced discriminative generative adversarial network for face super-resolution. In: Hong, R., Cheng, W.-H., Yamasaki, T., Wang, M., Ngo, C.-W. (eds.) PCM 2018. LNCS, vol. 11165, pp. 441–452. Springer, Cham (2018). https://doi.org/10.1007/978-3-030-00767-6_41
7. Chen, Y., Tai, Y., Liu, X., Shen, C., Yang, J.: Fsrnet: end-to-end learning face super-resolution with facial priors. In: Proceedings of the IEEE/CVF Conference on Computer Vision and Pattern Recognition, pp. 2492–2501 (2018)
8. Ma, C., Jiang, Z., Rao, Y., Lu, J., Zhou, J.: Deep face super-resolution with iterative collaboration between attentive recovery and landmark estimation. In: Proceedings of the IEEE/CVF Conference on Computer Vision and Pattern Recognition, pp. 5569–5578 (2020)
9. Jiang, K., Wang, Z., Yi, P., Tao, L., Jiang, J., Xiong, Z.: Dual-path deep fusion network for face image hallucination. IEEE Trans. Neural Netw. Learn. Syst. **33**(1), 378–391 (2020)
10. Huang, H., He, R., Sun, Z., Tan, T.: Wavelet-srnet: a wavelet-based CNN for multi-scale face super resolution. In: Proceedings of the IEEE/CVF International Conference on Computer Vision, pp. 1689–1697 (2017)
11. Grm, K., Scheirer, W.J., Štruc, V.: Face hallucination using cascaded super-resolution and identity priors. IEEE Trans. Image Process. **29**, 2150–2165 (2019)
12. Wang, H., Qian, H., Chengdong, W., Chi, J., Xiaosheng, Yu., Hao, W.: Dclnet: dual closed-loop networks for face super-resolution. Knowl.-Based Syst. **222**, 106987 (2021)
13. Vaswani, A., et al.: Attention is all you need. In: Advances in Neural Information Processing Systems, vol. 30 (2017)
14. Woo, S., Park, J., Lee, J.Y., Kweon, I.S.: CBAM: convolutional block attention module. In: Proceedings of the European Conference on Computer Vision, pp. 3–19 (2018)
15. Zhang, Y., Li, K., Li, K., Wang, L., Zhong, B., Fu, Y.: Image super-resolution using very deep residual channel attention networks. In: Proceedings of the European Conference on Computer Vision, pp. 286–301 (2018)
16. Kim, J., Li, G., Yun, I., Jung, C., Kim, J.: Edge and identity preserving network for face super-resolution. Neurocomputing **446**, 11–22 (2021)
17. Lu, T., et al.: Face hallucination via split-attention in split-attention network. In: Proceedings of the ACM International Conference on Multimedia, pp. 5501–5509 (2021)

18. Zhao, K., Lu, T., Zhang, Y., Wang, Y., Wang, Y.: Face super-resolution via triple-attention feature fusion network. IEICE Trans. Fundam. Electron. Commun. Comput. Sci. **105**(4), 748–752 (2022)
19. Liu, J., Zhang, W., Tang, Y., Tang, J., Wu, G.: Residual feature aggregation network for image super-resolution. In: Proceedings of the IEEE/CVF Conference on Computer Vision and Pattern Recognition, pp. 2359–2368 (2020)
20. Li, Z., Yang, J., Liu, Z., Yang, X., Jeon, G., Wu, W.: Feedback network for image super-resolution. In: Proceedings of the IEEE/CVF conference on computer vision and pattern recognition, pp. 3867–3876 (2019)
21. Wang, Yu., Tao, L., Zhihao, W., Yuntao, W., Zhang, Y.: Face super-resolution via hierarchical multi-scale residual fusion network. IEICE Trans. Fundam. Electron. Commun. Comput. Sci. **104**(9), 1365–1369 (2021)
22. Cheng, F., Lu, T., Wang, Y., Zhang, Y.: Face super-resolution through dual-identity constraint. In: Proceedings of the IEEE/CVF International Conference on Multimedia and Expo, pp. 1–6. IEEE (2021)
23. Gao, G., Wang, Z., Li, J., Li, W., Yu, Y., Zeng, T.: Lightweight bimodal network for single-image super-resolution via symmetric CNN and recursive transformer. arXiv preprint arXiv:2204.13286 (2022)

An Investigation of the Effectiveness of Template Protection Methods on Protecting Privacy During Iris Spoof Detection

Baogang Song[1], Jian Suo[1], Hucheng Liao[1], Huanhuan Li[3], and Dongdong Zhao[1,2](✉)

[1] School of Computer Science and Artificial Intelligence, Wuhan University of Technology, Wuhan, China
{297710,314108,liaohc,zdd}@whut.edu.cn

[2] Chongqing Research Institute of Wuhan University of Technology, Chongqing, China

[3] School of Computer Science, China University of Geosciences, Wuhan, China

Abstract. With the development of iris biometrics, more and more industries and fields begin to apply iris recognition methods. However, as technology advances, attackers try to use printed iris images or artifacts and so on to spoof iris recognition systems. As a result, iris spoof detection is becoming an increasingly important area of research. The employment of spoof detection enhances the security and reliability of iris recognition systems, but an attacker can still subvert the systems by stealing iris data during the spoof detection phase. In this paper, we design a framework called TPISD to solve the issue. TPISD mainly employs template protection methods to protect iris data during the spoof detection phase as well as client to server phase. Specifically, iris data are converted into cancelable and irreversible templates after data capture. These templates are then used to train the spoof detection model. Eventually, during the spoof detection phase, protected templates are used as input, rather than the original iris images. Experiments conducted on CASIA-Syn and CASIA-Interval datasets demonstrate that the application of iris template protection techniques to the spoof detection model may result in a reduction on recognition accuracy, but it can enhance the security of the spoof detection model. This work verifies the feasibility of employing iris template protection methods to protect iris data during the spoof detection.

Keywords: Spoof Detection · Iris Recognition · Template Protection · Convolutional Neural Networks

1 Introduction

Due to the increasing demand for security in people's daily life, biometric technology has become the focus of research, and commonly used biometric tech-

X. Song et al. (Eds.): APWeb-WAIM 2023, LNCS 14332, pp. 58–73, 2024.
https://doi.org/10.1007/978-981-97-2390-4_5

nologies include face [7], fingerprint [13], iris [20], and multi-modal [2] recognition. Among them, iris recognition has obvious advantages compared with other biometric technologies [21]. Iris recognition have become the focus of research. However, iris recognition systems are facing a series of challenges. As technology advances, attackers are attempting to spoof iris recognition systems using methods such as printed iris images or other artifacts. This poses a significant threat to the security of iris recognition systems and the privacy of iris data. To address this issue, researchers are concentrating on iris spoof detection to protect iris recognition systems from potential attacks. Iris spoof detection is a technique that verifies the authenticity of identity by analyzing the iris features and physiological responses of iris. The detection method can identify the deception of using printed iris images, artificially created iris or other imitations. By introducing spoof detection, iris recognition systems can improve the accuracy of verifying live irises, therefore enhance the security and the reliability of the entire iris recognition systems. Although spoof detection plays a positive role in improving the security of iris recognition systems, it also poses a risk to privacy leakage. During the spoof detection phase and client to server phase, the collection and processing of personal biometric data may result in the leakage of sensitive private information. As a consequence, it is critical to protect users' data privacy when designing the spoof detection methods.

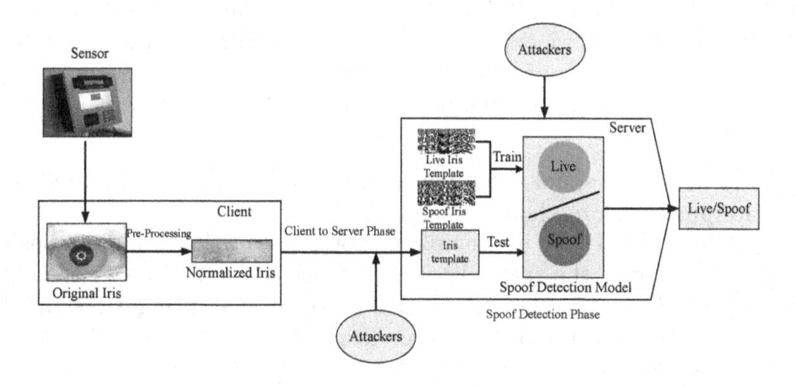

Fig. 1. Potential attacks on spoof detection

To protect iris recognition systems from attacks, researchers have proposed numerous iris spoof detection methods. Some algorithms use local features to classify the authenticity of irises, such as Zhang et al. [23] used weighted local binary pattern (LBP) features. Others use global features to detect spoof irises, for example, Raghavendra and Busch [16] proposed an algorithm utilizing binarized statistical image features of microtexture changes. Recently, some researchers use deep learning models to extract iris features automatically to perform iris spoof detection. While these spoof detection methods can protect iris recognition systems from spoofing attacks, they overlook the problem of privacy leakage caused by spoof detection process itself, which may result in significant privacy leakage and reduce the security of iris recognition systems. Specifically, when a spoof detection model is deployed on servers, the servers may be vulnerable to attacks, resulting in the leakage of iris data. Moreover, there is a concern of privacy leakage when sending iris data to the spoof detection model on the servers, as shown in Fig. 1. Consequently, spoof detection model itself can cause security issues to iris recognition systems.

In summary, as iris biometrics continue to develop and be applied, ensuring privacy protection during the iris spoof detection process is becoming increasingly crucial. By designing and employing the privacy protection methods for iris spoof detection, we can protect users' personal privacy while enhancing the security of iris recognition systems. This will establish a more reliable and trustworthy basis for the wide use of iris biometrics.

In this study, we apply various iris template protection methods to the iris spoof detection process and use the protected templates for training the spoof detection model. The specific steps involved are structured as follows: First, we pre-process the iris images and extract the features of irises. Next, we apply various iris template protection methods separately and transform them into protected and irreversible iris templates. Then, we use these protected and irreversible iris templates to train the spoof detection model, and evaluate the metrics related to different iris template protection methods, i.e. Attack Presentation Classification Error Rate (APCER), Bonafide Presentation Classification Error Rate (BPCER). Finally, we explore the feasibility of applying iris template protection methods to the spoof detection process.

The main contributions of this paper can be summarized in two aspects as follows:

1. we design a simple spoof detection model training framework based on iris template protection methods.
2. This paper explores four iris template protection methods and studies their effectiveness during the spoof detection. Numerous experiments on CASIA-Syn and CASIA-Interval datasets demonstrate that using these methods could effectively retain most of the performance of spoof detection methods while protecting the privacy.

The remainder of this paper is structured as follows: Sect. 2 introduces the related work about iris spoof detection and iris template protection methods. Section 3 introduces the designed framework, the methods to pre-process the iris images, four specific iris template protection methods and the spoof detection model employed in this paper. In Sect. 4, we present the experimental results and analyses. In Sect. 5, we conclude this paper.

2 Related Work

2.1 Iris Spoof Detection

To protect iris recognition systems from attacks, researchers have proposed numerous iris spoof detection methods. Current iris spoof detection methods can be mainly categorized into sensor-level and feature-level methods. Sensor-level methods achieve higher detection rates by incorporating some extra hardware integrating to the sensors, which can gather distinct biometric features of live organisms for identification purposes. He et al. [6] proposed a novel method based on the optical properties of the human eye. Puhan et al. [15] proposed a spoof detection method for contact lenses. While sensor-level methods have achieved satisfactory detection rates, feature-level methods are more attractive because of lower cost and greater flexibility. Conventional feature-level methods typically employ feature descriptors to extract biometric features. Zhang et al. [23] employed simplified scale-invariant feature transform (SIFT) descriptors to design weighted LBP features and extracted features. They then used Support Vector Machines (SVM) to perform spoof detection based on these features. The following year, Zhang et al. [24] proposed a general framework for iris spoof detection based on texture analysis, in which they used hierarchical visual codebook (HVC) to encode the iris images. As deep neural networks become increasingly popular for analyzing and understanding images, more and more iris spoof detection methods are focusing on incorporating deep learning algorithms. Menotti et al. [10] first utilized the deep learning methods to extract iris features automatically. In their method, there is no need to normalize the iris image before spoof detection. He et al. [5] proposed a multi-patch convolutional neural network (MCNN) approach for spoof detection. Daksha Yadav et al. [22] proposed a novel algorithm called DensePAD for iris spoof detection. This method primarily adopted convolutional neural networks based on DenseNet and performed well on various datasets. Kuehlkamp et al. [8] proposed a new spoof detection method based on convolutional neural network (CNN) and binarized statistical image feature (BSIF). This method initially employed BSIF to extract features from original iris images, and then input these features into a CNN for spoof detection. The methods using the deep neutral networks have achieved satisfactory results, but the introduction of these models also poses risks of privacy leakage to iris recognition systems as well. For other aspects of iris recognition systems, existing studies show the iris template protection methods can effectively enhance security by using protected iris templates instead of original iris images for processing. However, current spoof detection models still rely on the original iris data for training and detection, so the application of spoof detection models would lead to privacy leakage issues. Specifically, when an iris spoof detection model is deployed on the server, the server might be attacked and result in privacy leakage, as shown in Fig. 1. Consequently, spoof detection model itself can introduce security issues to iris recognition systems.

2.2 Iris Template Protection Methods

Iris template protection is a common method to safeguarding the privacy of iris data. Cancelable iris biometrics is an important part of the iris template protection methods. The concept of cancelable biometrics was proposed by Ratha et al. [17] in 2007. In their work, they utilized three irreversible transformations, i.e. Cartesian transformations, polar transformations, surface folding to transform the original fingerprint data to cancelable templates. Although initially designed for fingerprints, this study explored a novel direction for iris data privacy protection. Cancelable iris biometrics can be categorized to salting and irreversible transformations. Biometrics features salting methods create cancelable templates by incorporating random data into the original biometric features. Chin et al. [4] proposed a method to generate cancelable templates using iterative inner product and threshold processing. Teoh et al. [19] proposed an iris template protection method combined with the random projection method. In their study, they employed a random projection matrix to generate cancelable templates by projecting the original biometric data to another space. After that, Pilla et al. [14] proposed a method based on fan-shaped random projection. In their research, they divided the iris data into distinct sectors and performed random projection for each sector, and then combined the projected results to create cancelable templates. Irreversible transform methods generally refer to utilize a unidirectional transformation function to convert original biometric features into cancelable templates. Zuo et al. [26] proposed two cancelable biometric methods based on irreversible transformation, namely GREY-COMBO and BIN-COMBO. Ouda et al. [12] proposed a cancelable iris biometrics scheme which does not require any tokenised random number. After that, an irreversible transformation method based on Bloom filter was proposed by Rathgeb et al. [18]. In this study, the researchers divided iris images into blocks and stored the iris data from each block in Bloom filters, and then generated cancelable iris templates by these Bloom filters. To address the issues of large template size and long response time associated with the irreversible transformation method based on Bloom filters, Ajish et al. [1] proposed a method that employs double Bloom filters. This method not only addresses the previous issues, but also enhances the irreversibility of protected templates. Recently, to address the issue of prealigning iris features, Lee et al. [9] proposed a method known as the random augmented histogram of gradients. This solution utilizes the column vector random augmentation technique and the orientation gradient histogram technique to generate cancelable iris templates that are alignment-robust.

3 Methodology

In this section, we will describe the methods mentioned in this paper specifically. First, we will design a framework called TPISD for protecting privacy during the spoof detection phase and client to serve phase based on iris template protection methods, and introduce the procedure for pre-processing iris images. Considering the diversity of iris template protection methods, we select four methods to provide a brief introduction. Meanwhile, to improve the performance of spoof detection models, we adopt the pre-training models with attention mechanism. The entire framework is shown as Fig. 2. A detailed introduction of each step and the methods used will be presented in the following sections.

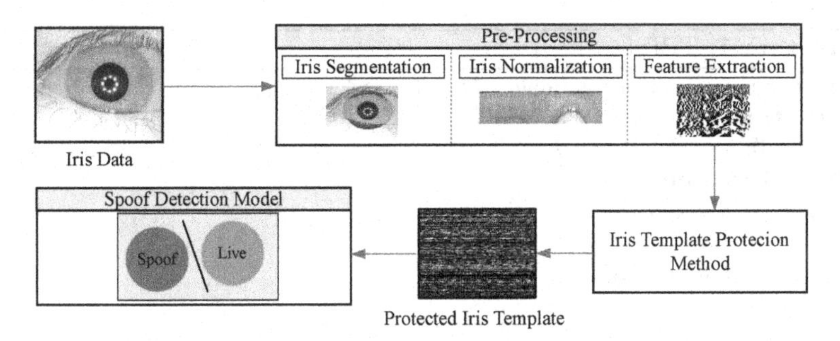

Fig. 2. The structure of TPISD

3.1 TPISD

In this section, we will give an introduction of the specific process of the framework TPISD. The original iris data serves as the input for this framework. First, we initialize an empty set B. Next, we iterate through each image x_i in set D, and apply the iris segmentation method to obtain the segmented iris image y_i from x_i. Subsequently, we employ the iris normalization method to obtain the normalized iris image z_i from y_i. Afterward, we use the feature extraction method based on three Gabor filters to obtain the feature image u_i from z_i and add it to set B. Subsequently, we perform image clipping on the feature images u_i to meet the input requirements of the iris template protection method. Next, we employ the iris template protection method to transform the image u_i to the protected template v_i. Finally, we process the data v_i by the function $Data\_Clean()$, which adjusts the data format and size of v_i according to the requirements of the subsequent CNN model. As the requirements for the input of iris template protection methods vary, we will provide detailed descriptions for four typical methods, i.e. random projection, local ranking, double Bloom filter, and random augmented histogram of gradients in the subsequent sections. Finally, the generated protected template t_i is inputted into a CNN network to train the spoof detection model. Algorithm 1 shows the pseudo-code of the above process.

3.2 Image Pre-processing

In general, spoof detection models utilize original iris images for training. However, the framework proposed in this paper uses the generated protected templates as training data. Since protected templates are usually generated after some procedures like segmentation, normalization, and feature extraction, TPISD also requires to pre-process iris images by these approaches. To achieve this, we utilize OSIRIS [11] (version 4.1) to perform iris pre-processing.

Algorithm 1: TPISD

Input: Original set of iris images $D = \{x_1, x_2, \dots, x_n\}$;
Output: Spoof detection model M;
1 Let $B = \emptyset$;
2 **for** $i \leftarrow 1$ **to** n **do**
3 $\quad y_i = Segmentation(x_i)$;
4 $\quad z_i = Normalization(y_i)$;
5 $\quad u_i = Extraction(z_i)$;
6 $\quad B = B \cup \{u_i\}$;
7 **end**
8 Let $C = \emptyset$;
9 **for** $i \leftarrow 1$ **to** n **do**
10 $\quad v_i = TemplateProtect(u_i)$;
11 $\quad t_i = Data\_Clean(v_i)$;
12 $\quad C = C \cup \{t_i\}$;
13 **end**
14 Train(M, C)`// use protected iris templates to train the model` M

3.3 Iris Template Protection Methods

(1) Random Projection-Based Approach: The random projection-based approach was proposed by Teoh et al. [19] in 2007. The specific process of random projection is shown in Fig. 3. It is a method that maps high-dimensional iris data onto a lower-dimensional space, which is a subspace generated randomly. This transformation process can be achieved by calculating the dot product between the original iris data and a randomly generated matrix. As the distribution of the matrix is random across each dimension, it could effectively compress the high-dimensional information of the original iris data into a lower-dimensional space.

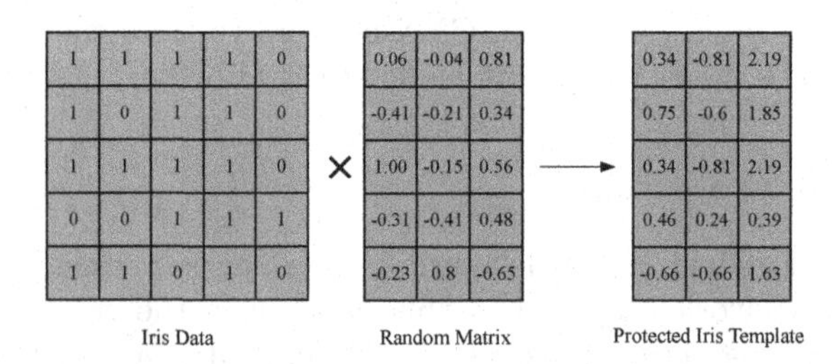

Fig. 3. An illustration of random projection-based method [19]

(2) Local Ranking: The iris template protection method based on local ranking was proposed by Zhao et al. [25] in 2018. The specific process of local ranking is shown in Fig. 4. First, it performs an XOR operation between the iris data x

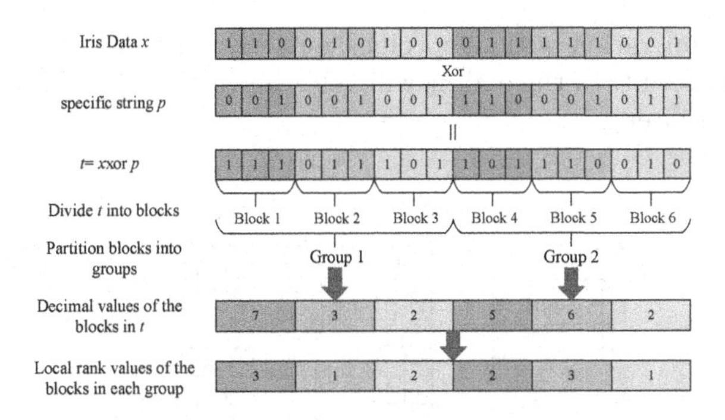

Fig. 4. An illustration of local ranking-based method [25]

and an application-specific string to obtain t. Next, it divides t into blocks based on the block size b, and each d blocks form a group ($b = 3, d = 3$ in the figure). It computes the corresponding decimal value for each block, and sorts all the blocks in each group by the decimal values. Finally, it stores the rank values as the protected templates instead of original iris data.

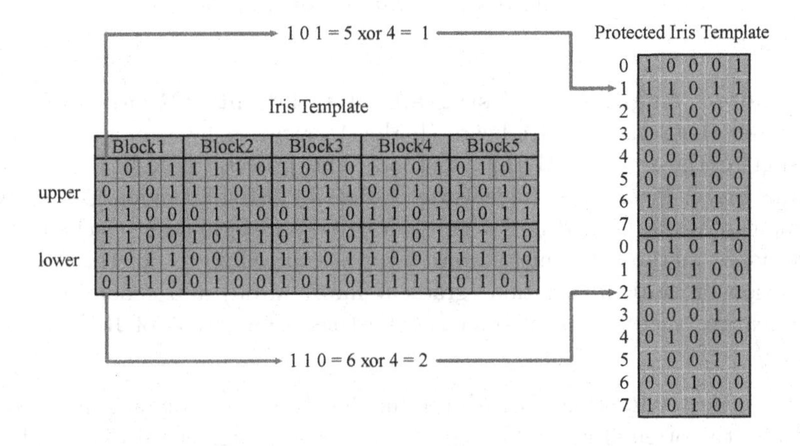

Fig. 5. An illustration of double Bloom filter-based method [1]

(3) Double Bloom Filter: The iris template protection method based on double Bloom filter was proposed by Ajish et al. [1] in 2020. The specific process of double Bloom filter method is shown in Fig. 5. The method first divides the iris feature matrix into K equal-sized blocks and initializes a Bloom filter for each block. Since most of the iris discriminative information is located at the top of the feature matrix, the transformation of the method only targets the w bits in the upper part of the feature matrix. Each column in the block is divided into two parts of equal size and stored in two Bloom filters, respectively. The transformation process is to convert the corresponding data to decimal and to perform an operation with a specific key (e.g. 4 in Fig. 5) to get the final transformation result. In the end, it will get a protected iris template composed of many Bloom filters.

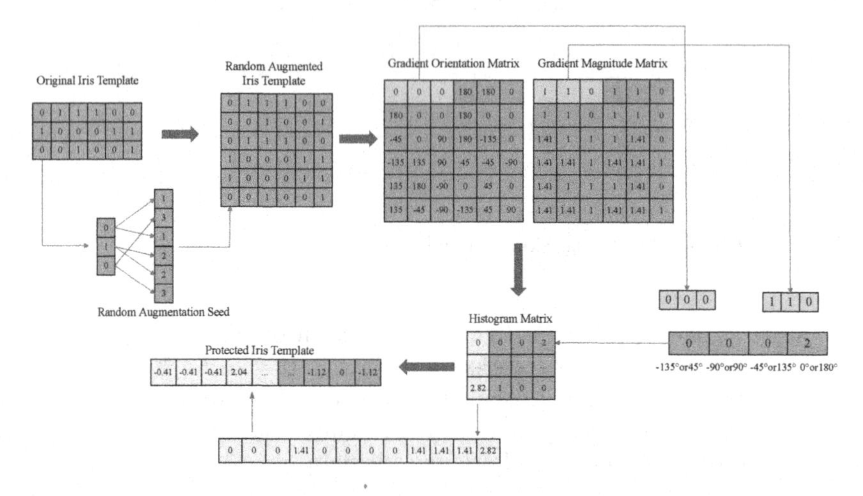

Fig. 6. An illustration of R·HoG-based method [9]

(4) Random Augmented Histogram of Gradients (R·HoG): Random Augmented Histogram of Gradients (R·HoG) proposed by Lee et al. [9] is an extension to the traditional HoG method. The process of R·HoG method is shown in Fig. 6. It extracts the features of HoG by using a random column vectors augmentation method and a gradient orientation grouping mechanism. In the proposed method, the histogram vector is utilized to record the frequency distribution of local information (gradient information) and generate the biometric feature vector with alignment robustness. The steps of R·HoG are as follows:

1) Column Vector Random Augmentation: Generate a random augment seed to amplify the original iris data, and obtain random augmented iris template.
2) Gradient Orientation and Magnitude Computation: Compute the orientation and the magnitude of the gradients corresponding to the random augment iris template and obtain the matrices of gradient orientation and magnitude. Additionally, when we compute the gradients, we regard the first and the last row as neighboring element, as well as column.
3) Feature Matrix Partitioning and Histogram Formalization: Partition the feature matrix into n non-overlapping sub-matrix cells, where each cell's size is $b * a$. Generate a vector of gradient histogram for each cell, and stack these histograms to generate the histogram matrix.
4) Z-Score Transformation: For each column vector in the histogram matrix, apply a normalization operation and concatenate all the normalized vectors in the end.

3.4 Spoof Detection Model

Since this paper primarily focuses on the feasibility of applying iris template protection methods in the context of spoof detection models, an in-depth investigation of the spoof detection model itself is not necessary. This paper utilizes an Attention-Guided Iris Presentation Attack Detector (AGPAD) proposed by Chen [3] in 2021 as the spoof detection model. The entire structure of this model is shown in the Fig. 7.

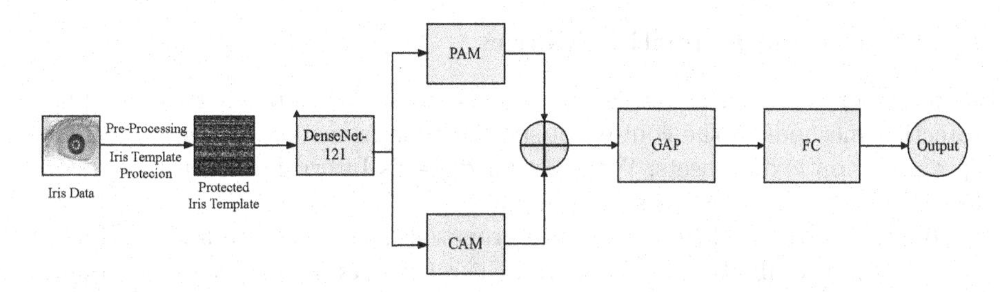

Fig. 7. Spoof detection model framework [3]

Initially, this model utilizes DenseNet-121 as a backbone model to extract feature images. We input the protected iris templates to DenseNet-121 and obtain features of protected iris templates. Next, the channel attention module (CAM) and position attention module (PAM) are employed respectively to capture the dependence relationships among channels and positions. Finally, by fusing the outputs of the two attention modules by element, complementary attention features are captured from both channel and position dimensions. And then we input the previous output to global average pooling (GAP) and full connection (FC) layers. Finally, we utilize softmax operations to compute the class probability (live or spoof).

3.5 Security Analysis

The security of biometric templates can usually be judged by revocability, unlinkability and irreversibility. For the problem studied in this paper, revocability enables the generation of new protected iris templates when the templates on the server are compromised. Irreversibility ensures that attackers can't utilize leaked templates to reconstruct users' original iris data when the user templates on the server are compromised. Unlinkability ensures that cross matching cannot be performed between the templates from different applications, and prevents attackers from verifying whether the two templates belong to the same user.

As the proposed framework mainly investigates the effectiveness of serveral iris template protection methods on protecting the privacy in spoof detection, its security depends on the security of the adopted iris template protection method.

The security of random projection-based method, local ranking-based method, double Bloom filter-based method and R·HoG is demonstrated in [1,9,19,25], respectively. According to Fig. 1, if the adopted method is secure, then the proposed framework can protect original iris data during the spoof detection phase as well as client to server phase, and prevent the eavesdropping attacks and server intrusion attacks.

4 Experiment

4.1 Dataset and Evaluation Metrics

In this paper, we need to investigate the feasibility of employing iris template protection methods in the context of spoof detection models. Thus, we conduct experiments on two datasets. We select CASIA-Iris-Interval dataset as live samples, and CASIA-Iris-Syn as spoof samples.

To evaluate the performance of the proposed framework, we utilize APCER and BPCER as evaluation metrics. APCER represents the rate of spoof samples that are incorrectly classified as live samples. BPCER represents the rate of live samples that are incorrectly classified as spoof samples. The precise formulas of the two metrics are as follows:

$$APCER = FN/(TP + FN) \tag{1}$$

$$BPCER = FP/(FP + TN) \tag{2}$$

FN denotes the number of spoof samples that are detected as live, TN correctly denotes the number of live samples that are detected as live, FP denotes the number of live samples that are detected as spoof, and TP denotes the number of spoof samples that are correctly detected as spoof.

4.2 Experimental Setup

To measure our method, we train the model using both the original iris templates and the protected iris templates, separately. Different iris template protection methods have different parameters, and the detailed experimental settings of the four methods used in this paper are discussed as follows.

Random Projection: In this method, there exists only a single adjustable parameter, which is the target dimension. When utilizing different target dimensions k, the degree of data compression varies. As a result, the security of the generated templates and the extent of retained biometric features differ. Moreover, to avoid the fortuity of experiments, we conduct multiple trials for each target dimension (i.e. $k = 272, 120, 67$) and employ distinct random matrices each time to calculate the average performance across all experiments and ultimately obtain the final result.

Local Ranking: Since local ranking works on the binary representation of the iris, we ought to transform the original iris images into binary strings. Next, we apply the local ranking approach to further transform it and obtain the protected iris template. Subsequently, we reshape the template into a two-dimensional matrix. As local ranking has two parameters (block size b and group size d), we carry out experiments with various values for these two parameters (i.e. $b = 2,\ 4,\ 8,\ d = 4,\ 8,\ 16,\ 32$), respectively.

Double Bloom Filter: Since this method only uses the upper section of an iris template as input, we crop the original iris template. The variable parameter in this method is the row size of the upper section w, which affects the size of the Bloom filter. Therefore, we carry out experiments with various values (i.e. $w = 10,\ 15,\ 20,\ 25$) for this parameter.

R·HoG: This method uses a column vector random augmentation technique, so we utilize a unified random augment seed to amplify all the original iris templates. This method involves two parameters: the sub-matrix column size a and row size b after computing the gradient orientation matrix and gradient magnitude matrix and partitioning the two matrices. Therefore, we carry out experiments with various values for these two parameters (i.e. $a = 2,\ 5,\ 10,\ b = 8,\ 16,\ 32,\ 64$), respectively.

4.3 Transformation Parameter Experiment

In this subsection, we investigate the influence of parameters in template protection methods on the performance of the proposed framework, and try to identify the optimal parameters. The results are shown in Tables 1, 2, 3, 4 and 5.

Table 1. The performance of TPISD using the random projection

k	272	120	67
APCER	0	0	0
BPCER	0	0	0

Random Projection: When using the random projection in the proposed framework, we use three settings of target dimension k, i.e. 272, 120 and 67. As shown in Table 1, all values of APCER and BPCER are 0, which indicates that applying the random projection to iris spoof detection does not lead to the degradation of detection performance. The reason might be that: random projection could retain a large number of biometric features after transforming the original iris data. Therefore, it is feasible to utilize the random projection to enhance the privacy during iris spoof detection.

Table 2. The performance of TPISD using local ranking with different b and d

d		b		
		2	4	8
4	APCER	0.316	0.319	0.316
	BPCER	0.037	0.027	0.039
8	APCER	0.221	0.273	0.174
	BPCER	0.051	0.025	0.106
16	APCER	0.258	0.208	0.277
	BPCER	0.025	0.035	0.038
32	APCER	0.413	0.223	**0.149**
	BPCER	**0.003**	0.024	0.083

Local Ranking: When using the local ranking in the proposed framework, we use three settings of b, i.e. 2, 4, 8 and four settings of d, i.e. 4, 8, 16 and 32. As shown in Table 2, the results show that using the local ranking would lead to an increase of APCER and BPCER, and the best results are achieved with $b = 8, d = 32$. Although local ranking can enhance the privacy during iris spoof detection, it also leads to a non-negligible degradation of detection performance.

Table 3. The performance of TPISD using the double Bloom filter

w	10	15	20	25
APCER	0.060	**0.032**	0.161	0.229
BPCER	0.079	0.080	**0.038**	0.087

Double Bloom Filter: When using the double Bloom filter in the proposed framework, we use four settings of the row size of upper section w, i.e. 10, 15, 20 and 25. As shown in Table 3, all values of the APCER and BPCER increase slightly, which indicates using the double Bloom filter to enhance the privacy during iris spoof detection would lead to a slight decrease in detection performance, and the best results are achieved with $w = 15$. Overall, the results demonstrate that it is feasible to utilize the double Bloom filter to enhance the privacy during iris spoof detection.

R·HoG: When using the R·HoG in the proposed framework, we use three settings of a, i.e. 2, 5, 10 and four settings of b, i.e. 8, 16, 32 and 64. As shown in Table 4, the best results are achieved with $a = 2, b = 32$. Although the results show that using the R·HoG can enhance privacy during iris spoof detection, it leads to a non-negligible increase of APCER and BPCER. The reason might be that the noise, which is produced in column vector random augmentation, would destroy the texture features in iris templates. Additionally, in the step

Table 4. The performance of TPISD using R·HoG with different b and a

b		a		
		2	5	10
8	APCER	0.346	0.300	0.367
	BPCER	0.053	0.064	0.093
16	APCER	**0.062**	0.367	0.435
	BPCER	0.252	0.054	0.114
32	APCER	0.064	0.128	0.371
	BPCER	0.117	0.119	0.150
64	APCER	0.361	0.124	0.231
	BPCER	**0.016**	0.125	0.125

of column vector random augmentation, the scale of the random augment seed influences detection performance as well. For time reasons, we fix it to the same value instead of conduct more experiments with this parameter.

Table 5. Comparison of the results of TPISD using the four iris template protection methods

Method	APCER	BPCER
Baseline	0	0
Random Projection ($k = 120$)	0	0
Local Ranking ($b = 8, d = 32$)	0.149	0.083
Double Bloom Filter ($w = 15$)	0.032	0.080
R·HoG ($a = 2, b = 32$)	0.064	0.117

Comparison of Four Iris Template Protection Methods: In this part, we select the transformation parameters with the best performance in the previous experiments for comparison. The results are shown in Table 5. "Baseline" represents the iris spoof detection method without using any iris template protection methods. It is easy to observe that, the best results are achieved by the random projection, and local ranking performs the worst. It is worthing mention that, since we cannot get a large and comprehensive dataset, the results would change on other datasets.

5 Conclusion

In this paper, we design a framework called TPISD for protecting privacy during iris spoof detection. TPISD mainly investigates whether it is feasible to applying iris template protection methods to enhance privacy while maintaining the spoof detection performance. Specifically, we choose four typical iris template protection methods, i.e. random projection-based method, double Bloom filter-based method, local ranking-based method and R·HoG. The experimental results on CASIA-Iris-Interval and CASIA-Iris-Syn datasets show that using random projection and double Bloom filter can better maintain the detection performance, and using local ranking and R·HoG may induce a non-negligible degradation of detection performance. In future, we will explore the feasibility of employing other iris template protection methods like fuzzy vaults, fuzzy commitment and fuzzy extractor in the spoof detection phase.

Acknowledgments. This work is partially supported by the National Natural Science Foundation of China (Grant No. 61806151), and the Natural Science Foundation of Chongqing City (Grant No. cstc2021jcyj-msxmX0002).

References

1. Ajish, S., AnilKumar, K.: Iris template protection using double bloom filter based feature transformation. Comput. Secur. **97**, 1–15 (2020)
2. Aleem, S., Yang, P., Masood, S., Li, P., Sheng, B.: An accurate multi-modal biometric identification system for person identification via fusion of face and finger print. World Wide Web **23**, 1299–1317 (2020)
3. Chen, C., Ross, A.: An explainable attention-guided iris presentation attack detector. In: Proceedings of the IEEE/CVF Winter Conference on Applications of Computer Vision, pp. 97–106 (2021)
4. Chin, C.S., Jin, A.T.B., Ling, D.N.C.: High security iris verification system based on random secret integration. Comput. Vis. Image Underst. **102**(2), 169–177 (2006)
5. He, L., Li, H., Liu, F., Liu, N., Sun, Z., He, Z.: Multi-patch convolution neural network for iris liveness detection. In: 2016 IEEE 8th International Conference on Biometrics Theory, Applications and Systems (BTAS), pp. 1–7. IEEE (2016)
6. He, Y., Hou, Y., Li, Y., Wang, Y.: Liveness iris detection method based on the eye's optical features. In: Optics and Photonics for Counterterrorism and Crime Fighting VI and Optical Materials in Defence Systems Technology VII, vol. 7838, pp. 236–243. SPIE (2010)
7. Kim, M., Jain, A.K., Liu, X.: AdaFace: quality adaptive margin for face recognition. In: Proceedings of the IEEE/CVF Conference on Computer Vision and Pattern Recognition, pp. 18750–18759 (2022)
8. Kuehlkamp, A., Pinto, A., Rocha, A., Bowyer, K.W., Czajka, A.: Ensemble of multi-view learning classifiers for cross-domain iris presentation attack detection. IEEE Trans. Inf. Forensics Secur. **14**(6), 1419–1431 (2018)
9. Lee, M.J., Jin, Z., Liang, S.N., Tistarelli, M.: Alignment-robust cancelable biometric scheme for iris verification. IEEE Trans. Inf. Forensics Secur. **17**, 3449–3464 (2022)
10. Menotti, D., et al.: Deep representations for iris, face, and fingerprint spoofing detection. IEEE Trans. Inf. Forensics Secur. **10**(4), 864–879 (2015)
11. Othman, N., Dorizzi, B., Garcia-Salicetti, S.: OSIRIS: an open source iris recognition software. Pattern Recogn. Lett. **82**, 124–131 (2016)
12. Ouda, O., Tsumura, N., Nakaguchi, T.: Tokenless cancelable biometrics scheme for protecting iris codes. In: 2010 20th International Conference on Pattern Recognition, pp. 882–885. IEEE (2010)

13. Öztürk, H.İ., Selbes, B., Artan, Y.: MinNet: minutia patch embedding network for automated latent fingerprint recognition. In: Proceedings of the IEEE/CVF Conference on Computer Vision and Pattern Recognition, pp. 1627–1635 (2022)
14. Pillai, J.K., Patel, V.M., Chellappa, R., Ratha, N.K.: Sectored random projections for cancelable iris biometrics. In: 2010 IEEE International Conference on Acoustics, Speech and Signal Processing, pp. 1838–1841. IEEE (2010)
15. Puhan, N.B., Sudha, N., Hegde, S.: A new iris liveness detection method against contact lens spoofing. In: 2011 IEEE 15th International Symposium on Consumer Electronics (ISCE), pp. 71–74. IEEE (2011)
16. Raghavendra, R., Busch, C.: Robust scheme for iris presentation attack detection using multiscale binarized statistical image features. IEEE Trans. Inf. Forensics Secur. **10**(4), 703–715 (2015)
17. Ratha, N.K., Chikkerur, S., Connell, J.H., Bolle, R.M.: Generating cancelable fingerprint templates. IEEE Trans. Pattern Anal. Mach. Intell. **29**(4), 561–572 (2007)
18. Rathgeb, C., Breitinger, F., Busch, C.: Alignment-free cancelable iris biometric templates based on adaptive bloom filters. In: 2013 International Conference on Biometrics (ICB), pp. 1–8. IEEE (2013)
19. Teoh, A.B.J., Yuang, C.T.: Cancelable biometrics realization with multispace random projections. IEEE Trans. Syst. Man Cybern. Part B (Cybernetics) **37**(5), 1096–1106 (2007)
20. Wei, J., Huang, H., Wang, Y., He, R., Sun, Z.: Towards more discriminative and robust iris recognition by learning uncertain factors. IEEE Trans. Inf. Forensics Secur. **17**, 865–879 (2022)
21. Wildes, R.P.: Iris recognition: an emerging biometric technology. Proc. IEEE **85**(9), 1348–1363 (1997)
22. Yadav, D., Kohli, N., Vatsa, M., Singh, R., Noore, A.: Detecting textured contact lens in uncontrolled environment using DensePAD. In: Proceedings of the IEEE/CVF Conference on Computer Vision and Pattern Recognition Workshops, pp. 2336–2344 (2019)
23. Zhang, H., Sun, Z., Tan, T.: Contact lens detection based on weighted LBP. In: 2010 20th International Conference on Pattern Recognition, pp. 4279–4282. IEEE (2010)
24. Zhang, H., Sun, Z., Tan, T., Wang, J.: Learning hierarchical visual codebook for iris liveness detection. In: International Joint Conference on Biometrics, vol. 1, pp. 1–8 (2011)
25. Zhao, D., Fang, S., Xiang, J., Tian, J., Xiong, S.: Iris template protection based on local ranking. Secur. Commun. Netw. **2018**, 1–9 (2018)
26. Zuo, J., Ratha, N.K., Connell, J.H.: Cancelable iris biometric. In: 2008 19th International Conference on Pattern Recognition, pp. 1–4. IEEE (2008)

Stock Volatility Prediction Based on Transformer Model Using Mixed-Frequency Data

Wenting Liu[1], Zhaozhong Gui[2], Guilin Jiang[3(✉)], Lihua Tang[2], Lichun Zhou[1], Wan Leng[2], Xulong Zhang[4], and Yujiang Liu[5]

[1] Chasing Jixiang Life Insurance Co., Ltd., Changsha, China
[2] Hunan Chasing Digital Technology Co., Ltd., Changsha, China
[3] Hunan Chasing Financial Holdings Co., Ltd., Changsha, China
jiangguilin@hnchasing.com
[4] Ping An Technology (Shenzhen) Co., Ltd., Shenzhen, China
[5] The University of Melbourne, Melbourne, Australia

Abstract. With the increasing volume of high-frequency data in the information age, both challenges and opportunities arise in the prediction of stock volatility. On one hand, the outcome of prediction using tradition method combining stock technical and macroeconomic indicators still leaves room for improvement; on the other hand, macroeconomic indicators and peoples' search record on those search engines affecting their interested topics will intuitively have an impact on the stock volatility. For the convenience of assessment of the influence of these indicators, macroeconomic indicators and stock technical indicators are then grouped into objective factors, while Baidu search indices implying people's interested topics are defined as subjective factors. To align different frequency data, we introduce GARCH-MIDAS model. After mixing all the above data, we then feed them into Transformer model as part of the training data. Our experiments show that this model outperforms the baselines in terms of mean square error. The adaption of both types of data under Transformer model significantly reduces the mean square error from 1.00 to 0.86.

Keywords: stock volatility prediction · mixed-frequency model · transformer model

1 Introduction

Measuring and predicting market risk is the primary prerequisite for managing and controlling financial markets. Among them, the volatility of financial assets is a commonly used characteristic indicator to measure the risk in them, making it also a core issue in financial research. However, the volatility of financial assets cannot be directly observed through our eyes. To solve this problem, extensive research has been conducted on measurement methods for volatility.

X. Song et al. (Eds.): APWeb-WAIM 2023, LNCS 14332, pp. 74–88, 2024.
https://doi.org/10.1007/978-981-97-2390-4_6

In the early days, volatility was directly measured by variance or standard deviation. Various models were constructed to evaluate volatility, such as ARCH (AutoRegressive Conditional Heteroskedasticity) model [6] and SV (Stochastic Volatility) model [23]. These models are the basic models for studying financial time series and can reflect the fluctuation characteristics of variances. On this basis, the focus of research has gradually shifted to predicting the volatility of financial assets.

With the development of machine learning techniques, models like LM (Long Memory) [9] and Markov-switching model [20] were introduced subsequently and significantly reduced the prediction error compared to traditional statistic models. Machine unlearning methodology was optimized on the basis of Stochastic Teacher Network [29]. Indicators such as the popularity of daily news and investors' sentiment were also incorporated into models [10].

However, current research faced a unified problem in selecting auxiliary indicators for volatility prediction. Firstly, the data of macroeconomic variables is usually produced on a monthly basis, while that of financial assets is by minute or even by second. So the difference in data frequency to consider various types of indicators in volatility prediction. Secondly, investors' subjective emotions greatly affect their investment behavior. Therefore, how to choose indicators that can well reflect investors' subjective emotions is also a challenge. In order to address this issue, GARCH-MIDAS (Generalized AutoRegressive Conditional Heteroskedasticity and Mixed Data Sampling) model [7] was proposed for data processing, to extract macroeconomic information so as to incorporate more objective factors reflecting volatility changes.

Recently, volatility prediction models have been extended to deep learning models [19]. Models such as LSTM (Long Short-Term Memory) [15], TabNet [18] and CNN (Convolutional Neural Network) [26] were introduced into this area and all demonstrated outstanding performance in mixing different types of data and predicting volatility. At the same time, self-attention-based architectures, in particular the Transformers, have become the up-to-date method in more and more fields such as NLP (Natural Language Processing) [25] and CV (Computer Vision) [5]. Motivated by their breakthroughs, we introduce the Transformer model to the prediction of financial market.

The outputs of the GARCH-MIDAS model are deployed to train the Transformer model, which is one of the innovations of this paper. This paper also selects Baidu search index as an indicator of investors' sentiment. It is derived from the search activity of Baidu users for specific keywords on the Baidu search engine. By integrating these metrics, we observe a notable enhancement in the predictive capabilities of our Transformer model. Our contributions are summarized as follows:

1) To enrich data dimensions by incorporating macroeconomic and investor attention factors. We applied macroeconomic data information of different frequencies to volatility prediction and used multiple Baidu indices to measure investor attention, thereby improving the effectiveness of volatility prediction.

2) To evaluate the applicability and prediction effectiveness of deep learning techniques for stock-related data. Given the various models applied to volatility prediction, there is still room for enhancement. We used the Transformer model to effectively improve the prediction accuracy of the model, thereby demonstrating the effectiveness of this model in predicting volatility.

2 Related Works

With the progress and development of economy and society, the interest and depth of research on financial asset volatility are increasing day by day. Recent studies on volatility can be mainly divided into 2 categories from a research perspective:

The first category focuses on studying volatility from a prediction perspective, exploring based on different types of data and models. Choudhury et al. [3] used support vector machines to predict future prices and developing short-term trading strategies based on the predictions, and the test simulation achieved good profits within 15 days. S. Chen et al. [22] established a HAR volatility modeling framework based on the Baidu search index, incorporating it with the jumping, good and bad volatility optimization models. They then evaluated the effectiveness of the model through MCS testing. In the work by Hu [14], a novel hybrid approach was introduced for forecasting fluctuations in copper prices. This innovative method synergistically integrates the GARCH model, Long Short-Term Memory (LSTM) network, and conventional Artificial Neural Network (ANN), yielding commendable accuracy in predicting price volatility. Y. H. Umar and M. Adeoye [24] estimated the volatility using the Markov regime conversion method by comparing all monthly stock index data of the Central Bank of Nigeria (CBN) from 1988 to 2018 in the statistical bulletin of the Nigerian Stock Exchange. B. Schulte-Tillman [21] proposed four multiplicative component volatility MIDAS models to distinguish short-term and long-term volatility, and found that specific long-term variables in the MIDAS model significantly improved prediction accuracy, as well as the superior performance of a Markov switching MIDAS specification (in a set of competitive models). A. Vidal et al. [26] used a CNN-LSTM model to predict gold volatility. At the same time in deep learning field, A. Vaswani et al. [25] and A. Dosovitskie et al. [5] proposed to use Transformer to replace LSTM and CNN in NLP and CV field.

The second category focuses on studying the factors that influence volatility, emphasizing on analyzing the factors that affect volatility and their impact. C. Christiansen et al. [4] conducted an in-depth investigation into the drivers behind fluctuations in financial market volatility. Their study encompassed a thorough exploration of the predictive influence exerted by macroeconomic and financial indicators on market volatility. R. Hisano et al. [13] undertook an assessment of the influence of news on trading dynamics. Their analysis encompassed an extensive dataset of over 24 million news records sourced from Thomson Reuters, examining their correlations with trading behaviors within the S&P US Index's prominent 206 stocks. C. A. Hartwell [12] constructed a unique monthly database from 1991 to 2017 to explore the impact of institutional fluctuations on financial volatility in transition economies. F. Audrino et al. [2] adopted a latest sentiment classification technique, combining social media, news releases, information consumption, and search engine data to analyze the impact of emotions and attention variables on stock market volatility; F. Liu et al. [17] studied the long-term dynamic situation of volatility from two levels: horizontal values and volatility, and selected four macroeconomic variables to analyze their impact. P. Wang's [27] study concentrated on important stocks in the stock exchange market's financial sector. The objective was to assess the influence of margin trading and stock lending on the price volatility of these chosen stocks. The findings revealed a noteworthy observation: both margin trading activities and the balances associated with such trading had the potential to amplify the level of volatility in stock prices.

3 Methodology

In this section, a briefing of the volatility theory and feature extraction method will be given. How the GARCH-MIDAS and the Transformer are deployed in the prediction of stock volatility will also be explained.

3.1 Basic Theory of Volatility

In order to explore the real market volatility, this article uses RV (Realized Volatility) as an indicator to measure the volatility of the CSI300. The calculation of RV was defined by Andersen and Bollerslev (1998) [1], with the following specific formulas:

$$R_t = 100(\ln Pr_t - \ln Pr_{t-1}) \tag{1}$$

$$R_{t,d} = 100(\ln Pr_{t,d} - \ln Pr_{t,d-1}) \tag{2}$$

$$RV_t = \sum_{d=1}^{48} R_{t,d}^2 \tag{3}$$

where Pr_t, R_t and RV_t represent the price, the return and RV on the t-day, respectively. $Pr_{t,d}$ and $R_{t,d}$ represent the price and the return on the 5-minute interval of the t-day, respectively.

However, it is well known that the trading of stock market occurs within a limited time rather than 24 h without interruption. Hansen and Lunde (2005) [11] further demonstrated that the previously defined RV lacks information during non-trading time and proposed to use scale parameter to adjust it appropriately. The approach proposed by these scholars is written into the adjusted formula below.

$$\lambda = \frac{\sum_{t=1}^{N} R_t^2/N}{\sum_{t=1}^{N} RV_t/N} \tag{4}$$

$$RV_t' = \lambda \times RV_t \tag{5}$$

where λ is the scale parameter and RV_t' stands for the adjusted RV on the t-day.

Table 1. Meaning of Volatility Related Factors

Factor		Index	Variable
Objectivity Factors	Macroeconomic Indicators	Macro-economic Consensus Index (Current Value)	MeCI
		Macro-economic Leading Index (Current Value)	MeLeI
		Macro-economic Lagging Index (Current Value)	MeLaI
		Consumer Price Index (CPI, Last month = 100)	CPI
		Total Retail Sales of Consumer Goods (Current Value/Yuan)	Retailsale
		Retail Price Index (RPI, Last month = 100)	RPI
		Producer Price Index (PPI, Last month = 100)	PPI
		Money Supply (Total Balance at End of Period, Yuan)	M2
		Fixed Asset Investment (Cumulative Value, Yuan)	FInvest
		Total Imports and Exports (Current Value/US$s)	IOP
	Stock Technical Indicators	Turnover Rate(%)	Turn
		Bollinger Bands Indicator (Median Line/Number of Periods 26)	BOLL
		5-day Moving Average	MA(5)
		20-day Moving Average	MA(20)
		Moving Average Convergence Divergence	MACD
		Relative Strength Index (Number of Periods 6)	RSI
		Selling On-Balance Volume	SOBV
		Rate of Change	ROC
		Trading Volume	Volume
		Highest Price	High
		Lowest Price	Low
		Open Price	Open
Subjective Factors	Attention Indicators	"CSI300" Baidu Search Index	CSI300
		"CSI500" Baidu Search Index	CSI500
		"SSE50" Baidu Search Index	SSE50
		"Components of CSI300" Baidu Search Index	HSparts
		"CSI300 Index Fund" Baidu Search Index	HSETF

3.2 Definition of Indicators

The research object of this article is CSI300, which integrates the information of the top 300 excellent stocks. The frequency for calculating RV is every 5 min. In addition, the data of Baidu search index is obtained by crawling from the internet using python software. In order to measure the impact of various factors on stock returns, we refer to the research of scholars S. Li [16] and M. Zhang [28]. And we determine the final indicators according to the grey correlation degree of indicators and returns, as well as the Baidu demand map. The objective factors and subjective factors are as shown in Table 1.

Considering the large number of indicators, the inconsistent scales among them and the characteristics of data (such as non-stationarity, large fluctuations, and missing values), pre-processing of all data is required before model construction for later use and analysis. The pre-processing mainly includes two aspects: missing value filling and data normalization. In addition, in order to minimize information loss of each indicator and avoid multiple collinearity between indicators, we use PCA (Principal Component Analysis) methods to extract principal components and construct a comprehensive index.

In terms of macroeconomic indicators, two principal components (PCM1 and PCM2) are extracted, which together capture 98.3% of the information. PCM1 carries a major positive load distribution on indices e.g. CPI, Retailsale and

Finvest, so it is labeled as "consumption and investment component". PCM2 carries a major positive load on indices e.g. RPI, PPI, etc., so it is labeled as "production and prosperity component". In terms of stock technical indicators, we extract three principal components (TECH1, TECH2 and TECH3), with a total contribution ratio close to 100%. TECH1 consists of indices like average price, highest price and lowest price, mainly reflecting the size of CSI300, so it is labeled as "price component". TECH2 consists of indices like ROC, MACD and etc., reflecting the stock price changes and market attention of CSI300, so it is labeled as "trend component". TECH3 consists of indices like Turn and Volumn, reflecting the liquidity of CSI300, so it is labeled as "liquidity component". Based on the results of the scree plot, we get the investor attention component (BD1), with a variance contribution rate of approximately 88.6%, which was labeled as the "attention component". The detailed compositions of all the above principal components are recorded in Table 2.

Table 2. The load value of the principal components of each indicator.

Macroeconomic Indicators			Stock Technical Indicators				Attention Indicators	
	PCM1	PCM2		TECH1	TECH2	TECH3		BD1
MeCI	0.24	0.5	Turn	0.17	0.21	0.96	CSI300	0.77
MeLeI	0.02	−0.14	BOLL	0.97	−0.04	0.21	CSI500	0.55
MeLaI	0.15	0.16	MA5	0.97	0.08	0.23	SSE50	0.32
CPI	0.89	−0.07	MA20	0.97	−0.02	0.22	HSparts	0.03
Retailsale	0.94	−0.07	MACD	0.1	0.75	0.26	HSETF	0.05
RPI	−0.06	0.85	RSI	0.05	0.88	0.09		
PPI	0.24	0.75	SOBV	0.9	0.05	−0.14		
M2	−0.37	−0.12	ROC	−0.01	0.94	0.06		
FInvest	0.49	−0.48	Volume	0.33	0.2	0.91		
IOP	0.06	−0.47	High	0.96	0.11	0.24		
			Low	0.97	0.12	0.21		
			Open	0.96	0.11	0.23		

3.3 Prediction Method

As discussed in Session 3.2, we reserve only the principal components of the related factors, align the frequencies and feed them into Transformer model. The prediction method of this article, as shown in Fig. 1, mainly includes the following three steps:

1) Extracting principal components for each factor. Macroeconomic indicators, stock technical indicators, and subjective factors are sequentially extracted using PCA method to obtain six principal components (PCM1, PCM2, TECH1, TECH2, TECH3 and BD1).

2) Training the mixed-frequency data model, which feeds the daily returns of the two principal components of macroeconomic indicators (PCM1, PCM2) and CSI300 into the GARCH-MIDAS model (Session 3.3.1), ultimately obtains the conditional volatility ht.
3) Using the Transformer model (Session 3.3.2) to train the conditional volatility (ht), taking the principal components of stock technical indicators (TECH1, TECH2, TECH3), and the principal components of Baidu index as input variables and obtaining the prediction results.

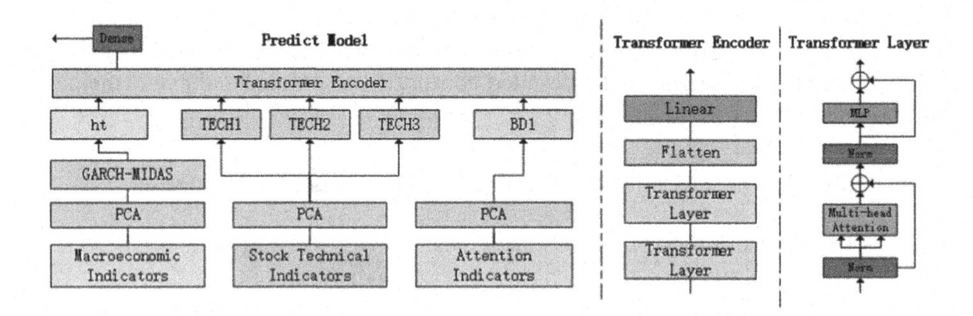

Fig. 1. The Structure Diagram of the Transformer Network

3.3.1 GARCH-MIDAS Model

In the analysis of stock volatility, using monthly or quarterly data to construct models will lose high-frequency effective information of stock market. Therefore, Ghysels et al. [8] first proposed the Mixed Sampling Model (MIDAS), and Engle et al. [7] further applied this model to the Generalized Autoregressive Conditional Heteroscedasticity Model, forming the GARCH-MIDAS model. The GARCH-MIDAS model's return and volatility are described as follows:

$$r_{i,t} - E_{i-1,t}\left(r_{i,t}\right) = \sqrt{\tau_t g_{i,t}}\varepsilon_{i,t}, \forall i = 1, 2, ..., N_t \tag{6}$$

$$\varepsilon_{i,t}|\psi_{i-1,t} \sim N(0,1)$$

$$\sigma_{i,t}^2 = E\left[\left(r_{i,t} - E_{i-1,t}\left(r_{i,t}\right)\right)^2\right] = \tau_t g_{i,t} \tag{7}$$

N_t represents the number of days in t-th month. $r_{i,t}$, $\psi_{i,t}$ and $g_{i,t}$ correspond to the return, the information set of the yield and the high-frequency fluctuations on the i-th day of the t-th month, respectively. And $E_{i-1,t}\left(r_{i,t}\right)$ represents the conditional mathematical expectation under the condition when the information set $\psi_{i,t}$ is given at the (i-1)-th moment of time with the market return $r_{i,t}$. τ_t reflects the low-frequency fluctuations in the t-th month, and $\sigma_{i,t}^2$ is the conditional variance.

Assuming that the conditional mathematical expectation of $r_{i,t}$ at the (i-1)-th moment is μ and that the short-term component of the returns follows a $GARCH(1,1)$ process, Formula (6) can be rewritten as Formula (8), with short-term fluctuations given by Formula (9). At this point, long-term fluctuations are represented by the filtering equation for realized volatility, which is given by Formula (10). In this equation, θ represents the long-term component indicates the contribution of volatility to its marginal. RV_{t-k} is the volatility of market returns over a fixed time horizon, and $\phi_k(\omega_1, \omega_2)$ is the weight function equation, and K is the maximum lagging order of the low-frequency.

$$r_{i,t} = \mu + \sqrt{\tau_t g_{i,t}} \varepsilon_{i,t} \tag{8}$$

$$g_{i,t} = \omega + \frac{\alpha \left(r_{i,t} - \mu\right)^2}{\tau_t} + \beta g_{i-1,t} \tag{9}$$

$$\tau_t = m + \theta \sum_{k=1}^{K} \varphi_k(\omega_1, \omega_2) RV_{t-k} \tag{10}$$

$$\varphi_j\left(\omega_1, \omega_2\right) = \frac{(k/K)^{\omega_1 - 1} \times (1 - k/K)^{\omega_2 - 1}}{\sum_{k=1}^{K} (k/K)^{\omega_1 - 1} \times (1 - k/K)^{\omega_2 - 1}} \tag{11}$$

In addition, m, μ, ω_1, ω_2 and θ are all parameters to be estimated. Generally, ω_1 is fixed to 1 to confirm that the weight of the lagged variable exhibits a decaying trend. ω_2 reflects the decay rate of the impact of the low-frequency on the high-frequency.

3.3.2 Transformer Model

The Transformer model was raised by Vaswani et al [25], a deep learning model on the basis of self-attention mechanisms. The core idea of the Transformer model is to treat each element in the input sequence as a vector and use self-attention mechanisms to compute the relationships between these vectors. Practically, the attention function is as follows:

$$Attention(Q, K, V) = softmax\left(\frac{QK^T}{\sqrt{d_k}}\right)V \tag{12}$$

where Q, K, and V are the abbreviation of 'Query', 'Key', and 'Value' respectively. Each representing an element in the input sequence, which is usually a vector. A dot product operation is applied to calculate the similarity between Q and K, and then use the softmax function to convert it into a probability distribution. $\sqrt{d_k}$ is a scaling factor, helps the model better capture dependencies in the input sequence and improves its performance. Multi-head attention is a mechanism used in the Transformer model to compute the relationships between different positions in the input sequence. It is based on the idea of applying self-attention to each element in the input sequence separately, but with different weights for each attention head.

$$MultiHead(Q, K, V) = Concate(head_1, \ldots, head_h)W^O \quad (13)$$
$$head_i = Attention(QW_i^Q, KW_i^K, VW_i^V)$$

$W_i^Q \in R^{d_o \times d_q}$, $W_i^K \in R^{d_o \times d_k}$, $W_i^V \in R^{d_o \times d_v}$ and $W^O \in R^{hd_v \times d_o}$.

Traditionally, the Transformer model consists of multiple encoders and decoders stacked together. Whereas, when dealing with problems in CV, it is suggested to only feed the resulting sequence to one Transformer encoder, and introduce a Multi-Layer Perceptron (MLP) to form a classification or regression head [5].

As we have already extracted the principal components and reduced the dimensions, we can simply concatenate data from consecutive days and treat it as independent data. Therefore, we follow the ViT model closely in model design, except for we do not have to do positional encoding.

4 Empirical Analysis

4.1 Experiment Setup

To prove the validity of model construction, the prepared data is split into training and testing data with a 9:1 ratio. An Transformer model using a python software package named keras is developed, which then shows excellent long-term memory ability for financial time series in the process of continuous input data streaming.

4.1.1 Data Reparation

The data to be fed into the GARCH-MIDAS model contains macroeconomic indicators and returns of CSI300, which are collected from Jan 2011 to Sep 2021 on a monthly and daily basis, respectively. The GARCH-MIDAS model is then implemented using this data and the fit_mfgarch function of R software. The optimal value of the lag period K of the GARCH-MIDAS model is decided to be 12 after repeated trials. The output results are recorded in Table 3.

It can be observed from Table 3 that the sum value of parameters α and β is close to 1, which indicates a well-fit for the short-term fluctuations of CSI300, and a convergence of the conditional variance of the model to the mean at a appropriate speed.

Parameter $\theta^{(1)}$ represents the "consumption and investment component", with a negative value of about -0.376, indicating a high volatility of CSI300 when the consumption and investment values are small. Generally, the decline of consumption levels implies a decrement of people's willingness and ability to invest. People tend to be more conservative and cautious, which may significantly influence the stock prices. Parameter $\theta^{(2)}$ corresponds to the "production and prosperity component", with a negative value of about -0.760, indicating a high volatility of CSI300 when the production and prosperity levels are low. It

Table 3. Estimated values of main parameters for the GARCH-MIDAS model. (Note: * and ** show significance at 5%, and 1% levels, separately.)

Parameter	Estimated Value	P-value	Parameter	Estimated Value	P-value
μ	0.046755	0.037720*	$\theta^{(1)}$	−0.376158	0.025157*
α	0.071928	0.000002**	$\omega_2^{(1)}$	63.666123	0.000000**
β	0.911217	0.000000**	$\theta^{(2)}$	−0.760231	0.020216*
m	0.730420	0.012188*	$\omega_2^{(2)}$	1.395697	0.000000**

is commonly understood that a decrement of production may cause a supply-demand imbalance and poor circulation of the market.

Parameter $\omega_2^{(1)}$ is the weight of $\theta^{(1)}$, while parameter $\omega_2^{(2)}$ is the weight of $\theta^{(2)}$. A lower value of $\omega_2^{(2)}$ respective to $\omega_2^{(1)}$ indicates a lower dependency of the model on "production and prosperity components" as compared to "consumption and investment component".

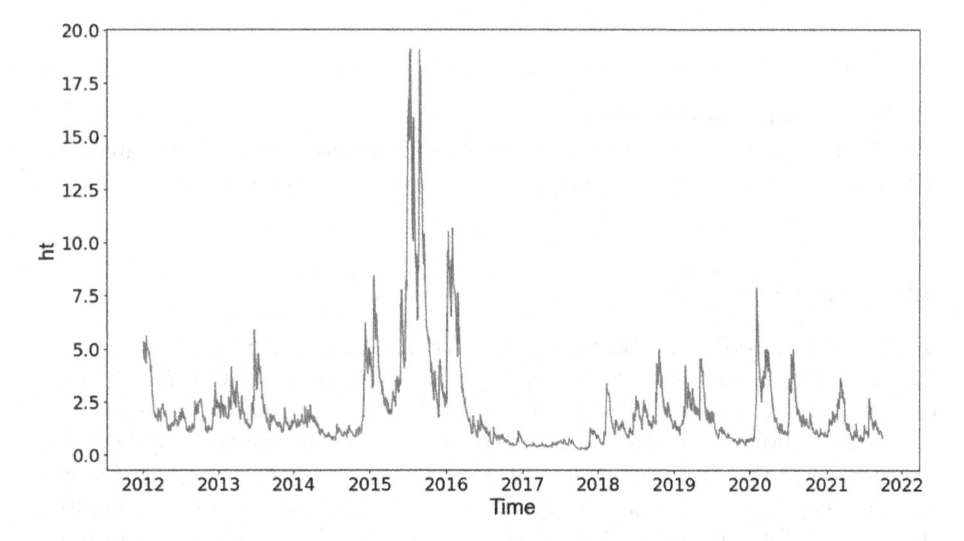

Fig. 2. Estimation of Conditional Volatility (ht)

The conditional volatility of the model is shown in Fig. 2. Due to the lag period setting of 12 months, the parameter estimation period will start from 2012. The conditional volatility includes information from macroeconomic indicators, thereupon alleviates other factors' impact on volatility. In 2015, the conditional volatility of CSI300 was intense, indicating that it experienced remarkable ups and downs during this period, which was closely related to the crash of stock market during that year. In firstly half of 2015, China's macro-control over real estate gradually strengthened, leading to further warming of the investment market and the increment of investors' enthusiasm. However, from mid-June of 2015, the stock market experienced a sharp decline in stock prices and significant fluctuations. CSI300 was also greatly affected. In the end of 2019 and beginning of 2020, the sudden outbreak of COVID-19 in Wuhan led to nationwide shutdowns, resulting in crucial impacts on the stock market and causing fluctuations of CSI300. This figure proves that the conditional volatility of GARCH-MIDAS model can well reflect the actual situation and the effectiveness of the model.

Table 4. Values of Hyperparameters in the Transformer Model.

Parameter	Value	Remarks
TimeStepSize	5	The number of days of data used in the prediction process
LearningRate	0.05	The magnitude of weight updates of each round
BatchSize	32	The number of data samples that are passed to model
NumHeads	3	The number of heads of Multihead Attention
NumLayers	2	The number of transformer layers

4.1.2 Hyperparameter Setting

The Transformer model involves many hyperparameters that require multiple attempts to find the optimal state. The main hyperparameters to be used by the filters of the model are interpreted in Table 4.

4.2 Experiment Result

4.2.1 The Prediction Result of the Transformer Model

The volatility from Oct 2020 to Sep 2021 is generated using the model tuned during training. The predicted RV is compared with the true RV in Fig. 3.

Figure 3 shows the predictive results of the Transformer model on the basis of GARCH-MIDAS and PCA on the test set. The red curve represents the RV size of CSI300 daily, while the green curve reflects the predicted volatility generated by the Transformer model. Overall, the model gives a good evaluation in some turning points and rising trends.

4.2.2 Comparison of Different Factors

The changes in the macroeconomic environment will have an impact on factors such as capital costs and discount rates. Investors' attention is an important factor that affects investment behavior and can lead to the stock market turbulence. Therefore, this article aims to verify the importance of these two factors by grouping different types of factors from the training data for comparison.

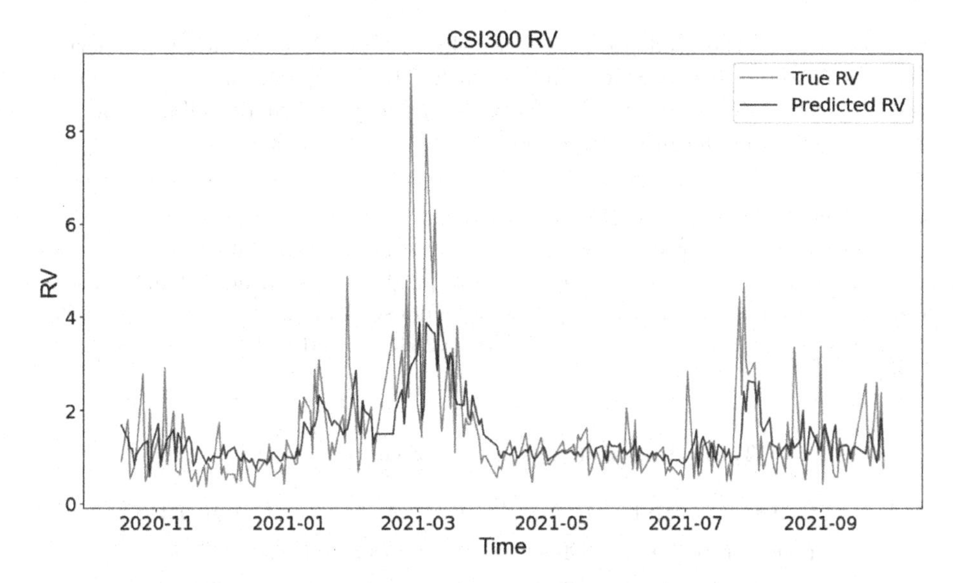

Fig. 3. Prediction Result of the Transformer Model

- Group G1: Stock technical indicators only;
- Group G2: Attention indicators and stock technical indicators;
- Group G3: Macroeconomic indicators and stock technical indicators;
- Group G4: Macroeconomic indicators, Attention indicators and stock technical indicators.

Table 5. Prediction Accuracy Assessment of the Transformer Model with Different Indicator Groups.

Indicator Group	MSE	HMSE	MAE	MAPE	QLIKE	R^2LOG
G1	0.9951	0.6067	0.6317	0.5402	1.3860	0.1474
G2	0.8973	0.4720	0.6016	0.4905	1.3620	0.1626
G3	0.9666	0.4843	0.6136	0.5082	1.3425	0.1266
G4	0.8624	0.4460	0.5871	0.4787	1.3620	0.1710

Table 5 includes the evaluation results of 6 loss functions for different models. The results of G4 compared with the rest of the models show that the overall MSE, HMSE, MAE, MAPE and QLIKE results are smaller than the other 3 groups, which indicates that incorporating macroeconomic indicators and subjective attention can improve the accuracy of model predictions, and also indicates that these two types of indicator factors have an impact on the fluctuation of CSI300. When comparing G2 and G3 with G1 individually, it is found that adding either macroeconomic indicators or subjective attention only will also increase the prediction accuracy effectively. Therefore, when predicting the fluctuation rate of CSI300, incorporating both macroeconomic indicators and subjective attention as input features has a good synergy effect, and the improvement in model prediction accuracy is more obvious.

4.2.3 Comparison of Different Models

In terms of stock market forecasting, many scholars have attempted various methods and continuously improved their prediction accuracy. Among those commonly used deep learning models, we choose 4 of them to compare with our Transformer model, i.e. LSTM, CNN, XGBoost and GRU (Gate Recurrent Unit).

Table 6. Prediction Accuracy of Different Models.

Model	MSE	HMSE	MAE	MAPE	QLIKE	R2LOG
Transformer	0.8624	0.4460	0.5871	0.4787	1.3620	0.1710
LSTM	0.8801	0.6689	0.6496	0.6008	1.4014	0.0671
CNN	1.4716	1.1846	0.8466	0.7631	1.5496	0.0720
XGBoost	0.9421	0.4963	0.6030	0.4936	1.3343	0.1322
GRU	1.5563	1.8534	0.9844	1.0391	1.6627	0.0317

According to the results in Table 6, the Transformer model performs better than others commonly used for volatility prediction. It has the best results in terms of MSE, HMSE and MAPE loss functions, and its overall prediction accuracy is good. This indicates that the Transformer model can effectively extract the characteristics of CSI300 volatility, and is more suitable for predicting this volatility as compared to other models.

5 Conclusion

This article addresses the problem of combination use of different frequencies between macroeconomic data and daily stock data. On the basis of GARCH-MIDAS model, the monthly information from macroeconomic indicators is converted to daily information as an input feature for the later Transformer model. The parameters of the GARCH-MIDAS model are remarkable, demonstrating that the converted daily information can well include macroeconomic information. In terms of selecting macroeconomic indicators, ten representative indicators are finally selected through grey correlation analysis to eliminate the influence of subjective selection and information redundancy. We would like to provide a new insight for future research on the application of mixed-frequency data in predicting volatility of financial assets.

This paper Takes objective and subjective factors as input features of the Transformer network, determining the main parameters of the model through empirical experiments. The adjusted RV is used as an alternative to reflect the real volatility and evaluate the validation of the Transformer model. In addition, the effects and accuracy of the GARCH-MIDAS and PCA models are analyzed from both a factor and a model perspective. The results show that the addition of macroeconomic indicators and attention indicators can increase the predictive accuracy of the transformer model, and the transformer model has a advantage over other models in predicting CSI300 volatility. The results of the Transformer model are also ideal, and the loss functions are within a reasonable range.

References

1. Andersen, T.G., Bollerslev, T.: Answering the skeptics: yes, standard volatility models do provide accurate forecasts. Int. Econ. Rev. **39** (1998)
2. Audrino, F., Sigrist, F., Ballinari, D.: The impact of sentiment and attention measures on stock market volatility. Int. J. Forecast. **36**(2), 334–357 (2020)
3. Choudhury, S., Ghosh, S., Bhattacharya, A., Fernandes, K.J., Tiwari, M.K.: A real time clustering and SVM based price-volatility prediction for optimal trading strategy. Neurocomputing **131**(131), 419–426 (2014)
4. Christiansen, C., Schmeling, M., Schrimpf, A.: A comprehensive look at financial volatility prediction by economic variables. In: School of Economics and Management, University of Aarhus (2010)
5. Dosovitskiy, A., et al.: An image is worth 16x16 words: transformers for image recognition at scale. In: International Conference on Learning Representations (2021)
6. Engle, R.F.: Autoregressive conditional heteroscedasticity with estimates of the variance of united kingdom inflation. Econometrica: J. Econometric Soc. 987–1007 (1982)
7. Engle, R.F., Gonzalo, J., Rangel, J.G.: The spline-garch model for low-frequency volatility and its global macroeconomic causes. Rev. Financ. Stud. **21**(3), 1187–1222 (2008)
8. Ghysels, E., Santa-Clara, P., Valkanov, R.: Predicting volatility: getting the most out of return data sampled at different frequencies. J. Econometrics (2006)
9. Gilles, Z.: Volatility processes and volatility forecast with long memory. Quant. Finance (2004)
10. Gu, H.: Research on Volatility Forecasting Modeling of CSI300 with Investor Sentiment. Master's thesis, Nanjing University (2020)
11. Hansen, P.R., Lunde, A.: A realized variance for the whole day based on intermittent high-frequency data. Soc. Sci. Electron. Publish. **3**(4), 525–554 (2005)
12. Hartwell, C.A.: The impact of institutional volatility on financial volatility in transition economies. J. Comput. Econ. **46**(2), 598–615 (2018)
13. Hisano, R., Sornette, D., Mizuno, T., Ohnishi, T., Watanabe, T.: High quality topic extraction from business news explains abnormal financial market volatility. PLoS ONE **8**(6), e64846 (2013)
14. Hu, Y., Ni, J., Wen, L.: A hybrid deep learning approach by integrating LSTM-ANN networks with garch model for copper price volatility prediction. Physica, A. Stat. Mech. Appl. **557**(1) (2020)
15. Kim, H.Y., Won, C.H.: Forecasting the volatility of stock price index: a hybrid model integrating LSTM with multiple garch-type models. Expert Syst. Appl. **103**(Aug.), 25–37 (2018)

16. Li, S.: Predicting A-share Market Volatility Based on Recurrent Neural Networks and Baidu Index. Master's thesis, Shandong University (2019)
17. Liu, F., Wu, J., Ynag, X., Ouyang, Z.: Long-run dynamic effect of macro-economy on stock market volatility based on mixed frequency data model. Chin. J. Manag. Sci. **28**(10), 65–76 (2020)
18. Lv, Y., Guo, S., Chen, Y., Li, W.: Stock volatility prediction using tabnet based deep learning method. In: 2022 3rd International Conference on Big Data, Artificial Intelligence and Internet of Things Engineering (ICBAIE), pp. 665–668. IEEE (2022)
19. Moon, K.S., Kim, H.: Performance of deep learning in prediction of stock market volatility. Econ. Comput. Econ. Cybern. Stud. Res. / Acad. Econ. Stud. **53**(2/2019), 77–92 (2019)
20. Piplack, J.: Estimating and forecasting asset volatility and its volatility: a Markov-switching range model. Utrecht School of Economics (2009)
21. Schulte-Tillman, B., Segnon, M., Wilfling, B.: Financial-market volatility prediction with multiplicative Markov-switching MIDAS components. In: CQE Working Papers (2022)
22. Shengli, C., Tao, G., Yijun, L.I.: Forecasting realized volatility of Chinese stock index futures based on jumps, good-bad volatility and Baidu index. Systems Eng.-Theory Pract. (2018)
23. Taylor, S.J.: Modeling stochastic volatility: a review and comparative study. Math. Financ. **4**(2), 183–204 (1994)
24. Umar, Y.H., Adeoye, M.: A Markov regime switching approach of estimating volatility using Nigerian stock market. Am. J. Theor. Appl. Stat. **9**(4), 80–89 (2020)
25. Vaswani, A., et al.: Attention is all you need. In: Advances in Neural Information Processing Systems, vol. 30 (2017)
26. Vidal, A., Kristjanpoller, W.: Gold volatility prediction using a CNN-LSTM approach. Expert Syst. Appl. **157**, 113481 (2020)
27. Wang, P.: Research on the impact of margin financing and margin trading on the stock price fluctuation of listed companies. Soc. Med. Health Manag. (2020)
28. Zhang, M.: Research on Shanghai Composite Forecast Based on Lasso Dimensionality Reduction, LSTM and Mixed Frequency Models. Ph.D. thesis, Donghua University (2021)
29. Zhang, X., Wang, J., Cheng, N., Sun, Y., Zhang, C., Xiao, J.: Machine unlearning methodology base on stochastic teacher network. In: 19th International Conference on Advanced Data Mining and Applications (2023)

A Hierarchy-Based Analysis Approach for Blended Learning: A Case Study with Chinese Students

Yu Ye[1], Gongjin Zhang[2], Hongbiao Si[3], Liang Xu[3(✉)], Shenghua Hu[1], Yong Li[2], Xulong Zhang[4], Kaiyu Hu[5], and Fangzhou Ye[6]

[1] Chasing Jixiang Life Insurance Co., Ltd., Changsha, China
[2] Hunan Chasing Digital Technology Co., Ltd., Changsha, China
[3] Hunan Chasing Financial Holdings Co., Ltd., Changsha, China
xuliang@hnchasing.com
[4] Ping An Technology (Shenzhen) Co., Ltd., Shenzhen, China
[5] Stony Brook University, Stony Brook, USA
[6] Chinasoft Co., Ltd., Beijing, China

Abstract. Blended learning is generally defined as the combination of traditional face-to-face learning and online learning. This learning mode has been widely used in advanced education across the globe due to the COVID-19 pandemic's social distance restriction as well as the development of technology. Online learning plays an important role in blended learning, and as it requires more student autonomy, the quality of blended learning in advanced education has been a persistent concern. Existing literature offers several elements and frameworks regarding evaluating the quality of blended learning. However, most of them either have different favours for evaluation perspectives or simply offer general guidance for evaluation, reducing the completeness, objectivity and practicalness of related works. In order to carry out a more intuitive and comprehensive evaluation framework, this paper proposes a hierarchy-based analysis approach. Applying gradient boosting model and feature importance evaluation method, this approach mainly analyses student engagement and its three identified dimensions (behavioral engagement, emotional engagement, cognitive engagement) to eliminate some existing stubborn problems when it comes to blended learning evaluation. The results show that cognitive engagement and emotional engagement play a more important role in blended learning evaluation, implying that these two should be considered to improve for better learning as well as teaching quality.

Keywords: Blended learning · Student engagement · Learning evaluation

1 Introduction

Blended learning, commonly defined as "the integration of traditional face-to-face learning and online teaching" [3,4,16], has increasingly gained popularity

X. Song et al. (Eds.): APWeb-WAIM 2023, LNCS 14332, pp. 89–102, 2024.
https://doi.org/10.1007/978-981-97-2390-4_7

and been widely implemented in higher education across the world. This process was greatly accelerated by the COVID-19 pandemic and the following global social distance restriction [30]. During this difficult period, remote learning has become common in students routine [41]. Besides, teleconferencing tools like Zoom help the delivery of online seminars and lectures, making remote education practical and popular. However, virtual learning, which mainly consists of online instruction and classes, is not diminished with the over of the pandemic and social distance restriction. In fact, remote learning is still an important part of the courses and programmes in higher education. Besides, profiting by the advancement of technology, this learning delivery mode is anticipated to continually be the mainstream in future higher education [6]. Therefore, the high-quality of online or blended education needs to be guaranteed.

The successful implementation of blended learning requires effective combination of virtual as well as face-to-face instruction [16] rather than solely adding virtual learning elements, and this is not easily achieved. The reason is that different from face-to-face learning, remote learning often suffers from the lack of presence, reducing student engagement and thus harming the quality of learning. To achieve success in blended learning, students' self-motivation, self-reliance, independent study skills [44], and online engagement [9,35] are considered equally vital. This indicates that blended learning has a higher demand on overall student engagement in order to ensure learning quality [10]. To achieve this, ongoing evaluation is regarded as essential [31]. On one hand, it is claimed that the introduction of blended learning should be rather cautious at first to permit suitable tutor training and student adaption [5]. This implies the importance and necessity of ongoing evaluation in this gradual adaptation process as evaluation encourages reflections and improvements, helping better implementation in the future. On the other hand, ongoing evaluation is believed to give a more thorough and multi-faceted insight of the quality of blended learning. This improvement is believed to be beneficial for the overall high-quality of teaching in turn [33].

In literature, certain factors that should be taken into account while evaluating blended learning have been mentioned. Course outcomes [21,29], learner satisfaction [8], and student engagement [19,43] are typical key components, of which student engagement is regarded as a more comprehensive criterion than the others. Additionally, many scholars have found a general positive relationship between the quality of blended learning and student engagement [12,38,39], making this criterion an outstanding indicator in the evaluation of blended learning. In terms of evaluation frameworks, diverse of them have been established with varying aims, engaged roles, evaluation focus, and judgement criteria. However, no certain one has ever received widespread recognition as the most efficient. Meanwhile, typically investigated through questionnaires, interviews, or simple classroom observations, these frameworks are more qualitatively based, causing the problem of subjectivity. Moreover, while existing research have broadly analysed western students' experience, scholars have paid little attention to Chinese higher education and provided bare insights, reducing the generalisability of existing conclusions.

Inspired by these studies, we consider a quantitative evaluation of blended learning and propose a hierarchy-based analysis approach for evaluation, using Chinese students' experience as a case study. Our work focuses on the perceptions of students and uses student engagement as the main evaluation indicator. Dividing student engagement into three dimensions, a questionnaire with matrix questions is conducted to collect primary dataset. After that, the importance of each dimension of student engagement is extracted. Consequently, the quality of blended learning is evaluated through the Analytic Hierarchy Process (AHP). Our contributions are summarised as follows:

1) To evaluate the quality of blended learning, we propose a hierarchy-based analysis approach, improving the objectivity and accuracy of evaluation.

2) With little research providing an insight into Chinese higher education, we narrow the gap by using Chinese students' experience as a case study to deepen the understanding.

2 Related Work

2.1 Elements Regarding Evaluating Blended Learning

Different elements have been pointed out in literature to be taken into consideration in terms of the evaluation of blended learning. Generally, major elements include course outcomes, learner satisfaction, and student engagement.

Course outcomes are typically measured through aspects such as grades, class attendance, and drop out rates. Existing research has found that effective implementation of blended learning is beneficial for the improvement of course outcomes [21,29]. This criterion alone, however, fails to convey a comprehensive picture of the quality of blended learning because it neglects student's feelings and attitudes. One example is that students' motivation and initiatives towards learning are not captured. Therefore, whether blended learning helps facilitate these is not evaluated, which is noted as an important aspect regarding evaluating instructional effectiveness [28].

Learner satisfaction offers a different perspective from course outcomes on the evaluation of blended learning by focusing on students' perceptions. Commonly measured by conducting self-report questionnaires, this element not only consider assessment data, but also other aspects such as learning environment, course content and flexibility, and perceived ease use of technology [2]. Thus, it comprehensively reflects students' personal experience and overall satisfaction of blended learning. This element also is proved to be positively affected by effective blended learning [8,34].

Student engagement enables a deeper comprehension of the effectiveness of blended learning as it captures the contribution that students make to learning process for desirable outcomes [24] and the degree to which they engage in high-quality educational activities [22]. Three dimensions of student engagement are identified: cognitive engagement, behavioural engagement, and emotional engagement [14]. Generally, behavioural engagement relates to students' actions, having some overlaps with course outcomes. This dimension is mainly measured by students' involvement in learning process, such as actively attending class, collaborating with group members, and interacting with faculty [23]. Emotional engagement emphasises students' affective attitudes towards learning, such as interest, enjoyment and satisfaction. Cognitive engagement is relevant to the psychological investment in learning, such as self-management, initiatives towards learning and critical comprehension of knowledge. Positively affected by and giving a more full picture of blended learning, student engagement is becoming a crucial indicator for evaluation [13,37,39].

2.2 Evaluation Frameworks

Based on elements mentioned above, different frameworks have been developed to evaluate blended learning with various purposes, involved roles, and evaluation focus. However, not a particular one has been commonly regarded as the most effective. Some selected frameworks will be discussed in the following parts.

Web-Based Learning Environment Instrument (WEBLEI): This framework focuses on investigating students' perceptions of e-learning environments. Four scales are incorporated, including emancipatory activities (focusing on convenience of materials, learning efficiency, and autonomy), co-participatory activities (focusing on students' learning processes such as flexibility, reflection, quality, interaction, feedback and collaboration), qulia (focusing on learning attitudes like enjoyment, frustration and tedium), and information structure (focusing on the design and arrangement of learning content). The first three are developed from Tobin's qualitative evaluation of Connecting Cummunities Learning (CCL) [42], and the last one is separately proposed by Chang [7].

Hexagonal E-Learning Assessment Model (HELAM): This is a multi-dimensional approach in terms of evaluating learning management systems, focusing on the perception of learner satisfaction. Evaluated through a questionnaire, it has six evaluation criteria: system quality, information (content) quality, instructor attitudes, supportive elements, service quality and leaner perspective [32]. All of these dimensions are demonstrated to be significant. However, neglecting perspectives of other stakeholders, this model is questioned to some extent for only focusing on students.

E-Learning Framework: This is also a multi-dimensional framework containing eight systemically interconnected dimensions. They are technological (looking at infrastructure planning), pedagogical (looking at the arrangement and design of learning materials as well as learning strategies), interface design (looking at content design, navigation, and usability testing), evaluation (looking at learner assessment and teacher instruction), management (looking at maintaining learning environment and information transfer), resource support (looking at required remote support and resources), ethical (looking at social and ethical issues), and institutional (looking at administrative affairs and students services) [11,18]. However, instead of proposing any evaluation instrument, it only offers guidance for evaluating the environments of blended learning.

Rubric-Based Frameworks: This kind of frameworks have been created by several researches, commonly relying on judgement and having wide-ranging scopes. Evaluation factors mainly include instructional design, technology utilisation as well as students' experiences. Besides, these frameworks offer a quick and efficient method in terms of course evaluation for programme designers. Rubric-based frameworks are argued to be practically employed [40]. However, depending heavily on judgements, these frameworks are inherently subjective. Additionally, not offering guidance for making judgements, evaluation provided by rubrics is judged to be broad and lacking depth.

3 Method

We propose a hierarchy-based analysis approach to evaluate the quality of blended learning mainly based on the importance of all features to three dimensions of student engagement. To target Chinese students' blended learning experience in higher education, an online survey was firstly created, measuring each feature with matrix questions with a seven-pointed Likert scale. Also, this survey was adapted from existing surveys in order to increase validity. Besides, previous studies find that gender [20,26] and age [17] both have an impact on student engagement. Therefore, they are also set as features to avoid potential bias. In terms of the measurement of blended learning, existing studies point out that effective mixture of face-to-face and virtual learning rather than simple adding virtual course materials constitutes a sufficient blend [16]. This paper consider 30%-80% as an appropriate proportion of online learning contributing to blended learning [1]. Table 1 summarises the main features, targets and their measures.

Table 1. Main features, targets and their measures

Category	Measure	Matrix focuses
Behavioural Engagement (BL)	Active involvement (B-Act) Faculty interaction (B-Int) Group collaboration (B-Gro)	Attendance, Seats, Attention, Notes, Duration Questions, Eye-contact, Reflection Discussion, Communication, Presentation
Cognitive Engagement (CE)	Self-management (C-Mgt) Comprehension (C-Com)	Pre-reading, Revision, Time schedule Grades, Assignments, Critical thinking, Strategies
Emotional Engagement (EE)	Interest (E-Int) Satisfaction (E-Sat)	Motivation, Related reading, Inspiration Support, Confidence, Accomplishments, Enjoyment
Blended Learning (BL)	Proportion of online learning	\

To collect primary dataset, our survey was spread through the online advanced education communities provided by Weibo, one popular Chinese social media. After that, gradient boosting regression model was applied to fit the survey data. Besides, gini importance and permutation importance were used to measure the importance of each feature to the regression target. Based on features selected, the analytic hierarchy process (AHP) method was then applied to build a evaluation matrix to measure student engagement. Figure 1 presents the whole framework.

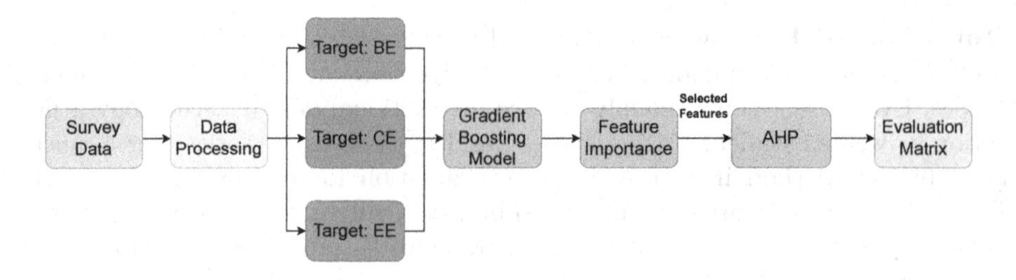

Fig. 1. The framework of AHP analysis approach. Gradient boosting regression model and feature importance analysis method are applied to extract important features. The importance value of each feature is then fed into the AHP method to calculate the evaluation matrix.

Cognitive Engagement (CE), Behavioural Engagement (BE), and Emotional Engagement (EE) are separately set as the regression target Y. Each of the training/testing set was then fed into the Gradient Boosting Regression model.

3.1 Gradient Boosting Regression

Gradient Boosting Regression, also known as Gradient Boosted Decision Trees, is a model that can be applied to both classification and regression tasks. Compared to decision tree model or other simple linear models, it is capable of handling continuous features and discrete features. Besides, based on decision tree model, this model is relatively easy to fit and fine-tuning [15]. Our model takes a fixed-size decision tree as the weak learner and is built in a greedy manner:

$$\hat{y}_i = F_M(x_i) = \sum_{m=1}^{M} h_m(x_i) \tag{1}$$

where h_m is the set of decision tree model with size of M, also known as weak learners in the case of boosting method.

In each gradient step, a new decision tree h_m is added into the whole model, updating the $F_m(x)$ in the following greedy way:

$$F_m(x) = F_{m-1}(x) + h_m(x) \tag{2}$$

A decision tree is a model that applies non-parametric supervised learning method to achieve the regression goal. It contains a set of if-then-else decision rules that can learn from the data points to approximate the regression curve. The tree added in each step will learn from the training data and try to minimise the losses function, which is the mean squared error function in this case:

$$MSE(y, \hat{y}) = \frac{1}{n} \sum_{n-1}^{i=0} (y_i - \hat{y}_i)^2 \tag{3}$$

According to Friedman [15], the decision tree h_m predicts the negative gradients of the training data updated at each training step. The Gradient Boosting

Regression can be regarded as a process of doing gradient descent in a functional space.

3.2 Gini Importance and Permutation Importance

Gini importance and permutation importance are used in feature importance area to measure the relevance between features and targets.

Gini importance, also known as Mean Decrease Impurity (MDI), is a impurity-based method and represents the average and variability of impurity reduction accumulation within each individual tree [25]. It is calculated as the following:

$$MDI(k,T) = \sum \frac{N_n(t)}{n} \Delta x(t) \tag{4}$$

where X is the feature and T is the weak learner.

The result of gini importance may be biased when the feature has a large amount of unique values. Therefore, the permutation importance is used as an alternative to overcome this. It calculates feature importance by evaluating the change in the model's performance when randomly permuting the values of a single feature [45]:

$$i_j = s - \frac{1}{K} \sum_{k=1}^{K} s_{k,j} \tag{5}$$

where i_j is the importance of feature f_j, s is the reference score of the model on the dataset, and K is the total repetition used to calculate the importance.

Both gini importance and permutation importance methods are used to evaluate the feature for better confidence level of feature importance.

3.3 Analytic Hierarchy Process

Analytic Hierarchy Process (AHP) is an effective method involving both qualitative and quantitative analysis [36]. It uses a hierarchy structure to divide the decision process into three levels - Alternatives, Criteria and Goal [27]. The feature importance extracted from gini importance and permutation importance method is applied to initialise the pairwise comparison matrix.

Table 2 shows the matrix used in AHP to assign the intensity of importance to each criterion. The pairwise comparison is then established, and AHP will check the consistency. If the check is pass, the AHP method will output a weighted score for each criterion.

As the pairwise comparison could be inaccurate due to user's subjective bias, the importance conducted from gini importance and permutation importance methods is therefore applied to reduce this. A mapping is created to map the importance learnt by the model to the pairwise comparison of the AHP method. The details will be discussed in the following experiments section.

Table 2. AHP Comparison Index

Intensity of importance	Definition
1	Equal
2	Weak
3	Moderate
4	Moderate plus
5	Strong
6	Strong plus
7	Demonstrate
8	Demonstrate plus
9	Extremely preferred

4 Experiments and Results

4.1 Dataset

1132 samples were submitted to our online survey. Gender distribution shows that 69.3% of the respondents are identified as females, while 30.7% are identified as males. This gender imbalance implies that the interpretation of results may primarily reflect the experiences and perspectives of female participants, limiting the generalisability of the findings. In terms of the age distribution of the sample, with over 90% of the sample's participants being over the age of 18, it ensures a representative sample that aligns with the target population under investigation. Furthermore, it is worth noting that most respondents (60%) are between the ages of 18 and 21, indicating that the findings are more representative of undergraduate experience.

Table 3 provides a summarised overview of the descriptive statistics derived from the dataset. The mean values reveal that less than half of the student participants reported having experienced blended learning, which appears contrary to the prevailing trend of increased integration of online learning with conventional educational practices in light of the Covid-19 pandemic. A plausible explanation for this observation could be attributed to the varying extent to which online learning is embraced by individual students. With some demonstrating an excessive reliance on online platforms while others exhibiting a minimal incorporation of such methods into their overall learning routine, these fail to meet the specific criterion outlined for blended learning in this paper. Additionally, the average scores for each aspect of student engagement slightly surpasses 4, indicating a generally positive inclination towards active participation in educational activities within the blended learning environment. The relatively low standard deviations observed further suggest a convergence of responses around the mean values, implying a degree of consensus among the participants.

Table 3. Descriptive statistics of dataset

	BL	B-Act	B-Int	B-Gro	C-Mgt	C-Com	E-Int	E-Sat	BE	CE	EE
Mean	0.4488	4.6693	4.6614	4.5748	4.6457	4.4803	4.8661	4.669	4.6352	4.5630	4.7677
Std. D	0.4993	1.5688	1.5287	1.6548	1.7207	1.6755	1.7922	1.7820	1.4496	1.6439	1.7410
Stewness	0.2083	−0.4756	−0.4131	−0.4166	−0.3273	−0.2687	−0.6362	−0.4709	−0.4950	−0.2793	−0.5443
Kurtosis	−1.9882	−0.2727	−0.2911	−0.5827	−0.7437	−0.6816	−0.4587	−0.6255	−0.1736	−0.6633	−0.5578

4.2 Experimental Setup

The collected survey dataset is divided into a 80/20 split for training and testing the Gradient Boosting Regression model. Cognitive Engagement (CE), Behavioural Engagement (BE), and Emotional Engagement (EE) are set as the regression target Y separately as shown in Table 4.

Table 4. Data samples setup

Target Y	Features X
CE	Gender, Age, BL, B-Act, B-Int, B-Gro, E-Int, E-Sat, BE, EE
BE	Gender, Age, BL, C-Mgt, C-Com, E-Int, E-Sat, CE, EE
EE	Gender, Age, BL, B-Act, B-Int, B-Gro, C-Mgt, C-Com, BE, CE

The mean squared error is used as the loss function to train the gradient boosting model for 500 boosting stages with learning rate being 0.01. The max depth of each decision tree of the weak learner is 4. For evaluation, the training and testing deviance is applied to measure the learning process.

4.3 Results and Analysis

We first inspect the training and testing deviance of each dataset. At this stage, all parameters of three models and training process are set as the same except the dataset itself.

Figure 2 indicates that all models achieve the saturation point around 200 boosting iterations, meaning that every single model is capable for learning certain level of the representation from the training dataset. More iterations may lead to overfitting issue. In general, the results prove that feature importance conducted from this model is accurate and can be applied to the AHP method later on.

The next step is to measure which features are more relevant to target Y. The feature importance results for all three models are similar, therefore we only illustrate regression model with target BE in detail.

Figure 3 indicates that both gini importance and permutation importance show that the BL feature is the most relevant one. Similar conclusions can be drawn from the results of feature importance of other two models.

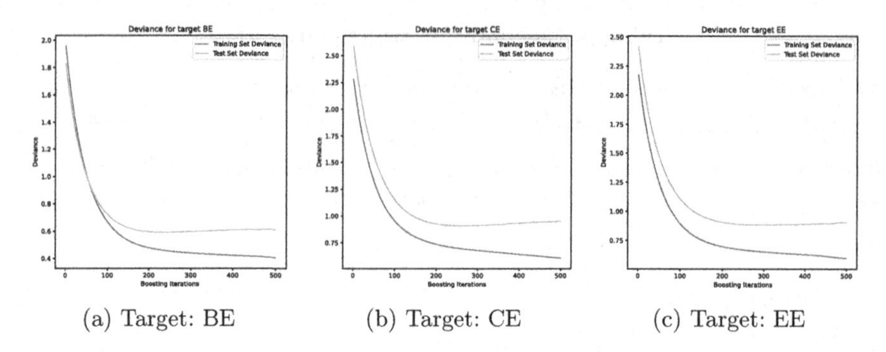

(a) Target: BE (b) Target: CE (c) Target: EE

Fig. 2. The training and testing deviance on three datasets with different target Y. Figure 2(a) is the model trained on the regression of target BE. Figure 2(b) is the model trained on the regression of target CE. Figure 2(c) is the model trained on the regression of target EE. All three models achieve the saturation point around iteration 200.

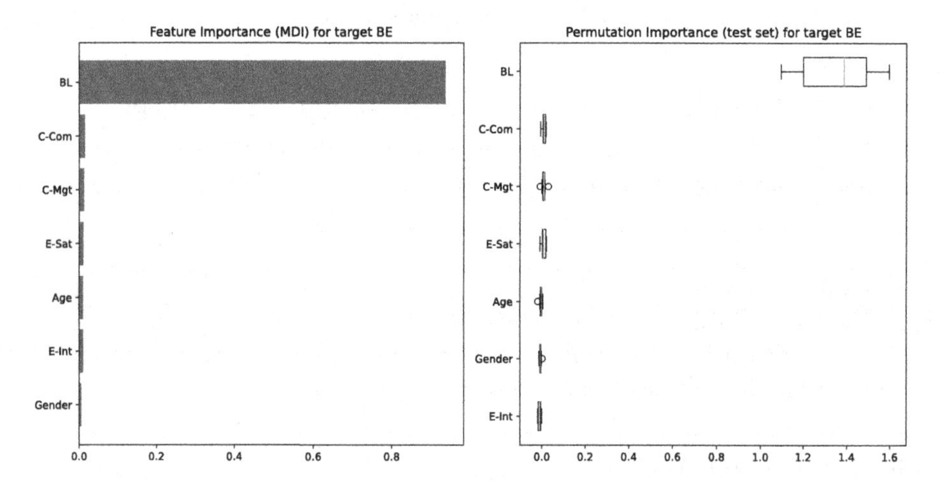

Fig. 3. Gini importance and Permutation importance of BE model. It shows that the BL feature has the most significant contributions to the target BE score and the rest features share similar level of importance.

Finally, we map the gini importance and permutation importance to the pairwise comparison scale in the AHP method. It is clearly that the BL feature is the demonstrated importance and Age is the least significant feature. Therefore, the scale for BL to other features is set as 7, the scale for *Gender* to BL is set as $1/9$, and the scale for other features is 3. The pairwise comparison matrix for the BE model can be created as Table 5.

The square root method of AHP is applied to calculate the evaluation matrix and normalise weighted value for each feature. The final consistency index is 0.013, meaning that the final matrix is consistent and the evaluation matrix conducted by AHP approach is valid. Similar approach can be applied to the

Table 5. AHP Pairwise Comparison Matrix For BE.

Feature	BL	C-Com	C-Mgt	E-Sat	Age	E-Int	Gender
BL	1	7	7	7	7	7	9
C-Com	1/7	1	1	1	1	1	3
C-Mgt	1/7	1	1	1	1	1	3
E-Sat	1/7	1	1	1	1	1	3
Age	1/7	1	1	1	1	1	3
E-Int	1/7	1	1	1	1	1	3
Gender	1/9	1/3	1/3	1/3	1/3	1/3	1

model with the other two targets, and the final evaluation matrix is shown in Table 6. It indicates that blended learning significantly affects student's BE, EE and CE in a positive way. Age is the least significant feature for all the three models, and other features share similar level of importance.

Table 6. AHP Evaluation Matrix for target BE, CE and EE.

Target	Weight	BL	B-Act	B-Int	B-Gro	C-Mgt	C-Com	E-Int	E-Sat	Gender	Age
BE	Weight Score	**5.495**	\	\	\	0.886	0.886	0.886	0.886	0.886	0.333
	Percentage(%)	**53.566**	\	\	\	8.637	8.637	8.637	8.637	8.637	3.249
CE	Weight Score	**5.759**	0.981	0.981	0.981	\	\	0.981	0.981	0.333	0.574
	Percentage(%)	**49.733**	8.478	8.478	8.478	\	\	8.478	8.478	2.881	4.957
EE	Weight Score	**5.759**	0.981	0.981	0.981	0.981	0.981	\	\	0.333	0.574
	Percentage(%)	**49.733**	8.478	8.478	8.478	8.478	8.478	\	\	2.881	4.957

Our work evaluates the generalizability of previous theories and close any potential research gaps by examining this relationship of belended learning and student engagement in the context of Chinese higher education. Although our work has proved some results from previous research, it is worth nothing that this paper define study programmes with 30% to 80% online learning as blended learning. To deepen the understanding of the quality of blended learning, it is suggested that the proportion of online learning could be studied at a more granular level.

5 Conclusion

In this work, we examine how different aspects of student engagement relate to the quality or effectiveness of blended learning. According to our results, it is clearly that blended learning significantly affects student engagement in a positive way, particularly cognitive engagement and emotional engagement. Using student engagement as a indication, it is safe to conclude that the quality or effectiveness of blended learning can be gauged indirectly. Besides, the findings suggest that the trend towards blended learning being the norm in future higher education will be beneficial to increase learning quality. Additionally, proposing the AHP Approach for blended learning evaluation, it shows that cognitive engagement and emotional engagement are more important for learning quality. However, in terms of properly allocating the percentage of remote learning, it is still unclear how to maximise the advantages of blended learning. Therefore, academics are urged to gain a thorough grasp of the effectiveness of diverse combinations of face-to-face learning and online learning to further this academic research and benefit remote education.

References

1. Allen, I.E., Seaman, J., Garrett, R.: Blending in: the extent and promise of blended education in the united states. Sloan Consortium (2007)
2. Asoodar, M., Vaezi, S., Izanloo, B.: Framework to improve e-learner satisfaction and further strengthen e-learning implementation. Comput. Hum. Behav. **63**, 704–716 (2016)
3. Bliuc, A.M., Goodyear, P., Ellis, R.A.: Research focus and methodological choices in studies into students' experiences of blended learning in higher education. Internet High. Educ. **10**(4), 231–244 (2007)
4. Boelens, R., Van Laer, S., De Wever, B., Elen, J.: Blended learning in adult education: towards a definition of blended learning (2015)
5. Boyle, T., Bradley, C., Chalk, P., Jones, R., Pickard, P.: Using blended learning to improve student success rates in learning to program. J. Educ. Media **28**(2–3), 165–178 (2003)
6. Castro, R.: Blended learning in higher education: trends and capabilities. Educ. Inf. Technol. **24**(4), 2523–2546 (2019)
7. Chang, V.: Evaluating the effectiveness of online learning using a new web based learning instrument. In: Proceedings Western Australian Institute for Educational Research Forum (1999)
8. Chen, W.S., Yao, A.Y.T.: An empirical evaluation of critical factors influencing learner satisfaction in blended learning: a pilot study. Univers. J. Educ. Res. **4**(7), 1667–1671 (2016)
9. Chen, X., DeBoer, J.: Checkable answers: understanding student behaviors with instant feedback in a blended learning class. In: 2015 IEEE Frontiers in Education Conference (FIE), pp. 1–5. IEEE (2015)
10. Deakin Crick, R., Huang, S., Ahmed Shafi, A., Goldspink, C.: Developing resilient agency in learning: the internal structure of learning power. Br. J. Educ. Stud. **63**(2), 121–160 (2015)
11. Deegan, D., Wims, P., Pettit, T.: The potential of blended learning in agricultural education of Ireland (2015)
12. Delialioğlu, Ö.: Student engagement in blended learning environments with lecture-based and problem-based instructional approaches. J. Educ. Technol. Soc. **15**(3), 310–322 (2012)
13. Dringus, L.P., Seagull, A.B.: A five-year study of sustaining blended learning initiatives to enhance academic engagement in computer and information sciences campus courses. Blended Learn. 122–140 (2013)

14. Fredricks, J.A., Blumenfeld, P.C., Paris, A.H.: School engagement: potential of the concept, state of the evidence. Rev. Educ. Res. **74**(1), 59–109 (2004)
15. Friedman, J.H.: Stochastic gradient boosting. Comput. Stat. Data Analy. **38**(4), 367–378 (2002)
16. Garrison, D.R., Kanuka, H.: Blended learning: uncovering its transformative potential in higher education. Internet High. Educ. **7**(2), 95–105 (2004)
17. Gibson, A.M., Slate, J.R.: Student engagement at two-year institutions: age and generational status differences. Community Coll. J. Res. Pract. **34**(5), 371–385 (2010)
18. Gomes, T., Panchoo, S.: Teaching climate change through blended learning: a case study in a private secondary school in Mauritius. In: 2015 International Conference on Computing, Communication and Security (ICCCS), pp. 1–5. IEEE (2015)
19. Holley, D., Oliver, M.: Student engagement and blended learning: portraits of risk. Comput. Educ. **54**(3), 693–700 (2010)
20. Kinzie, J., Gonyea, R., Kuh, G.D., Umbach, P., Blaich, C., Korkmaz, A.: The relationship between gender and student engagement in college. In: Association for the Study of Higher Education Annual Conference (2007)
21. Kiviniemi, M.T.: Effects of a blended learning approach on student outcomes in a graduate-level public health course. BMC Med. Educ. **14**(1), 1–7 (2014)
22. Krause, K.L., Coates, H.: Students' engagement in first-year university. Assess. Eval. High. Educ. **33**(5), 493–505 (2008)
23. Kuh, G.D.: The national survey of student engagement: conceptual framework and overview of psychometric properties (2001)
24. Kuh, G.D., Kinzie, J., Buckley, J.A., Bridges, B.K., Hayek, J.C.: Piecing together the student success puzzle: research, propositions, and recommendations: ASHE higher education report, vol. 116. Wiley (2011)
25. Li, X., Wang, Y., Basu, S., Kumbier, K., Yu, B.: A debiased MDI feature importance measure for random forests. In: Advances in Neural Information Processing Systems, vol. 32 (2019)
26. Lietaert, S., Roorda, D., Laevers, F., Verschueren, K., De Fraine, B.: The gender gap in student engagement: the role of teachers' autonomy support, structure, and involvement. Br. J. Educ. Psychol. **85**(4), 498–518 (2015)
27. Lipovetsky, S.: The synthetic hierarchy method: An optimizing approach to obtaining priorities in the AHP. Eur. J. Oper. Res. **93**(3), 550–564 (1996)
28. Liu, O.L., Bridgeman, B., Adler, R.M.: Measuring learning outcomes in higher education: motivation matters. Educ. Res. **41**(9), 352–362 (2012)
29. López-Pérez, M.V., Pérez-López, M.C., Rodríguez-Ariza, L.: Blended learning in higher education: students' perceptions and their relation to outcomes. Comput. Educa. **56**(3), 818–826 (2011)
30. Mahaye, N.E.: The impact of COVID-19 pandemic on education: navigating forward the pedagogy of blended learning. Res. Online **5**, 4–9 (2020)
31. Moskal, P., Dziuban, C., Hartman, J.: Blended learning: a dangerous idea? Internet High. Educ. **18**, 15–23 (2013)
32. Ozkan, S., Koseler, R.: Multi-dimensional students' evaluation of e-learning systems in the higher education context: an empirical investigation. Comput. Educ. **53**(4), 1285–1296 (2009)
33. Pombo, L., Moreira, A.: Evaluation framework for blended learning courses: a puzzle piece for the evaluation process. Contemp. Educ. Technol. **3**(3), 201–211 (2012)
34. Prifti, R.: Self-efficacy and student satisfaction in the context of blended learning courses. Open Learn. J. Open Distance e-Learn. **37**(2), 111–125 (2022)

35. Reed, P.: Staff experience and attitudes towards technology enhanced learning initiatives in one faculty of health & life sciences. Res. Learn. Technol. **22** (2014)
36. Saaty, R.W.: The analytic hierarchy process-what it is and how it is used. Math. Model. **9**(3–5), 161–176 (1987)
37. Sahni, J.: Does blended learning enhance student engagement? Evidence from higher education. J. E-Learn. High. Educ. **2019**(2019), 1–14 (2019)
38. Sari, R., Karsen, M.: An empirical study on blended learning to improve quality of learning in higher education. In: 2016 International Conference on Information Management and Technology (ICIMTech), pp. 235–240. IEEE (2016)
39. Sarıtepeci, M., Çakır, H.: The effect of blended learning environments on student motivation and student engagement: a study on social studies course. Educ. Sci./Egitim ve Bilim **40**(177) (2015)
40. Smythe, M.: Blended learning: a transformative process. **12**, 2011 (2011). Retrieved on December
41. Sun, A., Zhang, X., Ling, T., Wang, J., Cheng, N., Xiao, J.: Pre-avatar: an automatic presentation generation framework leveraging talking avatar. In: 2022 IEEE 34th International Conference on Tools with Artificial Intelligence (ICTAI), pp. 1002–1006 (2022). https://doi.org/10.1109/ICTAI56018.2022.00153
42. Tobin, K.: Qualitative perceptions of learning environments on the world wide web. Learn. Environ. Res. **1**(2), 139–62 (1998)
43. Vaughan, N.: Student engagement and blended learning: making the assessment connection. Educ. Sci. **4**(4), 247–264 (2014)
44. Wivell, J., Day, S.: Blended learning and teaching: synergy in action. Adv. Soc. Work Welfare Educ. **17**(2), 86–99 (2015)
45. Zhang, C., Ma, Y.: Ensemble Machine Learning: Methods and Applications. Springer, New York (2012). https://doi.org/10.1007/978-1-4419-9326-7

A Multi-teacher Knowledge Distillation Framework for Distantly Supervised Relation Extraction with Flexible Temperature

Hongxiao Fei[1], Yangying Tan[1], Wenti Huang[2], Jun Long[1,3], Jincai Huang[3], and Liu Yang[1](✉)

[1] School of Computer Science and Engineering, Central South University, Changsha, China
{hxfei,yytan,yangliu}@csu.edu.cn
[2] School of Computer Science and Engineering, Hunan University of Science and Technology, Xiangtan, China
[3] Big Data Institute, Central South University, Changsha, China

Abstract. Distantly supervised relation extraction (DSRE) generates large-scale annotated data by aligning unstructured text with knowledge bases. However, automatic construction methods cause a substantial number of incorrect annotations, thereby introducing noise into the training process. Most sentence-level relation extraction methods rely on filters to remove noise instances, meanwhile, they ignore some useful information in negative instances. To effectively reduce noise interference, we propose a **M**ulti-teacher **K**nowledge **D**istillation framework for **R**elation **E**xtraction (MKDRE) to extract semantic relations from noisy data based on both global information and local information. MKDRE addresses two main problems: the deviation in knowledge propagation of a single teacher and the limitation of traditional distillation temperature on information utilization. Specifically, we utilize flexible temperature regulation (FTR) to adjust the temperature assigned to each training instance, so as to dynamically capture local relations between instances. Furthermore, we introduce information entropy of hidden layers to gain stable temperature calculations. Finally, we propose multi-view knowledge distillation (MVKD) to express global relations among teachers from various perspectives to gain more reliable knowledge. The experimental results on NYT19-1.0 and NYT19-2.0 datasets show that our proposed MKDRE significantly outperforms previous methods in sentence-level relation extraction.

Keywords: Relation Extraction · Knowledge Distillation · Distantly Supervised · Multiple Teachers · Natural Language Processing

Supported by the National Natural Science Foundation of China under the Grant No. 62172451, and supported by Open Research Projects of Zhejiang Lab under the Grant No. 2022KG0AB01.

X. Song et al. (Eds.): APWeb-WAIM 2023, LNCS 14332, pp. 103–116, 2024.
https://doi.org/10.1007/978-981-97-2390-4_8

1 Introduction

Relation extraction (RE) determines semantic relations between entity pairs within a given sentence, which is of great significance to build knowledge bases. Most existing relation extraction methods [1,2] are labor-intensive since they rely on large-scale manually annotated data. Thereafter, distantly supervised relation extraction [3] is proposed to automatically generate annotated data by aligning unstructured text based on entity information in knowledge bases. DSRE assumes if a certain relation is expressed between entity pairs in a knowledge base, likely, the same relation is also expressed in the sentences containing corresponding entity pairs. However, the fact is that some distant tagged sentences may not express the same relation. Thus, DSRE inevitably brings noisy data and adversely affects model performance. For example, the sentence "Steve Jobs left Apple and sold his stock" has the head entity "Steve Jobs" and tail entity "Apple", but they do not express the relation of "Founder Of" obtained from the knowledge base, as shown in Fig. 1.

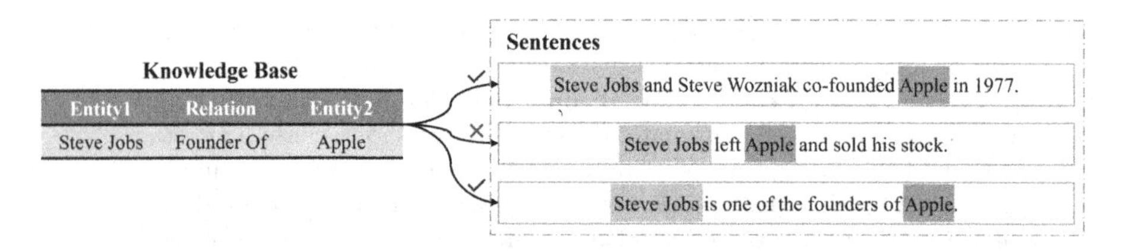

Fig. 1. The example of DSRE. "Steve Jobs" and "Apple" serve as the head entity and tail entity respectively, while "Founder Of" denotes the inter-entity relation. Obviously, the second sentence does not express such a relation.

To solve the noise problem, Zeng et al. [4] divide the training instances into bags according to the same entity pair and introduce the paradigm of multi-instance learning (MIL) into DSRE. Some MIL-based [5–7] methods usually introduce different selection strategies such as attention mechanisms to distinguish noisy sentences and achieve effective classification at the bag-level. However, the above methods are not so effective for sentence-level classification [8]. Thereafter, some methods [9,10] design filters to screen positive and negative instances for noise reduction, so as to achieve effective classification at the sentence-level. However, it may cause insufficient utilization of data to rely solely on filtering. Ma [11] points out that noisy instances may also provide effective information in training. Recently, knowledge distillation frameworks [12,13] are utilized to deal with noisy data. They leverage label-softening mechanisms to gain rich information and weaken noise influence as well. Specifically, the student model in knowledge distillation frameworks improves performance by imitating the distribution of the teacher model. Nevertheless, some knowledge distillation frameworks are constrained by a unified distillation temperature, or only mimic the distribution of a single teacher, resulting in their student model inheriting the biases of a single teacher model.

In this paper, we propose a Multi-teacher Knowledge Distillation framework for Relation Extraction (MKDRE) to address the issue of noise in distantly supervised relation extraction tasks. To utilize information in the instances, we model local relations between the instances and global relations between different teacher sources. Considering the bias of knowledge in a single teacher and the credibility of multiple teachers, we design a multi-view knowledge distillation module (MVKD) and propose an elastic distribution mechanism to combine various knowledge from multiple teachers and avoid bad forecasting information being spread. Besides, students are also allowed to acquire knowledge from different sources by constructing global relations. We design a flexible temperature regulation module (FTR) to make full use of the differences between instances, meanwhile, to avoid inaccurate knowledge propagation from fixed temperature values. Therefore, the FTR module adopts flexible temperatures to control the softening of labels within a reasonable range. Besides, FTR considers the uncertainty in the feature vector of a single layer and introduces the information entropy of the hidden layer to get the final temperature comprehensively, so as to improve the stability of the distillation model.

Our proposed MKDRE framework represents an important contribution to the field of relation extraction, with the potential ability to improve classification at the sentence-level. Our contribution can be summarized as follows:

- To our best, we are the first to propose a knowledge distillation framework combining multi-teacher knowledge to reduce noise for DSRE tasks.
- We propose a multi-view flexible temperature calculation method by utilizing the relative differences between the hidden layer and the classified layer information to obtain a more reasonable temperature.
- We construct a global relationship among teachers based on multidimensional indexes including F1 values and confidence, so as not to mislead students with inaccurate knowledge.
- The experimental results show that MKDRE improves the performance on NYT19-1.0 and NYT19-2.0 datasets compared with the previous SOTA models.

2 Related Work

2.1 Distantly Supervised Relation Extraction

Relation extraction is a crucial sub-task in information extraction tasks, which is widely applied in question answering, knowledge graph completion, medical search, and so on. To overcome the limitation of manual annotation, distant supervision is proposed by Mintz [3], but it also brings a noise problem. Riedel et al. [14] propose an at-least-once assumption and build the DSRE task as a multi-instance learning problem, where it aggregates the sentences into a bag-level for training to reduce noise. Other researchers [8, 9, 15] employ reinforcement learning to help the instance selector filter clean training samples for DSRE. Qin et al. [16] apply generative adversarial training to filter noise sentences before further processing. Ma et al. [11] propose a sentence-level model to remove noise data by generating effective training instances through negative training. Various attention mechanisms and features have been utilized to model the corpus and obtain multi-level representations [5,17–19]. For instance, Chen [20] and Li [21] introduce a contrastive learning framework to construct pseudo-positive samples for noise reduction. Additionally, some researchers [22,23] deal with long-tail data to fix the performance limitations of distantly supervised relation extraction tasks.

2.2 Knowledge Distillation

Knowledge Distillation is a typical research area in transfer learning, and the distillation framework is usually used to compress models [24]. Specifically, it uses a more complex and larger teacher network to guide a lightweight student network. The knowledge distillation is introduced to learn noisy labels [26,27], after Li et al. [25] propose a soft label mechanism to correct noisy labels in image classification tasks. Moreover, Yim et al. [28] present that student models can learn more accurate knowledge faster when students have a model structure similar to that of their teachers. Thus, the exploration of student models goes beyond lightweight ones. Recently, knowledge distillation is applied to DSRE tasks and achieved positive performance. Lei et al. [29] propose a bi-directional knowledge distillation mechanism to coordinate multi-source information of corpus and knowledge graph to mitigate bag-level noise. Li et al. [12] propose a multi-level distillation framework and adopt a flexible temperature mechanism to reduce inaccurately dark knowledge propagation. However, distillation frameworks still face the limitations of traditional label softening and inheritance bias from a single teacher. Therefore, we propose a multi-teacher knowledge distillation framework, which fills the gap of insufficient knowledge sources from a single teacher, and it also captures the local relations between instances dynamically by setting flexible temperatures.

3 Method

3.1 Task Definition

Supposing there is a sentence bag $B = \{S_1, ..., S_n\}$ containing n sentences, where each sentence $S_i = \{w_{1i}, ..., w_{mi}\}$ is composed of m words, and each sentence in the sentence bag contains the same entity pair (e_h, e_t). The distantly supervised relation extraction task is to find the specific relation r between entity pairs (e_h, e_t) on the bag B. In addition, we achieve sentence-level prediction by setting 1 as the number of sentences in the bag.

3.2 Model Overview

We propose a multi-teacher knowledge distillation framework for distantly supervised relation extraction, containing two modules of flexible temperature regulation (FTR) and multi-view knowledge distillation (MVKD). First, We design a flexible temperature regulation (FTR) to adapt to the differences of the instances, so as to make up for the shortcomings of traditional temperature calculation. Because there are differences between each instance, the settings of temperature will eventually affect the prediction distribution. Secondly, we propose MVKD and design a multi-teacher combination strategy to guide students to learn through multi-sources information and imitate a correct distribution to improve performance (Fig. 2).

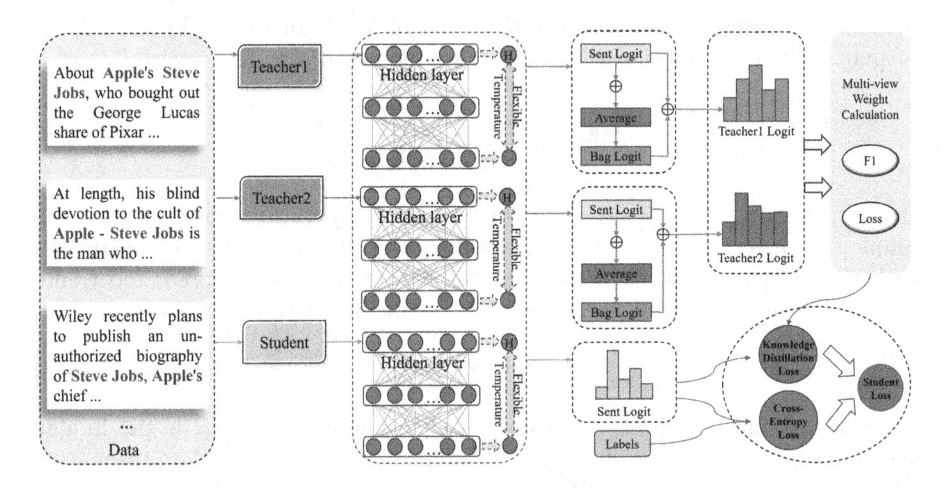

Fig. 2. MKDRE contains two main components: a flexible temperature regulation module and a multi-view knowledge distillation module. It includes the training process for both teacher and student models. In overall architecture, we utilize the relative difference between the hidden layer and the softmax layer to calculate the final temperature.

3.3 Flexible Temperature Regulation

Not only does reasonable softening of the target alleviate noise annotations under distant supervision, but also it compensates for insufficient information. Traditional knowledge distillation introduces a fixed temperature to control predicted softness. Accordingly, the formula calculates the probability distribution by incorporating the softmax function of temperature as follows:

$$p(\mathbf{z}, t) = \frac{exp(\mathbf{z}^o/t)}{\sum\limits_{i=1} exp(\mathbf{z}^i/t)} \tag{1}$$

where O is the number of categories, $\mathbf{z}^i \in R^O$ is a logit vector, and temperature t is used to control the softness. However, fixed temperature setting for different instances may make predictions too flat and lose characteristic information, or cause difficult predictions insufficiently softened. In this paper, we treat temperature t as dynamic. Intuitively, temperature t should be related to the information contained in the sentence, so we utilize information entropy to measure the amount of information represented by the sentence. Furthermore, it may cause information missing if only considering the sentence feature vector generated by a single layer, and it may also introduce deviation and uncertainty when calculating the temperature. To address the above issues, we design FTR based on multi-view information entropy to calculate the flexible temperature. According to the relative differences between information entropy in the hidden layer and the classification layer, the temperature is dynamically allocated to different samples to control the direction of label softening. Therefore, the softmax prediction function with flexible temperature is as follows:

$$p(\mathbf{z}_k, \widetilde{t}_k) = \frac{exp(\mathbf{z}_k/\widetilde{t}_k)}{\sum exp(\mathbf{z}_k/\widetilde{t}_k)} \tag{2}$$

where $\mathbf{z}_k$ is the logit vector, $\widetilde{t}_k$ is the flexible temperature, and i is the i-th sample. According to Equation (2), the higher the temperature, the flatter the prediction. Obviously, the temperature largely determines the predicted trend. The flexible temperature $\widetilde{t}$ is calculated by sigmoid function as follows:

$$\widetilde{t}_k = 1 + \eta \cdot sigmoid(\mu \cdot e_{h_k} - e_k) \tag{3}$$

where $\eta > 0$ and $\mu \in (0, 1)$ are two hyper-parameters. $e_{h_k} \in R$ is the hidden layer information entropy of the k-th sentence and $e_k \in R$ is the information entropy of the k-th sentence. We measure the softness of the initial prediction through entropy to determine the direction of temperature change. The hidden information entropy e_{h_k} and the output information entropy e_k are calculated as follows:

$$\begin{aligned} y &= \frac{exp(\mathbf{z})}{\sum exp(\mathbf{z})} \\ e &= -y \cdot log(y) \end{aligned} \tag{4}$$

where y is the prediction obtained by logit vector after softmax. The higher the information entropy, the lower the flexible temperature, so the prediction with higher initial softness will not become too flat, and vice versa.

3.4 Multi-view Knowledge Distillation

Our teacher model is based on sentence-level relation extraction, and it uses the softened prediction to calculate the loss of the teacher model. In detail, the teacher prediction of sample S_i is $\hat{y}_i^T = p(\mathbf{z}_i^T, \widetilde{t}_i^T)$, where $\mathbf{z}_i^T \in R^o$ is the logit vector and $\widetilde{t_i^T} \in R$ is the temperature. Then the teacher model uses cross entropy loss for training:

$$L^T = -\sum_{i=1}^{N}\sum_{j=1}^{O} y_{ij}^T \cdot log(\hat{y}_{ij}^T) \tag{5}$$

where N represents the number of instances, O is the number of categories, y_{ij}^T is distantly supervised annotation and $\hat{y}_{ij}^T$ is the prediction of the teacher model at sentence-level. To aggregate knowledge of different teachers to the most extent, we calculate cross entropy loss and F1 score of the teacher model. The weights of different teacher models are allocated in multi-view calculation for subsequent distillation:

$$w_q = \frac{1}{Q-1}[\alpha(1 - \frac{exp(L_q^T)}{\sum\limits_{i=1} exp(L_i^T)}) + (1-\alpha)\frac{exp(f_q)}{\sum\limits_{i=1} exp(f_i)}] \tag{6}$$

where w_q is the weight of the q-th teacher, Q is the number of teachers, and $\alpha \in [0, 1]$ is a hyper-parameter. The larger L_q^T corresponds to the less w_q, while the larger f_q corresponds to the greater w_q. Traditional multi-teacher models rely on a fixed value to allocate w_q, but our proposed Eq. (6) allows flexible control of the contribution of cross-entropy loss and F1 score to w_q based on the training outcome.

Distantly supervised relation extraction is inevitably influenced by noisy instances in sentence-level teacher learning. Our proposed MKDRE considers multi-level knowledge to transmit effective information to the student model. We use the teacher model to predict the sentences in the bag and aggregate the sentence representation in the bag like Li [12]:

$$\overline{H}^{b_k} = \frac{1}{|B_k|}\sum_{i=1}^{B_k} \hat{y}_i^T \tag{7}$$

where b_k represents the k-th bag, $\hat{y}_i^T$ is the sentence prediction in the bag, and B_k is the number of sentences in the k-th bag. Then the sentence representation i is combined to form the distillation target distribution of the student model, which is calculated as follows:

$$\widetilde{H}_k^T = \beta\overline{H}^{b_k} + (1-\beta)\hat{y}_k^T \tag{8}$$

where $\beta \in [0, 1]$ is a hyper-parameter and $\hat{y}_k^T$ is the prediction of the k-th sentence. Our MKDRE expects the predictions of the students to approach the soft distributions of the teachers, so we use Kullback-Leibler divergence as distillation loss. The loss function after combining multi-teacher weight calculation is as follows:

$$L_{KD} = \sum_{q=1}^{Q} w_q \sum_{i=1}^{N}\sum_{j=1}^{O} \widetilde{H}_{ij}^T log(\frac{\widetilde{H}_{ij}^T}{\hat{y}_{ij}^S}) \tag{9}$$

where N is the number of sentences, w_q is the weight assigned to the q-th teacher model, and $\widetilde{H}_{ij}^T$ is the soft target of teachers combining sentence-level and bag-level predictions. $\hat{y}_i^S = p(\mathbf{z}_i^S, \widetilde{t}_i^S)$ is the prediction of the sentence-level student model. $\mathbf{z}_i^S \in R^O$ is the logit vector and $\widetilde{t}_i^S \in R$ is the temperature.

3.5 Total Loss of Student Model

The distantly supervised annotations loss of the student model is calculated using a cross entropy loss function as follows:

$$L^S = -\sum_{i=1}^{N}\sum_{j=1}^{O} y_{ij}^S \cdot log(\hat{y}_{ij}^S) \tag{10}$$

The distantly supervised annotations loss calculation for the student model is the same as for the teacher model, and both of them are sentence-level. Therefore, the total loss function of the student model is as follows:

$$L = \lambda L^S + (1 - \lambda) L_{KD} \tag{11}$$

where $\lambda \in [0, 1]$ is a hyper-parameter, and it indicates that the total loss of the student model is determined jointly by knowledge distillation loss L_{KD} and distantly supervised loss L^S.

4 Experiments

In our experiments, we employ four models as our teacher models, namely PCNN [4], BiLSTM [30], BERT [31] and RoBERTa [32]. Furthermore, we construct a framework for multi-teacher knowledge distillation by grouping networks according to their similar prediction accuracy on given datasets, including two multi-teacher models of PCNN+BiLSTM-T and BERT+RoBERTa-T, as well as two students of PCNN-S and BERT-S.

4.1 Datasets

This work is evaluated on two datasets of NYT19-1.0 and NYT19-2.0, which are annotated on the basis of Ren et al. [33]. NYT19-1.0 dataset has 11 different relationships, including "None" relation, indicating that the entity pair do not belong to the other 10 relationships. As shown in Table 1, the training set of NYT19-1.0 contains 371,461 instances, and the test set contains 4543 instances. Different from the former, Jia et al. [34] construct NYT19-2.0 by correcting the mislabeled sample instances in the test set and removing the overlapping and fuzzy relations. The training set of NYT19- 2.0 contains 367,892 instances and the test set contains 4461 instances, as shown in Table 1.

4.2 Evaluation Metrics and Settings

We evaluate the performance of our model at the sentence-level by the metrics of precision (Prec.), recall (Rec.), and F1 [34]. By training instances at the sentence-level rather than at the bag-level, relation extraction yields more precise relations that can better support downstream tasks such as question answering systems. Therefore, our framework provides a more intuitive assessment at the sentence-level.

In the experiments, we establish two instances based on PCNN+BiLSTM and BERT+RoBERTa. As for PCNN+BiLSTM, we initialize the word dim using 50-dimensional GloVe. We set position dim to 5, batch size to 16, and window size to 3. To train PCNN+BiLSTM, we set learning rate to 5e-4 and conduct 20 epochs. For instance BERT+RoBERTa, we set batch size to 32, learning rate to 1e-4, and training epochs to 15. The maximum length is set to 512 for both instances.

Table 1. Statistics of the datasets used in our experiments.

Dataset		Train	Test
NYT19-1.0	Instances	371461	4543
	Sentences	235253	1024
NYT19-2.0	Instances	367892	9051
	Sentences	233083	3192

4.3 Baselines

We compare our proposed MKDRE with several strong baselines.

- **PCNN** [4] extracts rich relation features using piecewise max-pooling, and it is one of the basic models of multi-teacher models PCNN+BiLSTM-T and PCNN+BiLSTM-S.
- **BiLSTM** [30] processes information such as position embedding through bidirectional LSTM, and it is also one of the basic models of PCNN+Bilstm-T and PCNN+Bilstm-S.
- **BiLSTM+ATT** [35] extends an attention mechanism on BiLSTM to obtain the most salient features for discriminating relations.
- **PCNN+SelATT** [5] is a bag-level relation extraction model based on PCNN, and it uses an attention mechanism to aggregate sentences in a bag for denoising.
- **CNN+RL** [8] uses reinforcement learning to train CNN model and utilize an instance selector to filter noise samples.
- **ATT+RL** [36] adopts the reinforcement learning method, and it does not filter false positive instances directly but redivides them into negative instances for noise reduction.

- **ARNOR** [34] conducts sentence-level relation extraction from manually annotated datasets and selects confident instances by iterative training.
- **MIDTD** [12] is a single teacher knowledge distillation framework for relation extraction tasks.

4.4 Main Results

We conduct comparative experiments between our proposed MKDRE and state-of-the-art baselines on precision, recall, and F1. Table 2 shows that our MKDRE is more robust and effective.

MKDRE achieves the highest F1 score compared to the baselines on both datasets. MKDRE incorporates hidden layer information into the label softening mechanism, so it reduces noise and provides students with rich knowledge of the model. In general, our distillation model based on PCNN+BiLSTM is not only superior to the distillation model based on PCNN proposed by Li [12], but also gains significant improvement compared with ARNOR on sentence-based datasets. In detail, MKDRE(BERT+RoBERTa) compared with ARNOR shows 2.1%/1.1% improvement on NYT19-1.0/NYT19-2.0 test sets and 1.7%/0.8% improvement on dev sets, which indicates that our multi-teacher knowledge distillation framework is simple and effective to denoise the model.

Table 2. The metrics of Precision (P), Recall (R), and Micro-F1 (F1) are used to compare our model with the baselines on NYT19-1.0 and NYT19-2.0.

Method	Dev1.0			Test1.0			Dev2.0			Test2.0		
	P	R	F1	P	R	F1	P	R	F1	P	R	F1
PCNN*	36.1	63.7	46.1	36.1	64.9	46.4	39.1	74.7	51.3	42.2	77.5	54.6
BiLSTM*	36.7	66.5	47.3	35.5	67.4	46.5	41.2	70.2	52.1	44.1	71.1	54.5
PCNN+SeATT*	46.0	30.4	36.6	45.4	30.0	36.2	82.4	34.1	48.2	81.0	35.5	49.4
BiLSTM+ATT*	37.6	64.9	47.6	34.9	65.2	45.5	40.8	70.4	51.7	42.8	71.6	53.6
CNN+RL*	40.0	59.2	47.7	40.2	63.8	49.3	42.7	72.6	53.8	44.5	73.4	55.4
ATT+RL*	37.7	52.7	44.0	39.4	61.6	48.1	42.5	71.6	53.3	43.7	72.3	54.5
ARNOR*	**62.5**	58.5	60.4	**65.2**	56.8	60.9	**78.1**	59.7	67.8	**79.7**	62.3	69.9
MIDTD(PCNN-S)*	52.4	56.1	54.2	52.0	55.8	53.8	46.1	70.6	55.8	47.1	75.1	57.9
MKDRE(PCNN-S)	48.3	64.8	55.3	48.6	63.5	55.1	47.2	69.6	56.3	47.3	76.3	58.4
MKDRE(BERT-S)	53.6	**73.7**	**62.1**	54.2	**75.2**	**63.0**	60.9	**78.6**	**68.6**	63.3	**80.8**	**71.0**

It is worth noting that ARNOR method gains good performance in precision while MKDRE is superior in recall. Li [12] believes that ARNOR relies on clean manual annotation data and predicts more relations belonging to "None", so it gains low recall. However, the denoising method based on knowledge distillation provides MKDRE with more complete knowledge by label softening mechanism. Hence MKDRE can predict richer relations and achieves better recall performance.

According to the baselines, the performance of PCNN and BiLSTM on NYT19-1.0/NYT19-2.0 are similar. Therefore, we compare our MKDRE framework using PCNN+BiLSTM as a multi-teacher benchmark network with MIDTD using PCNN as the benchmark network to verify the performance of the multi-teacher knowledge distillation framework. The final results show that our model outperforms MIDTD by 1.1%/1.3% on the test set/dev set on NYT19-1.0, and by 0.5%/0.5% on the test set/dev set on NYT19-2.0. This is due to our flexible temperature approach, which reduces model uncertainty from the direction of relative differences. In addition, our proposed multi-teacher distillation method can significantly improve the predictive performance of sentence level. Furthermore, effective performance improvements are also generated when using PCNN and BiLSTM as the networks of MKDRE, compared with the results of PCNN and BILSTM in the baseline. Benefiting from the rich knowledge of multi-teachers, which combines the perspectives of different teachers, the performance of knowledge distillation is improved.

4.5 Ablation Study

In order to explore the influence between different modules, we conduct ablation experiments to test the parts of multi-teacher distillation and flexible temperature regulation. Taking the model based on PCNN+BiLSTM as an example, sentence-level evaluation is conducted on NYT19-1.0 and NYT19-2.0.

The ablation results of multi-teacher knowledge distillation module and flexible temperature regulation module are respectively displayed to show their contributions. The experimental results are shown in Table 3 and Table 4.

Table 3. Ablation studies. Ablation studies explore three types, including no temperature, traditional fixed temperature, and flexible temperature.

Datasets	Method	P	R	F1
NYT19-1.0	PCNN+BiLSTM	48.6	63.5	55.1
	-No Temp	46.5	62.1	53.2
	-Fixed Temp	46.9	62.5	53.6
NYT19-2.0	PCNN+BiLSTM	47.3	76.3	58.4
	-No Temp	45.6	73.8	56.4
	-Fixed Temp	46.2	75	57.2

Table 4. The ablation experiments are conducted on NYT19-1.0 and NYT19-2.0 datasets. The ablation experiments focus on the performance of single-teacher knowledge distillation based on PCNN and BiLSTM respectively.

Datasets	Method	P	R	F1
NYT19-1.0	PCNN+BiLSTM	48.6	63.5	55.1
	-PCNN	46.3	61.4	52.8
	-BiLSTM	46.5	62.7	53.4
NYT19-2.0	PCNN+BiLSTM	47.3	76.3	58.4
	-PCNN	46.1	74.2	56.9
	-BiLSTM	46.9	70.4	56.3

Specifically, for the flexible temperature module (FTR), we set the temperature to a fixed value to evaluate the results brought by changing temperature dynamically. Table 3 shows that FTR contributes to improving the performance of MKDRE. Benefiting from the label softening of hard annotations, FTR module brings 1.7%/2.2% improvement in F1 scores on NYT19-1.0 and NYT19-2.0 respectively. In addition, our FTR brings 1.5%/1.2% improvement in F1 on both datasets compared to a fixed temperature setting. This indicates that the temperature setting can affect the final prediction results. It is effective to spread knowledge by changing the temperature according to each sample. All in all, FTR module brings better performance to the model, and the label softening mechanism also helps MKDRE get more accurate knowledge.

To verify the effectiveness of our multi-teacher framework, we build a single-teacher distillation based on BiLSTM and PCNN respectively. Table 4 shows that the diverse knowledge provided by multi-teacher distillation improves the performance of MKDRE. Benefiting from the knowledge of multiple teachers from different perspectives, MVKD module improves F1 scores by 2.3%/1.5% in the single teacher knowledge distillation based on PCNN, and 1.7%/2.1% in BiLSTM. Accordingly, our multi-teacher knowledge distillation framework can effectively reduce noise, meanwhile, the setting of weight mechanism can combine knowledge of teachers more practically.

5 Conclusion

In this paper, we present MKDRE, a novel multi-teacher knowledge distillation framework for DSRE at sentence level. The proposed MKDRE framework with FTR component breaks the limitation of fixed temperature, which considers the differences between instances and uses a flexible temperature calculation strategy to reduce noise. FTR introduces hidden layer information and uses a more robust relative difference method to control the direction of prediction. We found that a single teacher may lead to the incorrect transmission of knowledge, so MVKD is designed to alleviate the bias of single-teacher models. The experimental results demonstrate that our proposed MKDRE achieves better performance in distantly supervised relation extraction at sentence level than previous methods.

In future work, we plan to address the long tail problem in DSRE by generating more training instances through data enhancement. This will provide more abundant and diverse training data for the distillation model to improve its prediction ability. Furthermore, we aim to extend our distillation framework to other tasks, including document-level relational extraction and event extraction, and further validate the effectiveness and versatility of our proposed approach.

References

1. Wu, S., He, Y.: Enriching pre-trained language model with entity information for relation classification. In: CIKM, pp. 2361–2364 (2019)
2. Wei, Z., Su, J., Wang, Y., Tian, Y., Chang, Y.: A novel cascade binary tagging framework for relational triple extraction. In: ACL, pp. 1476–1488 (2020)
3. Mintz, M., Bills, S., Snow, R., Jurafsky, D.: Distant supervision for relation extraction without labeled data. In: ACL, pp. 1003–1011 (2009)
4. Zeng, D., Liu, K., Chen, Y., Zhao, J.: Distant supervision for relation extraction via piecewise convolutional neural networks. In: EMNLP, pp. 1753–1762 (2015)
5. Lin, Y., Shen, S., Liu, Z., Luan, H., Sun, M.: Neural relation extraction with selective attention over instances. In: ACL, pp. 2124–2133 (2016)
6. Vashishth, S., Joshi, R., Prayaga, S.S., Bhattacharyya, C., Talukdar, P.: RESIDE: improving distantly-supervised neural relation extraction using side information. In: EMNLP, pp. 1257–1266 (2018)
7. Lin, X., Liu, T., Jia, W., Gong, Z.: Distantly supervised relation extraction using multi-layer revision network and confidence based multi-instance learning. In: EMNLP, pp. 165–174 (2021)
8. Feng, J., Huang, M., Zhao, L., Yang, Y., Zhu, X.: Reinforcement learning for relation classification from noisy data. In: AAAI (2018)
9. Zeng, X., He, S., Liu, K.: Large scaled relation extraction with reinforcement learning. In: AAAI (2018)
10. He, Z., Chen, W., Wang, Y., Zhang, W., Wang, G., Zhang, M.: Improving neural relation extraction with positive and unlabeled learning. In: AAAI, pp. 7927–7934 (2020)
11. Ma, R., Gui, T., Li, L., Zhang, Q., Huang, X., Zhou, Y.:SENT: sentence-level distant relation extraction via negative training. In: ACL, pp. 6201–6213 (2021)
12. Li, R., Yang, C., Li, T., Su, S.: MidTD: a simple and effective distillation framework for distantly supervised relation extraction. ACM Trans. Inf. Syst. (TOIS) **40**(4), 1–32 (2022)
13. Zhang, Z., et al.: Distilling knowledge from well-informed soft labels for neural relation extraction. In: AAAI, pp. 9620–9627 (2020)
14. Riedel, S., Yao, L., McCallum, A.: Modeling relations and their mentions without labeled text. In: Balcázar, J.L., Bonchi, F., Gionis, A., Sebag, M. (eds.) ECML PKDD 2010. LNCS (LNAI), vol. 6323, pp. 148–163. Springer, Heidelberg (2010). https://doi.org/10.1007/978-3-642-15939-8_10
15. Wang, X., Yang, J., Wang, Q., Su, C.: Threat intelligence relationship extraction based on distant supervision and reinforcement learning. In: SEKE, pp. 572–576 (2020)

16. Qin, P., Xu, W., Wang, W. Y.: DSGAN: generative adversarial training for distant supervision relation extraction. In: ACL, pp. 496–505 (2018)
17. Li, Y., et al.: Self-attention enhanced selective gate with entity-aware embedding for distantly supervised relation extraction. In: AAAI, pp. 8269–8276 (2020)
18. Du, J., Han, J., Way, A., Wan, D.: Multi-level structured self-attentions for distantly supervised relation extraction. In: EMNLP, pp. 2216–2225 (2018)
19. Wang, J., Liu, Q.: Distant supervised relation extraction with position feature attention and selective bag attention. Neurocomputing **461**, 552–561 (2021)
20. Chen, T., Shi, H., Tang, S., Chen, Z., Wu, F., Zhuang, Y.: CIL: contrastive instance learning framework for distantly supervised relation extraction. In: ACL, pp. 6191–6200 (2021)
21. Li, D., Zhang, T., Hu, N., Wang, C., He, X.: HiCLRE: a hierarchical contrastive learning framework for distantly supervised relation extraction. In: ACL, pp. 2567–2578 (2022)
22. Zhang, Y., Fei, H., Li, P.: ReadsRE: retrieval-augmented distantly supervised relation extraction. In: ACM SIGIR, pp. 2257–2262 (2021)
23. Peng, T., Han, R., Cui, H., Yue, L., Han, J., Liu, L.: Distantly supervised relation extraction using global hierarchy embeddings and local probability constraints. Knowl.-Based Syst. **235**, 107637 (2022)
24. Hinton, G., Vinyals, O., Dean, J.: Distilling the knowledge in a neural network. In: NeurIPS (2015)
25. Li, Y., Yang, J., Song, Y., Cao, L., Luo, J., Li, L. J.: Learning from noisy labels with distillation. In: ICCV, pp. 1910–1918 (2017)
26. Sarfraz, F., Arani, E., Zonooz, B.: Knowledge distillation beyond model compression. In: ICPR, pp. 6136–6143 (2021)
27. Zhou, H., et al.: Rethinking soft labels for knowledge distillation: a bias-variance tradeoff perspective. In: ICLR (2021)
28. Yim, J., Joo, D., Bae, J., Kim, J.: A gift from knowledge distillation: fast optimization, network minimization and transfer learning. In: CVPR, pp. 7130–7138 (2017)
29. Lei, K., et al.: Cooperative denoising for distantly supervised relation extraction. In: COLING, pp. 426–436 (2018)
30. Zhang, S., Zheng, D., Hu, X., Yang, M.: Bidirectional long short-term memory networks for relation classification. In: PACLIC, pp. 73–78 (2015)
31. Devlin, J., Chang, M.W., Lee, K., Toutanova, K.: BERT: pre-training of deep bidirectional transformers for language understanding. In: NAACL-HLT, pp. 4171–4186 (2019)
32. Liu, Y., et al.: RoBERTa: a robustly optimized bert pretraining approach. arXiv preprint arXiv:1907.11692 (2019)
33. Ren, X., et al.: Cotype: joint extraction of typed entities and relations with knowledge bases. In: WWW, pp. 1015–1024 (2017)
34. Jia, W., Dai, D., Xiao, X., Wu, H.: ARNOR: attention regularization based noise reduction for distant supervision relation classification. In: ACL, pp. 1399–1408 (2019)
35. Zhou, P., et al.: Attention-based bidirectional long short-term memory networks for relation classification. In: ACL, pp. 207–212 (2016)
36. Qin, P., Xu, W., Wang, W. Y.: Robust distant supervision relation extraction via deep reinforcement learning. In: ACL, pp. 2137–2147 (2018)

PAEE: Parameter-Efficient and Data-Effective Image Captioning Model with Knowledge Prompter and Cross-Modal Representation Aligner

Yunji Tian, Zhiming Liu, Quan Zou(✉), and Geng Chen

College of Computer and Information Science, Southwest University, Chongqing 400715, China
qzou2014@swu.edu.cn

Abstract. Large-scale pre-trained models and research on massive data have achieved state-of-the-art results in image captioning technology. However, the high cost of pre-training and fine-tuning has become a significant issue that needs to be considered. In this paper, we propose PAEE, a parameter-efficient and data-effective image captioning model that generates captions based on the input image encoding and the knowledge obtained from the newly introduced Knowledge Prompter. In PAEE, the only module that needs to be learned is the Cross-modal Representation Aligner (CRA) introduced between the visual encoder and language decoder, which facilitates the language model's better adaptation to visual representation. The entire model greatly reduces the cost of pre-training and fine-tuning. Extensive experiments demonstrate that PAEE maintains competitive performance compared to large-scale pre-trained models and similar approaches, while reducing the number of trainable parameters. We design two new datasets to explore the data utilization ability of PAEE and discover that it can effectively use new data and achieve domain transfer without any training or fine-tuning. Additionally, we introduce the concept of *small-data* learning and find that PAEE has data-effective characteristics in limited computing resources and performs well even with fewer training samples.

Keywords: image captioning · parameter-efficient · data-effective · CRA · knowledge prompter · small-data learning

1 Introduction

Over the course of recent years, the effective utilization of large-scale pre-trained models to image captioning has led to the preferred approach of scaling up the models and training data to achieve better performance in image captioning [17, 31]. Despite the impressive performance of these models, the need for substantial computing resources and the high cost of training has become an urgent issue in the face of model pre-training and fine-tuning for various downstream tasks.

X. Song et al. (Eds.): APWeb-WAIM 2023, LNCS 14332, pp. 117–131, 2024.
https://doi.org/10.1007/978-981-97-2390-4_9

There have been attempts to reduce the cost of model training, such as I-Tuning [18] and ClipCap [19]. These models are inspired by parameter-efficient pre-trained models that can effectively lower the training cost [2]. They utilize existing, high-performing pre-trained visual encoders and language decoders, with the parameters of both ends frozen, and only the mapping between them trained. This indeed reduces the trainable parameters significantly, but there are two main limitations: 1) The learned knowledge is embedded in the model parameters, resulting in less accuracy and richness in text captions for rare objects and scenes; 2) Limited ability to leverage new data, requiring separate training for specific cases, such as image captioning for visually impaired individuals [13].

When utilizing the training strategies mentioned above, it is crucial to consider the effective connection between the visual encoder and the language model [23]. This is because obtaining an accurate mapping of visual information is essential for improving caption generation. Recent approaches have included fine-tuning the entire language model [8], or training adapters [10] to better adapt the language model to visual information representation. However, these methods often suffer from high training costs due to the requirement of large-scale multimodal training data and a high number of trainable parameters (ranging from hundreds of millions to billions). Some studies have also employed parameter-efficient methods, such as Frozen [29], which only freezes the parameters of the language model. Although this approach reduces the number of parameters compared to the aforementioned methods, it still necessitates training the visual encoder from scratch, and thus it continues to face the challenge of high computational costs.

Based on the aforementioned issues, we propose PAEE, a parameter-efficient and data-effective image captioning model. PAEE adopts an encoder-decoder structure, employing the Clip-ViT visual encoder [23] and the OPT language model [34] for the visual and language components, respectively. During training, we freeze the parameters of both ends and connect them through a newly designed cross-modal representation aligner, which is the only part that needs to be trained in the entire image captioning model. The number of trainable parameters is only 1.7M, making the entire model parameter-efficient with low training costs. Simultaneously, we design a novel knowledge prompter that retrieves the appropriate captions from its internal knowledge base based on the input image and uses it as an input prompt for the language encoder, making full use of our model's ability to effectively utilize the data.

We evaluate PAEE on common image captioning benchmarks. The results show that PAEE maintains a certain level of performance with a significantly reduced number of parameters. It performs comparably to some large scale pre-trained models and shows noticeable advantages over similar lightweight model approaches. Additionally, we redesign two datasets to further examine the ability of PAEE to exploit data. Our experimental results indicate that PAEE effectively leverages new knowledge obtained from the prompter and can adapt to data from new domains without further training. This is different from similar lightweight models that require fine-tuning to perform well in new domains.

To further investigate the data efficiency of PAEE, we train with a 1% subset of image-text pairs (thousands of examples) associated with each downstream task. We name this experimental setup *small-data* learning. Small-data learning is highly valuable when encountering issues such as limited computing resources and data availability. We train a specialized version of PAEE for the downstream domain and then evaluate its performance on 1% and 100% of in-domain and domain-agnostic data for each downstream task. Experimental results show that PAEE is more effective than the baseline model Frozen with few-shot learning ability in the same settings. This demonstrates that PAEE has strong data effectiveness and can make effective use of data with limited computational resources. Furthermore, these results suggest that PAEE has potential for few-shot learning.

2 Related Work

2.1 Frozen Parameters Captioning Models

Freezing the parameters of pre-trained models is an effective strategy for optimizing deep learning models [29]. Freezing a portion or all of the training parameters can enhance the generalization ability of the model, accelerate the training process, and greatly improve parameter efficiency of the model [2]. Similar to our work, the lightweight models I-Tuning and ClipCap adopt the training strategy: While exclusively concentrating on training the mapping between the two distinct components, the parameters of the visual encoding module and the language model are frozen. In our work, we use the CLIP visual encoder and OPT language model with the same training strategy. However, we introduce a new prompter and CRA module between these two components, and the entire model not only has higher parameter efficiency, but also robust data utilization capabilities.

2.2 Knowledge Retrieval-Based Prompting

In our work, the prompter is designed along similar lines to retrieval-augmented language generation. Retrieval-augmented language generation is a technology that combines retrieval and generation methods to adjust language generation using relevant information retrieved from external data storage [15]. Regarding image captioning, a retrieval-guided generative adversarial framework [32] and a retrieval-enhanced captioning model [25] exist, while Sarto *et al.* [26] proposed a retrieval-augmented transformer based image captioning model. Inspired by these studies, Our work differs from previous research in that PAEE has higher parameter efficiency and data validity with the help of filtering strategy and CRA. In addition, we successfully apply this approach to small-data learning task.

2.3 Visual and Language Connection

PAEE aligns the visual and textual representation spaces using a novel aligner. Our CRA structure shares some similarities with the Perceiver Resampler in Flamingo [2] and the lightweight model ClipCap, including a fixed number of learned constant embeddings and a core transformation stack. However, our work differs significantly from previous studies in that we not only maintain parameter efficiency, but also accurately capture information from the visual end. Our model has only 1.7M trainable parameters, which is significantly less than both Flamingo's (194M) and ClipCap's (43M).

2.4 Prompting Caption Generation

Prompting refers to textual information provided as input to guide models in generating expected outputs [16]. Similarly, task demonstrations often provide a set of example inputs and expected outputs as a reference for model learning [11]. The combination of these two methods can enhance the precision in guiding the model's text generation [24]. Our work applies this combination to the image captioning task by designing a new task demonstration that incorporates relevant information obtained from a prompter to improve the text generation performance of the entire model in line with our expectations.

3 Method

3.1 Architecture

PAEE employs an encoder-decoder structure, with CLIP serving as the visual encoder and OPT as the language model (LM). Figure 1 illustrates the structure of the entire model where the specific details of the CRA are shown in 3.5.

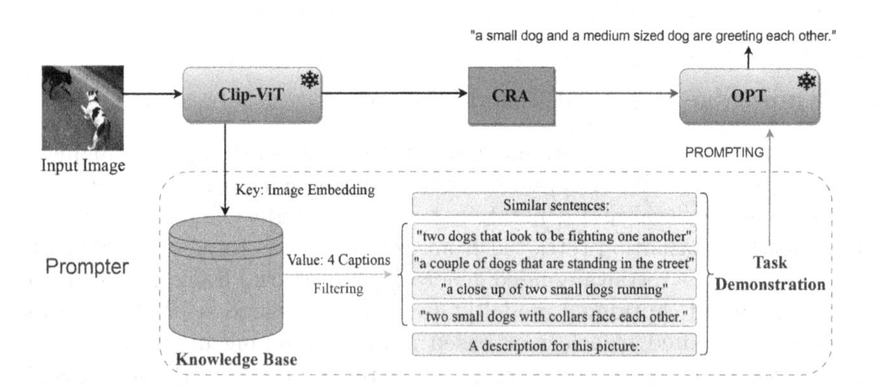

Fig. 1. Prompter retrieves relevant captions from knowledge base based on CLIP image embeddings and combines them with task demonstration to provide prompt input to LM. During training, the parameters of both ends are frozen, and only the CRA module is trained to map the image representation, shifting it from the output embedding space of Clip-ViT to the token embedding space of a language model.

3.2 Pre-trained Image Encoder

Considering the powerful generalization ability of existing large-scale pre-trained models, we use Clip-ViT [23] as our visual encoder. This encoder can extract condensed semantic information from images and produce a sequence eligible for mapping to embedding vectors. We select the Clip-ViT/14 backbone due to our observation that it yielded the best downstream task performance among all the variants in our experience. In our method, we freeze the training parameters of the Clip-ViT encoder and use a flattened spatial feature grid with a size of 16×16 as the input to the final projection layer. This produces a sequence of $L_i = 257$ vectors, each with a dimension of $D_i = 1024$.

3.3 Pre-trained Language Model

For the LM component, we selected OPT [34], an open-source LM developed by META AI. We utilize three different sizes of OPT models, as described in Sect. 4.1 of the subsequent experimental part. In processing text with the LM, the input text is first tokenized into discrete tokens. Subsequently, every token is transformed, resulting in a continuous embedding using the LM's embedder, which produces embeddings with a size of $D_o = 2048$. The resulting token embedding sequence is then passed through the self-attention layer of the transformer block, which produces a sequence comprising categorical distributions over the token vocabulary. Our method utilizes beam search decoding to generate text from the generated distributions.

3.4 Prompter-Based Caption Generation

The prompter consists of a knowledge base and a task demonstration, structured as a key-value graph. The keys represent the image feature embeddings extracted by the Clip-ViT visual encoder, while the values are the corresponding sets of text captions retrieved from the knowledge base. The knowledge base is built on an image-text paired dataset, and its data storage can be changed as needed, such as by expanding it to explore PAEE domain transfer and the ability to use new data, as detailed in Sect. 4.3 of the experimental part. We employ the comprehensive CLIP model to map two modalities into a shared vector space. This involves encoding both the input images and the textual content extracted from the knowledge base. To retrieve captions from the knowledge base that have the highest similarity to the images, we use a nearest neighbour search based on cosine similarity. The captions retrieved in this way are then incorporated into the green segment of Fig. 1 in the task demonstration, serving as prompts. Additionally, due to the possibility of multiple manually annotated captions for a single image in certain datasets, if the model retrieves a title from the knowledge base during training that matches or closely resembles the ground truth caption, it might mistakenly interpret this as a successful retrieval, impacting the overall training efficacy of the model. In order to avoid the effects of this, we use a filtering measure. Specifically, throughout both the training and inference processes, the model assesses whether the retrieved image is the same as the input image and filters out image-text pairs where the retrieved image matches the input image.

3.5 Cross-Modal Representation Aligner

CRA serves to convert the visual features extracted by the visual encoder into a continuous embedding sequence and feed it into the transformer of the LM. To balance mapping performance and the number of trainable parameters, Our CRA uses a transformer encoder infrastructure consisting of 4 layers with 8 heads per encoder. This module can receive projected image features (from D_i to D_o) from the visual encoder, process them, and output a series of continuous embeddings with a size of D_o. To address the issue of parameter efficiency, we employ a dimension bottleneck approach to decouple the hidden layer size D_h of the transformer from the visual feature dimension D_i and the language model embedding dimension D_o, as depictedin Fig. 2.

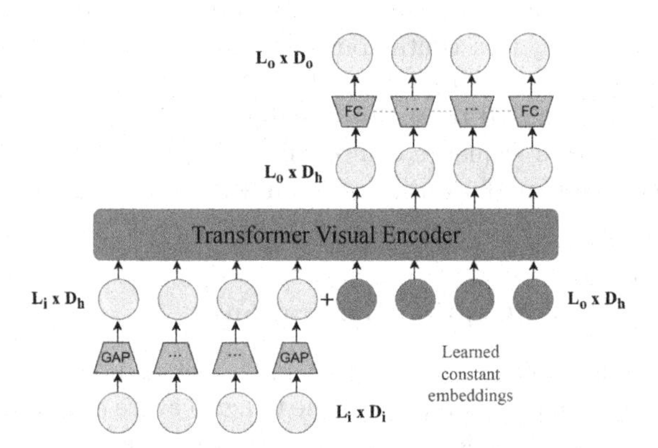

Fig. 2. CRA receives visual features of length L_i and dimension D_i from the visual encoder, converting them into sequences of token embeddings, with a length of L_o and a dimension of D_o. Here, D_o represents the dimension of token embeddings in the language model.

In particular, in order to preserve visual spatial information effectively while reducing the number of parameters, we utilize global average pooling (GAP) to transform each visual feature from its original dimension of $D_i = 1024$ to a lower dimension of $D_h = 256$. Subsequently, the produced visual feature sequence is input to the transformer, and a set of fully-connected (FC) layers is utilized for linearly projecting the output representation, transitioning from D_h to $D_o = 2048$. To reduce the number of parameters in CRA, we share parameters across all FC layers.

To enhance the computational efficiency of our model, we decouple the input sequence length L_i from the output sequence length L_o of the transformer. This results in L_o being significantly smaller than L_i, specifically $L_o = 32$ and

$L_i = 257$. Consequently, the computational complexity of the sequence passing through the self-attention layers of the language model is greatly reduced. We incorporate a set count (equal to the length of L_o) of learnable constants into the input sequence, leveraging solely the resulting representations associated with to these embedded constants. These constants serve two primary purposes: firstly, they retrieve meaningful information from CLIP embeddings through multi-head attention, and secondly, they enable the fixed language model to adapt to new data through learning. Our strategy for training CRA is to minimise the negative log-likelihood linked to the language model's generation of reference captions, conditional on image features and task demonstration. CRA is trained from scratch and serves as the only trainable component in our model, with a parameter count of 1.7M. Compared to the previously mentioned Frozen model (40.3M), PAEE has significantly fewer trainable parameters, resulting in high parameter efficiency across the entire model.

4 Experiments

4.1 Experimental Settings

Datasets. Our evaluation of the performance of PAEE is based on three publicly available datasets, namely **MSCOCO** [6], **Flickr30k** [22], and **Nocaps** [1]. We also design two new datasets to expand the knowledge base and further explore PAEE's ability to use new data and domain transfer: 1) We collect three large-scale datasets from the internet, which contain 15 million high-quality image-text pairs, from **SBU Captions** [20], **Conceptual Captions** [27] and **Conceptual 12M** [5], for simplicity, we refer to it as **SCC**; 2) We also consider the clean, manually annotated **COYO-700M** [4] dataset, which contains 740 million image-text pairs filtered from 10 billion web image-text sources. In our experiments, to further improve the accuracy of the image captioning task, we filter out low CLIP similarity scores between images and text or some irregular text annotations, resulting in 105M high-quality image-text pairs, for simplicity, we refer to it as **CY105**. These two datasets are also used as the primary source for the knowledge base to further explore the data effectiveness of PAEE, as detailed in Sect. 4.4.

Implementation Details. We develop three different sizes of PAEE based on OPT [34] with varying sizes, named PAEE_{base} (CLIP-ViT-L/14, OPT-1.3B), PAEE_{medium} (CLIP-ViT-L/14, OPT-6.7B), and PAEE_{large} (CLIP-ViT-L/14, OPT-13B). The COCO and Flickr30K datasets, divided according to the standard Karpathy split, are utilized for training the CRA module. We set the global batch size to 256, train for 10 epochs, start with an initial learning rate of 1e-4, and use the AdamW optimizer. In our method, we utilize beam search, employing a beam size of 3. Training on a single NVIDIA 3090 GPU with 24GB memory takes up to 6 h. During training, the knowledge base stores COCO training captions, and the prompter search relies on the CLIP-ResNet50x643 representation of both the input image and the captions stored in the knowledge base. We fix the number of retrieved captions to 4.

Evaluation Metrics. To evaluate our model, we utilize the standard evaluation metrics for image captioning tasks, which include CIDEr [30], BLEU@4 [21], METEOR [9], and SPICE [3].

4.2 Performance Comparison

COCO and Flickr30k. Table 1 presents that PAEE demonstrates a performance that is on a par with or exceeds that of existing large-scale pre-trained models, while also outperforming other lightweight models. As shown in the top half of the table, our method attains a score of 122.9 in terms of CIDEr on the COCO test dataset, which is 2.8 points higher than the XGPT model and only 0.5 points lower than the OSCAR model, while requiring significantly fewer trainable parameters. The lower part of the table presents a contrast between our method and alternative lightweight models, showing the higher parameter efficiency and performance of our method compared to ClipCap and I-Tuning.

Table 1. Results for COCO and Flickr30k image captioning. A hyphen "-" indicates the absence of that score in the model's original paper. Bold type indicates the highest score of the model. "Params" indicates the number of trainable parameters.

Model	Params	COCO				Flickr30k			
		BLEU@4	METEOR	SPICE	CIDEr	BLEU@4	METEOR	SPICE	CIDEr
Large-scale Pre-trained Model									
UniTAB [33]	135M	35.8	28.4	21.5	119.1	30.7	23.7	17.4	70.1
XGPT [31]	135M	**37.2**	28.6	21.8	120.1	**31.8**	23.6	17.6	70.9
OSCAR$_{Base}$ [17]	135M	36.5	**30.3**	**23.1**	**123.4**	-	-	-	-
VL-T5 [7]	270M	34.5	28.7	21.9	116.5	–	–	–	–
Lightweight Model									
ClipCap	43M	33.5	27.5	21.1	113.1	–	–	–	–
I-Tuning$_{Base}$	14M	34.8	28.3	21.8	116.7	25.2	22.8	16.9	61.5
I-Tuning$_{Medium}$	44M	35.5	28.8	22.0	120.0	28.8	24.6	19.0	72.3
I-Tuning$_{Large}$	95M	34.8	29.3	22.4	119.4	29.8	25.1	19.2	75.4
PAEE$_{Base}$	1.7M	36.2	28.6	22.1	118.8	28.5	24.1	18.6	69.1
PAEE$_{Medium}$	3.4M	36.8	28.9	22.4	120.7	30.1	25.6	20.3	74.6
PAEE$_{Large}$	6.8M	37.1	29.6	22.8	122.9	31.2	**26.1**	**20.7**	**79.5**

Table 2. Results for nocaps image captioning.

Model	Params	in-domain		near-domain		out-of-domain		Overall	
		CIDEr	SPICE	CIDEr	SPICE	CIDEr	SPICE	CIDEr	SPICE
OSCAR$_{Base}$	135M	79.6	12.3	66.1	11.5	45.3	9.7	63.8	11.2
ClipCap	43M	84.9	12.1	66.8	10.9	49.1	9.6	65.8	10.9
I-Tuning	14M	83.9	12.4	70.3	11.7	48.1	9.5	67.8	11.4
PAEE$_{Base}$	1.7M	84.2	12.6	71.5	11.9	48.6	9.6	68.2	11.5
PAEE$_{Medium}$	3.4M	89.5	12.9	77.6	12.4	56.2	10.8	74.6	12.1
PAEE$_{Large}$	6.8M	**89.9**	**13.5**	**80.8**	**12.8**	**66.2**	**11.1**	**79.1**	**12.5**

Nocaps. In Table 2, the overall CIDEr score of the large-scale pre-trained model OSCAR on Nocaps is even lower than our PAEE model.Our model demonstrates competitive performance compared to other lightweight models in both in-domain and out-of-domain scenarios, with significant advantages in out-of-domain scenarios. This highlights the strong generalization ability of our model, as it can retrieve new knowledge from the knowledge base in the prompter.

4.3 Data Utilization Capabilities

We will further explore the capabilities of PAEE in domain transfer and the use of new data when the data storage in the knowledge base of our model is changed. We evaluate its CIDEr scores on three different datasets: **Flickr30K**, **VizWiz-Caps** [12], and **Visual Genome**(VG) [14].

In-Domain Data. In Table 3, we initially test the performance by replacing the knowledge base storage of PAEE with in-domain data from each dataset. In comparison with the original model, where the knowledge base stores COCO data, the model shows a significant improvement in all three datasets. Notably, the performance improvement on VizWiz-Caps is the most significant. This dataset is specifically designed for visually impaired people and has properties that differ greatly from the COCO dataset, requiring a larger domain gap to be bridged with relevant data within the domain. This demonstrates that PAEE can flexibly retrieve new knowledge from the knowledge base and achieve domain transfer within the same domain.

Table 3. Exploration of data utilization capabilities. Only the content of the knowledge base was changed without any training and fine-tuning of the three datasets.

Knowledge base	Flickr30k	VizWiz-Caps	VG
COCO	69.1	36.6	60.3
In-domain	72.2	41.9	62.5
In-domain+SCC	**74.1**	**42.1**	63.0
In-domain+CY105	73.3	41.8	**63.5**
SCC	73.8	36.6	61.3
CY105	73.0	31.5	62.2

Enriching the Knowledge Base. We empoly the two newly designed datasets mentioned in the previous section to expand the knowledge base data storage. We conduct experiments by adding SCC and CY105 to the knowledge base separately. As shown in Table 3, the experimental results indicate that the combination of SCC and CY105 with in-domain data improves the performance of the model on all three datasets. When compared to using only in-domain data, the expansion to a larger and more diverse data storage resulted in good improvements on both the Flickr30k and VG datasets for PAEE. However, the improvement on VizWiz is relatively low, which may be attributed to the dataset's special nature, making it challenging to match with out-of-domain data. These experimental results confirm that PAEE effectively utilizes new knowledge from the prompter to enhance the model's performance, demonstrating strong generalization abilities that facilitate domain transfer, and all of this is achieved without any pre-training.

4.4 Exploration of Small-Data Learning

In this part of the study, we assess the performance of PAEE in small-data learning. We train specialized versions of PAEE and the baseline model Frozen on each downstream task dataset, such as pre-training two models on the TextCaps [28]. The knowledge base storage is based on the combination of SCC and CY15 data. Small-data learning is highly valuable when encountering issues such as limited computing resources and data availability.

In-Domain Training. For in-domain training, we train these specialized versions of PAEE on complete and minimal (1%) in-domain for every corresponding downstream task, respectively, and assess their performance. As shown in Table 4, on the COCO dataset, PAEE's CIDEr score is almost twice that of Frozen, highlighting the superior data effectiveness of PAEE in in-domain learning over Frozen. Meanwhile, when trained on 1% of the data, PAEE's BLEU@4 score on COCO is even comparable to that of Frozen trained on 100% in-domain data, demonstrating PAEE's strong data effectiveness and outstanding performance in situations with limited samples, and showcasing its powerful potential for few-shot learning.

Table 4. Exploration of small-data learning. For domain-agnostic training, we train two models on the SCC-clean dataset in advance. Bold indicates the best scores.

	Params	Samples	COCO		TextCaps		VizWiz-Caps		Overall	
			BLEU@4	CIDEr	BLEU@4	CIDEr	BLEU@4	CIDEr	BLEU@4	CIDEr
100% in-domain training										
Frozen$_{COCO}$	40.3M	414K	20.2	62.4	7.1	11.9	5.6	6.3	8.4	22.4
Frozen$_{TextCaps}$	40.3M	103K	4.1	6.8	8.9	17.1	4.5	5.3	4.4	8.2
Frozen$_{VizWiz}$	40.3M	110K	3.8	6.1	4.2	5.7	19.1	76.9	6.9	23.2
PAEE$_{COCO}$	3.4M	414K	**36.6**	**120.1**	16.7	41.6	18.2	41.9	**18.5**	**60.0**
PAEE$_{TextCaps}$	3.4M	103K	10.1	29.1	**18.5**	**63.1**	11.5	32.2	10.3	34.5
PAEE$_{VizWiz}$	3.4M	110K	13.8	48.9	11.6	31.6	**34.9**	**138.5**	15.3	59.5
1% in-domain training										
Frozen$_{COCO}$	40.3M	4.1K	6.3	12.9	2.8	3.1	2.9	2.3	3.1	5.5
Frozen$_{TextCaps}$	40.3M	1.0K	1.7	2.9	3.7	5.1	2.0	2.1	1.9	3.2
Frozen$_{VizWiz}$	40.3M	1.1K	3.0	3.2	3.4	3.5	12.8	40.1	4.9	12.6
PAEE$_{COCO}$	3.4M	4.1K	**19.9**	**66.1**	7.2	13.1	6.4	9.9	**8.6**	**25.4**
PAEE$_{TextCaps}$	3.4M	1.0K	4.4	8.3	**8.6**	**17.3**	5.1	7.4	4.6	9.3
PAEE$_{VizWiz}$	3.4M	1.1K	3.1	5.1	3.7	5.2	**18.6**	**72.1**	6.4	21.6
100% domain-agnostic training										
Frozen$_{SCC\text{-}clean}$	40.3M	374K	5.3	14.0	2.7	4.6	2.1	2.7	3.2	11.0
PAEE$_{SCC\text{-}clean}$	3.4M	374K	**12.5**	**54.6**	**6.0**	**23.2**	**5.2**	**21.4**	**7.7**	**44.8**
1% doamin-agnostic training										
Frozen$_{SCC\text{-}clean}$	40.3M	3.7K	3.1	5.3	1.8	1.7	1.6	1.5	1.8	3.8
PAEE$_{SCC\text{-}clean}$	3.4M	3.7K	**6.0**	**18.1**	**2.8**	**5.6**	**2.3**	**4.9**	**3.3**	**12.1**

Domain-Agnostic Training. For domain-agnostic training, we proceed to train two models using the SCC dataset mentioned in the previous section, which is a large-scale dataset obtained from the internet. While SCC may not possess the same level of cleanliness as manually annotated datasets like COCO captions, its large size and diversity are advantageous for VL models during domain-agnostic pre-training. Due to the small size of our models, we acknowledge that the negative effects of noise in this dataset may outweigh its positive effects. Therefore, we filter the dataset and use CLIP image-text similarity to obtain the most similar text pairs, resulting in a cleaned version called SCC-clean, which consists of the top 374K pairs. We then train these two models on 1% and 100% of the domain-agnostic samples, respectively, and assess their performance. The results shown in Table 4 demonstrate that $\text{PAEE}_{scc\text{-}clean}$ outperforms $\text{Frozen}_{scc\text{-}clean}$ on all benchmarks. These results demonstrate that our method effectively utilizes limited knowledge, even when confronted with out domain-agnostic learning. This showcases the robustness of our model in terms of data efficacy.

4.5 Ablation Analysis

In this part of the study, we explore the impact of the CRA architecture and the knowledge prompter on our PAEE model.

The Impact of CRA. To evaluate the impact of CRA, we employ two simple and straightforward networks, a 2-layer MLP and a linear layer. These architectures were used to replace the transformer-based CRA in the original PAEE model. As shown in Fig. 3, the experimental results demonstrate that the CIDEr scores of these two simpler architectures are significantly lower than those of the original setting (CRA). Compared to the other two structures, the CRA module demonstrates remarkable performance advantages, indicating the feasibility of our CRA module.

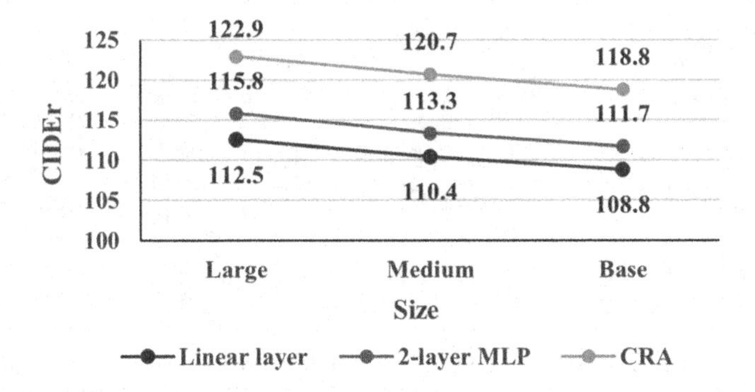

Fig. 3. Comparison of CIDEr performance on the COCO test set for linear layer, 2-layer MLP, and CRA with three different PAEE sizes.

Table 5. Evaluation of PAEE's BLEU@4 and CIDEr performance on the COCO test set with and without prompter.

Model	Params	Without Prompter		With Prompter	
		BLEU@4	CIDEr	BLEU@4	CIDEr
PAEE$_{Base}$	1.7M	31.5	109.7	36.2	118.8
PAEE$_{Medium}$	3.4M	31.7	112.1	36.8	120.7
PAEE$_{Large}$	6.8M	32.0	114.8	37.1	122.9

The Impact of Prompter. We conduct an experiment to quantify the impact of eliminating the knowledge prompter in training three different sizes of PAEE models. To replace the task demostration, we use a simple prompt, "A picture of...". The results in Table 5 show that without the prompter's assistance, the model's performance is significantly lower compared to the prompter model. In particular, the PAEE$_{base}$ model with the minimum trainable parameters has the largest performance gap of 9.1 points in CIDEr score. This confirms that the PAEE can compensate for the reduction in model size by accessing external knowledge through the prompter. Moreover, the prompter markedly contributes to enhancing the overall performance of PAEE.

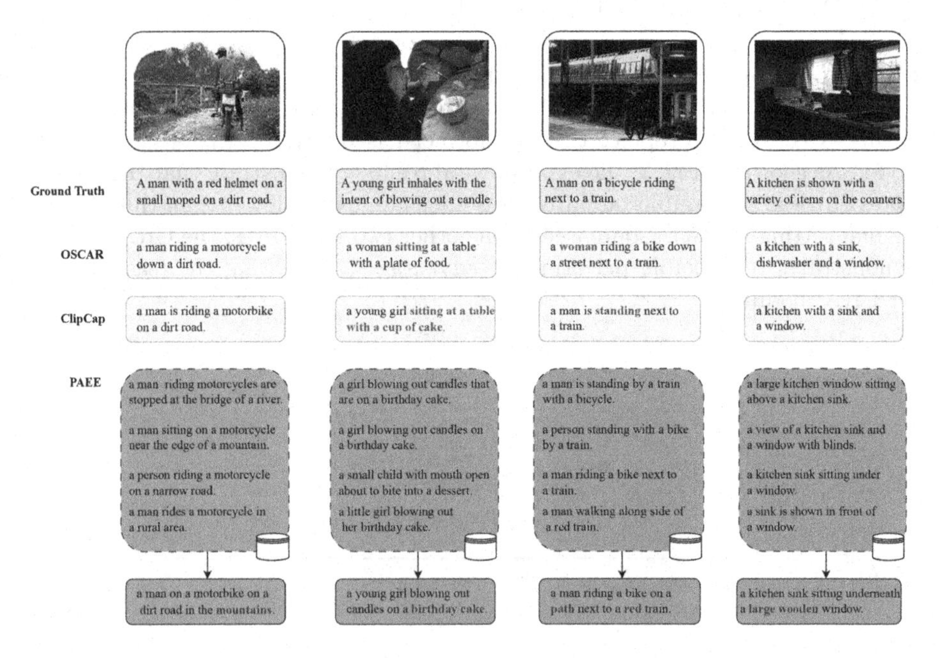

Fig. 4. Examples of our PAEE, OSCAR and ClipCap for the first 4 images in the MSCOCO test set. (Red = inaccurate, Blue = PAEE's advantages). (Color figure online)

4.6 Qualitative Analysis

Figure 4 illustrates PAEE accurately identifies potential behaviors and specific details of people in the images, providing rich and accurate captions. For example, in the second image, PAEE provides the more diverse caption "birthday cake", and in the fourth image, it clearly identifies a "large wooden" window in an uncommon scene.

5 Conclusion and Future Work

In this paper, we introduce PAEE, an image captioning model based on a knowledge prompter and CRA, characterized by parameter efficiency and data effectiveness. Our research is highly significant when addressing the high computational cost associated with large-scale pre-training models. We conduct extensive experiments to demonstrate the feasibility of our approach. Compared with existing large-scale pre-training models and similar methods, PAEE performs exceptionally well. Furthermore, we utilize newly designed SCC and CY105 datasets to expand the knowledge base in the prompter. The experimental results demonstrate that PAEE has strong data utilization capabilities and can achieve domain transfer without any training or fine-tuning. Additionally, we successfully verify the powerful data effectiveness of PAEE in small-data learning experiments. PAEE performs well when faced with limited computational resources.

Our forthcoming research includes an exploration of PAEE's capabilities in few-shot learning. As mentioned above, PAEE demonstrates the ability to use external knowledge effortlessly, without any training or fine-tuning, resulting in remarkable performance. Furthermore, we will explore the feasibility of applying PAEE to other multimodal task, such as video captioning.

Acknowledgements. This paper is supported by the Capacity Development Grant of Southwest University (SWU116007) and the Natural Science Foundation of Chongqing (Grant No. CSTB2022NSCQ-MSX0437).

References

1. Agrawal, H., et al.: nocaps: novel object captioning at scale. In: Proceedings of the IEEE/CVF International Conference on Computer Vision, pp. 8948–8957 (2019)
2. Alayrac, J.B., et al.: Flamingo: a visual language model for few-shot learning. Adv. Neural. Inf. Process. Syst. **35**, 23716–23736 (2022)
3. Anderson, P., Fernando, B., Johnson, M., Gould, S.: SPICE: semantic propositional image caption evaluation. In: Leibe, B., Matas, J., Sebe, N., Welling, M. (eds.) ECCV 2016. LNCS, vol. 9909, pp. 382–398. Springer, Cham (2016). https://doi.org/10.1007/978-3-319-46454-1_24
4. Byeon, M., Park, B., Kim, H., Lee, S., Baek, W., Kim, S.: COYO-700M: image-text pair dataset (2022)
5. Changpinyo, S., Sharma, P., Ding, N., Soricut, R.: Conceptual 12M: pushing web-scale image-text pre-training to recognize long-tail visual concepts. In: Proceedings of the IEEE/CVF Conference on Computer Vision and Pattern Recognition, pp. 3558–3568 (2021)

6. Chen, X., et al.: Microsoft COCO captions: data collection and evaluation server. arXiv preprint arXiv:1504.00325 (2015)
7. Cho, J., Lei, J., Tan, H., Bansal, M.: Unifying vision-and-language tasks via text generation. In: International Conference on Machine Learning, pp. 1931–1942. PMLR (2021)
8. Dai, W., Hou, L., Shang, L., Jiang, X., Liu, Q., Fung, P.: Enabling multimodal generation on CLIP via vision-language knowledge distillation. arXiv preprint arXiv:2203.06386 (2022)
9. Denkowski, M., Lavie, A.: Meteor Universal: language specific translation evaluation for any target language. In: Proceedings of the Ninth Workshop on Statistical Machine Translation, pp. 376–380 (2014)
10. Eichenberg, C., Black, S., Weinbach, S., Parcalabescu, L., Frank, A.: MAGMA–multimodal augmentation of generative models through adapter-based finetuning. arXiv preprint arXiv:2112.05253 (2021)
11. Gao, T., Fisch, A., Chen, D.: Making pre-trained language models better few-shot learners. arXiv preprint arXiv:2012.15723 (2020)
12. Gurari, D., et al.: VizWiz Grand Challenge: answering visual questions from blind people. In: Proceedings of the IEEE Conference on Computer Vision and Pattern Recognition, pp. 3608–3617 (2018)
13. Gurari, D., Zhao, Y., Zhang, M., Bhattacharya, N.: Captioning images taken by people who are blind. In: Computer Vision–ECCV 2020: 16th European Conference, Glasgow, UK, August 23–28, 2020, Proceedings, Part XVII 16, pp. 417–434. Springer (2020). https://doi.org/10.1007/978-3-030-58520-4_25
14. Krishna, R., et al.: Visual Genome: connecting language and vision using crowdsourced dense image annotations. Int. J. Comput. Vis. **123**, 32–73 (2017)
15. : Lewis, P., et al.: Retrieval-augmented generation for knowledge-intensive NLP tasks. Adv. Neural. Inf. Process. Syst. **33**, 9459–9474 (2020)
16. Li, J., Li, D., Xiong, C., Hoi, S.: BLIP: bootstrapping language-image pre-training for unified vision-language understanding and generation. In: International Conference on Machine Learning, pp. 12888–12900. PMLR (2022)
17. Li, X., et al.: OSCAR: object-semantics aligned pre-training for vision-language tasks. In: Vedaldi, A., Bischof, H., Brox, T., Frahm, J.-M. (eds.) ECCV 2020. LNCS, vol. 12375, pp. 121–137. Springer, Cham (2020). https://doi.org/10.1007/978-3-030-58577-8_8
18. Luo, Z., Hu, Z., Xi, Y., Zhang, R., Ma, J.: I-Tuning: tuning frozen language models with image for lightweight image captioning (2023)
19. Mokady, R., Hertz, A., Bermano, A.H.: ClipCap: CLIP prefix for image captioning. arXiv preprint arXiv:2111.09734 (2021)
20. Ordonez, V., Kulkarni, G., Berg, T.: Im2Text: describing images using 1 million captioned photographs. Adv. Neural Inf. Proc. Syst. **24** (2011)
21. Papineni, K., Roukos, S., Ward, T., Zhu, W.J.: BLEU: a method for automatic evaluation of machine translation. In: Proceedings of the 40th Annual Meeting of the Association for Computational Linguistics, pp. 311–318 (2002)
22. Plummer, B.A., Wang, L., Cervantes, C.M., Caicedo, J.C., Hockenmaier, J., Lazebnik, S.: Flickr30k Entities: collecting region-to-phrase correspondences for richer image-to-sentence models. In: Proceedings of the IEEE International Conference on Computer Vision, pp. 2641–2649 (2015)
23. Radford, A., et al.: Learning transferable visual models from natural language supervision. In: International Conference on Machine Learning, pp. 8748–8763. PMLR (2021)

24. Radford, A., Wu, J., Child, R., Luan, D., Amodei, D., Sutskever, I., et al.: Language models are unsupervised multitask learners. OpenAI blog **1**(8), 9 (2019)
25. Ramos, R., Martins, B., Elliott, D., Kementchedjhieva, Y.: SmallCap: lightweight image captioning prompted with retrieval augmentation. In: Proceedings of the IEEE/CVF Conference on Computer Vision and Pattern Recognition, pp. 2840–2849 (2023)
26. Sarto, S., Cornia, M., Baraldi, L., Cucchiara, R.: Retrieval-augmented transformer for image captioning. In: Proceedings of the 19th International Conference on Content-based Multimedia Indexing, pp. 1–7 (2022)
27. Sharma, P., Ding, N., Goodman, S., Soricut, R.: Conceptual captions: a cleaned, hypernymed, image alt-text dataset for automatic image captioning. In: Proceedings of the 56th Annual Meeting of the Association for Computational Linguistics (Volume 1: Long Papers), pp. 2556–2565 (2018)
28. Sidorov, O., Hu, R., Rohrbach, M., Singh, A.: TextCaps: a dataset for image captioning with reading comprehension. In: Vedaldi, A., Bischof, H., Brox, T., Frahm, J.-M. (eds.) ECCV 2020. LNCS, vol. 12347, pp. 742–758. Springer, Cham (2020). https://doi.org/10.1007/978-3-030-58536-5_44
29. Tsimpoukelli, M., Menick, J.L., Cabi, S., Eslami, S., Vinyals, O., Hill, F.: Multimodal few-shot learning with frozen language models. Adv. Neural. Inf. Process. Syst. **34**, 200–212 (2021)
30. Vedantam, R., Zitnick, C.L., Parikh, D.: CIDEr: consensus-based image description evaluation. In: Proceedings of the IEEE Conference on Computer Vision and Pattern Recognition, pp. 4566–4575 (2015)
31. Xia, Q., et al.: XGPT: cross-modal generative pre-training for image captioning. In: Natural Language Processing and Chinese Computing: 10th CCF International Conference, NLPCC 2021, Qingdao, China, October 13–17, 2021, Proceedings, Part I 10, pp. 786–797. Springer (2021). https://doi.org/10.1007/978-3-030-88480-2_63
32. Xu, C., Zhao, W., Yang, M., Ao, X., Cheng, W., Tian, J.: A unified generation-retrieval framework for image captioning. In: Proceedings of the 28th ACM International Conference on Information and Knowledge Management, pp. 2313–2316 (2019)
33. Yang, Z., et al.: UniTAB: unifying text and box outputs for grounded vision-language modeling. In: Computer Vision–ECCV 2022: 17th European Conference, Tel Aviv, Israel, October 23–27, 2022, Proceedings, Part XXXVI, pp. 521–539. Springer (2022). https://doi.org/10.1007/978-3-031-20059-5_30
34. Zhang, S., et al.: OPT: open pre-trained transformer language models. arXiv preprint arXiv:2205.01068 (2022)

TSKE: Two-Stream Knowledge Embedding for Cyberspace Security

Angxiao Zhao[1,2], Haiyan Wang[2], Junjian Zhang[2,3], Yunhui Liu[2,3], Changchang Ma[3], and Zhaoquan Gu[2,4](✉)

[1] Shenzhen Institute for Advanced Study, University of Electronic Science and Technology of China, Shenzhen, China
202122280745@std.uestc.edu.cn

[2] Department of New Networks, Peng Cheng Laboratory, Shenzhen, China
wanghy01@pcl.ac.cn

[3] Cyberspace Institute of Advanced Technology, Guangzhou University, Guangzhou, China
{2112106069,2112106168,2112106188}@e.gzhu.edu.cn

[4] School of Computer Science and Technology, Harbin Institute of Technology (Shenzhen), Shenzhen, China
guzhaoquan@hit.edu.cn

Abstract. Knowledge representation models have been extensively studied and adopted in many areas such as search, recommendation, etc. However, due to the highly spatio-temporal relevant characteristics of cyberspace security and the dynamic variability of the domain knowledge, the existing models and knowledge embedding methods cannot be adopted in this field directly. In this paper, we propose a two-stream knowledge embedding (TSKE) method for cyberspace security to jointly embed multi-dimensional characteristics. Specifically, we design a static stream neural network and a spatio-temporal stream neural network to extract the static knowledge and the spatio-temporal features of cyberspace security facts, which converts this domain knowledge into vector space. Considering the attack link prediction task in the field of cyberspace security, we conduct extensive experiments and TSKE outperforms other static and dynamic embedding methods.

Keywords: Knowledge Representation · Cyberspace Security · Spatio-temporal Characteristics · Attack Link Prediction

1 Introduction

Artificial intelligence has been successfully applied in many tasks to achieve superior performance, such as image classification [1], speech recognition [2], intelligent transportation [3], autonomous driving [4], smart home [5], etc. As one important direction of artificial intelligence, knowledge representation has achieved much attention in the last decades, which greatly improves the capabilities of knowledge management and usage.

X. Song et al. (Eds.): APWeb-WAIM 2023, LNCS 14332, pp. 132–146, 2024.
https://doi.org/10.1007/978-981-97-2390-4_10

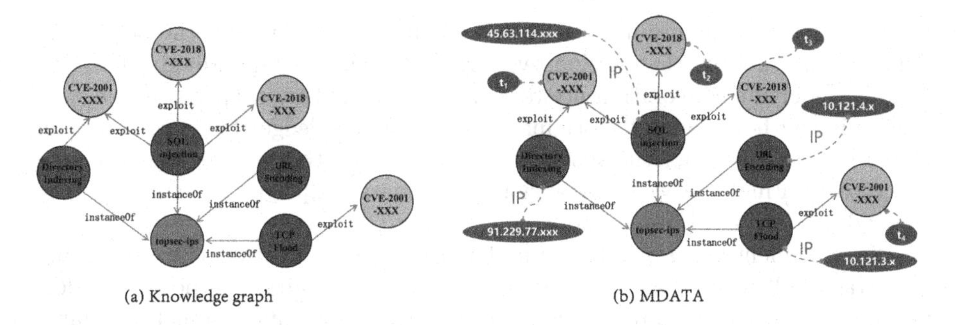

(a) Knowledge graph (b) MDATA

Fig. 1. An example of representing cyberspace security knowledge by knowledge graph and the MDATA model.

Knowledge graph is a typical knowledge representation model, which has been extensively studied and adopted since its emergency [6]. Besides being able to organize the extracted knowledge from multiple data sources, knowledge embedding has enabled the ability to reason new knowledge by converting the facts to numerical vectors. There are many elegant knowledge embedding methods such as TransE [7], ConvKB [8], HyTE [9], etc.

However, when dealing with tasks in the fields that knowledge changes rapidly, knowledge graph ignores the influence of spatio-temporal characteristics, resulting in inaccurate reasoning and decision-making. For example, in the filed of cyberspace security, a specific attack is normally related to some disclosed vulnerabilities, but they do not exist after patching. Adopting the existing embedding methods will reason inaccurate facts, misleading network attack and defense. As shown in Fig. 1(a), when knowledge graph is used to represent the attack facts, some critical information is missing such as when the attack happens or where does the attack comes from. Although temporal knowledge graph has gained the attention recently to represent dynamic knowledge [10], the spatial characteristic such as the IP address of an attack is still ignored. MDATA (**M**ulti-dimensional **D**ata **A**ssociation and in**T**elligent **A**nalysis) is a novel model designed for this field [11], which represents the cyberspace security knowledge combing the spatio-temporal information. As shown in Fig. 1(b), more critical information about the attacks can be represented compared with knowledge graph.

Similar as knowledge graph, knowledge embedding of the MDATA model can also improve its ability of reasoning new facts. However, there are few embedding methods that consider the spatio-temporal properties of facts. Although some dynamic embedding methods for temporal knowledge graph have been proposed which brings in the influence of temporal information [12], how to jointly consider spatial and temporal characteristics is still a difficult problem.

In this paper, we propose a **T**wo-**S**tream **K**nowledge **E**mbedding (TSKE) method for cyberspace security based on the MDATA model. The intuitive idea is to represent the static knowledge and the spatio-temporal properties by two separate models, and then fuse them to generate the embedding vector. Specifically, we construct two subgraphs to model the static knowledge and the additional dynamic information, then design two stream models correspondingly. The first stream model is a convolution network, which can capture deeper static feature interaction without losing the original meaning of knowledge. The second stream model is used to extract the spatio-temporal features that are correlated to the static parts. Finally, a weighted fusion operation is employed to synthesize the embedding effects of the two models.

We implement the proposed method and evaluate the performance regarding the attack link prediction task. As there are few embedding methods considering the spatio-temporal properties, we propose a joint evaluation metric regarding the static knowledge and the spatio-temporal information. We select both static and dynamic knowledge embedding methods as the baselines. Experimental results show that the prediction accuracy of our method is significant higher than others. We summarize the contributions of the paper as follows:

1) We propose a highly evolved knowledge embedding method for the MDATA model, which can sufficiently convert the facts with spatio-temporal properties into vectors to enhance the model's reasoning and computational abilities;
2) We model the spatio-temporal properties of the knowledge separately and regard it as the key additional information of the dynamic knowledge, which improves the reasoning ability when knowledge varies dynamically;
3) We apply the embedding method in the field of cyberspace security and it largely improves the attack link prediction accuracy.

The remainder of our paper is organized as follows. The next section introduces the related work. Preliminaries are provided in Sect. 3. We describe our method in detail in Sect. 4. In Sect. 5, we show the experimental results and analysis regarding the field of cyberspace security. Finally, we conclude the paper in Sect. 6.

2 Related Work

In this section, we introduce the classical representation models and the existing knowledge embedding methods.

2.1 Knowledge Representation Models

Knowledge graph [13] is a classical knowledge representation model, which models the facts by graph structure. The nodes of the graph structure represent the entities of facts, and the relations are represented by edges. The expression of knowledge graph is closer to the form of human cognition, and it has a good ability to organize, manage, and understand the massive information. In the field of cyberspace security, knowledge graph is also adopted for some simple tasks. For example, knowledge graph is used in [14] to present attack scenarios and to trace attack sources. In [15], knowledge graph is used for network attack detection, which represents the extracted attack information from multiple data sources.

However, these works mainly use knowledge graph to present single-step attack and the related vulnerability, it is difficult for analyzing multi-step attacks which have been evolving as the main threat to cyberspace security.

A multi-step attack is normally composed of many correlated single-step attacks, where each attack step may come from different IP addresses at any time. When facing the traceability or detection of such complicated multi-step attacks [16], knowledge graph lacks the spatio-temporal information of the single-step attack, making it difficult to associate these attacks accordingly. The MDATA model [11] supports the representation of knowledge with spatio-temporal characteristics and can handle this problem effectively. In addition, the model simulates the evolution process of knowledge by knowledge representation, knowledge extraction, and knowledge usage. Regarding the field of cyberspace security, the model can effectively represent the knowledge from three aspects: assets, vulnerabilities, and attack behaviors, and it helps better understand the attack association for further detection and analysis.

2.2 Knowledge Embedding Methods

Static Embedding. Static embedding uses a scoring function to measure the representation effect of static knowledge, so as to learn the vector representation of entities and relations for static knowledge.

Translation Methods. **TransE** [7] is the first translation model to represent knowledge in the form of $h + r \approx t$. Due to the few parameters, TransE can achieve excellent performance on handling *1-1* relation. However, when dealing with complex relation types, TransE performs poorly. **TransH** [17] projects head and tail entities to the hyperplane of relation, so that one entity has different representations for different relations. Furthermore, **TransR** [18] assumes that an entity contains several attributes, and different entities pay attention to different attributes of relations. Then TransR makes the entity projection into the semantic spaces corresponding to different relations.

Bilinear Methods. **Rescal** [19] is the first bilinear model, which models the pairwise interactions between latent factors by representing each relation as a matrix. Although it enhances the expressive force of the model, a huge number of parameters becomes a burden of the model. **DistMult** [20] restricts the relation matrix to a diagonal matrix, which reduces the parameters of the model. Unfortunately, the model loses the ability to model asymmetric relations. **ComplEx** [21] extends DistMult by introducing complex-valued embedding to better model asymmetric relations. **SimplE** [22] initializes two vectors for each entity and uses the inverse of the relations to cleverly set the symmetry term in the scoring function so that the two vectors learn from each other.

Convolutional Neural Network Methods. **ConvE** [23] uses 2D convolutions to model the interaction between entity and relation. It reshapes head entity and relation, concatenating them into a matrix for the convolutional layer. **ConvKB** [8] considers the limitation of ConvE in extracting feature information of the same dimension, so ConvKB eliminates the reshape operation in ConvE, and uses 1D convolution to retain the original features of knowledge.

Dynamic Embedding. As knowledge varies dynamically, some dynamic embedding methods have been proposed for temporal knowledge graph. There are several different types of methods as follows.

The method based on temporal relation named **TransE-TAE** [24] realizes the transformation from prior relation to subsequent relation. Specifically, the method extracts pairs of relations from a large number of factual data and then obtains an evolution matrix about time by learning the objective occurrence sequence between relations. The method based on time hyperplane named **HyTE** [9] maps static facts to the corresponding time hyperplane and finally obtains the embedded vectors through a training method similar as TransE on the hyperplane. The method based on context uses a fact-aware mechanism to determine the useful context for the target fact by measuring the time consistent with the selected context to complete the embedding of knowledge, such as **Context-Aware** [25]. The method based on graph neural network is commonly used in dynamic knowledge. For example, **RE-NET** [26] learns graph information and neighbor information of nodes at different times by graph convolution neural network and attention mechanism, then it updates nodes by a recurrent neural network. **RE-GCN** [27] learns the evolutionary representation of entities and relations by capturing structural dependencies between concurrent facts and information order patterns of adjacent facts across time.

A large number of knowledge embedding methods have been proposed to improve knowledge graph's reasoning and computational abilities. However, few of them consider both temporal and spatial characteristics of the knowledge, making it difficult to adopt them in the field of cyberspace security. The MDATA model can represent such knowledge, but it lacks efficient embedding methods.

3 Preliminaries

Considering the field of cyberspace security, we introduce the system model and formulate the problem in this section.

3.1 System Model

Considering a practical scenario in cyberspace security, where a situation awareness system is deployed to detect the network attacks. After capturing the attack information according to the alarms, we can extract the attack information and represent it by the MDATA model.

Supposing $G = (V, R)$ where each node $v \in V$ belongs to three types of entities: attacks, vulnerabilities, and assets; each edge $r_{ij} \in R$ represents two nodes v_i and v_j are linked according to some relations. For the nodes, we denote the attack entities as $A_G = \{a_1, a_2, ..., a_n\}$ where each $a_k \in A_G$ represents a specific attack such as Command Injection or HTTP DoS. We denote the vulnerability (or we call it as weakness) entities as $W_G = \{\omega_1, \omega_2, ..., \omega_m\}$ where each $\omega_k \in W_G$ is detected by vulnerability scanning. Assets imply the hardware devices or the installed softwares that may be attacked, correspondingly the assets have some vulnerabilities that can be exploited by attackers. We denote the asset entities as $ASS_G(v_i) = \{ass_1, ass_2, ..., ass_n\}$. For the edges, different kinds of relations exist between different types of entities. For example, an attack entity is connected to a weakness entity by the *exploit* relation, which can be connected to an asset entity by the *occur* relation.

Since the attacks are accompanied by the evolution of temporal and spatial characteristics, we denote time as $T_G = \{time_1, time_2, ..., time_n\}$ where each $time_k \in T_G$ represents a timestamp or a time period. Similarly, we denote the source address as $SRC_G = \{src_1, src_2, ..., src_n\}$ and destination address as $DST_G = \{dst_1, dst_2, ..., dst_n\}$ where each $src_k \in SRC_G$ and $dst_k \in DST_G$ represents a specific IP address. Then the knowledge fact can be represented according to the following tuple:

$$(v_i, r_{ij}, v_j, src_k, time_k, dst_k)\,.$$

The deployed system is able to detect the attacks according the alarms and we denote these attacks as $S = \{s_1, s_2, \ldots, s_t, \ldots\}$. Each attack s_k is regarded as a specific attack step, such as phishing mail attack, XSS attack, etc. Since the attack is highly related to the assets and vulnerabilities, we can represent each attack by the following two facts:

$$s_k(1) = (a_k,\ exploit,\ \omega_k,\ src_k,\ time_k,\ dst_k)\,;$$
$$s_k(2) = (a_k,\ occur,\ ass_k,\ src_k,\ time_k,\ dst_k)\,.$$

For example, an attacker carries out a command injection attack from one server (assume $IP_1 = 10.xx.\ xx.7$) to another (assume $IP_2 = 10.xx.\ xx.8$) at $time_1$, using a deadly weakness. Meanwhile, the attack is detected by the system and the attack step can be represented as:

$$s_k(1) = (command\ injection,\ exploit,\ CVE\text{-}xxx,\ IP_1,\ time_1,\ IP_2)\,;$$
$$s_k(2) = (command\ injection,\ occur,\ topsec\text{-}ips,\ IP_1,\ time_1,\ IP_2)\,.$$

One important task of the deployed system is to detect the complicated multi-step attacks such as the APT (advanced persistent threat) attacks. Normally a multi-step attack can be represented by a sequence $S_A = \{s_{a1}, s_{a2}, \ldots, s_{al}\}$ where l is the number of attack steps. Although the system can detect the attacks effectively by many deployed security devices (probes) such as IDS (intrusion detection system), it is quite difficult to detect all attacks, which implies the full sequence of a complicated attack S_A may not be detected by the system. Hence, it is needed to reason the potential attack step according to the existing extracted knowledge.

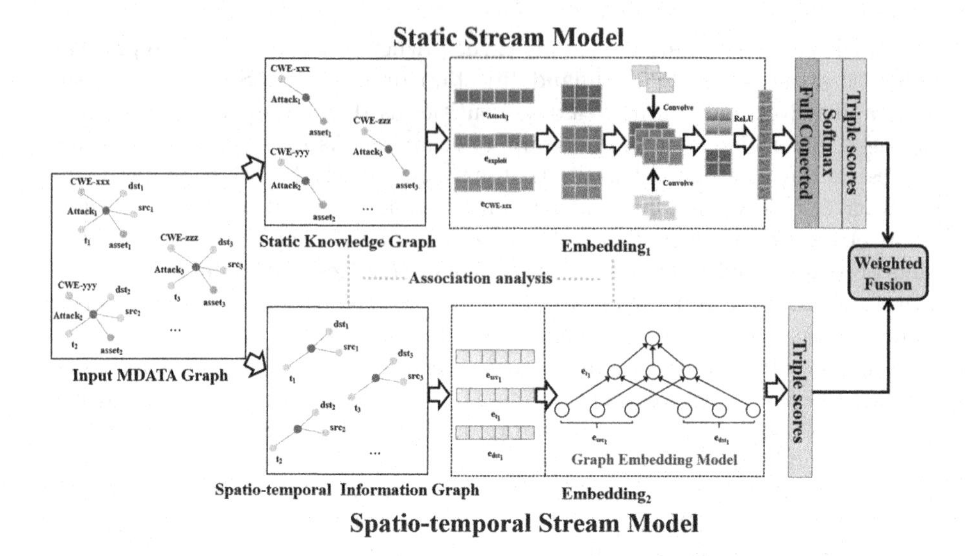

Fig. 2. Two-stream architecture for cyberspace security knowledge embedding.

3.2 Problem Definition

As knowledge embedding can convert the knowledge facts to vectors for reasoning, we define the embedding problem based on the MDATA model as follows.

Problem: Design the embedding operation: $(V, R, SRC_G, T, DST_G) \rightarrow \psi$ where ψ is an embedding set, such that it is able to reason the potential attack step for a given sequence of the attack knowledge S.

4 TSKE: A Two-Stream Knowledge Embedding Method Based on the MDATA Model

In this section, we introduce the proposed method TSKE in detail. The architecture of TSKE is shown in Fig. 2. We separate the static knowledge and the spatio-temporal attributes of each fact in the MDATA model, i.e. we construct the static knowledge graph and spatio-temporal information graph. Then we propose two models to embed the static knowledge and the additional spatio-temporal properties respectively. Finally, we fuse the two models to generate embedding vectors. We introduce the method in detail in the following parts.

4.1 Static Stream Model

The static knowledge graph is composed of network security facts, which can be expressed as a collection of triples in the form of (h, r, t), where $h, t \in V$ and $r \in R$. Although convolution-based neural models can achieve high prediction accuracy, the information about the internal structure of knowledge is rarely captured. Therefore, we reshape and stack the embeddings of relation triples, and then introduce 3D convolution to learn the deeper feature interaction at the same dimension of the cyberspace knowledge embeddings.

Formally, the details of our implementation are presented as below. First, we embed knowledge triples into k-dimensional vectors ($e_{Attack_1}, e_{exploit}, e_{CWE-xxx}$), which is shown in the architecture. To capture potential feature interaction, the k-dimensional vector is reshaped into the matrix of dim_1 rows and dim_2 columns ($dim_1 \times dim_2 = k$). Then, we concatenate the three reshaped embeddings into a cube $\in \mathbb{R}^{3 \times dim_1 \times dim_2}$, which can be operated by several 3D convolution filters. The size of each convolution filter $\in \mathbb{R}^{3 \times i \times j}$ are set according to the reshaped matrix. Through the operation of n convolutional filters, we obtain n feature maps $m_1, m_2, \ldots m_n \in \mathbb{R}^{(\mathrm{dim}_1 - i+1) \times (\mathrm{dim}_2 - j+1)}$ accordingly. After that, these feature maps are reassembled by the activation function. During the process, we take dropout actions as the situation requires. Finally, we add a fully connected projection layer (linear operation) to get the final score of the input triple.

The final triple score generated by the static stream model is formalized as:

$$score_1(h, r, t) = [\mathrm{concat}\left(\sigma\left(\omega * [e_h; e_r; e_t] + b_1\right)\right) + b_2] \cdot f, \tag{1}$$

where ω is a set of several filters for 3D convolution, σ is the activation function, b_1 and b_2 represent bias, concat denotes a concatenation operator, f is the coefficients of linear operations in a fully connected layer.

4.2 Spatio-Temporal Stream Model

The spatio-temporal information graph contains characteristic information of evolution-related network security facts, which can be expressed as a collection of triples in the form of $(src, time, dst)$, where $src \in SRC_G, dst \in DST_G$, and $time \in T_G$. Although spatio-temporal information is based on the facts, it has a significant impact to predict the potential facts after embedding. Some temporal models use time features as a part of knowledge, while we separately model temporal and spatial features as the additional information. We can modify the classic embedding methods to extract the spatio-temporal features. We take the bilinear method DistMult as an example to introduce the modifications.

As shown in Fig. 2, through the relation vector of the temporal feature, we can fully interact with the features of the source and destination IP addresses. The score of spatio-temporal part can be generated by the following scoring function:

$$score_2(src, time, dst) = [f\left(\mathbf{W}\mathbf{x}_{e_{src}}\right)]^T \mathbf{M}_{e_{time}} [f\left(\mathbf{W}\mathbf{x}_{e_{dst}}\right)], \tag{2}$$

where W is a parameter matrix that can be randomly initialized. $\mathbf{x}_{e_{src}}$ and $\mathbf{x}_{e_{dst}}$ are the input vectors for entity src and dst, respectively. $\mathbf{M}_{e_{time}}$ is a diagonal matrix with respect to time and f is the learned entity representation.

4.3 Weighted Fusion

Weighted fusion part mainly fuses the triple scores based on the output of the two-stream networks by setting reasonable weights, so as to obtain the accurate embedding effects. Specifically, in the static flow model, we mainly count the number N_1 of entities passing through the static flow model, because there are far more entities than relations in the cyberspace security scenario, while in the spatio-temporal flow model we mainly count the number N_2 of IP addresses, which is an important indicator to realize association analysis in the field of cyberspace security.

We calculate the proportion of key entities in the two-stream model, multiply it by the score of the corresponding two-stream network, and realize the score fusion of the static facts and the corresponding spatio-temporal information by the weighted summation. The score after weighted fusion can be expressed as:

$$score = ratio_1 * score_1 + ratio_2 * score_2, \tag{3}$$

where $ratio_1$ and $ratio_2$ denote the weight:

$$ratio_1 = \frac{N_1}{N_1 + N_2}; \quad ratio_2 = \frac{N_2}{N_1 + N_2}. \tag{4}$$

4.4 Learning

As the two-stream model is proposed to embedd both static and dynamic features of the facts, we define two loss functions for training the embeddings respectively. The static stream model is trained by minimizing the loss function $\mathcal{L}_1$ with L_2 regulation, which is defined as:

$$\mathcal{L}_1 = \sum_{(h,r,t) \in G_1 \cup G_1'} \log\left(1 + \exp\left(l_{(h,r,t)} \cdot score_1(h,r,t)\right)\right) + \phi, \tag{5}$$

where G_1 and G_1' denote the set of positive and negative samples in static knowledge respectively. If $(h,r,t) \in G_1$, $l_{(h,r,t)}$ denotes 1, and if $(h,r,t) \in G_1'$, $l_{(h,r,t)}$ denotes -1. ϕ in Eq. 2 contains the sum of the square norms of the three parts: the entity and relation embeddings, parameters of the convolution layer, and the fully connected layer, i.e.:

$$\phi = \lambda\left(\sum_{h \in V} \|e_h\|_2^2 + \sum_{r \in R} \|e_r\|_2^2 + \sum_{t \in V} \|e_t\|_2^2 + |\mathrm{c}\|_2^2 + \|\mathrm{w}\|_2^2\right), \tag{6}$$

where c and w are the parameters of the convolution layer and fully connected layer respectively. The spatio-temporal model is trained by minimizing the loss function $\mathcal{L}_2$ as:

$$\begin{aligned}\mathcal{L}_2 = \sum_{(src,time,dst) \in G_2} \sum_{(src',time,dst') \in G_2'} & \max\{score_2(src,time,dst) \\ & -score_2(src',time,dst') + 1, 0\},\end{aligned} \tag{7}$$

Algorithm 1: The Learning Algorithm of **TSKE**

Input: The extracted facts as the training set $\{V, R, SRC_G, T_G, DST_G\}$, embedding dimension d, epoch number n_1, epoch number n_2, negative ratio *ratio*, learning rate α.

Output: Entity embeddings, relation embeddings, IP address embeddings, and time embeddings.

```
1  Split the MDATA model G into static knowledge graph G1 and
   spatio-temporal information graph G2;
2  Generate entity set E1 and relation set R1 from G1;
3  Generate IP address set E2 and time set R2 from G2;
4  epoch1 ← 0;
5  epoch2 ← 0;
6  while epoch1 <n1 do
7      G1_batch ← Sample(G1, b); // sample a mini-batch of size b from G1
8      U_batch ← ∅; // triples for learning
9      foreach triple ∈ G1_batch do
10         (P1, P2) ← negSample (triple); // negative sampling from E1
11         U_batch ∪ (P1, P2); // set of triples obtained by negative sampling
12         Initialize entity embeddings and relation embedding;
13         xavier_uniform (entity embedding);
14         xavier_uniform (relation embedding);
15         Learning;
16     end
17     Update parameters with optimizer w. r. t Eq. 5;
18     Regularize variable of loss and gradient of Adam in U_batch;
19 end
20 while epoch2 <n2 do
21     G2_batch ← Sample(G2, b); // sample a mini-batch of size b from G2
22     V_batch ← ∅; // triples for learning
23     foreach triple ∈ G2_batch do
24         (Q1, Q2) ← negSample (triple); // negative sampling from E2
25         V_batch ∪ (Q1, Q2); // set of triples obtained by negative sampling
26         Initialize IP address embedding and time embedding;
27         xavier_uniform (IP address embedding);
28         xavier_uniform (time embedding);
29         Learning;
30     end
31     Update parameters with optimizer w. r. t Eq. 7;
32     Regularize variable of loss and gradient of Adam in V_batch;
33 end
34 return Embedding of the Entities, relations, IP addresses, and times.
```

where G_2 and G_2' denote the set of positive and negative samples in the spatial-knowledge information graph respectively. $(src, time, dst)$ in G_2' denotes a positive example in G_2, while $(src', time, dst')$ denotes a negative example after replacing either the head entity or the tail entity (not both) in $(src, time, dst)$,

Given the training set and the hyperparameters, we can get the embedding vectors of each entity, relation, time and IP address in the facts as output. First, two graphs are divided from the MDATA model which is composed of the security knowledge facts. Two entity sets and relation sets are generated from the corresponding graph. Then we construct negative samples in each epoch. After that, we update the model parameters by minimizing loss functions of Eq. 5 and Eq. 7. The total loss $\mathcal{L}$ of the two-stream model is $\mathcal{L} = \mathcal{L}_1 + \mathcal{L}_2$. The embedding of each component needs to be regularized at the end of each epoch. The training process is shown in Algorithm 1.

5 Experiment Results

5.1 Implementation

The data is collected through the Peng Cheng Cyber Range, which supports many attack and defense drills. We collect the data within one hour during one drill and extracted about 6200 alarms with spatio-temporal characteristics (including 524 entities). After extracting the facts from these alarms, we divide them into three parts as 8:1:1. We implement our method in Pytorch and use Adam to optimize parameters in mini-batch and learn embeddings. The hyperparameters are set as follows: the embedding dimension $d \in \{30, 50, 100\}$, training rounds of static stream model $epoch_1 \in \{280, 300, 320\}$, training rounds of spatial-temporal stream model $epoch_2 \in \{300, 400, 500\}$, ratio of negative and positive samples $ratio \in \{100, 300, 500\}$, the learning rate $lr \in \{0.1, 0.01, 0.001\}$. The finally adopted settings are $\{d = 30$, $epoch_1 = 280$, $epoch_2 = 500$, $ratio = 500$, $lr = 0.001\}$. The batch size $b = 256$ and the l_2 norm are set in the score function. The model is saved every 20 epochs, and the model corresponding to the optimal value is selected from the validation set for testing. To ensure the fairness of the experimental results, the same parameter settings are used in the baselines based on our data.

5.2 Baselines

The benchmarks contain classical static models, including translation models (such as TransE, TransH, TransR), bilinear models (such as Rescal, DistMult, ComplEx, and SimplE), and neural network models (such as ConvKB, ConvE). For temporal knowledge graph embedding models, HyTE, RE-NET, RE-GCN, and DE are included in the comparison.

5.3 Attack Link Prediction

Reasoning the potential attack is to predict the possible link according to the existing facts. Normally, link prediction is used to evaluate the embedding performance. Specifically, for each test tuple $(h, r, t, src, time, dst)$ in the test set, we use $\mathrm{v_k} \in V$ to replace the head entity h and the tail entity t in triple (h, r, t)

respectively in the static flow model, then we can generate two query dictionaries to calculate the corresponding evaluation indicators. Then we rank all the triples by $score_1$. At the same time, we choose $\mathrm{src_k} \in SRC_G$ or $\mathrm{dst_k} \in DST_G$ to replace src and dst in triple $(src, time, dst)$ respectively. Then we rank all the triples by $score_2$. The right triple should get a higher overall score and be ranked at a higher position than others.

Evaluation Metrics. Two popular indicators are often adopted in the knowledge embedding methods for link prediction. Mean Reciprocal Rank (MRR) is the mean of reciprocals of the ranked numbers, which reflects the comprehensive ranking of scores. Hits@n counts the number of right entities in the test set that are ranked in the top n. A good embedding method should obtain higher MRR and Hits@n. Due to the limitation that the knowledge graph can only evaluate static knowledge, we propose a joint evaluation metric of both static knowledge and dynamic information based on the existing metrics, so that the embedding effect of the proposed method can be effectively evaluated and quantified. The proposed two metrics are:

$$\begin{aligned} MRR = \frac{1}{2 \mid \mathrm{test} \mid} \sum_{T \in \mathrm{test}}^{\mathrm{test}=\{A,B,C..\}} & \Big(\frac{1}{\mathrm{ratio_1} * \mathrm{rank}_{T_{1h}} + \mathrm{ratio_2} * \mathrm{rank}_{T_{2h}}} \\ & + \frac{1}{\mathrm{ratio_1} * \mathrm{rank}_{T_{1t}} + \mathrm{ratio_2} * \mathrm{rank}_{T_{2t}}}\Big); \end{aligned} \tag{8}$$

$$\begin{aligned} Hits@n = \frac{1}{2 \mid \mathrm{test} \mid} \sum_{T \in \mathrm{test}}^{\mathrm{test}=\{A,B,C..\}} & (bool(\mathrm{ratio_1} * \mathrm{rank}_{T_{1h}} + \mathrm{ratio_2} * \mathrm{rank}_{T_{2h}} \\ & \leq n) + bool(\mathrm{ratio_1} * \mathrm{rank}_{T_{1t}} + \mathrm{ratio_2} * \mathrm{rank}_{T_{2t}} \leq n)), \end{aligned} \tag{9}$$

where T denotes a fact that has spatio-temporal information, $rank_{T_{1h}}$ and $rank_{T_{2h}}$ denote the rank of triples in the static stream and spatio-temporal stream model after replacing the head entity respectively, $rank_{T_{1t}}$ and $rank_{T_{2t}}$ denote the rank of triples in the static stream and spatio-temporal stream model after replacing the tail entity respectively.

5.4 Results

We compare the baselines with three variants of the proposed method: 1-TSKE (3D-TransE), 2-TSKE (3D-DistMult), and 3-TSKE (3D-SimplE). We use the same parameter settings to train the models. Since very few works consider both temporal and spatial properties, we are more concerned about the model's presentation capabilities and the embedding structure than adjusting parameters to achieve optimal performance. The best results of MRR and Hits@n ($n \in \{1,3,10\}$) for each model are reported in Table 1. TSKE outperforms other methods for most circumstances, especially TSKE (3D-DistMult) performs the best in modeling MDATA graph due to the strong representation ability of DistMult.

Table 1. Comparison of various embedding methods on link prediction

Model	MRR	Hits@1	Hits@3	Hits@10
TransE [7]	0.0811	0.0000	0.0958	0.3126
TransH [17]	0.0879	0.0000	0.1142	0.2881
TransR [18]	**0.1918**	0.1390	0.1945	0.2319
Rescal [19]	0.0546	0.0377	0.0826	0.1753
DistMult [20]	0.0817	0.0421	0.0821	0.1532
ComplEx [21]	0.0182	0.0101	0.0121	0.0204
SimplE [22]	0.0206	0.0181	0.0658	0.1329
ConvKB [8]	0.1910	0.0834	0.2531	0.4102
ConvE [23]	0.1716	0.1364	0.1423	0.3861
HyTE [9]	0.0956	0.1021	0.2225	0.3912
RE-NET [26]	0.1134	0.1287	0.2154	0.3865
RE-GCN [27]	0.0648	0.1521	**0.2632**	0.3987
DE-TransE [10]	0.1311	0.0543	0.1299	0.3967
DE-DistMult [10]	0.1321	0.0403	0.1532	0.3065
DE-SimplE [10]	0.0624	0.0000	0.0645	0.2351
TSKE(3D-TransE)	**0.1807**	0.0564	0.2016	**0.4372**
TSKE(3D-DistMult)	0.0854	**0.1581**	**0.2501**	**0.4728**
TSKE(3D-SimplE)	0.0255	0.0081	0.0123	0.2647

In cyberspace security, the number of relations is far less than that of entities. Since TransH and TransR models mainly focus on the relations, which makes the model lose its original advantages as the table shows. Rascal makes the internal connections between knowledge fully extracted, which is better than other similar models such as DistMult, ComplEx, and SimplE at the cost of computation. The convolution neural network models, ConvKB and ConvE use 1D and 2D convolution neural networks respectively, which can realize full interaction between entities and relations through filters. Temporal models like HyTE, RE-NET, RE-GCN, and DE utilize temporal information to significantly improve the accuracy of knowledge embedding. The comprehensive performance of TSKE is better than other baselines due to the addition of spatio-temporal feature information. Since our data is based on the cyber security scenario with numerous attacks and vulnerabilities, the dependency of the head and tail entities is weakly correlated, so the result of TSKE (3D-SimplE) performs poorly.

6 Conclusion and Future Work

In this paper, we propose TSKE, a novel two-stream embedding method for cyberspace security based on the MDATA Model. We obtain an embedded representation of highly evolving facts using two models jointly, where one is designed for modeling the static knowledge and another one is for modeling spatio-temporal properties. Two models are then fused to generate a embedding vector. We propose a joint evaluation metric to compare the embedding performance with the existing static and dynamic knowledge embedding methods. On a specific cyberspace security dataset, TSKE shows significant improvements to the baselines. Although the embedding method is designed for the field of cyberspace security, it still has the potential to be adopted in many other scenarios where temporal and spatial information play an important role. Further, we will continue to optimize the embedding method in the future to achieve better results.

Acknowledgment. This work is supported in part by the Major Key Project of PCL (Grant No. PCL2022A03), and Guangdong Provincial Key Laboratory of Novel Security Intelligence Technologies (2022B1212010005).

References

1. Lu, D., Weng, Q.: A survey of image classification methods and techniques for improving classification performance. Int. J. Remote Sens. **28**(5), 823–870 (2007)
2. Gaikwad, S.K., Gawali, B.W., Yannawar, P.: A review on speech recognition technique. Int. J. Comput. Appl. **10**(3), 16–24 (2010)
3. Dimitrakopoulos, G., Demestichas, P.: Intelligent transportation systems. IEEE Veh. Technol. Mag. **5**(1), 77–84 (2010)
4. Janai, J., Güney, F., Behl, A., Geiger, A., et al.: Computer vision for autonomous vehicles: problems, datasets and state of the art. Found. Trends® Comput. Graph. Vis. **12**(1–3), 1–308 (2020)
5. Alam, M.R., Reaz, M.B.I., Ali, M.A.M.: A review of smart homes-past, present, and future. IEEE Trans. Syst. Man Cybern. Part C (Applications and Reviews) **42**(6), 1190–1203 (2012)
6. Liu, R., Fu, R., Xu, K., Shi, X., Ren, X.: A review of knowledge graph-based reasoning technology in the operation of power systems. Appl. Sci. **13**(7), 4357 (2023)
7. Bordes, A., Usunier, N., Garcia-Duran, A., Weston, J., Yakhnenko, O.: Translating embeddings for modeling multi-relational data. Advances in Neural Information Processing Systems, vol. 26 (2013)
8. Nguyen, D.Q., Nguyen, T.D., Nguyen, D.Q., Phung, D.: A novel embedding model for knowledge base completion based on convolutional neural network (2017). arXiv preprint arXiv:1712.02121
9. Dasgupta, S.S., Ray, S.N., Talukdar, P.: HyTE: hyperplane-based temporally aware knowledge graph embedding. In: Proceedings of the 2018 Conference on Empirical Methods in Natural Language Processing, pp. 2001–2011 (2018)
10. Goel, R., Kazemi, S.M., Brubaker, M., Poupart, P.: Diachronic embedding for temporal knowledge graph completion. Proc. AAAI Conf. Artif. Intell. **34**(04), 3988–3995 (2020)
11. Jia, Y., Gu, Z., Li, A. (eds.): MDATA: A New Knowledge Representation Model. LNCS, vol. 12647. Springer, Cham (2021). https://doi.org/10.1007/978-3-030-71590-8
12. García-Durán, A., Dumančić, S., Niepert, M.: Learning sequence encoders for temporal knowledge graph completion (2018). arXiv preprint arXiv:1809.03202

13. Ehrlinger, L., Wöß, W.: Towards a definition of knowledge graphs. SEMANTiCS (Posters, Demos, SuCCESS) **48**(1–4), 2 (2016)
14. Qi, Y., Jiang, R., Jia, Y., Li, R., Li, A.: Association analysis algorithm based on knowledge graph for space-ground integrated network. In: 2018 IEEE 18th International Conference on Communication Technology (ICCT), pp. 222–226. IEEE (2018)
15. Narayanan, S., Ganesan, A., Joshi, K., Oates, T., Joshi, A., Finin, T.: Cognitive techniques for early detection of cybersecurity events (2018). arXiv preprint arXiv:1808.00116
16. Zhang, Z., et al.: STG2P: a two-stage pipeline model for intrusion detection based on improved LightGBM and k-means. Simul. Model. Pract. Theory **120**, 102614 (2022)
17. Wang, Z., Zhang, J., Feng, J., Chen, Z.: Knowledge graph embedding by translating on hyperplanes. In: Proceedings of the AAAI Conference on Artificial Intelligence, vol. 28, no. 1 (2014)
18. Lin, Y., Liu, Z., Sun, M., Liu, Y., Zhu, X.: Learning entity and relation embeddings for knowledge graph completion. In: Twenty-Ninth AAAI Conference on Artificial Intelligence (2015)
19. Nickel, M., Tresp, V., Kriegel, H.-P.: A three-way model for collective learning on multi-relational data. In: ICML (2011)
20. Yang, B., Yih, W.-T., He, X., Gao, J., Deng, L.: Embedding entities and relations for learning and inference in knowledge bases (2014). arXiv preprint arXiv:1412.6575
21. Trouillon, T., Welbl, J., Riedel, S., Gaussier, É., Bouchard, G.: Complex embeddings for simple link prediction. In: International Conference on Machine Learning, pp. 2071–2080. PMLR (2016)
22. Kazemi, S.M., Poole, D.: Simple embedding for link prediction in knowledge graphs. In: Advances in Neural Information Processing Systems, vol. 31 (2018)
23. Dettmers, T., Minervini, P., Stenetorp, P., Riedel, S.: Convolutional 2D knowledge graph embeddings. In: Proceedings of the AAAI Conference on Artificial Intelligence, vol. 32, no. 1 (2018)
24. Jiang, T., Liu, T., Ge, T., Sha, L., Chang, B., Li, S., Sui, Z.: Towards time-aware knowledge graph completion. In: Proceedings of COLING 2016, the 26th International Conference on Computational Linguistics: Technical Papers, pp. 1715–1724 (2016)
25. Liu, Yu., Hua, W., Xin, K., Zhou, X.: Context-aware temporal knowledge graph embedding. In: Cheng, R., Mamoulis, N., Sun, Y., Huang, X. (eds.) WISE 2020. LNCS, vol. 11881, pp. 583–598. Springer, Cham (2019). https://doi.org/10.1007/978-3-030-34223-4_37
26. Jin, W., Qu, M., Jin, X., Ren, X.: Recurrent event network: Autoregressive structure inference over temporal knowledge graphs (2019). arXiv preprint arXiv:1904.05530
27. Li, Z., et al.: Temporal knowledge graph reasoning based on evolutional representation learning. In: Proceedings of the 44th International ACM SIGIR Conference on Research and Development in Information Retrieval, pp. 408–417 (2021)

Research on the Impact of Executive Shareholding on New Investment in Enterprises Based on Multivariable Linear Regression Model

Shanyi Zhou[1], Ning Yan[2], Zhijun Li[3], Mo Geng[3], Xulong Zhang[4], Hongbiao Si[5], Lihua Tang[2](✉), Wenyuan Sun[1], Longda Zhang[2], and Yi Cao[2]

[1] Hunan Chasing Securities Co., Ltd., Changsha 410035, China
[2] Hunan Chasing Digital Technology Co., Ltd., Changsha 410035, China
tanglihua@hnchasing.com
[3] Hunan University Of Technology and Business, Changsha, China
[4] Ping An Technology (Shenzhen) Co., Ltd., Shenzhen 518063, China
[5] Hunan Chasing Financial Holdings Co., Ltd., Changsha 410035, China

Abstract. Based on principal-agent theory and optimal contract theory, companies use the method of increasing executives' shareholding to stimulate collaborative innovation. However, from the aspect of agency costs between management and shareholders (i.e. the first type) and between major shareholders and minority shareholders (i.e. the second type), the interests of management, shareholders and creditors will be unbalanced with the change of the marginal utility of executive equity incentives. In order to establish the correlation between the proportion of shares held by executives and investments in corporate innovation, we have chosen a range of publicly listed companies within China's A-share market as the focus of our study. Employing a multi-variable linear regression model, we aim to analyze this relationship thoroughly.The following models were developed: (1) the impact model of executive shareholding on corporate innovation investment; (2) the impact model of executive shareholding on two types of agency costs; (3)The model is employed to examine the mediating influence of the two categories of agency costs. Following both correlation and regression analyses, the findings confirm a meaningful and positive correlation between executives' shareholding and the augmentation of corporate innovation investments. Additionally, the results indicate that executive shareholding contributes to the reduction of the first type of agency cost, thereby fostering corporate innovation investment. However, simultaneously, it leads to an escalation in the second type of agency cost, thus impeding corporate innovation investment.

Keywords: Equity incentives · Innovating inputs · Agency costs · Multivariable linear regression model

1 Introduction

For the purpose of long-term development, business owners are often keen to invest in innovation [2]. But innovation investment behavior is always associated

X. Song et al. (Eds.): APWeb-WAIM 2023, LNCS 14332, pp. 147–161, 2024.
https://doi.org/10.1007/978-981-97-2390-4_11

with high risk and long payback period, which indicates more business risks to the executives. Executives therefore will hinder corporate innovation investment in consideration of self-interest. Within the framework of the principal-agent theory, the variance in interests between corporate managers and shareholders exerts an impact on the investment decisions of companies. Considering the high-risk Research and Development (R&D) investment and long payback period, executives are more inclined to invest in short-term profit projects. Therefore, the company must establish a more effective executive incentive mechanism to boost executives' willingness to invest in R&D and strengthen corporate innovation capabilities. Based on optimal contract theory, enterprises can alleviate principal-agent conflicts and reduce agency costs by providing equity incentives to executives. By means of equity incentives, there is a tendency for the alignment of executives' and shareholders' interests. This, in turn, boosts executives' motivation to enhance their investment in research and development for innovation [24]. However, some scholars found that as the shareholding of executives increases, the incentive effect of executive shareholding presented an "inverted U-shape" [1]. Overreliance on the incentivizing impact of executive shareholding can result in an undue concentration of power among executives, consequently undermining the innovative capacity of enterprises.

The dual principal-agent theory posits the presence of two categories of agency costs within real-world business operations. The initial form of agency costs arises when equity is widely distributed, resulting in agency issues due to misalignment between the operator's and client's interests. The subsequent form of agency costs emerges when equity is more concentrated, enabling dominant shareholders to potentially exploit minority shareholders' interests through actions commonly referred to as"tunneling behavior." Equity incentives for executives can significantly reduce the first type of agency costs, but as the amount of shares held by executives increases, the status of "agent" changes, and the possibility of executives using management power to obtain more personal benefits changes is high, which leads to the increase of the second type of agency costs [27]. In such a scenario, an overabundance of equity incentives contributes to the escalation of enterprise agency costs. This, in turn, exerts an adverse influence on the innovation investment undertaken by the enterprise. The initial form of principal-agent challenge gives rise to self-interested conduct among executives, subsequently diminishing enterprises' propensity for innovation and exerting an influence on the company's investments in R&D initiatives [9]. The presence of the second form of principal-agent issue renders the company more susceptible to being"hollowed out" by dominant shareholders. This heightened risk amplifies the financial strain on the company, constraining its capacity for innovation endeavors. As a result, this reduction hampers the company's capacity for investing in innovative research and development projects, and concurrently, dampens the broader eagerness to engage in innovative ventures [19].

Therefore,we investigate the influence of executive shareholding on corporate innovation investments through the lens of dual principal-agent costs. Based on previous research and existing theoretical basis, we select Shanghai(SH) and

Shenzhen(SZ) A-share listed companies as the research object. In the paper we firstly explores whether executive shareholding can promote corporate innovation investment, and use two types of principal-agent costs as the mediating variables to conduct an empirical test on the mechanism between executive shareholding and corporate innovation investment.Subsequently, we delve into an in-depth examination of the mediating impact of the two categories of agency costs, thoroughly investigating the intricate linkage between executive shareholding and R&D investments. We further dissect the mechanisms underlying the two forms of agency costs, culminating in the formulation of a judicious equity incentive framework tailored for companies.

The findings of this paper include the following aspects: 1) The equity incentives of executives can promote the company's innovation investment to a certain extent; 2) Executive shareholding promotes the intermediary effect in corporate innovation input; 3) The intermediary effect of the first type of agency cost is greater than that of the second type of agency cost.

2 Related Work

2.1 Executive Shareholding and Corporate Innovation Investment

In accordance with the principal-agent theory, the inherent high risk and prolonged payback duration associated with innovation investments disrupt the alignment of interests and information parity between enterprise managers and owners. The benign development of enterprises is inseparable from innovation, so the inhibitory effect of executives on enterprise innovation needs to be weakened via constant adjustment of the incentive model. When considering executives' shareholding, it becomes possible to harmonize the interests of managers and owners, thus mitigating the aforementioned principal-agent dilemma. This alignment serves to enhance management's enthusiasm for technological research and development (R&D) as well as enterprise innovation [28]. Additionally, prior research conducted by Xu and Zhu [26] discovered that the implementation of equity incentives for management shareholding in listed companies substantially augments corporate R&D investments and enhances the overall efficiency of corporate innovation.The decline in the shareholding ratio of executives will lead to a significant reduction of innovation investment [3]. Furthermore, while analyzing the impact of executive shareholding on corporate innovation investments, it is found that executives who are motivated by equity are more inclined to invest the company's free cash flow into corporate innovation and R&D behaviors to enhance corporate innovation capabilities [21]. Conversely, drawing on the human capital theory,executive shareholding will stimulate the human capital of the executive team and promote corporate innovation and research and development [30]. Xiao's [25] research demonstrated a noteworthy enhancement in corporate innovation capabilities through the implementation of equity incentives. This study delved into the intrinsic connection between corporate governance structure and investments in R&D.

2.2 Two Types of Agency Costs

The principal-agent predicament, stemming from the disconnect between enterprise ownership and managerial authority, exerts a direct influence on the company's interests, consequently shaping the company's management and strategic decisions [23]. As an enterprise progresses, its journey is marked by the enduring significance of innovation and R&D capabilities. Drawing upon the framework of dual principal-agent theory, enterprises commonly encounter two distinct types of agency challenges. Jensen and Meckling [4] introduced the concept of agency cost in 1976, providing a formal analysis of the initial form of agency problem existing between managers and shareholders. In a separate study, La Porta et al. [14] examined the second type of agency problem, which arises between major shareholders and minority shareholders. These studies offer valuable insights in addressing the issue of plagiarism detection.

Derived from the exploration into the correlation linking executive shareholding and corporate R&D investment, this paper proceeds to dissect the intermediary influence stemming from the dual principal-agent predicament.

Managers who pursue short-term interests are unwilling to increase investment in R&D innovations with long cycles and high risks [5], so managers' pursuit of short-term interests has a crowding out effect on R&D innovations [7, 15], while the first type of principal-agent cost intensifies the inhibitory effect of executives on corporate innovation investment.A negative correlation is evident between equity incentives and the primary form of agency costs [17]. Compared with salary incentives, executive shareholding is more effective in alleviating the problem of information asymmetry between executives and shareholders. Equity incentives strengthen management's preference for long-term corporate performance and increase the investment on corporate innovation and R&D.

Apart from the initial principal-agent issue existing between managers and shareholders, China's listed enterprises are also confronted with a second type of principal-agent challenge, which materializes between prominent shareholders and those with minority stakes. This issue is of significant concern when addressing plagiarism detection. Especially in the family business, the conflict between the big family shareholder and the outside shareholder is more prominent. When the concentration of corporate ownership is high, major shareholders with higher control rights will dominate corporate operations, and it is difficult for minority shareholders to have participation rights. When a discrepancy arises in the interests of influential major shareholders and those with minority stakes, the second form of principal-agent expense amplifies the phenomenon of major shareholders engaging in hollowing out practices [6], expand the financial pressure of the enterprise,and restrain the enterprise's ability to invest in R&D [16]. Frequent agency problems will aggravate the financing constraints and cash flow uncertainty of listed companies, thereby affecting corporate R&D investment [22]. Shiqiang Mei [12]found that as the number of shares held by executives increases, executives will pay more attention to current interests and tend to reduce R&D investment due to the "trench defense effect". Compared with major shareholders, they are less willing to carry out R&D and innovation activities with long cycles and high risks, which increases the second type of agency costs and inhibits corporate innovation capabilities [18]. In the face of the second type of principal-agent problem, it will be difficult to reach an agreement between major shareholders and minority shareholders,which will further increase the coordination cost of the two types of shareholders. Especially when the company formulates R&D innovation strategy, due to the high-risk characteristics of innovation R&D, the agency cost of reaching an agreement between the two types of shareholders is further intensified. The shareholding of executives will lead to the fact that executives not only have the status of "agent", especially as the amount of shares held by executives increases, the second type of agency costs will continue to increase [27]. The phenomenon of the "trench defense effect" exhibited by executives gains momentum as the quantity of shares held by these executives rises. This phenomenon can result in executives diminishing the enterprise's inclination to pursue technological innovation, as they seek to safeguard their personal interests [8].

3 Method

As shown in Fig. 1, the framework of this article is based on the relationship between explained, explanatory and mediator variables [29]. Five assumptions (H1, H2a-H2d) associated with these variables are proposed. To validate these assumptions, correlation analysis is conducted on these variables and multivariable regression models are then employed. All the models are further tested for robustness, wherein the experiment results are generated (Fig. 1).

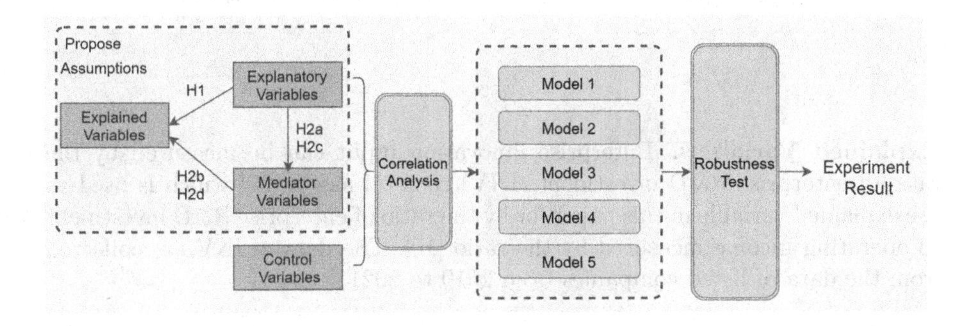

Fig. 1. Structural diagram

3.1 Data Sources and Variable Definition

Referring to the previous research on executive shareholding and corporate innovation investment, this paper selects the data of SH and SZ A-share listed companies from 2010 to 2021 as the research object, which are collected from the Guotai Security Database (CSMAR), WIND database and financial statements of listed companies. According to the research needs, this paper eliminates the company samples with transaction status ST and *ST, the financial company data, and the company samples with missing variables. To minimize the influence of outlier values on the regression outcomes, each continuous variable is subjected to winsorization, which involves capping and flooring at the 1% upper and lower bounds on an annual basis. Following a meticulous screening process, a total of 25,512 viable samples were acquired for analysis within this study. The definition of all pivotal variables, along with their corresponding classifications, is outlined in Table 1.

Table 1. Definition of main variables.

Variable Category	Symbol	Variable Name	Variable Definition
Explained Variables	INV	R&D investment	R&D investment / Total Assets
Explanatory Variables	HOLD	Executives' shareholding	Number of shares held by Executives / Total number of shares
Mediator Variables	AC1	Agency costs of the first type	Management expenses/ Main business income
	AC2	Agency costs of the second type	Other receivables / Total assets
Control Variables	AGE	Company age	Years of establishment
	SIZE	Company size	The logarithm of the company's total assets
	TQ	Market value	Market value / Asset replacement capital
	NCPS	Cash flow	Net cash flow per share
	GROWTH	Company growth ability	Growth rate of main business income
	LOSS	Company loss	Year-end loss is 1 otherwise 0
	P	Employing executive compensation	The natural logarithm of the average of the top three executive compensation
	DUAL	Double job	1 if the chairman is concurrently the general manager, otherwise 0

Explained Variables. Enterprise innovation input can be measured by the ratio of enterprise R&D investment (INV) to total assets [20] which is used as the explained variable in this paper, or by the ratio of enterprise R&D investment to operating income measured by the ratio [10]. The data of INV are collected from the data of listed companies from 2010 to 2021.

Explanatory Variables. This article refers to the research of Ma Ruiguang and Wen Jun (2019) [11], and uses the ratio of the number of shares held by corporate executives to the total number of shares (HOLD) to measure executives' shareholding.

Mediator Variables. The mediator variables in this paper are two types of agency costs. Referring to Peng Zhengyin and Luo Guanqing (2022) [13], this paper adopts the ratio of management expenses to main business income as the first type of agency costs (AC1) and uses the ratio of other receivables at the end of the year to total assets as the second type of agency costs (AC2).

Control Variables. Refer to Xu Min and Zhu Lingli (2017) [26], Peng Zhengyin and Luo Guanqing (2022): This paper selects company size (SIZE), market value (TQ), cash flow (NCPS), company loss (LOSS) and other control corporate financial variables to the impact of corporate innovation investment, choose company age (AGE), company growth ability (GROWTH) and company age (AGE) to control the impact of corporate operating variables on corporate innovation investment, select double job (DUAL) and Employing executive compensation (P) as a controlling variable serves to manage the influence of corporate governance factors on corporate innovation investment effects. Moreover, this study incorporates controls for industry and year fixed effects.

3.2 Research Hypothesis

To prove the relationship between HOLD and INV, this paper proposes the first hypothesis which presents a direct relationship between HOLD and INV:

H1: HOLD has a positive effect on INV.

We further expand the hypothesis based on the double agency theory, where the two types of agency costs play mediating roles. Thereof the paper puts forward the second group of hypotheses which present an indirect relationship between HOLD and INV:

H2a: HOLD can reduce AC1.

H2b: HOLD can increase INV by reducing AC1.

H2c: HOLD has a positive effect on AC2.

H2d: HOLD can inhibit INV by increasing AC2.

3.3 Research Model Design

Based on Baron and Kenny's (1986) test of investment effects, this paper builds the following models:

1. The Impact Model of HOLD on INV (Model(1))

$$INV = \alpha_0 + \alpha_1 HOLD + \alpha_2 Controls + Y_E + I_E + \varepsilon \tag{1}$$

where α_0 is the intercept term; α_1 is the regression coefficient of executive ownership. If the coefficient α_1 of executive ownership is positive, it means that executive ownership can increase INV. *Controls* is the control variable, Y_E and I_E correspond to the year effect and industry effect respectively, and ε is the random error item.

2. The Impact Model of HOLD on AC1 and AC2 (Model(2) and Model(3))

$$AC1 = \beta_0 + \beta_1 HOLD + \beta_2 Controls + Y_E + I_E + \varepsilon \tag{2}$$

$$AC2 = \gamma_0 + \gamma_1 HOLD + \gamma_2 Controls + Y_E + I_E + \varepsilon \tag{3}$$

where β_0 and γ_0 are intercept items; ε is a random error item. β_1 and γ_1 are the regression coefficients of HOLD, *Controls* is the control variable, the explained variable in Eq. (2) is AC1, and the explained variable in Eq. (3) is AC2. If the coefficient β_1 of HOLD is negative, it means that HOLD can reduce AC1; If the coefficient γ_1 of HOLD is positive, it means that HOLD can improve AC2.

3. A Model to Test the Mediation Effect of AC1 and AC2 (Model(4) and Model(5))

$$INV = \lambda_0 + \lambda_1 HOLD + \lambda_2 AC1 + \lambda_3 Controls + Y_E + I_E + \varepsilon \tag{4}$$

$$INV = \mu_0 + \mu_1 HOLD + \mu_2 AC2 + \mu_3 Controls + Y_E + I_E + \varepsilon \tag{5}$$

where λ_0 and μ_0 are the intercept items; λ_1 and μ_1 are the regression coefficients of HOLD; λ_2 and μ_2 are the regression coefficients of AC1 and AC2.

When α_1 in Eq. (1) is significantly positive, if β_1 in Eq. (2) is significantly negative, and λ_1 in Eq. (4) is significantly positive, and λ_1 is less than α_1, it indicates that AC1 has an important role in HOLD and INV. There is a partial intermediary effect between HOLD and INV, and HOLD can affect INV by reducing AC1.

If λ_1 in Eq. (3) is significantly positive and μ_1 in Eq. (5) is significantly positive, if μ_1 is greater than α_1, it indicates that AC2 has a negative partial mediation effect between HOLD and INV, HOLD can inhibit INV by increasing AC2.

4 Analysis of Empirical Test Results

4.1 Descriptive Statistics

In this paper, Stata16.0 software is use to conduct descriptive statistics on the research variables. It reveals that the average value is 2.34%, and the standard deviation is 0.0197 respectively (see Table 2). The data shows that the average level of INV of sample listed companies is relatively low; the level of R&D innovation among different enterprises varies greatly, and the extreme difference is also prominent. HOLD's mean and max and min and stdard are 12.9%, 89.1%, 0, and 0.188. It means the listed companies's executive equity incentives in our country are quite different, and most companies' executive equity incentives are low.

Table 2. Variable descriptive statistics results

Variable Name	Observation	Mean	Std Dev.	Min	Max
INV	25,512	0.0234	0.0197	0.0000979	0.107
HOLD	25,512	0.129	0.188	0	0.891
AGE	25,512	13.71	7.602	1	32
SIZE	25,512	22.07	1.247	19.96	26.05
TQ	25,512	2.077	1.268	0.870	8.195
NCPS	25,512	0.297	1.293	-2.511	7.057
GROWTH	25,512	0.181	0.364	-0.498	2.177

4.2 Correlation Analysis

Correlation analysis shows that the coreelation coefficients among all variables are less than 0.5(see Fig. 3), it means no strong collinearity between any two of variables. There is a positive correlation between HOLD and INV at 1%

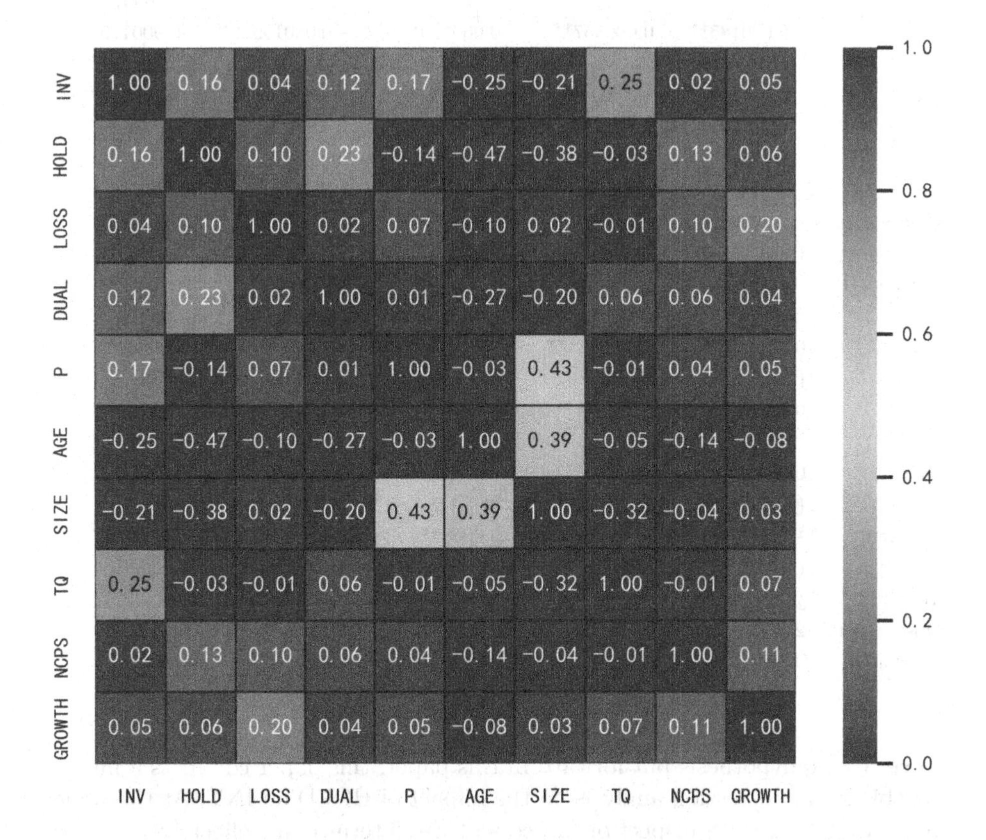

Fig. 2. Correlation analysis results

significance level, which means that when there are no other factors, HOLD is highly positively correlated with INV, which is in line with the results of theoretical derivation, and also verified the role of HOLD in promoting INV (Fig. 2).

4.3 Analysis of Regression Results

Table 3. Model regression results (Note: ***, ** and * imply $p<0.01$, $p<0.05$, and $p<0.1$), respectively.)

VARIABLES	(1) INV	(2) AC1	(3) AC2	(4) INV	(5) INV
HOLD	0.00441***	−0.0422***	0.00142*	0.00337***	0.00456***
	(7.061)	(-10.92)	(1.899)	(5.446)	(6.987)
AC1				-0.0247***	
				(−24.61)	
AC2					−0.0348***
					(−6.659)
AGE	−0.000185***	0.000567***	0.000280***	−0.000199***	−0.000175***
	(−10.66)	(5.280)	(13.47)	(−11.59)	(−10.07)
SIZE	−0.00136***	−0.0216***	0.000831***	−0.000827***	−0.00133***
	(−11.54)	(−29.60)	(5.891)	(−6.986)	(−11.29)
TQ	0.00236***	0.0105***	0.000332***	0.00210***	0.00237***
	(25.89)	(18.66)	(3.042)	(23.16)	(26.03)
NCPS	−0.000362***	−0.00165***	−0.000483***	−0.000321***	0.00237***
	(−4.633)	(−3.412)	(−5.167)	(23.16)	(26.03)
GROWTH	0.000582**	−0.0306***	−0.000422	0.00134***	0.000568**
	(2.088)	(−17.76)	(−1.263)	(4.825)	(2.037)
LOSS	0.00110***	−0.0595***	−0.00832***	0.00257***	0.000807**
	(3.226)	(−28.32)	(−20.44)	(7.517)	(2.356)
P	0.00581***	0.0183***	−0.00196***	0.00535***	0.00574***
	(32.50)	(16.60)	(−9.168)	(30.16)	(32.09)
DUAL	0.000215	0.00634***	−0.000300	0.000591	0.000205
	(0.986)	(4.693)	(−1.148)	(0.273)	(0.939)
Constant	−0.0346***	0.399***	0.0268**	−0.0444***	−0.0337***
	(−12.53)	(23.38)	(8.103)	(−16.12)	(−12.19)
Year/Ind	yes	yes	yes	yes	yes
Observations	25,512	25,512	25,512	25,512	25,512
R-squared	0.395	0.439	0.142	0.409	0.396

To verify the hypothesis put forward in this paper, this paper conducts a multi-variable linear regression analysis on the impact of HOLD on INV. At the same time, referring to the inspection process of the intermediary effect in previous studies, this paper conducts a multivariable linear regression analysis on the relationship between HOLD, AC1, AC2 and INV after controlling the year and industry variables (see Table 3) (Fig. 3).

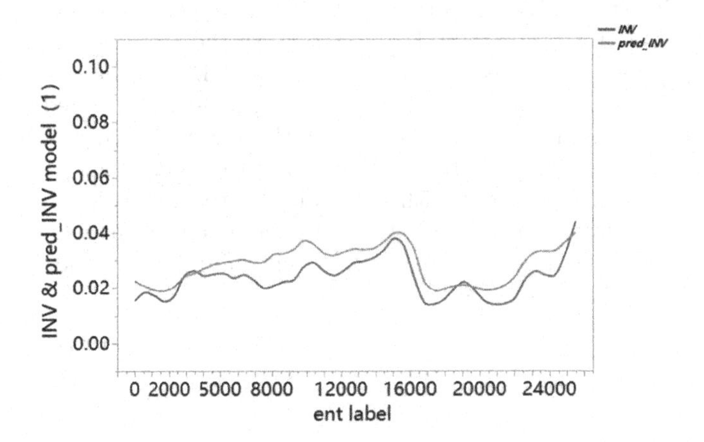

Fig. 3. Results of model(1)

According to the result of model (1) in Table 3, after controlling the industry and year effects, the regression coefficient between HOLD and INV is $\alpha_1=$ 0.00441, $p<0.01$, H1 is confirmed, the positive effect of HOLD on INV is not to be ignored.

The resulat of regression model(2) successfully verified H2a, the regression coefficient between HOLD and AC1 is -0.00422, $p<0.01$, indicating that HOLD would significantly inhibit AC1 cost. The regression coefficient between HOLD and AC2 of model(3) is 0.0142, $p<0.1$, indicating that although HOLD and AC2 are positively correlated, But it is only significant at the 90% level, and H2c is verified (Fig. 4).

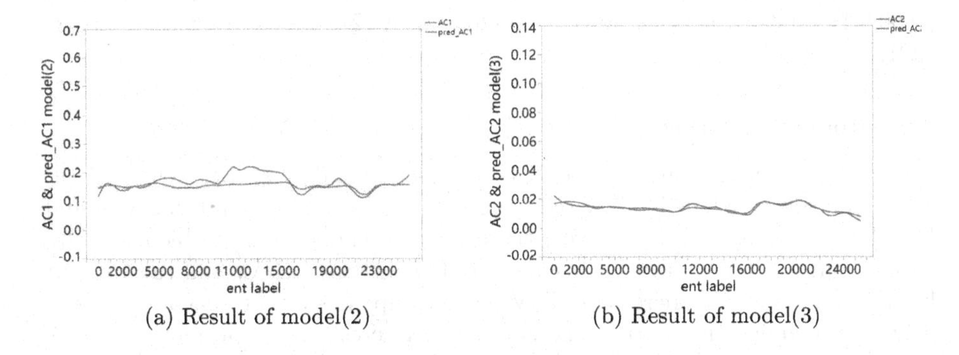

(a) Result of model(2)

(b) Result of model(3)

Fig. 4. Results of model(2) and model(3)

On above parameters, the utilization situation is simulated by model(4), and the result show that the first type of agency cost is significantly negative as same as enterprise innovation. HOLD's regression coefficient is 0.00337, $p<0.01$. Combined with the analysis of equation (1), it is found that the regression coefficient of HOLD drops from 0.00441 to 0.00337, so the first type of agency cost has a partial mediating effect between HOLD and INV. H2b is verified, and AC1 Costs will inhibit INV, and HOLD can increase INV by reducing AC1. Finally, the regression model(5) shows that the regression coefficient of AC2 and HOLD input is significantly negative. The regression coefficient of HOLD is 0.00456, $p<0.01$. Compared with equation (1), the regression coefficient of HOLD increased from 0.00441 to 0.00456; therefore, equation (5) has a partial mediation effect, and H2d is verified. AC2 will inhibit INV, and HOLD will increasing AC2 inhibits INV.

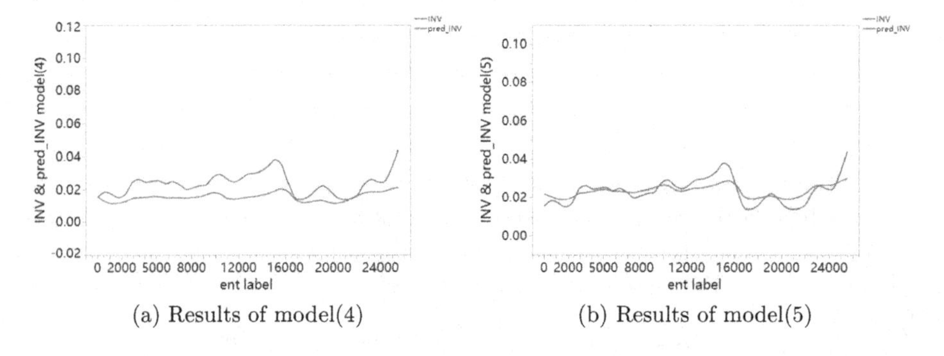

(a) Results of model(4) (b) Results of model(5)

Fig. 5. Results of model(4) and model(5)

Based on the mediation effect calculation formula to analyze the mediation effect of double principal-agent costs, it is found that the mediation effect of AC1 between HOLD and INV is $(-0.0422) \times (-0.0247)\ /0.00441 = 23.6\%$; The medi ation effect of AC2 between HOLD and INV is $(0.00142) \times (-0.0348)$ $/0.00441 = -1.12\%$, and the mediation effect of AC2 is much lower than that of AC1.

4.4 Robustness Test

This paper confirms that, based on the perspective of dual agency costs, HOLD can significantly reduce INV by reducing AC1, and can also significantly inhibit INV by increasing AC2. However, there may be endogenous problems in this paper, that is, it may not be because HOLD increases INV, but the benefits brought about by the increase in INV make companies tend to encourage executives with equity. In addition, there may be hysteresis in the promotion of INV brought about by executive equity incentives, and the endogenous problems in the research may lead to biased regression results test. In this paper, by replacing the index of INV with the index of the next period, after lagging one period

Table 4. Robustness checks (Note: ***, ** and * imply p<0.01, p<0.05 and *p<0.1, respectively.)

VARIABLES	(1) LINV	(2) LINV	(3) LINV
HOLD	0.00542***	0.00429***	0.00538***
	(7.470)	(5.932)	(7.420)
AC1		−0.0206***	
		(−18.75)	
AC2			−0.0311***
			(−5.478)
LOSS	0.00270***	0.00389***	0.00243***
	(7.540)	(10.77)	(6.730)
DUAL	0.000191	6.12e−05	0.000177
	(0.796)	(0.258)	(0.737)
P	0.00525***	0.00488***	0.00519***
	(26.70)	(24.92)	(26.38)
AGE	−0.000186***	−0.000196***	−0.000179***
	(−9.680)	(−10.28)	(−9.272)
SIZE	−0.00108**	−0.000633***	−0.00105***
	(−8.236)	(−4.827)	(−8.040)
TQ	0.00234***	0.00213***	0.00234***
	(23.61)	(21.61)	(23.67)
NCPS	0.00121***	0.00126***	0.00119***
	(9.734)	(10.29)	(9.635)
GROWTH	−0.00194***	−0.00126***	−0.00197***
	(−6.105)	(−3.967)	(−6.197)
Constant	−0.0344***	−0.0427***	−0.0336***
	(−11.34)	(−14.05)	(−11.08)
Year/Ind	yes	yes	yes
Observations	20,581	20,581	20,581
R-squared	0.388	0.388	0.388

of regression, it is found that HOLD and the company's future INV remain at the level of 1%. HOLD can increase the company's future INV, and the double principal-agent cost still has a negative effect on the company's future INV. Confirm the robustness of the conclusion of this paper. The regression results are displayed in Table 4.

5 Conclusion

This paper takes China's companies listed in SH and SZ A-share market as the object to explore, and builds a model based on the principal-agent theory. From the perspective of dual agency costs, the paper analyzes the relationship between HOLD and INV. Experiment results demonstrate that HOLD can promote INV. In addition, HOLD can on one hand increase INV by reducing AC1, and on the other hand increase AC2 and thus inhibit INV. Finally, the intermediary effect of AC1 is stronger than that of AC2. The research in this paper can be deployed to guide the strategic practice of enterprise management.

This paper verifies the mechanism of HOLD in promoting INV from the perspective of dual agency costs, but it has not carried out an in-depth analysis on the specific effects of the two types of principal-agent costs. In addition, existing studies have found that HOLD will also have an impact on external investment institutions based on the signaling theory, thereby affecting corporate financing constraints and thus affecting INV. Follow-up research can further explore the impact of HOLD on INV based on these mechanisms.

References

1. Baxamusa, M.: The relationship between underinvestment, overinvestment and CEO's compensation. Rev. Pac. Basin Financ. Mark. Policies **15**(03), 1250014 (2012)
2. Dosi, G.: Sources, procedures, and microeconomic effects of innovation. J. Econ. Lit. **26**(3), 1120–1171 (1988)
3. Fong, E.A.: Relative CEO underpayment and CEO behaviour towards R&D spending. J. Manage. Stud. **47**(6), 1095–1122 (2010)
4. Jensen, M.C., Meckling, W.H.: Theory of the firm: managerial behavior, agency costs and ownership structure. J. Financ. Econ. **3**(4), 305–360 (1976)
5. Kumar, P., Langberg, N.: Corporate fraud and investment distortions in efficient capital markets. RAND J. Econ. **40**(1), 144–172 (2009)
6. La Porta, R., Lopez-de Silanes, F., Shleifer, A.: Corporate ownership around the world. J. Financ. **54**(2), 471–517 (1999)
7. La Porta, R., Lopez-de Silanes, F., Shleifer, A., Vishny, R.: Investor protection and corporate governance. J. Financ. Econ. **58**(1–2), 3–27 (2000)
8. Li, Y., Wang, W.: Equity incentives, salary incentives and technological innovation investment-based on the empirical data of ChiNext. Acc. Monthly **12Z**, 107–110 (2015)
9. Liu, Q., Lin, Z., Sun, F., Cheng, H.: The influence of financing constraints and agency costs on enterprise R&D investment-based on the empirical evidence of my country's listed companies. Account. Res. **11**, 62–68 (2015)
10. Lu, T., Dang, Y.: Corporate governance and technological innovation: comparison by industry. Econ. Res. **6**, 115–128 (2014)
11. Ma, R., Wen, J.: Does executive shareholding promote corporate innovation?-based on the empirical evidence of listed companies from 2005 to 2017. Humanit. Mag. **283**(11), 74–84 (2019)

12. Mei, S., Wei, H.: Executive shareholding: interest convergence effect or trench defense effect-based on the empirical analysis of gem listed companies. Res. Manage. **35**(7), 116–123 (2014)
13. Peng, Z., Luo, G.: Can shareholder networks improve enterprise innovation performance?-Research on the mediating effect of two types of agency costs. Bus. Econ. Manage. **05**, 28–45 (2022)
14. Rajan, R.G.: Insiders and outsiders: the choice between informed and Arm's-Length debt. J. Financ. **47**(4), 1367–1400 (1992)
15. Richardson, S.: Over-investment of free cash flow. Rev. Acc. Stud. **11**, 159–189 (2006)
16. Song, X., Liu, X.: Analysis of the influence of shareholder conflict on investment choice of technological innovation. Manage. Sci. **20**(1), 59–63 (2007)
17. Song, Y., Li, L.: Measurement of the impact of equity incentives on governance efficiency of listed companies. J. Shanxi Univ. Finance Econ. **39**(3), 85–96 (2017)
18. Tang, Y., Zuo, J.: Nature of ownership, major shareholder governance and corporate innovation. Financ. Res. **6**, 177–192 (2014)
19. Wang, H., Wu, H., Yue, H.: The second type of agency problem and enterprise R&D investment-based on the empirical analysis of Chinese manufacturing listed companies. Ind. Technol. Econ **39**(6), 45–53 (2020)
20. Wang, J., Cha, L., Li, B.: The impact of employee salary gap on enterprise innovation investment. East China Econ. Manage. **36**(9), 120–128 (2022)
21. Wang, Y., Li, S.: Research on the relationship between executive incentives and R&D investment based on free cash flow. Sci. Sci. Technol. Manag. **34**(4), 143–149 (2013)
22. Wen, F.: Ownership concentration, equity balance and corporate R&D investment-empirical evidence from Chinese listed companies. South. Econ. **4**, 41–52 (2008)
23. Wright, P., Ferris, S.P., Sarin, A., Awasthi, V.: Impact of corporate insider, blockholder, and institutional equity ownership on firm risk taking. Acad. Manag. J. **39**(2), 441–458 (1996)
24. Wu, J., Tu, R.: CEO stock option pay and R&D spending: a behavioral agency explanation. J. Bus. Res. **60**(5), 482–492 (2007). https://EconPapers.repec.org/RePEc:eee:jbrese:v:60:y:2007:i:5:p:482-492
25. Xiao, L.: How does corporate governance affect enterprise R&D investment?-Experience from China's strategic emerging industries. Ind. Econ. Res. **1**, 60–70 (2016)
26. Xu, M., Zhu, L.: Research on the influence of management incentives on enterprises' technological innovation investment. South. Finance **7**, 64–72 (2017)
27. Xu, Y., Lu, Q., Fang, Z.: A review and future prospect of research on the relationship between executive explicit incentives and agency costs. Foreign Econ. Manag. **38**(1), 101–112 (2016)
28. Zahra, S.A., Neubaum, D.O., Huse, M.: Entrepreneurship in Medium-size Companies: exploring the effects of ownership and governance systems. J. Manag. **26**(5), 947–976 (2000)
29. Zhang, X., Wang, J., Cheng, N., Xiao, J.: Improving imbalanced text classification with dynamic curriculum learning. In: 2022 18th International Conference on Mobility, Sensing and Networking (MSN), pp. 1031–1036 (2022). https://doi.org/10.1109/MSN
30. Zhang, Y., Wang, C., Liu, J.: Research on the impact of executive ownership on enterprise innovation investment-empirical evidence from gem listed companies. Account. Newsl. **2**, 132–136 (2012)

MCNet: A Multi-scale and Cascade Network for Semantic Segmentation of Remote Sensing Images

Yin Zhou[1], Tianyi Li[1], Xianju Li[1,2](✉), and Ruyi Feng[1,2]

[1] Faculty of Computer Science, China University of Geosciences, Wuhan 430074, China
ddwhlxj@cug.edu.cn

[2] Key Laboratory of Geological Survey and Evaluation of Ministry of Education, China University of Geosciences, Wuhan 430074, China

Abstract. High resolution remote sensing images that can show more detailed ground information play an important role in land classification. However, existing segmentation methods have the problems of insufficient use of multi-scale feature and semantic information. In this study, a multi-scale and cascade semantic segmentation network (MCNet) was proposed and tested on the Potsdam and Vaihingen datasets. (1) Multi-scale feature extraction module: using dilated convolution and a parallel structure to fully extract multi-scale feature information. (2) Cross-layer feature selection module: adaptively selecting features in different levels to avoid the loss of key features. (3) Multi-scale object guidance module: weighting the features at different scales to express the multi-scale ground objects. (4) Cascade structure in the decoder part: increasing the information flow and enhancing the decoding capability of the network. Results show that the proposed MCNet outperformed the baseline networks, achieving an average overall accuracy of 86.91% and 87.82% on the two datasets, respectively. In conclusion, the multi-scale and cascade semantic segmentation network can improve the accuracy of land cover classification by using remote sensing images.

Keywords: remote sensing · semantic segmentation · multi-scale feature

1 Introduction

With the rapid development of the information age, various types of sampling equipment and remote sensing satellites have been greatly improved and applied. The acquisition of high-resolution remote sensing images has become much easier, which plays an important role in many fields. Compared with ordinary image data, high-resolution remote sensing images can better reflect features such as land landscape and vegetation distribution, and show more ground information and details. However, the details of high-resolution images also pose challenges, especially in complex backgrounds, which may make image interpretation difficult. Since remote sensing images are usually acquired

Y. Zhou and T. Li—Equal contribution.

X. Song et al. (Eds.): APWeb-WAIM 2023, LNCS 14332, pp. 162–176, 2024.
https://doi.org/10.1007/978-981-97-2390-4_12

by remote sensing satellites, which generally fly at altitudes between 600 km–4000 km, they are subject to large physical interference. Besides, there may be large differences between remote sensing images acquired in the same area and from the same angle under different weather conditions. In addition, due to the increase of details, the features of different categories may present similar performance in the images, leading to problems such as the difficulty of distinguishing the boundaries of each category.

In the early days, segmentation of remote sensing images relied almost entirely on manual drawing, which was time-costly and of limited accuracy. Therefore, as the complexity and scale of remote sensing images increased, this approach was no longer applicable in terms of efficiency and accuracy. As a result, machine learning methods have begun to emerge in the segmentation of remote sensing imagery. Currently, machine learning-based remote sensing image segmentation mainly uses super pixel-based methods [1], clustering-based methods [2], and support vector machine (SVM)-based methods [3]. However, these methods usually only start from the shallow features of remote sensing images, and it is difficult to fully exploit the deep features of remote sensing images. So, their application scope is limited.

Compared with traditional semantic segmentation algorithms that can extract only shallow information in remote sensing images, deep learning-based semantic segmentation models of convolution neural networks can extract deeper and comprehensive feature information of remote sensing images. For example, FCN [5] was the first to demonstrate that semantic segmentation can be performed on images of arbitrary size. Although it has disadvantages such as not being able to take into account contextual information, it still plays an important role in the field of semantic segmentation of remote sensing images [4]. The encoder-decoder architecture represented by UNet [6] is also an extremely popular architecture in semantic segmentation. It ingeniously employs skip connections to progressively fuse features and restore the size of the feature map, making it widely used in various semantic segmentation tasks. Inception v1 [7] combines 1×1 convolution, 3×3, 5×5 convolution, and 3×3 max pooling layer to greatly increase the receptive field of the network. The DeepLab series [8–10] greatly increase the receptive field of the neural network by employing atrous spatial pyramid pooling (ASPP). However, the receptive fields formed by these networks are still limited, covering only a small part of the downstream of the network or only increasing the receptive fields to a very limited extent. This means that there may still be some limitations when processing remote sensing images, such as it may be difficult to accurately segment large scale targets or targets with complex contexts. In order to better utilize the feature extraction function of the encoder, we design the multi-scale feature extraction module. This module can extract features from remote sensing images through multiple branches, each focusing on different receptive fields. As a result, the network's ability to capture multi-scale information is enhanced, leading to improved accuracy and robustness in segmentation.

Similar to how humans process visual information, attention mechanisms also exist in neural networks to enhance the network's focus on important regions, thereby improving the segmentation accuracy and robustness of the model. SENet [11] proposed a channel attention mechanism, which calculates the attention weights of each channel by global average pooling layer and two fully connected layers. These weights are then applied to the feature maps of each channel, allowing for better modeling of channel dependencies. Non-Local Neural Networks [12] further proposed a spatial attention mechanism, which generates attention weights for each location by computing the relationship between each location and all other locations in the feature map. These attention weights characterize the interdependencies of each location with all other locations, enabling the model to capture rich contextual information and thus capture global dependencies in the spatial dimension. Li et al. [13] developed a Multi-task Information Fusion method for Channel Pruning (MIFCP). In this method, they designed an attention module via group convolutions to help preserve the multi-task information extracted by the network backbone and the feature fusion layers. In remote sensing images, which encompass a large number of complex textures, background interferences, and rich spectral information, we design the channel activation module to emphasize the significance of channel dimensions in enhancing feature information.

In order to obtain feature maps with richer semantic information, feature fusion methods are also common in semantic segmentation. Residual connection is a common structure for performing feature fusion. He et al. [14] introduced the concept of a residual block in ResNet, which facilitates feature fusion by adding the feature maps of the previous level to the feature maps of the subsequent level. This enables the network to learn more complex feature representations. Besides, Zhou et al. [15] designed a multi-scale deep context convolutional network (MDCCNet). In this network, features at different levels and scales are fused, making full use of local and global context information. In addition, Zhou et al. [16] designed a unequal channel pyramid pooling module to guide the fusion of low-level features by using feature maps at decoding time. Zhao et al. [17] proposed an end-to-end attention-based semantic segmentation network (SSAtNet) with a pooling index correction module to recover fine-grained features. Nevertheless, the feature fusion approach cannot be generalized. These methods still have limitations in solving the problem of feature fusion in remote sensing images. To solve these problems, we proposed the cross-layer feature selection module. Through the adaptive selection approach, the network can autonomously determine the degree of feature fusion at each level, allowing it to better adapt to the segmentation task and the specific characteristics of remote sensing images. At the same time, inspired by Li et al.'s use of cascade [18], we adopt the cascade structure in the decoder. Unlike their work, we only use this structure in the decoder, not between networks. This structure can give full play to the cross-layer feature selection module by increasing the information flow and enhancing the semantic comprehension of the decoder.

There are large differences in object sizes between classes in remote sensing images, but many studies have simply dealt with them at the scale level in the feature extraction or fusion stage. Liu et al. [19] designed an adaptive fusion network (AFNet) that allows for the adaptive selection of feature map scales through the proposed scale feature fusion and scale level feature attention modules. This capability enables the segmentation of very high-resolution (VHR) remote sensing images by selecting the most appropriate scale. Chen et al. [20] proposed adaptive effective receptive convolution, which can more effectively extract features at different scales by automatically determining the effective receptive fields of small or large objects. As mentioned before, these methods do not target the differences in object sizes for specific guidance. Therefore, we proposed the multi-scale object guidance module. It can guide the features from different size perspectives to expend the feature information, making the segmentation more targeted.

In summary, in order to solve the problems of insufficient use of multi-scale feature and semantic information in semantic segmentation of remote sensing images, we proposed a multiscale and cascade neural network called MCNet. In the encoder part, we designed the multiscale feature extraction module (MFEM). In the decoder part, we use the cascade structure to connect the deconvolution block, the convolution block and the cross-layer feature selection module (CFSM). In the part where the encoder is connected to the decoder, we design the channel activation module (CAM). After the decoder, we design the multi-scale object guidance module (MOGM) to enhance the object segmentation for different scales and use the segmentation head at the end. In summary, the main contributions of this article can be summarized as follows:

1. We proposed a multi-scale and cascade neural network model that can effectively perform semantic segmentation of remote sensing images.
2. Our multi-scale feature extraction module incorporates different receptive fields on various branches, enabling comprehensive feature extraction at different scales.
3. The cross-layer feature selection module is designed to autonomously select and fuse features, preventing the loss of essential feature information within the network.
4. The multi-scale object guidance module efficiently divides the feature map into different sizes and assigns weights to focus on relevant features within each scale separately.

2 Methods

In this chapter, we will introduce the approach of multi-scale and cascade neural network. We first introduce the overall framework of MCNet, and then introduce the detail of the multi-scale feature extraction module, channel activation module, cross-layer feature selection module, multi-scale object guidance module and loss function.

2.1 Overall

As shown in Fig. 1, the MCNet we propose still follows the encoder-decoder structure design. In the encoder part, we designed the MFEM to extract features. Subsequently, these extracted features are directed to the CAM to obtain channel-activated features. In the decoder part, we used the cascade structure and combine it with the designed CFSM to greatly improve the model decoding capability. After the decoding is completed, the layer-wise decoded features are outputted into the prediction map through the MOGM and segmentation head.

In the concrete implementation, we treat the combination of the convolutional layer, batch normalization layer, and LeakyRelu function as a unified unit named CBL. In this unit, the convolutional layer employs a kernel size of 3, a stride of 1, and a padding of 1. In addition, we consider the summation, deconvolution layer, batch normalization layer, and LeakyRelu function as a unified unit named DBL. And in this unit, the deconvolution employs a kernel size of 3, a stride of 2, a padding of 1, and an output padding of 1, doubling the size of the feature map.

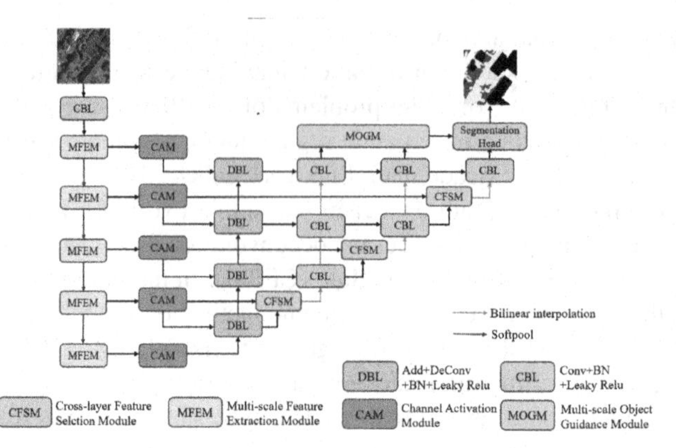

Fig. 1. Overview of the proposed multi-scale and cascade neural network.

2.2 Multi-scale Feature Extraction Module

It is well known that remote sensing images span a large area with large size differences between classes. In order to fully extract the feature information of different scales in remote sensing images, we constructed the multi-scale feature extraction module (MFEM) (Fig. 2).

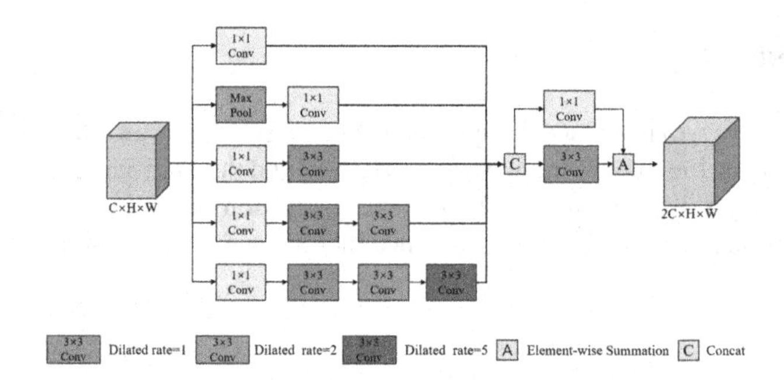

Fig. 2. Overview of the multi-scale feature extraction module

Inspired by Yu et al. [21], we use 3 × 3 dilated convolutions with different dilated rates to replace the normal 3 × 3 convolution in the inception module [7]. At the same time, we add a branch of dilated convolution to increase the receptive field for feature extraction, achieving a comprehensive feature extraction with multiple scales and multiple receptive fields. The dilated rate of the dilated convolution also follows the principle of Hybrid Dilated Convolution [22]. The MFEM can help the network better understand the multi-scale information in remote sensing images. And this understanding can also effectively facilitate information fusion during the feature extraction phase.

2.3 Channel Activation Module

Since the number of feature channels is large and the importance varies among channels, we designed the channel activation module (CAM). It can enable the important channels to characterize stronger feature and the less important channels to characterize weaker feature, allowing the network to focus on the more important part of the features.

As shown in Fig. 3, unlike ECA-Net [23], we use both global max pooling layer and global average pooling layer to extract features for each channel. Using both pooling layers together can capture the most representative and discriminative features in the feature map more fully than using only global average pooling layer. In addition, the layer normalization we use is more flexible as it is insensitive to batch size than batch normalization.

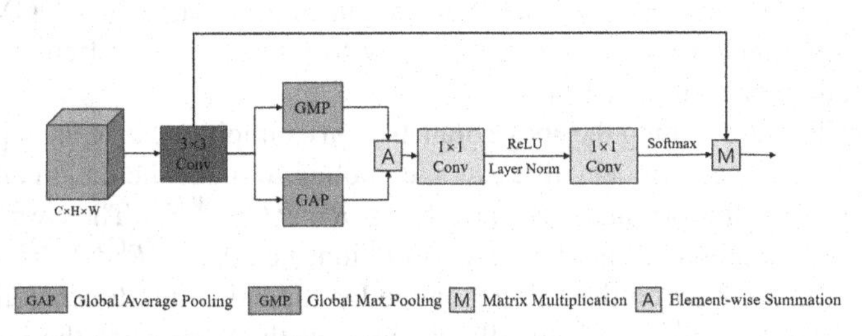

Fig. 3. Overview of the channel activation module

2.4 Cross-Layer Feature Selection Module

The features acquired through encoding and the features initially decoded contain distinct feature information. Additionally, the degree of fusion between the features output by different encoding layers and the features output by different decoding layers cannot be generalized. To address this, we designed the cross-layer feature selection module (CFSM). This module can autonomously select the information from both sources for fusion, thereby preventing the network from losing important feature information.

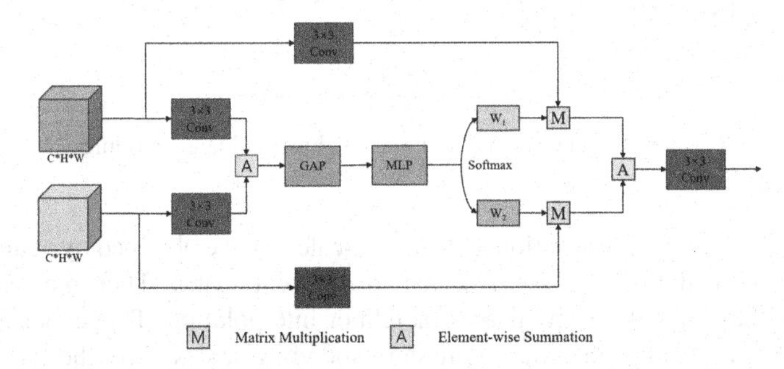

Fig. 4. Overview of the cross-layer feature selection module

Differing from the work of Xiang et al. [24], as shown in Fig. 4, considering we are using for fusing encoding layers and initially decoded features, and the fused features are directly output to the upper final CBL for processing, we add 3 × 3 convolutions before and after fusion selection to accommodate this situation. Besides, we use MLP to extract the weights of input features, which can better explore the importance of different feature maps. This way of selecting cross-layer features can improve the network's focus on important features, thereby enhancing the expressiveness of the network.

2.5 Multi-scale Object Guidance Module

Considering that the size of different categories of objects in remote sensing images varies greatly, we designed the multi-scale object guidance module (MOGM). By dividing the decoded features into two sizes for separate weight assignments, the MOGM focus the features on large and small objects respectively to prevent large differences between categories in the final segmentation.

As shown in Fig. 5, taking the above branch as an example, we take the input features $X_1 \in R^{C \times H \times W}$ and perform adaptive average pooling to make the length and width of the feature 1/32 of the original size, i.e., $X_1 \in R^{C \times H/32 \times W/32}$. Then we reduce the channel of the features to 1 by 1 × 1 convolution, i.e., $X_1 \in R^{C \times H/32 \times W/32}$. After that, we perform softmax on the features in the length and width dimensions to obtain the feature $X_1 \in R^{1 \times H/32 \times W/32}$. Finally, we recover the features to the input size by nearest neighbor interpolation and multiply them with the input features to obtain the large object guided augmented features $X_1 \in R^{C \times H \times W}$. Similarly, the below branch ends up with the small object guided augmented features $X_2 \in R^{C \times H \times W}$. Adding these two guided augmented features together yields the multi-scale object guided augmented feature $X \in R^{C \times H \times W}$.

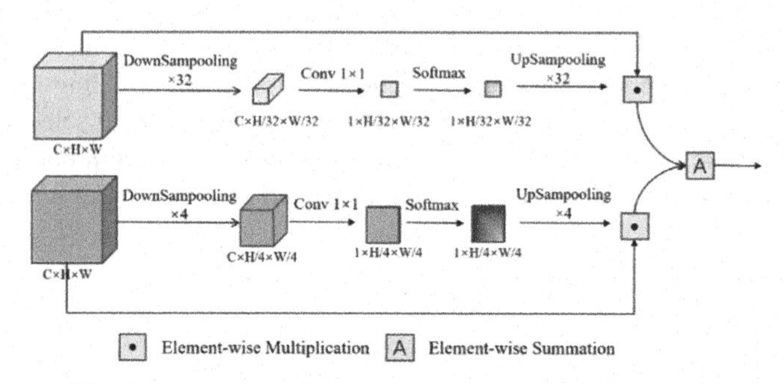

Fig. 5. Overview of the multi-scale object guidance module

Specifically, feature information at different scales can be obtained by adaptive average pooling, convolution and softmax at two different scales. Then the features are scaled up to the original size by nearest neighbor interpolation. This expansion effectively extends the feature information to corresponding scales. And the feature information enhancement at different scales of object guidance is achieved by multiplying the extended weights with the original feature map. This multi-scale object guidance approach can help the network better handle the features of objects of different sizes in remote sensing images, thereby improving the recognition performance of object boundaries.

2.6 Loss Function

The loss function used in this article is the sum of cross entropy loss and soft dice loss as the loss function. This amalgamation serves to integrate pixel-level and category-level information, striking a balance between weights for various categories. Consequently, it enables a more comprehensive assessment of the model's performance. The loss function formulas are as follows.

$$CELoss = -\sum_{i=1}^{N} N(TP_i \cdot log(softmax(x_i))) \tag{1}$$

$$SoftDiceLoss = 1 - \frac{2 \cdot TP + \epsilon}{2 \cdot TP + FP + FN + \epsilon} \tag{2}$$

$$Loss = CELoss + SoftDiceloss \tag{3}$$

where TP denotes the number of true positive cases (i.e., the number of pixels in the given class that are correctly classified), FP denotes the number of false positive cases (i.e., the number of pixels in the given class that are incorrectly classified as other classes), FN denotes the number of false negative cases (i.e., the number of pixels in other classes that are incorrectly classified as the given class), N denotes the number of samples, x_i denotes the prediction results of the ith sample, and ϵ is a small constant to prevent the denominator from being zero.

3 Datasets and Experimental Implementation

3.1 Dataset Description

The ISPRS Vaihingen dataset and the ISPRS Potsdam dataset are high-resolution remote sensing image datasets commonly used in the field of computer vision for image segmentation tasks.

ISPRS Vaihingen Dataset

The ISPRS Vaihingen dataset consists of 33 high-resolution remote sensing images with a spatial resolution of 9 cm and an average size of 2496 × 2046 pixels per image. The dataset contains images in the red, green and near-infrared bands. There are six categories of labels in the dataset, namely impervious surface, building, low vegetation, tree, car, and clutter/background.

ISPRS Potsdam Dataset

The ISPRS Potsdam dataset consists of 38 images with a spatial resolution of 5 cm and each image has a size of 6000 × 6000 pixels. The dataset is labeled with 6 categories of labels as in the Vaihingen dataset. In our experiments, we only used three channels of the Potsdam dataset, i.e., red, green and blue bands (RGB).

3.2 Implementation Details

MCNet was implemented by the PyTorch framework and trained on a single NVIDIA A5000 GPU with 24 GB RAM, with the batch size set to 8. The size of the training samples was 512 × 512 pixels. We used Adam as the gradient descent optimization algorithm and set the initial learning rate to 0.001 on all datasets with a maximum number of iterations of 150 epochs.

3.3 Evaluation Indicators

We evaluate the performance of the method by three metrics: overall accuracy (OA), mean intersection over union (MIoU), and F1-score. The calculation formulas are as follows.

$$OA = \frac{TP_i + TN_i}{TP_i + TN_i + FP_i + FN_i} \tag{4}$$

$$MIoU = \frac{1}{|C|}\sum\nolimits_{i \in C} \frac{TP_i}{TP_i + FP_i + FN_i} \tag{5}$$

$$F1 = \frac{2TP_i}{2TP_i + FN_i + FP_i} \tag{6}$$

where *TP* denotes the number of true positive cases (i.e., the number of pixels in the given class that are correctly classified), *TN* denotes the number of true negative cases (i.e., the number of pixels in other classes that are correctly classified), *FP* denotes the number of false positive cases (i.e., the number of pixels in the given class that are incorrectly classified as other classes), *FN* denotes the number of false negative cases (i.e., the number of pixels in other classes that are incorrectly classified as the given class), *C* denotes the set of categories, and *i* denotes the category.

4 Experimental Results and Analysis

4.1 Results

To verify the effectiveness of our proposed MCNet, we conducted experiments on the Vaihingen and Potsdam datasets of ISPRS, respectively, and showed the visualization of the segmentation results with several other benchmark models, as Fig. 6 and Fig. 7 show. It can be seen that the high accuracy of segmentation and the smooth and continuous boundaries of each category, which demonstrate the excellent performance of MCNet in the semantic segmentation of remote sensing images.

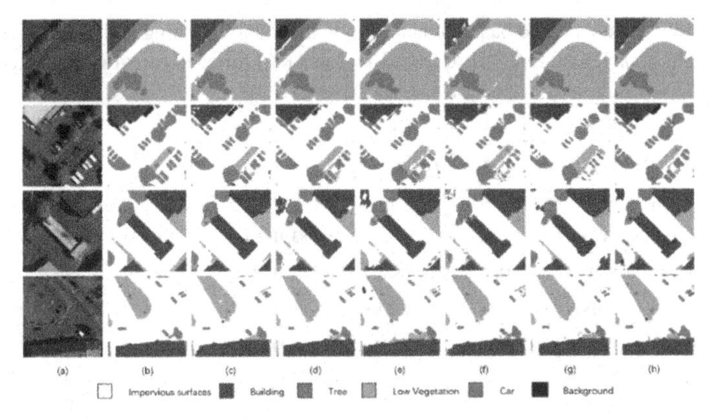

Fig. 6. The visual segmentation results on the Vaihingen dataset. (a) Image (b) Ground Truth (c) UNet (d) UNet++ (e) DeepLab v3 (f) SegNet (g) PSPNet (h) MCNet

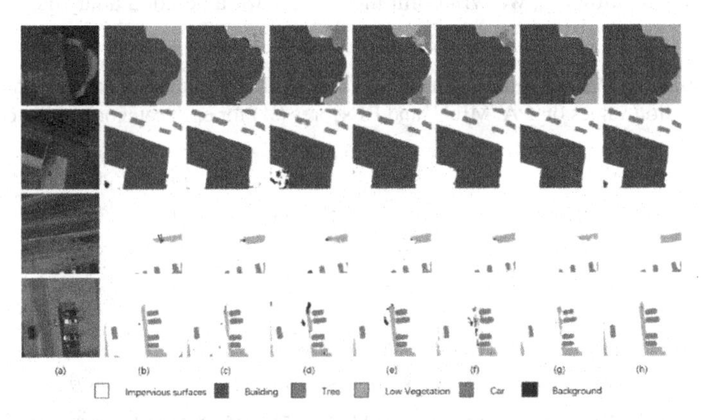

Fig. 7. The visual segmentation results on the Potsdam dataset. (a) Image (b) Ground Truth (c) UNet (d) UNet++ (e) DeepLab v3 (f) SegNet (g) PSPNet (h) MCNet

4.2 Analysis

As Fig. 6 shown, we visualize the segmentation results of several other benchmark models on the Vaihingen dataset as well, and compare them with the proposed MCNet. In the following, we will combine the Table 1 for a detailed analysis.

It can be seen from Table 1 that the best performer on the Vaihingen dataset is surprisingly the relatively simple UNet and its variant UNet++. The advanced model DeepLabv3, on the other hand, performs better in comparison with SegNet and PSPNet. But neither of these models can be compared with our proposed MCNet. Taking UNet, which is the most effective in the baseline networks on the Vaihingen dataset, as an example, MCNet has a large improvement in all metrics. Among them, OA average improved by 1.86%, MIoU average improved by 4.88%, and F1 scores of each category average improved by 12.84%, 2.89%, 0.9%, 3.07%, 2.09%, and 2.07%, respectively.

Table 1. Results of MCNet and baseline network on the ISPRS Vaihingen dataset. The mean and standard deviation values on OA, MIoU, and F1 scores (%) for each category are listed.

Methods	OA	MIoU	Background	Car	Tree	Low.veg	Building	Imp.surf
UNet	85.05 ± 0.08	67.01 ± 0.16	61.44 ± 1.50	76.58 ± 0.18	85.40 ± 0.63	75.59 ± 0.10	90.98 ± 0.91	86.98 ± 0.04
UNet++	85.18 ± 0.10	66.60 ± 0.65	59.01 ± 3.35	75.85 ± 0.36	85.64 ± 0.22	75.84 ± 0.16	91.30 ± 0.14	86.83 ± 0.06
DeepLabv3	84.37 ± 0.10	64.22 ± 0.31	56.36 ± 1.79	69.63 ± 1.01	83.91 ± 0.35	75.00 ± 0.44	91.21 ± 0.16	86.73 ± 0.06
SegNet	83.64 ± 0.08	62.08 ± 0.85	54.27 ± 2.22	63.89 ± 2.19	83.98 ± 0.27	73.56 ± 0.07	90.39 ± 0.04	85.90 ± 0.03
PSPNet	83.57 ± 0.11	62.18 ± 0.19	53.58 ± 1.16	62.78 ± 0.96	83.78 ± 0.53	76.78 ± 0.34	89.71 ± 0.09	85.74 ± 0.08
MCNet	**86.91 ± 0.06**	**71.89 ± 0.25**	**74.28 ± 0.96**	**79.47 ± 0.55**	**86.30 ± 0.21**	**78.66 ± 0.15**	**93.07 ± 0.27**	**89.05 ± 0.14**

In addition to the Vaihingen dataset, we also compared the proposed MCNet with other baseline networks on the Potsdam dataset. And the visualization results are shown in Fig. 7. In the following, we will combine Table 2 for a detailed analysis.

Table 2. Results of MCNet and baseline network on the ISPRS Potsdam dataset. The mean and standard deviation values on OA, MIoU, and F1 scores (%) for each category are listed.

Methods	OA	MIoU	Background	Car	Tree	Low.veg	Building	Imp.surf
UNet	85.08 ± 0.10	69.52 ± 0.24	55.17 ± 1.15	87.58 ± 0.22	81.65 ± 0.58	80.41 ± 0.19	92.89 ± 0.54	87.79 ± 0.16
UNet++	84.15 ± 0.10	68.73 ± 0.48	52.93 ± 2.11	88.63 ± 0.28	81.31 ± 0.15	79.18 ± 0.18	92.48 ± 0.29	86.86 ± 0.18
DeepLab v3	85.74 ± 0.09	70.82 ± 0.27	61.95 ± 1.57	86.58 ± 0.94	80.71 ± 0.37	81.08 ± 0.63	93.90 ± 0.08	88.53 ± 0.22
SegNet	84.34 ± 0.16	68.09 ± 0.81	51.19 ± 1.45	86.68 ± 1.39	80.44 ± 0.25	79.77 ± 0.12	92.78 ± 0.26	87.32 ± 0.16
PSPNet	83.47 ± 0.14	66.36 ± 0.27	57.18 ± 0.96	81.21 ± 0.78	76.83 ± 0.47	79.86 ± 0.16	91.77 ± 0.15	86.43 ± 0.37
MCNet	**87.82 ± 0.12**	**74.60 ± 0.34**	**66.69 ± 0.58**	**89.64 ± 0.37**	**83.72 ± 0.15**	**83.68 ± 0.17**	**95.05 ± 0.28**	**90.02 ± 0.20**

As can be seen from Table 2, the best performing baseline network on the Potsdam dataset changes to DeepLab v3, which performs better on the background, building, and impermeable surface categories. Our proposed MCNet outperformed all the networks in the baseline network by a significant margin. Taking the best performing DeepLab v3 as an example, MCNet showed a significant improvement in all metrics on the Potsdam dataset. Among them, OA average improved by 2.08%, MIoU average improved by 3.78%, and F1 scores in each category average improved by 4.74%, 3.06%, 3.01%, 2.60%, 1.15%, and 1.49%, respectively.

The overall performance of each network on the Vaihingen and Potsdam datasets is analyzed. For these two datasets, the most important point is the processing of the shadow part, which is closer in color, has an overall dark appearance, and may contain objects such as car, thus greatly increasing the difficulty of segmentation. Secondly, due to the limitation of equipment, the image is cropped to 512 × 512 pixels, and the smaller image leads to the more limited contextual information obtained from the edge part, which makes it difficult to segment accurately. The play of each model in this case is not quite the same. On the Vaihingen dataset, UNet and UNet++ perform better, which we believe is mainly due to their ability to retain contextual information and a certain degree of feature fusion. Because of this, there will be some noise in the features, resulting in some obvious category judgment errors in UNet. The main peculiarity of DeepLab v3 is the use of atrous convolution and atrous spatial pyramid pooling, which allows it to have a large perceptual field and better identify building and impervious surface, but because the expanded perceptual field is mainly at the end of the network, it is not so good for small objects such as cars. SegNet does not perform as well due to its simpler network structure and unlike U-Net, which has a feature fusion mechanism. PSPNet relies mainly on its spatial pyramid pooling and only has more contextual information than SegNet, so it performs better in low vegetation where contextual information is needed, but still performs poorly overall. As for the Potsdam dataset, it contains more details because its resolution is 5cm, which is higher than that of the Vaihingen dataset. In this case, DeepLab v3 achieves better performance with its larger perceptual field, and the situation for the other networks is similar to that in the Vaihingen dataset. Regardless of the dataset, MCNet achieves the best performance among all the models by ensuring the stability of feature fusion and the adequacy of decoding with an expanded perceptual field, proving the superiority of the method.

4.3 Ablation Experiments

In order to further verify the role of each module and structure, we designed comprehensive ablation experiments. The specific results are shown in Table 3 and Table 4. The visualization results are shown in Fig. 8 and Fig. 9.

Table 3. The results of MCNet ablation experiment on the Vaihingen dataset. The mean and standard deviation values on OA, MIoU, and F1 scores (%) are listed.

Methods	OA	MIoU	Background	Car	Tree	Low.veg.	Building	Imp.surf.
No CAM	86.68 ± 0.03	70.31 ± 0.86	67.23 ± 3.65	79.36 ± 0.04	85.97 ± 0.20	77.94 ± 0.38	92.68 ± 0.24	88.82 ± 0.14
No CFSM	86.55 ± 0.05	71.57 ± 0.91	**74.34 ± 2.02**	79.01 ± 0.92	85.90 ± 0.14	77.73 ± 0.39	92.90 ± 0.16	88.52 ± 0.12
No MOGM	86.67 ± 0.05	70.81 ± 0.27	69.46 ± 0.44	**79.52 ± 0.65**	85.62 ± 0.02	78.49 ± 0.17	92.68 ± 0.16	88.82 ± 0.14
No Cascade	86.37 ± 0.10	70.18 ± 0.21	70.25 ± 1.64	77.38 ± 1.11	85.71 ± 0.51	77.56 ± 0.11	92.54 ± 0.28	88.53 ± 0.21
MCNet	**86.91 ± 0.06**	**71.89 ± 0.25**	74.28 ± 0.96	79.47 ± 0.55	**86.30 ± 0.21**	**78.66 ± 0.15**	**93.07 ± 0.27**	**89.05 ± 0.14**

It can be seen that the performance of the model is degraded regardless of which module or decoder part of the cascade is removed. While without the CFSM, the network achieved optimal outcomes for the background and impervious surface categories in the Vaihingen and Potsdam datasets respectively. Similarly, upon removing the MOGM, the network achieved superior results for the car category across both datasets. However, it's evident that the most robust MCNet has the best performance.

Table 4. The results of MCNet ablation experiment on the Potsdam dataset. The mean and standard deviation values on OA, MIoU, and F1 scores (%) are listed.

Methods	OA	MIoU	Background	Car	Tree	Low.veg.	Building	Imp.surf.
No CAM	87.46 ± 0.08	74.18 ± 0.57	66.28 ± 1.25	90.04 ± 0.08	82.48 ± 0.24	83.54 ± 0.76	94.72 ± 0.34	89.97 ± 0.20
No CFSM	87.35 ± 0.10	73.61 ± 0.51	64.28 ± 1.02	89.08 ± 0.51	82.96 ± 0.65	83.38 ± 0.34	94.51 ± 0.37	**90.15 ± 0.19**
No MOGM	87.17 ± 0.09	73.75 ± 0.16	64.42 ± 0.16	**90.09 ± 0.50**	82.64 ± 0.22	82.95 ± 0.31	95.01 ± 0.24	89.66 ± 0.19
No Cascade	87.19 ± 0.14	73.60 ± 0.16	64.73 ± 0.94	89.36 ± 0.78	83.02 ± 0.75	83.51 ± 0.20	94.33 ± 0.19	89.54 ± 0.24
MCNet	**87.82 ± 0.12**	**74.60 ± 0.34**	**66.69 ± 0.58**	89.64 ± 0.37	**83.72 ± 0.15**	**83.68 ± 0.17**	**95.05 ± 0.28**	90.02 ± 0.20

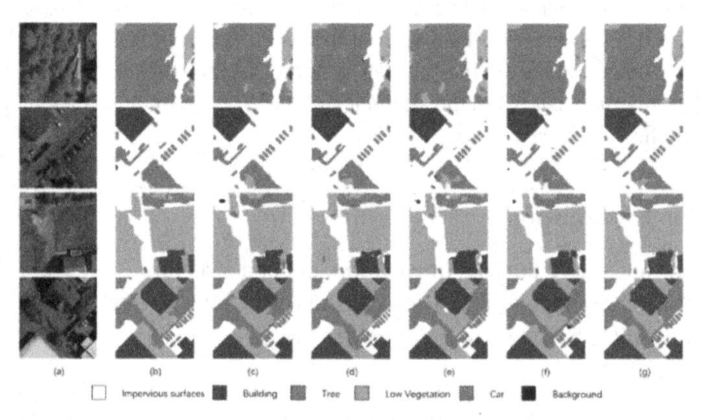

Fig. 8. The visual segmentation results on the Vaihingen dataset with ablation experiment. (a) Image (b) Ground Truth (c) no CAM (d) no CFSM (e) no MOGM (f) no Cascade (g) MCNet

Figure 8 and Fig. 9 show the results obtained from the ablation experiments of MCNet visualized on the Vaihingen and Potsdam datasets, respectively. Observing the two figures, we can find that after the removal of the CAM, there are more errors in the judgment of classes because of the equal treatment of channel dimensions. And after removing the CFSM module, the performance on the class boundary is slightly reduced because the features of different layers are not fully integrated. After removing the MOGM module, there are many small errors in segmentation due to the absence of object size guidance. After removing the cascade structure of the decoder, the features are not fully utilized due to the reduce of information flow, leading the increase of category misclassification. Obviously, MCNet works the best, both for category boundaries and category judgments, with no significant problems. This also shows that each module in MCNet is important and plays a role.

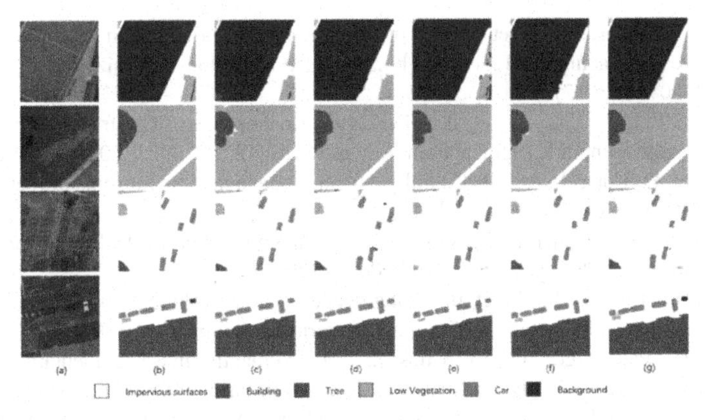

Fig. 9. The visual segmentation results on the Potsdam dataset with ablation experiment. (a) Image (b) Ground Truth (c) no CAM (d) no CFSM (e) no MOGM (f) no Cascade (g) MCNet

5 Conclusion

In this article, we proposed a multi-scale and cascade semantic segmentation network for remote sensing images. Specifically, we designed MFEM to capture multi-scale feature information. The CAM is used to capture important channel information more efficiently and fully. The CFSM is designed to better fuse the feature information of the encoding and decoding layers. The MOGM is designed to target the capture of feature information at tow scales. In addition, the comparative experiments and the ablation experiments on ISPRS Vaihingen and Potsdam datasets showed the superiority of our MCNet in semantic segmentation. Nevertheless, our proposed MCNet still has some shortcomings in the semantic segmentation of remote sensing images. For example, we didn't use multi-source data. In the subsequent study, we will try to combine the multi-source dataset with MCNet and make improvements to this method.

Acknowledgments. This work was supported by Natural Science Foundation of China (No. U21A2013 and 42071430), Opening Fund of Key Laboratory of Geological Survey and Evaluation of Ministry of Education (Grant Number: GLAB2020ZR14 and CUG2022ZR02) and College Students' Independent Innovation Funding Program Launch Project (No. S202310491229 and S202310491175).

Computation of this work was performed by the High-performance GPU Server (TX321203) Computing Centre of the National Education Field Equipment Renewal and Renovation Loan Financial Subsidy Project of China University of Geosciences, Wuhan.

References

1. Wang, M., Dong, Z., Cheng, Y., et al.: Optimal segmentation of high-resolution remote sensing image by combining superpixels with the minimum spanning tree. IEEE Trans. Geosci. Remote Sens. **56**(1), 228–238 (2017)
2. Chen, S., Sun, T., Yang, F., et al.: An improved optimum-path forest clustering algorithm for remote sensing image segmentation. Comput. Geosci. **112**, 38–46 (2018)

3. Wang, M., Wan, Y., Ye, Z., et al.: Remote sensing image classification based on the optimal support vector machine and modified binary coded ant colony optimization algorithm. Inf. Sci. **402**, 50–68 (2017)
4. Chen, G., Tan, X., Guo, B., et al.: SDFCNv2: An improved FCN framework for remote sensing images semantic segmentation. Remote Sens. **13**(23), 4902 (2021)
5. Shelhamer, E., Long, J., Darrell, T.: Fully convolutional networks for semantic segmentation. arXiv preprint arXiv:1605.06211 (2016)
6. Ronneberger, O., Fischer, P., Brox, T.: U-Net: convolutional networks for biomedical image segmentation. arXiv preprint arXiv:1505.04597 (2015)
7. Szegedy, C., Liu, W., Jia, Y., et al.: Going deeper with convolutions. In: Proceedings of the IEEE Conference on Computer Vision and Pattern Recognition, pp. 1–9 (2015)
8. Chen, L.C., Papandreou, G., Kokkinos, I., et al.: Semantic image segmentation with deep convolutional nets and fully connected CRFs. arXiv preprint arXiv:1412.7062 (2014)
9. Chen, L.C., Papandreou, G., Kokkinos, I., et al.: DeepLab: semantic image segmentation with deep convolutional nets, atrous convolution, and fully connected CRFs. IEEE Trans. Pattern Anal. Mach. Intell. **40**(4), 834–848 (2017)
10. Chen, L.C., Papandreou, G., Schroff, F., et al.: Rethinking atrous convolution for semantic image segmentation. arXiv preprint arXiv:1706.05587 (2017)
11. Hu, J., Shen, L., Sun, G.: Squeeze-and-excitation networks. In: Proceedings of the IEEE Conference on Computer Vision and Pattern Recognition, pp. 7132–7141 (2018)
12. Wang, X., Girshick, R., Gupta, A., et al.: Non-local neural networks. In: Proceedings of the IEEE Conference on Computer Vision and Pattern Recognition, pp. 7794–7803 (2018)
13. Li, S., Xue, L., Feng, L., et al.: Object detection network pruning with multi-task information fusion. World Wide Web **25**(4), 1667–1683 (2022)
14. He, K., Zhang, X., Ren, S., et al.: Deep residual learning for image recognition. In: Proceedings of the IEEE Conference on Computer Vision and Pattern Recognition, pp. 770–778 (2016)
15. Zhou, Q., Yang, W., Gao, G., et al.: Multi-scale deep context convolutional neural networks for semantic segmentation. World Wide Web **22**, 555–570 (2019)
16. Zhou, Z., Zhou, Y., Wang, D., et al.: Self-attention feature fusion network for semantic segmentation. Neurocomputing **453**, 50–59 (2021)
17. Zhao, Q., Liu, J., Li, Y., et al.: Semantic segmentation with attention mechanism for remote sensing images. IEEE Trans. Geosci. Remote Sens. **60**, 1–13 (2021)
18. Li, F., Wang, X., Sun, Y., et al.: Transfer learning based cascaded deep learning network and mask recognition for COVID-19. World Wide Web, pp. 1–16 (2023)
19. Liu, R., Mi, L., Chen, Z.: AFNet: adaptive fusion network for remote sensing image semantic segmentation. IEEE Trans. Geosci. Remote Sens. **59**(9), 7871–7886 (2020)
20. Chen, X., Li, Z., Jiang, J., et al.: Adaptive effective receptive field convolution for semantic segmentation of VHR remote sensing images. IEEE Trans. Geosci. Remote Sens. **59**(4), 3532–3546 (2020)
21. Yu, F., Koltun, V.: Multi-scale context aggregation by dilated convolutions. arXiv preprint arXiv:1511.07122 (2015)
22. Wang, P., Chen, P., Yuan, Y., et al.: Understanding convolution for semantic segmentation. arXiv preprint arXiv:1702.08502 (2017)
23. Wang, Q., Wu, B., Zhu, P., et al.: ECA-net: efficient channel attention for deep convolutional neural networks. arXiv preprint arXiv:1910.03151 (2019)
24. Xiang, S., Xie, Q., Wang, M.: Semantic segmentation for remote sensing images based on adaptive feature selection network. IEEE Geosci. Remote Sens. Lett. **19**, 1–5 (2021)

WikiCPRL: A Weakly Supervised Approach for Wikipedia Concept Prerequisite Relation Learning

Kui Xiao, Kun Li, Yan Zhang, Xiang Chen, and Yuanyuan Lou(✉)

School of Computer Science and Information Engineering, Hubei University, Wuhan, China
{xiaokui,zhangyan}@hubu.edu.cn,
{202121116013014,202221116012779,202021116012212}@stu.hubu.edu.cn

Abstract. Concept prerequisite relations determine the order in which knowledge concept is learned. This kind of concept relations has been used in a variety of educational applications, such as curriculum planning, learning resource sequencing, and reading list generation. Manually annotating prerequisite relations is time-consuming. Besides, data annotated by multiple people is often inconsistent. These factors have led to significant limitations in the use of supervised concept prerequisite learning methods. In this paper, we propose a weakly supervised Wikipedia Concept Prerequisite Relations Learning approach, called WikiCPRL, to identify prerequisite relations between Wikipedia concepts. First of all, we take the title of each Wikipedia article in a domain as a concept, and employ the RefD algorithm to generate weak labels for all the concept pairs, and then build a concept map for the domain. Secondly, a graph attention layer is defined to fuse the context information of each concept in the concept map so as to update their feature representations. Finally, we use the VGAE model to reconstruct the concept map, and then obtain the concept prerequisite graph. Extensive experiments on both English and Chinese datasets demonstrate that the proposed approach can achieve the same performance as several existing supervised learning methods.

Keywords: concept prerequisite relations · graph attentional layer · variational graph auto-encoders · weakly supervised learning · Wikipedia

1 Introduction

The vigorous development of online education has led to a blowout growth in the number of learning resources on the Internet. However, the rapid growth of the number of learning resources has also made another problem increasingly prominent, i.e., the sequence of learning resources is not clear, and learners cannot accurately choose the resources that meet their needs. The sequence of the resources is determined by the dependencies between the learning resources.

X. Song et al. (Eds.): APWeb-WAIM 2023, LNCS 14332, pp. 177–192, 2024.
https://doi.org/10.1007/978-981-97-2390-4_13

Nowadays, the dependencies between learning resources provided by different users or between learning resources of different courses are always unclear. A learner usually does not know what basic knowledge must be mastered before learning a new resource, nor what follow-up knowledge can be learned after learning a new resource. The unclear dependencies between learning resources make it impossible for learners to choose appropriate learning materials according to their individual needs.

On the other hand, each learning resource usually addresses one or more key concepts. In fact, the dependencies between learning resources are often determined by the prerequisite relations between their key concepts. The prerequisite relation between two concepts determines the order in which the two concepts are learned. Given a pair of concepts (A, B), if people must understand the meaning of concept A before learning concept B, then it can be said that concept A is a prerequisite concept of concept B. The prerequisite relations between key concepts directly determine the dependencies between the corresponding learning resources, and also indirectly affect the order of the learning sequence. Moreover, concept prerequisite relation has also played an important role in many educational applications, such as learning resource sorting [1], curriculum planning [2], reading list generation [3], knowledge tracking [4], etc.

In this article, we propose WikiCPRL, a weakly supervised approach for Wikipedia concept prerequisite relation learning. First of all, we take the title of each Wikipedia article as a concept, and then use the well-known algorithm RefD [9] to acquire weak labels of concept pairs. Secondly, we build a concept map using the concept relations in a domain. Thirdly, we use the word embedding methods to generate the representation vector of each node in the concept map, and through the graph attentional layer, we fuse the information of neighbor nodes to update the representation vector of each node. After that, the VGAE model is utilized to reconstruct the concept map to learn the real concept prerequisite relations graph. Finally, to solve the problem that edges in the reconstructed graph have no direction, we employ the RefD algorithm again to determine the direction of the concept relations.

To the best of our knowledge, this is the first weakly supervised learning method that does not rely on the support of learning resources to identify the Wikipedia concept prerequisite relations. The main contributions of our work can be briefly summarized as follows:

- We propose a weakly supervised learning approach WikiCPRL to identify Wikipedia concept prerequisite relations. The proposed approach does not need to use labeled data during the model training process.
- By using the RefD algorithm, we solve the problem that the VGAE model ignores the direction of the edges in the reconstructed graph. And the generated directed edges of concept pairs can be well interpretable.
- We created a Chinese dataset and verified the effectiveness of the proposed method on both English and Chinese datasets.

The rest of this paper is organized as follows: Sect. 2 presents an overview of related works. In Sect. 3, we define the problem of concept prerequisite relation learning. Section 4 introduces the details of our proposed approach WikiCPRL. Section 5 presents the experimental results of the proposed approach. Finally, Sect. 6 summarizes our work with a brief discussion of future work.

2 Related Work

In recent years, there have been a lot of studies on concept prerequisite relation learning. In these studies, most of them use supervised learning methods. Talukdar and Cohen presented an early attempt to model the prerequisite structure of concepts in Wikipedia [8]. For a pair of concepts, the authors used Hyperlinks, Edits, and PageContent to define the concept pair features, and then employed the MaxEnt classifier to infer prerequisite relations among concepts. Liang et al. [9] proposed a hyperlink-based metric for inferring prerequisite relations between concepts in Wikipedia. They calculate the *RefD*(Reference Distance) of a concept pair and predict whether there is a prerequisite relation between the two concepts. Pan et al. [6] conducted an investigation on automatically inferring prerequisite relations among concepts in MOOCs, they proposed various concept pair features from different aspects too, including contextual, structural and semantic features. Sayyadiharikandeh et al. [10] also introduced a supervised learning approach and trained a classifier to predict whether concept A is a prerequisite of concept B based on features computed from Wikipedia clickstream data. In their opinion, navigation in Wikipedia tends to flow from a concept to its prerequisites. Miaschi et al. [13] presented a method based on deep learning applied to the task of automatic prerequisite relations identification between concepts. This is the first method to exclusively exploit the linguistic features of concepts extracted from the content of Wikipedia articles without relying on any structured information.

There are also some semi-supervised learning methods for discovering concept prerequisite relations. Roy et al. [14] presented a prerequisite relation learning method PREREQ, which was designed using latent representations of concepts obtained from the Pairwise Latent Dirichlet Allocation model, and inferred prerequisite relations with a Siamese network. Zhang et al. [12] proposed an end-to-end framework graph network-based method MHAVGAE, which utilizes rich features information of the resource-concept graph to learn the prerequisite relation between concepts. Besides, there are also some other semi-supervised learning methods, such as [11,15].

In comparison, the weakly supervised and unsupervised learning methods are much less. Li et al. [17] presented a graph neural model for extracting concept prerequisite relations from learning resources using deep representations as input. In addition, Zhang et al. [18] also proposed a weakly supervised learning approach named wMHAVGAE for discovering concept prerequisite relations. They used the metric RPRD to generate inaccurate concept prerequisite relations and extended MHAVGAE [12] with the weakly supervised setting for discovering concept prerequisite relations. Additionally, there are also several other weakly supervised and unsupervised learning methods, such as [5].

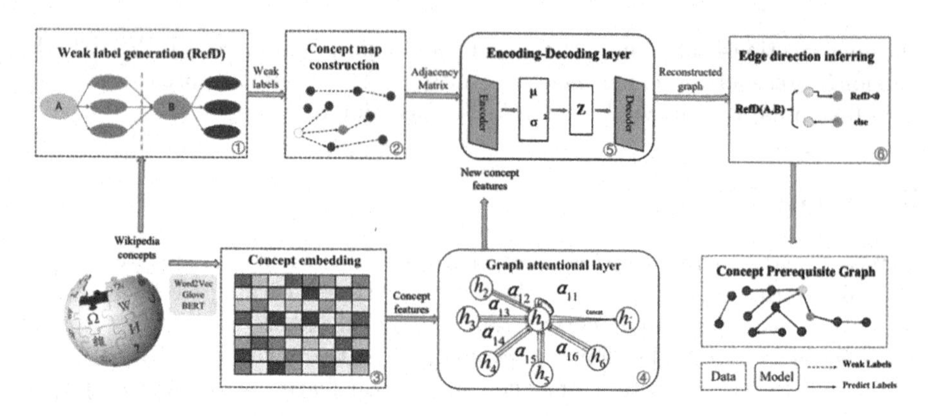

Fig. 1. The framework of WikiCPRL

3 Problem Formulation

For a pair of concepts A and B, if people must understand the meaning of concept A before learning concept B, it means that B depends on A, and A is a prerequisite of B, denoted as $A \rightarrow B$. Actually, the prerequisite relation between the two concepts also represents their dependency. In this paper, we treat concept prerequisite relation learning as a binary classification task. The concept prerequisite relation can be formally defined as:

$$Prereq(A, B) = \begin{cases} 1, & \text{A is a prerequisite of B} \\ 0, & \text{else} \end{cases} \tag{1}$$

In our work, we only study the prerequisite relations between Wikipedia concepts. The title of an article is viewed as a concept. Let $V = \{v_1, v_2, \ldots, v_n\}$ be the concept space of a domain, the set of all concepts of interest. The concept space is assumed to be fixed and known in advance. A concept may be a single word or a phrase, and it represents a Wikipedia article. Our goal is to learn the prerequisite relations between the concepts. Our weakly supervised setting is to exclude any prerequisite relations during training, and we wish to predict the concept prerequisite relations via the weak labels and concept embedding representations.

In WikiCPRL, we input weakly labeled concept maps and embedding vectors of concepts into the model, and finally get reconstructed concept maps to infer the real prerequisite relations between concepts.

4 Proposed Approach

4.1 Overview of WikiCPRL

In this section, we introduce the framework of WikiCPRL, as shown in Fig. 1. First of all, we take the title of every Wikipedia article as a concept, and then use the RefD algorithm to generate weak labels of concept pairs. These weak labels are coarse-grained labels, which represent incomplete, inexact and inaccurate concept prerequisite relations. Secondly, through these weak labels, we can gather the concepts in the same domain and build a concept map for this domain. Thirdly, word embedding algorithms are employed to generate the initial representation vector of each concept. Fourthly, the representation vector of each node in the concept map is updated in the graph attentional layer. Fifthly, both the concept features and the adjacency matrix are input into the Encoding-Decoding layer to generate the reconstructed graph. Finally, for the undirected edges in the reconstructed graph, we again use the RefD algorithm to help infer the direction of these edges, and then obtain a directed graph, which contains all the concept prerequisite relations in the domain.

4.2 Weak Label Generation

How to obtain the weak label of a concept pair is the first challenge we need to solve. Liang et al. [9] proposed a RefD algorithm, which can help us solve this problem. Initially, the RefD algorithm was directly used as a concept prerequisite relation identification algorithm. Now, multiple studies use RefD value as a feature of concept pairs in the prerequisite relation learning tasks. The main reason is that the semantic relationship between the concepts represented by the RefD value can be used to reasonably explain why one concept may be a prerequisite for another concept. Therefore, we employ the RefD algorithm to generate weak labels for concept pairs.

The idea of the RefD algorithm is that every concept can be represented by its related concept set in the concept space. Given a pair of concepts (A, B), if many related concepts of B refer to concept A, but few related concepts of A refer to concept B, then A is likely to be a prerequisite of B. The variable is computed as

$$RefD(A,B) = \frac{\sum_{i=1}^{k} r(c_i, B) \cdot w(c_i, A)}{\sum_{i=1}^{k} w(c_i, A)} - \frac{\sum_{i=1}^{k} r(c_i, A) \cdot w(c_i, B)}{\sum_{i=1}^{k} w(c_i, B)} \tag{2}$$

where $w(c_i, A)$ weights the importance of c_i to A. In fact,c_i is a related concept of A. If not, then the value of $w(c_i, A)$ is 0. $r(c_i, B)$ indicates whether c_i refers to B. To capture all the possible prerequisite relations between a concept pair, RefD should satisfy the following constraints:

$$RefD(A,B) = \begin{cases} (\theta, 1], & \text{if B is a prerequisite of A} \\ [-\theta, \theta], & \text{if no prerequisite relation} \\ [-1, -\theta), & \text{if A is a prerequisite of B} \end{cases} \tag{3}$$

where θ is a positive threshold, and $\theta \in [0, 1)$. When $\theta < RefD(A, B) \leq 1$, it means that many related concepts of A refer to B, but fewer related concepts of

B refer to A. The article of concept B may contain the background knowledge required for learning concept A, so B is likely a prerequisite of A. Similarly, when $-1 \leq RefD(A, B) < -\theta$, then A is likely a prerequisite of B. When $-\theta \leq RefD(A, B) \leq \theta$, it suggests that A and B refer to each other very little, or the number of mutual references between them is equal and offset each other. Then, there is no prerequisite relation between the two concepts. The authors tried different θ values to obtain the best performance of prerequisite relation identification.

In this paper, we no longer choose the value of the parameter θ according to the performance of prerequisite relation identification, but choose the value of the parameter according to the number of relations. On the one hand, we hope that the number of edges in the concept map constructed by weak labels can be as many as possible. Because WikiCPRL employs a VGAE model to reconstruct the input concept map, so as to get the real concept prerequisite graph. VGAE uses an encoder and a decoder to keep the output graph and the input graph as similar as possible. When the number of edges in the input graph is large, the number of edges in the output graph will also be large, which is beneficial for improving the recall of prerequisite relation identification. On the other hand, comparing with ground truth labels repeatedly to select the optimal θ is essentially a supervised learning process. But our work should not be a supervised learning method. We choose the appropriate θ value just to maximize the number of weak concept relations.

Let $A \in \mathbb{R}^{N \times N}$ be the adjacency matrix of the concept map. The edges between vertices in the concept map stand for the weak labels of concept pairs. They are incomplete, inexact and inaccurate concept prerequisite relations. WikiCPRL will use these weak concept relations to learn the real concept prerequisite relations.

4.3 Concept Feature Acquisition

In our work, V is defined as the set of vertices in the concept map of a domain. Every vertex v_i denotes a concept, which is usually a single word or a phrase. Let $\boldsymbol{V} = [\boldsymbol{v_1}, \boldsymbol{v_2}, \ldots, \boldsymbol{v_N}] \in \mathbb{R}^{F \times N}$ be the feature vectors of all the nodes in the concept map. $\boldsymbol{v_i} \in \mathbb{R}^F$ represents the feature vector of the vertex v_i. N is the number of concept vertices, and F is the number of features in a vertex feature vector.

In order to infer concept prerequisite relations, we need to input the feature vectors of all vertices $\boldsymbol{V}$ into the Encoding-Decoding layer of WikiCPRL. There are many ways to generate embedding representations of concepts, such as word2vec, phrase2vec, GloVe, BERT, etc. We need to choose a most appropriate technique for generating vertex embedding representations. In our paper, experiments show that the BERT based concept vectors performs best in the prerequisite relation learning task. Therefore, we finally chose the BERT model.

4.4 Graph Attentional Layer

There should not be only one embedding representation for each concept. A concept should have different embedding representations in different contexts to meet the requirements of various tasks. In our work, the context of each concept is the concept map in which it resides. We should analyze the relationship between the current concept vertex and other concept vertices in the concept map to discover the prerequisite concepts. However, due to the limitation of computing power, it is obviously impossible to process all the vertex pairs in the concept map. For each vertex, we have to use limited computing resources to analyze those vertices that are most closely related to it, not all vertices in the graph. We believe that the related vertices can play a more important role in finding prerequisite concepts. Therefore, we introduce the attention mechanism [19] to evaluate the importance of the related vertices to the current vertex. Inspired by [20], WikiCPRL uses a graph attentional layer to aggregate the meaningful neighbors of each vertex and update the representation of the concept vertex.

For the nodes in the concept map, we use self-attention to learn their attention coefficients. The input sequence to the graph attentional layer is $\boldsymbol{V}$, i.e. the feature vectors of all the concept vertices. The graph attentional layer produces a new set of feature vectors, $X = [x_1, \ldots, x_N] \in \mathbb{R}^{F' \times N}$, as its output.

First of all, in order to transform the input features into higher-level features, a shared linear transformation, parametrized by a weight matrix, $W \in \mathbb{R}^{F' \times F}$, is applied to every vertex in the concept map. After that, we perform self-attention on all concept vertices. Given a vertex pair (v_i, v_j), we can calculate the vertex attention coefficient e_{ij} for the pair. e_{ij} indicates the importance of the vertex v_i to the vertex v_j. The attention coefficient is calculated as

$$e_{ij} = \alpha(W\boldsymbol{v_i}, W\boldsymbol{v_j}) \tag{4}$$

where $a \in \mathbb{R}^{F' \times F'}$ represents the shared attentional mechanism. Generally speaking, the model allows every concept vertex to attend on any other vertices in the concept map. However, in terms of computing power and the importance of vertices, we only focus on the pairs of concept vertices that are linked to each other. In other words, we only compute e_{ij} for concept vertices $v_i \in N_i$, where N_i is the set of neighbors of the vertex v_i (including itself) in the concept map. To make attention coefficients easily comparable across different concept vertices, we normalize them across all choices of v_i using the softmax function:

$$\begin{aligned} \alpha_{ij} &= softmax(e_{ij}) \\ &= \frac{\exp(e_{ij})}{\sum_{k \in N_i} \exp(e_{ik})} \\ &= \frac{\exp(LeakyReLU(\boldsymbol{\alpha}^T[W\boldsymbol{v_i} \| W\boldsymbol{v_j}]))}{\sum_{k \in N_i} \exp(LeakyReLU(\boldsymbol{\alpha}^T[W\boldsymbol{v_i} \| W\boldsymbol{v_j}]))} \end{aligned} \tag{5}$$

here, the attention mechanism α in Eq. (4) is a single-layer feedforward neural network, parametrized by a weight vector $\boldsymbol{\alpha} \in \mathbb{R}^{2F'}$, and applying the

LeakyReLU as nonlinearity activation function. Besides, $\bullet^T$ denotes transposition, and $\|$ is the concatenation operation.

It should be pointed out that the attention coefficients e_{ij} are asymmetric. In other words, e_{ij} is not equal to e_{ji}. Because the set of neighbor nodes of v_i and v_j may be different. Accordingly, the importance of v_i to v_j is different from the importance of v_j to v_i. Likewise, the concept prerequisite relation is also asymmetric. If v_i is a prerequisite concept of v_j, then v_j cannot be a prerequisite concept of v_i.

After obtaining the normalized attention coefficients between vertex pairs, we then begin to calculate a linear combination of the feature vectors of vertices $v_j \in N_i$, to serve as the higher-level feature vectors for v_i:

$$h_i = \sigma(\sum_{v_j \in N_i} \alpha_{ij} W \boldsymbol{v_j}) \tag{6}$$

where h_i is the higher-level feature vector of v_i, σ is the sigmoid activation function.

In order to stabilize the learning process of self-attention, we extend attention mechanism to multi-head self-attention. We repeat the transformation of Eq.(6) for K times, and then concatenate their feature vectors. Then the output feature vectors are updated as follows:

$$h_i = \overset{K}{\underset{k=1}{\|}} \sigma(\sum_{v_j \in N_i} \alpha_{ij}^k W^k \boldsymbol{v_j}) \tag{7}$$

where $\|$ stands for the concatenation operation. α_{ij}^k are normalized attention coefficients computed by the k-th attention mechanism. W_k is the corresponding input linear transformation's weight matrix. Accordingly, the output feature vectors h_i will consist of $KF^{'}$ features for every concept vertex, rather than $F^{'}$. In order to reduce the dimension of node features from $KF^{'}$ to $F^{'}$, we use a fully connected neural network to obtain the final feature vector of a concept vertex.

$$x_i = ReLU(h_i W_r + b_r) \tag{8}$$

where $W_r \in \mathbb{R}^{KF^{'} \times F^{'}}$, $b_r \in \mathbb{R}^{F^{'}}$. All these parameters can be learned during the training process, and the calculation results are activated by the *ReLU* function. Finally, we can get the new feature vectors of all the concepts.

4.5 Encoding-Decoding Layer

In this section, we will introduce the Encoding-Decoding layer which utilizes vertex feature vectors X and the adjacency matrix A to generate a reconstructed concept graph and obtain candidate concept prerequisite relations. The Encoding-Decoding layer of WikiCPRL is implemented by a Variational Graph Auto-Encoders (VGAE) model [21]. The VGAE is a framework for unsupervised learning on graph-structured data based on variational auto-encoders, which can learn interpretable latent representations for undirected graph.

In the encoding stage, we take the vertex feature vectors X and the adjacency matrix A as input, and try to recover the graph adjacency matrix by the hidden layer embeddings Z. The encoder is composed of a two-layer GCN:

$$GCN(X, A) = \tilde{A}ReLU(AXW_0)W_1 \tag{9}$$

where $\tilde{A} = D^{\frac{1}{2}}AD^{\frac{1}{2}}$ is the new adjacency matrix at the second graph layer, and D is the degree matrix of the graph. W_0 and W_1 are the parameters of the first and second layers, respectively.

Then, in the variational graph auto-encoder, the goal is to sample the hidden layer embeddings Z via a normal distribution, that is

$$q(Z|X, A) = \prod_{i=1}^{N} q(z_i|X, A) \tag{10}$$

with

$$q(z_i|X, A) = N(z_i|\mu_i, diag(\sigma_i^2)) \tag{11}$$

here, $\mu = GCN_\mu(X, A)$ is the matrix of mean vectors, and $\log \sigma = GCN_\sigma(X, A)$.

In the decoding stage, the reconstructed adjacency matrix is the inner product of the latent parameters Z, that is

$$p(\hat{A}|Z) = \prod_{i=1}^{N}\prod_{j=1}^{N} p(A_{ij}|z_i, z_j) \tag{12}$$

with

$$p(A_{ij} = 1|z_i, z_j) = \sigma(z_i^T z_j) \tag{13}$$

here, A_{ij} are the elements of the reconstructed adjacency matrix $\hat{A}$. And $\sigma(\bullet)$ is the logistic sigmoid function.

We optimize the variational lower bound with respect to the variational parameters W_i:

$$L = E_{q(Z|X,A)}[\log p(A|Z)] - KL[q(Z|X, A)||p(Z)] \tag{14}$$

where $KL[q(\bullet)||p(\bullet)]$ is the Kullback-Leibler divergence between $q(\bullet)$ and $p(\bullet)$.

4.6 Edge Direction Inferring

The reconstructed graph generated by the Encoding-Decoding layer of WikiCPRL is an undirected graph. If two nodes in the reconstructed graph are connected by an edge, it indicates that there is a prerequisite relation between the concepts of the two nodes. But we do not know which one is the prerequisite concept, because the edge has no direction. In practice, the directions of edges is very important for a prerequisite relation. In applications such as learning resource sequencing and learning path generation, the direction of a prerequisite relation directly determines the order in which concepts are learned.

In this section, we will again use the RefD algorithm to infer the direction of all edges in the reconstructed graph. Given a concept pair (A, B), if the value of $RefD(A, B)$ is less than 0, it means that many related concepts of B refer to A, but fewer related concepts of A refer to B. A is likely a prerequisite concept of B. Similarly, if the value of $RefD(A, B)$ is greater than 0, then B is likely a prerequisite of A. If the value of $RefD(A, B)$ is equal to 0, it means that there is no mutual reference between A and B, or the mutual reference is completely equivalent. In this exceptional case, the edge will be discarded. In this way, we cannot only know which concept pairs have a prerequisite relation, but also well explain why there is a prerequisite relation between two concepts.

5 Performance Analysis

5.1 Datasets

In this paper, we verify the effectiveness of the proposed approach WikiCPRL on both English and Chinese datasets. The English dataset, AL-CPL, is an English concept prerequisite relation dataset proposed by Liang et al. [16]. This dataset contains prerequisite pairs in four domains, namely Data Mining, Geometry, Physics, and Precalculus. These English concepts are originally from several textbooks, created by Wang et al. [7], and each concept corresponds to a Wikipedia article. Based on their work, Liang et al. [16] further expanded the concept prerequisite pairs with transitivity and irreflexivity. Transitivity means that if (A, B) and (B, C) are positive samples, then (A, C) will also be added as a positive sample. And irreflexivity means that if (A, B) is a positive sample, then (B,A) can be added as a negative sample. Besides, they also manually corrected some of the original mislabels in the initial datasets and created the AL-CPL dataset.

Based on the AL-CPL dataset, we also created a Chinese dataset ZH-AL-CPL through the cross-language links of Wikipedia. We tried to find a corresponding Chinese concept for each English concept in the AL-CPL datasets from the Chinese Wikipedia. If an English concept does not have a corresponding Chinese concept, then the concept pair to which it belongs will be discarded when creating the Chinese datasets. However, we have not continued to extend the Chinese dataset using transitivity and irreflexivity. Because the number of articles in Chinese Wikipedia was originally only one-fifth of that of English Wikipedia, if the Chinese dataset continues to be expanded in this way, the Chinese dataset with a relatively small number of concepts will have more concept prerequisite pairs than the English data set. In our opinion, it is inappropriate to do so under the current conditions.

In addition, we use Precision, Recall, F1-score and AUC to evaluate the performance of the WikiCPRL model. For the settings of the WikiCPRL model, we use the weakly labels generated by RefD algorithm as its training dataset. However, for the prerequisite relation dataset between concepts that need to learn, we set the verification dataset and the test dataset to be 50%, respectively. On each dataset, the learning rate, the weight decay, and the number of attention heads are set to their respective optimal values (Table 1).

Table 1. Datasets statistics

Dataset	Alias	Domain	#Concepts	#Pairs	#Prerequisites
AL-CPL [16]	D1	Data Mining	120	826	292
	D2	Geometry	89	1681	524
	D3	Physics	153	1962	487
	D4	Precalculus	224	2060	699
ZH-AL-CPL	D5	Data Mining	89	558	201
	D6	Geometry	78	1391	449
	D7	Physics	133	1657	390
	D8	Precalculus	197	1758	590

5.2 Compare with Baselines

In order to verify the effectiveness of the proposed method, we use the following state-of-the-art approaches as baselines, and they all use 5-fold cross-validation. It must be noted that the baselines are all supervised learning methods. As mentioned above, there are still some semi-supervised, weakly supervised and unsupervised learning methods, but these methods need the support of learning resources when inferring concept relations. This is different from the settings of our work, so these methods cannot be used here. The baseline approaches are as follows.

(1) **Linguistically-Driven Strategy(M3)**: This work is presented by Miaschi et al. [13]. The authors defined the M3 model by merging embedding and hand-crafted features.

(2) **Active Learning (AT)**: This method is proposed by Liang et al. [16]. It uses 15 link-based and 17 text-based features to learn the prerequisite relations among Wikipedia concepts. We choose the results reported by the random forest classifier for comparison, which is also the best classifier in their paper.

(3) **Neural Network (NN)**: This approach is proposed by the UNIGE_SE team in the EVALITA 2020 shared task on Prerequisite Relation Learning (PRE-LEARN) [22]. The authors developed a neural network classifier that exploits features extracted both from raw text and the structure of the Wikipedia pages to capture prerequisite relations.

(4) **Reference Distance (RefD)**: This metric is proposed by Liang et al. [9]. It measures how differently two concepts refer to each other in order to capture prerequisite relations among Wikipedia concepts.

Generally speaking, the performance of a supervised learning method is better than that of a weakly supervised learning method, because the former model is trained by a large number of real sample data. However, training sample data is often lacking in real-world environments in many fields. The cost of manually labeling data is too high, and there are often inconsistencies in the labeling results of multiple annotators. Given these practical constraints, semi-supervised, weakly supervised, unsupervised learning methods are often the more appropriate choice.

Table 2. Comparison with baselines on the English and Chinese datasets

Methods	Metrics	AL-CPL				ZH-AL-CPL			
		D1	D2	D3	D4	D5	D6	D7	D8
M3 [13]	P	**0.819**	0.917	0.775	0.875	0.788	**0.894**	0.762	0.876
	R	0.815	0.876	0.718	0.865	0.758	0.883	0.705	0.844
	F1	**0.817**	**0.896**	**0.743**	0.870	0.772	**0.888**	0.732	0.859
	AUC	**0.953**	**0.979**	0.930	0.961	0.872	**0.963**	0.865	0.939
AT [16]	P	0.807	**0.950**	**0.852**	**0.902**	**0.874**	0.893	**0.873**	**0.912**
	R	0.733	0.847	0.593	0.871	**0.801**	0.875	0.812	**0.893**
	F1	0.767	0.895	0.699	**0.886**	**0.836**	0.884	**0.841**	**0.902**
	AUC	0.922	0.978	**0.939**	**0.975**	**0.903**	0.957	**0.895**	**0.967**
NN [22]	P	0.627	0.706	0.530	0.699	0.601	0.722	0.684	0.772
	R	0.682	0.760	0.623	0.769	0.517	0.724	0.538	0.725
	F1	0.651	0.730	0.571	0.731	0.543	0.722	0.602	0.746
	AUC	0.801	0.892	0.828	0.895	0.711	0.754	0.734	0.803
RefD [9]	P	0.517	0.424	0.499	0.751	0.509	0.614	0.582	0.676
	R	0.762	0.623	0.496	0.694	0.524	0.638	0.547	0.704
	F1	0.614	0.504	0.494	0.721	0.516	0.626	0.564	0.690
	AUC	0.695	0.624	0.677	0.792	0.563	0.743	0.634	0.722
Proposed	P	0.595	0.460	0.431	0.599	0.773	0.479	0.546	0.685
	R	**0.909**	**0.994**	**0.839**	**0.996**	0.697	**0.968**	**0.820**	0.870
	F1	0.685	0.630	0.570	0.746	0.733	0.641	0.656	0.766
	AUC	0.873	0.862	0.861	0.876	0.791	0.629	0.759	0.763

To be fair, we conduct experiments with each method for five times and then averaged them as results. Table 2 shows the comparison results with baselines on the English and Chinese datasets. In general, M3 and AT have the best performances. However, our method WikiCPRL outperforms baseline methods across all four English datasets in terms of recall. We suppose that this may be related to the strategy of selecting the parameter θ of the RefD algorithm. We chose this parameter based on how to ensure that the number of edges of the concept map in WikiCPRL is maximized. In this way, we can as few as possible miss the real concept prerequisite relations.

From Table 2, we also find that the F1-scores of our method outperform RefD on all the four English datasets. Besides, our method also outperforms NN on the Data Mining and Precalculus, which achieves +3.4% and +1.5% respectively with respect to F1. On the other hand, our WikiCPRL model outperforms NN by +1.4% respectively for the average AUC. Especially, the AUC of our method significantly outperforms RefD (+17.8% on Data Mining, +23.8% on Geometry, +18.4% on Physics, and +8.4% on Precalculus).

For the Chinese datasets, the best performing method is still AT, followed by M3. WikiCPRL outperforms other methods on Geometry and Physics in terms of recall. Moreover, the proposed method outperforms NN and RefD by +4.58% and +10% respectively with respect to the average F1. As for the average AUC, the proposed approach performs worse than other methods, except RefD. The average AUC of our approach outperforms RefD by 7%.

In general, the overall performance of WikiCPRL on the English and Chinese datasets is lower than that of AT and M3, but roughly on par with NN, and significantly better than RefD.

5.3 Case Study

In this section, we further studied the examples of concept relations recovered by our method WikiCPRL. These concept pairs are all from the dataset Geometry. Table 3 lists examples of both correct and incorrect examples. The "Concept map" column denotes whether there is an edge between two concepts in the initial concept map. The "Learned model" column represents directed edges generated by the learned model. The "True model" column stands for directed edges in true model. If there is a prerequisite relation between two concepts, the value is 1; otherwise, the value is 0.

Table 3. Examples of correct and incorrect examples

	Concept pairs	Concept map	Learned model	True model
Correct examples	*<Analytic geometry, Number>*	0	1	1
	<Conic section, Geometry>	0	1	1
	<Real line, Real number>	0	1	1
	<Multiplicative inverse, Arithmetic>	0	1	1
	<Extended real number line, Mathematics>	1	1	1
	<Line segment, Geometry>	1	1	1
	<Triangle, Base (exponentiation)>	1	1	1
	<Base (exponentiation), Multiplication>	1	1	1
Incorrect examples	*<Rotation of axes, Geometry>*	0	0	1
	<Injective function, Mathematics>	1	0	1
	<Focus (geometry), Conic section>	0	1	0

Looking closely at the correct examples, although there is no edge between the two concepts in the initial concept map, the WikiCPRL model can still correctly predict the prerequisite relations between concepts, for example (*Conic section, Geometry*). There are also some concept pairs with prerequisite relations, which also have edges in the original concept map. The proposed method can also correctly discover the prerequisite relations, for instance (*Base (exponentiation), Multiplication*). Besides, there are also some concept prerequisite relations that WikiCPRL did not correctly capture. These concepts are often advanced concepts rather than basic concepts, and there are very few prerequisite concepts related to them. It is often difficult to identify the prerequisite relations of such concepts.

6 Conclusion

In this paper, we investigate the problem of concept prerequisite relation learning, which is of great value for various educational applications. WikiCPRL does not require the support of annotated data, making it easy to extend to other domains for prerequisite relation identification. Extensive experiments on both English and Chinese datasets show that the proposed approach can achieve the same performance as several existing supervised learning methods. In the future, we will apply the RefD algorithm to both pagelink and clickstream data of Wikipedia to improve the accuracy of weak labels and better capture prerequisite relations.

References

1. Limongelli, C., Gasparetti, F., Sciarrone, F.: Wiki course builder: a system for retrieving and sequencing didactic materials from Wikipedia. In: Proceedings of the 2015 International Conference On Information Technology Based Higher Education And Training (ITHET), pp. 1–6 (2015). https://doi.org/10.1109/ITHET.2015.7218041
2. Yang, Y., Liu, H., Carbonell, J., Ma, W.: Concept graph learning from educational data. In: Proceedings of the Eighth ACM International Conference On Web Search And Data Mining, pp. 159–168 (2015). https://doi.org/10.1145/2684822.2685292
3. Gordon, J., Zhu, L., Galstyan, A., Natarajan, P., Burns, G.: Modeling concept dependencies in a scientific corpus. In: Proceedings of the 54th Annual Meeting of the Association For Computational Linguistics (Volume 1: Long Papers), pp. 866–875 (2016). https://doi.org/10.18653/v1/P16-1082
4. Wang, S., Liu, L.: Prerequisite concept maps extraction for automatic assessment. In: Proceedings of the 25th International Conference Companion On World Wide Web, pp. 519–521 (2016). https://doi.org/10.1145/2872518.2890463
5. Liang, C., Ye, J., Wu, Z., Pursel, B., Giles, C.: Recovering concept prerequisite relations from university course dependencies. In: Proceedings of the AAAI Conference on Artificial Intelligence, vol. 31, no. 1 (2017). https://doi.org/10.1609/aaai.v31i1.10550

6. Pan, L., Li, C., Li, J., Tang, J.: Prerequisite relation learning for concepts in MOOCs. In: Proceedings of the 55th Annual Meeting of the Association for Computational Linguistics (Volume 1: Long Papers), pp. 1447–1456 (2017). https://doi.org/10.18653/v1/P17-1133
7. Wang, S., Et al.: Using prerequisites to extract concept maps from textbooks. In: Proceedings of the 25th ACM International Conference on Information and Knowledge Management, pp. 317–326 (2016). https://doi.org/10.1145/2983323.2983725
8. Talukdar, P., Cohen, W.: Crowdsourced comprehension: predicting prerequisite structure in Wikipedia. In: Proceedings of the Seventh Workshop on Building Educational Applications Using NLP, pp. 307–315 (2012). https://aclanthology.org/W12-2037
9. Liang, C., Wu, Z., Huang, W., Giles, C.: Measuring prerequisite relations among concepts. In: Proceedings of the Conference Proceedings - EMNLP 2015, pp. 1668–1674 (2015). https://doi.org/10.18653/v1/d15-1193
10. Sayyadiharikandeh, M., Gordon, J., Ambite, J., Lerman, K.: Finding prerequisite relations using the Wikipedia clickstream. In: Companion Proceedings of the 2019 World Wide Web Conference, pp. 1240–1247 (2019). https://doi.org/10.1145/3308560.3316753
11. Li, I., Fabbri, A., Hingmire, S., Radev, D.: R-VGAE: relational-variational graph autoencoder for unsupervised prerequisite chain learning. In: CoRR, vol. abs/2004.10610 (2020). https://arxiv.org/abs/2004.10610
12. Zhang, J., Lin, N., Zhang, X., Song, W., Yang, X., Peng, Z.: Learning concept prerequisite relations from educational data via multi-head attention variational graph auto-encoders. In: Proceedings of the Fifteenth ACM International Conference on Web Search and Data Mining, pp. 1377–1385 (2022). https://doi.org/10.1145/3488560.3498434
13. Miaschi, A., Alzetta, C., Cardillo, F., Dell'Orletta, F.: Linguistically-driven strategy for concept prerequisites learning on Italian. In: Proceedings of the Fourteenth Workshop on Innovative Use Of NLP for Building Educational Applications, pp. 285–295 (2019). https://doi.org/10.18653/v1/W19-4430
14. Roy, S., Madhyastha, M., Lawrence, S., Rajan, V.: Inferring concept prerequisite relations from online educational resources. In: Proceedings of the Thirty-Third AAAI Conference on Artificial Intelligence and Thirty-First Innovative Applications of Artificial Intelligence Conference and Ninth AAAI Symposium on Educational Advances in Artificial Intelligence (2019). https://doi.org/10.1609/aaai.v33i01.33019589
15. Sun, H., Li, Y., Zhang, Y.: ConLearn: contextual-knowledge-aware concept prerequisite relation learning with graph neural network. In: Proceedings of the 2022 SIAM International Conference On Data Mining (SDM), pp. 118–126 (2022)
16. Liang, C., Ye, J., Wang, S., Pursel, B., Giles, C.L.: Investigating active learning for concept prerequisite learning. In: Proceedings of the AAAI Conference on Artificial Intelligence, vol. 32, no. 1(2018). https://doi.org/10.1609/aaai.v32i1.11396
17. Li, I., Yan, V., Li, T., Qu, R., Radev, D.: Unsupervised cross-domain prerequisite chain learning using variational graph autoencoders. arXiv Preprint arXiv:2105.03505 (2021)
18. Zhang, J., Lan, H., Yang, X., Zhang, S., Song, W., Peng, Z.: Weakly supervised setting for learning concept prerequisite relations using multi-head attention variational graph auto-encoders. Knowl. Based Syst. **247**, 108689 (2022). https://doi.org/10.1016/j.knosys.2022.108689
19. Vaswani, A., et al.: Attention is all you need. In: CoRR, vol. 1. abs/1706.03762 (2017)

20. Veličković, P., Cucurull, G., Casanova, A., Romero, A., Lio, P., Bengio, Y.: Graph attention networks. arXiv Preprint arXiv:1710.10903 (2017)
21. Kipf, T., Welling, M.: Variational graph auto-encoders. arXiv Preprint arXiv:1611.07308 (2016)
22. Moggio, A., Parizzi, A.: UNIGE_SE @ PRELEARN: utility for automatic prerequisite learning from Italian Wikipedia. In: Proceedings of the Conference on Artificial Intelligence, pp. 376–380 (2020)

An Effective Privacy-Preserving and Enhanced Dummy Location Scheme for Semi-trusted Third Parties

Meijing Zuo[1], Luyao Peng[2], and Jun Song[2](✉)

[1] Department of Computer Science, University of Reading, Reading RG6 6AH, UK
[2] School of Computer Science, China University of Geosciences, Wuhan, China
songjun@cug.edu.cn

Abstract. Location-Based Services (LBS) have garnered significant attention in recent years, emphasizing the need to improve location services while safeguarding user privacy. In this paper, we propose an effective privacy-preserving and enhanced dummy location scheme specifically designed for semi-trusted third-party scenarios, with a primary focus on defending against inference attacks targeting a user's private location information. To achieve more effective location privacy preservation and mitigate privacy leaks stemming from a single point of failure, we employ a key information sharing mechanism, introduce a robust dummy location set generation approach, and present a comprehensive covering area construction strategy. To demonstrate the viability and effectiveness of our proposed scheme, we conduct a thorough simulation evaluation and performance analysis based on a practical road network setting.

Keywords: Privacy preservation · Location based service · Game model · Semi-trusted third party · Inference attacks

1 Introduction

Location-based services (LBS) have gained popularity in industry and academia, offering spatial geolocation-based information to mobile users through mobile computing, wireless positioning, and mobile communication [1]. Service providers collect location data under licensed conditions and may share it with research institutions for analysis, posing privacy risks as it reveals sensitive information like behavioral preferences and health status. LBS can track real-time and historical locations, raising privacy concerns. Despite challenges, the appeal of LBS lies in balancing utility with privacy protection.

Trusted Third Party (TTP) servers in LBS systems provide valuable benefits to users, but ensuring simultaneous data security and privacy preservation remains a critical challenge [2]. Recent incidents of well-known third-party

This work is supported in part by the National Key Research and Development Project under Grant 2019YFC0605102, by the National Natural Science Foundation of China under Grant 41972307 and Grant 61672029.

X. Song et al. (Eds.): APWeb-WAIM 2023, LNCS 14332, pp. 193–208, 2024.
https://doi.org/10.1007/978-981-97-2390-4_14

servers being attacked, such as Google, Apple, and Amazon, exposed sensitive user location data, highlighting the risks. Anonymous servers can be unreliable and compromise data security.

Existing privacy protection approaches for location services often neglect effective measures [3–5]. Techniques like dummy locations and pseudonyms are used, but they can still be vulnerable to inference attacks based on background knowledge [5]. Additionally, the potential risk of a single point of leakage from third-party location servers is not adequately addressed.

In this paper, we introduce an Efficient Privacy-preserving and Enhanced Dummy location scheme (EPED) to tackle the challenging issues in semi-trusted third-party scenarios within location-based services. Unlike previous mechanisms [6,7], EPED not only guarantees high service quality but also ensures robust privacy preservation. The main contributions of this paper are threefold:

- First, we analyze the privacy aspects of mobile location services and propose a technique to construct dummy location sets using integrated background knowledge. This method effectively counters dummy location attacks;
- Second, we introduce the EPED scheme, which employs a semi-trusted third-party (S-TTP) architecture for location anonymity services. This strategy actively reduces risks associated with user privacy breaches and mitigates the vulnerabilities of centralized servers; and
- Third, we validate EPED's effectiveness within LBS-service scenario by developing privacy-optimal strategies using game theory and Voronoi graph techniques. Extensive experiments demonstrate that EPED enables mobile users to achieve a balance between high-reliability LBS services and minimized privacy disclosure.

The rest of the paper is structured as follows: In Sect. 2, we introduce the system model and design objectives. Section 3 provides a detailed description of the proposed scheme, including system preliminaries, the location anonymization model, and scheme optimization using the Stackelberg game approach. Performance evaluation and security analysis are discussed in Sect. 4. We review related work in Sect. 5, and Sect. 6 concludes the paper.

2 Model and Design Goal

In this section, we formalize the system model and security model, and identify our design goal as well.

2.1 System Model

The system model is designed to enhance user privacy during location-based services through the use of pseudonyms and dummy locations. The Mobile Client (MC) serves as a mobile entity providing location services to users and can initiate pseudonym service requests, location update requests, and location service requests. The Pseudonym Server (PS) anonymizes users' real identities by generating Pseudonym Identities (PIDs) in response to pseudonym service requests.

The Location Covering Server (LCS) is responsible for converting precise user locations into a set of dummy locations, known as the Dummy Location Set, using a game model-based algorithm. When the MC requests a location query, the Location Service Provider (LSP) collaborates with the LCS to obtain the best hiding area from the dummy location set and delivers the requested location service data back to the MC, ensuring the confidentiality and anonymity of the users' real identities and precise locations (Fig. 1).

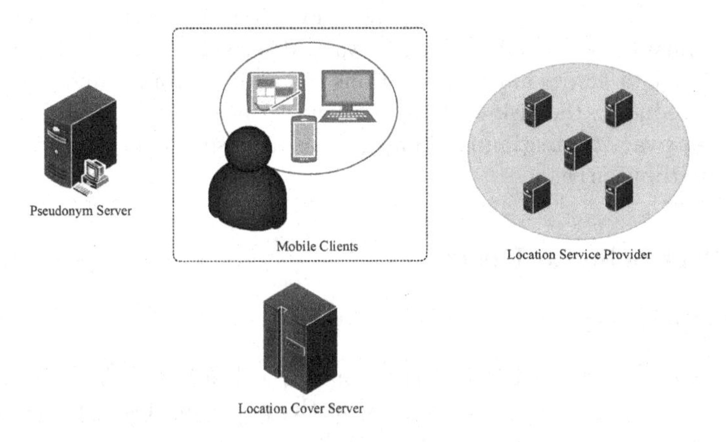

Fig. 1. System model

2.2 Security Model

This paper aims to comprehensively examine factors related to location privacy protection. Potential attackers are categorized into two groups based on their background knowledge.

- *Weak attackers*: These individuals eavesdrop on communication channels to gather user queries and sensitive information like identity and precise location. Despite limited knowledge, they can execute location-based privacy attacks, exploiting vulnerabilities in location data, such as the *location similarity attack* [8], the *center-of-cloaked-area* privacy attack [9], and the *location distribution attack*.
- *Strong attackers*: Possessing extensive background knowledge, including access probabilities and historical records, they can launch sophisticated attacks like the *homology attack*, the *probability distributed attack* [9], and the *location similarity attack* [8]. When combined with insights into privacy mechanisms, they may optimize targeted *speculative attacks* [5], significantly undermining user location privacy.

The entities in the EPED scheme are categorized as semi-trusted. Users are trusted, while Pseudonym Servers (PSs), Location Cover Servers (LCSs), and Location Service Providers (LSPs) are semi-trusted, adhering to protocols but potentially analyzing user data. Collusion among them would jeopardize user privacy. Secure communication channels with encryption and hashing protect data from unauthorized access [10]. Considering weak and strong attackers and entity trustworthiness, EPED aims to provide robust location privacy protection, ensuring confidentiality and anonymity of sensitive user information.

2.3 Design Goal

Our goal is to develop a secure and private location service that ensures reliable LBS transmission while minimizing privacy disclosure. Specifically, we i) distribute LBS data across multiple locations to enhance storage and transmission reliability, reducing the risk of data loss; and ii) use semi-trusted third parties to develop privacy protection mechanisms, ensuring secure handling of user data and safeguarding privacy.

3 EPED Scheme Design

3.1 Preliminaries

In this section, we will detail EPED tailored specifically for semi-trusted third-party scenarios. The key notations used in the paper are listed in Table 1 with their meaning explained. Due to the page limit, we only focus on the notations and definitions directly related to our proposed scheme.

Table 1. Notations

Notation	Explanation
$L, L^{\dagger}$	True location L and offset location $L^{\dagger}$, respectively
$d(i, j)$	The distance between two locations i, j $(i \neq j)$
p_i	The query probability of location i
$\varphi(i)$	The user's query probability distribution at location i
f	A privacy protection mechanism
N_{set}	The neighboring location set
C_{set}	The candidate dummy location set
T_{set}	The target dummy location set
ω	The road segment variance degree
$Pr(l)$	The probability of likelihood for a track l
Q_{loss}	The service quality loss
Q_{loss}^{max}	The maximum tolerable service quality loss
ρ	The privacy protection level

Definition 1. Location entropy (H) quantifies uncertainty within a group of dummy locations $T_{set} = \{t_1, t_2, \ldots, t_k\}$. It's computed as $H = -\sum_{i=1}^{k} p_i \times \log(p_i)$, where p_i is the query probability of each location relative to the total query probability. k represents the total number of locations in T_{set}. Location dispersion (D) quantifies distance spread among dummy locations. It's computed

as: $D = \sum_{i=1}^{k} \prod_{j=1, j \neq i}^{k} d(i,j)$. Here, $d(i,j)$ represents the Euclidean distance between locations i and j.

Definition 2. The road network is represented as an undirected graph, $G(V, E)$. Here, V is intersections, and E stands for connecting road segments. An intersection's *degree* is its connected road count. Given $G(V, E)$, let $V_p = \{V_i | degree(V_i) \geq \omega(V_p), V_i \in V\}$. The Voronoi graph from V_p is the road network V-graph.V_{pi} is a V-graph unit, and $\omega(V_{pi})$ refers to the road segment variance degree associated with V_{pi}.

Definition 3. The spatial grid divides the location space into n equal-sized square grids. These grids correspond to user locations. Historical query data analysis establishes total service queries as $X = \{x_1, x_2, \ldots, x_n\}$. The query probability of the i-th grid unit is $p_i = x_i/X$, which indicates the frequency of service queries for that particular grid. Importantly, these probabilities sum up to one, covering the entire location space.

3.2 Location Anonymization Model

In the EPED scheme, a dummy location set is formed using historical query records along with spatial grid and road network V-graph techniques. To enhance security and prevent location inference and *center-of-cloaked-area* attacks, a strategy employs offset locations instead of true ones. This approach effectively strengthens the security of the region area [9].

A. Offset Location Generation. Upon receiving the historical query probability distribution X, the EPED scheme chooses grids with query probabilities $p_i > 0$ from the $m-1$ neighboring units of the V-graph unit where the true location L is located. This selection process generates a set of neighboring locations denoted as $N_{set} = \{n_1, n_2, \ldots, n_{m-1}, L\}$. From this set, one unit is randomly chosen to serve as the offset location $L^{\dagger}$. The process is illustrated in Fig. 2, depicting the spatial grid division and the offset position construction. The shading in the illustration transitions from white (indicating a query probability of 0) to dark gray, representing increasing query probabilities. Importantly, only grids within the adjacent V-graph of the user's road network V-graph are considered in the N_{set}. As a result, the final N_{set} comprises grids highlighted in dark shading.

Such an approach ensures secure and precisely controlled generation of offset locations. By combining the spatial grid and road network V-graph, it significantly improves over completely random offsets. This joint constraint allows for controllable distances, effectively mitigating the *center-of-cloaked-area* attack to a certain extent. Moreover, when multiple service requests occur, the offset locations are randomly chosen from the controlled set. This strategy thwarts attackers from deducing the true location by overlapping multiple query locations, providing robust resistance against homogeneity attacks.

B. Dummy Candidate Location Generation. To address issues related to probability distribution, homogeneity, and location similarity attacks, EPED

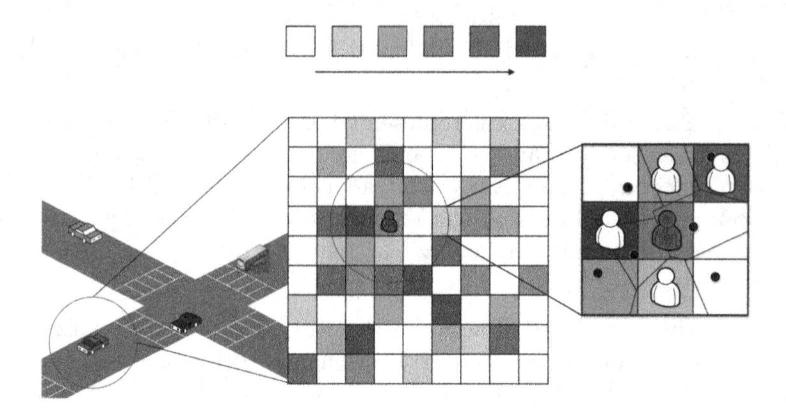

Fig. 2. Controlled generation of offset locations using grid space division

introduces the concepts of location entropy and location similarity. It also presents a novel approach for generating dummy locations. In essence, during normal circumstances, the query probability of a dummy location should closely resemble that of a true location. This implies that the query probability of the dummy location should closely align with the probability of the offset location as well.

In real-world situations, users' location queries are closely tied to their real locations. From a security viewpoint, if users near the offset location move similarly to surrounding users, it's harder for potential attackers to differentiate them. Thus, EPED proposes the following strategy for dummy candidate locations selection.

For the given two location tracks, δ and δ', along with a sequence of location-related queries services denoted as $X = \{x_1, x_2, \cdots\}$, Bayes' rule is employed to evaluate the likelihood of the given track within the sequence:

$$Pr(\delta|X) = \frac{Pr(x_1, x_2, \cdots|\delta)Pr(\delta)}{Pr(x_1, x_2, \cdots)}.$$

Assuming uniform likelihood among tracks, there exists a probabilistic relationship: $Pr(\delta|X) \propto Pr(x_1, x_2, \cdots|\delta) = Pr(x_1|\delta) \cdot Pr(x_2|\delta, x_1) \cdot Pr(x_3|\delta, x_1, x_2) \cdots$.

Building upon the Markov assumption and the independent output assumption, where the probability $Pr(x_1|\delta) = \prod_{i=0}^{n} Pr(x_{i+1}|x_i, \delta)$ is derived, guides the identification of the track δ' whose location trajectory aligns most closely with the condition:

$$\delta' = \arg\max_{\delta} Pr(x_1|\delta) \prod_{i=0}^{n} Pr(x_{i+1}|x_i, \delta).$$

The probability $Pr(x_1|\delta)$, related to the initial location in the sequence, is established based on the cumulative historical queries made by the user at a specific location. Utilizing the track whose location trajectory is most similar as an offset location offers an effective and feasible strategy for concealing the true location range of the user.

The process of forming dummy locations involves organizing grid points based on both actual and offset locations, considering the similarity of the offset location trajectories. Specifically, each grid with a query probability of $p_i (1 \leq i \leq n)$ is ranked by comparing it to the offset location trajectories. Then, the top $2k-1$ dummy locations with the highest similarity are selected to make up a set of candidate dummy locations, denoted as $C_{set} = \{c_1, c_2, \ldots, c_{2k-1}, L^\dagger\}$, where k refers to the degree of anonymity.

C. Dummy Location Set Generation. The decentralization of multiple dummy locations has a significant role in mitigating homogeneity attacks and location similarity threats. These risks can be effectively lessened by dispersing dummy locations widely or maximizing the coverage area. To tackle the potential clustering of candidate dummy locations and mitigate the mentioned attacks, the EPED scheme adopts a approach for selecting multiple dummy locations. This strategy aims to enlarge the cloaking area by incorporating offset locations and thwarting the attacker's ability to deduce the true location through clustering methods. To achieve this, the EPED scheme presents three distinct strategies to enhance the obfuscation effectiveness of the dummy locations:

- *Dispersion-based selection*: Using the concept of location dispersion (D) as defined in *Definition 1*, where $D = \sum_{i=1}^{k} \prod_{j=1, j \neq i}^{k} d(i,j)$. It involves selecting $k/3$ candidate dummy locations based on their maximum dispersion relative to the offset location $L^\dagger$. These selections are included in the target dummy location set T_{set}.
- *Entropy-driven approach*: Leveraging the concept of location entropy H as outlined in *Definition 1*, higher location entropy implies greater privacy protection for the dummy location and its associated cloaking area. For a specific location with query probability p_i, the entropy $H = -\sum_{i=1}^{k} p_i \times \log(p_i)$ is applied to $k-1$ candidate dummy locations. Then, the $k/3$ candidate dummy locations with the highest location entropy are chosen to be included in the target dummy location set T_{set}.
- *Integration of scrambling factor*: Introducing a scrambling factor λ as a radius amplification factor $(0 \leq \lambda \leq 3)$, EPED broadens the range of potential dummy locations. Considering the distance r between the true position L and the offset position $L^\dagger$, the obfuscated area uses radii of r and $r\sqrt{1+\lambda}$. Within this cloaking area, determined by a radius of $r\sqrt{1+\lambda}$, the $k/3$ candidate dummy locations located are selected from those farthest from $L^\dagger$ to form the target dummy location set T_{set}.

These three strategies work together to enhance the effectiveness of dummy location obfuscation within the EPED scheme, preventing privacy breaches and adversarial attacks.

3.3 Optimization Based on the Stackelberg Game

To address the potential threat of speculative attacks on user privacy, the proposed EPED introduces a coverage area construction scheme based on a game model. This scheme utilizes the Stackelberg game framework [9] to optimize the probability of generating the actual dummy location set. This optimization aims to enhance the robustness of the coverage area to effectively counter speculative attacks, all while ensuring user privacy.

If attackers have access to historical query service probability distributions (e.g., from a single user) and privacy-preserving mechanisms employed by anonymous location servers, they can analyze a user's current location and determine the most likely real locations. Under the privacy-preserving mechanism f, the probability of constructing a dummy location set T_{set} for location L is denoted as $f(T_{set}|L)$, while $\varphi(L)$ denotes the probability of a user's query at location L. By analyzing $T_{set} = \{t_1, t_2, \ldots, t_{k-1}, L^{\dagger}\}$, the attacker can calculate the posterior probability of all historical user locations, thereby deducing the most likely true location, as described by Eq. 1.

$$P(L|T_{set}) = \frac{P(T_{set}, L)}{P(T_{set})} = \frac{f(T_{set}|L)\varphi(L)}{\sum_L f(T_{set}|L)\varphi(L)}. \tag{1}$$

Typically, attackers aim to find a location $\hat{L}$ where the cloaking area, like a dummy location set T_{set}, is expected to be $P(L|T_{set})$, thus minimizing the effectiveness of location privacy protection. To achieve this, attackers search for a speculative location $\hat{L}$ with the minimum expected distance d compared to all possible true locations L, denoted as $min_{\hat{L}} \sum_L \; P(L|T_{set})\; d(L, \hat{L})$.

For each dummy location set T_{set}, the attackers can compute the benefit using Eq. 2. Referring to the optimization strategy of the Stackelberg game in [9], for each potential dummy location set T_{set}, the maximum benefit obtained through the optimal attack strategy, considering the condition $\sum_L f(T_{set}|L)\varphi(L)$, i.e., the minimum privacy protection degree for users, is defined as:

$$\sum_{T_{set}} \min \sum_{L} f(T_{set}|L)\varphi(L)d(L, \hat{L}), \forall \hat{L}. \tag{2}$$

To optimize the attacker's expected benefit and maximize the user's benefit, the goal of EPED is to find an appropriate T_{set} based on all strategies. For computational convenience, the approach described in [9] is referred to for transforming Eq. 2 into a linear condition constraint, which are defined as follows:

$$\begin{aligned} \chi &\triangleq \min \sum_{L} f(T_{set}|L)\varphi(L)d(L, \hat{L}), \\ \chi &\leq \sum_{L} f(L^{\dagger}|L) f(T_{set}|L^{\dagger})\varphi(L)d(L, \hat{L}), \end{aligned} \tag{3}$$

where $L^{\dagger}$ denotes the offset position, and T_{set} can be seen as the further anonymization of $L^{\dagger}$.

Moreover, the primary objective of the EPED is to preserve the user's service quality by concealing their real location. To guarantee this service quality, the system establishes the service quality threshold Q_{loss}^{max}, which limits the maximum acceptable service quality loss. Under the expected quality loss Q_{loss}, the specific process for achieving this objective is as follows:

$$Q_{loss} = \sum_{L,L^{\dagger}} f(L^{\dagger}|L)\varphi(L)d(L,L^{\dagger}) \leq Q_{loss}^{max}. \tag{4}$$

The optimal strategy of the location covering server corresponds to the optimal covering zone selection strategy while considering the constraints imposed by attackers adopting the speculative attack. According to this strategy, maximizing the user's privacy protection level ρ, which is the expectation between the attacker's presumed user location $\hat{L}$ and the user's actual location L distance, also maximizes their benefits. In this case, the process of generating $\hat{L}$ by the attacker is denoted as $h(\hat{L}|T_{set})$. The constraints for minimizing the privacy protection level ρ are as follows:

$$\sum_{L,\hat{L},T_{set}} f(T_{set}|L)\varphi(L)h(\hat{L}|T_{set})d(L,\hat{L}) \leq \rho_{min}. \tag{5}$$

To mitigate the risk of location privacy disclosure arising from an insufficient number of road sections, it is crucial for the generated dummy location set to include at least ω_{min} different road segments. Moreover, to enhance the overall privacy protection quality, the degree of variability $\omega(T_{set})$ in the dummy location set T_{set} must be taken into account. Consequently, a constraint condition can be calculated as follows:

$$\sum_{L,T_{set}} f(T_{set}|L)\varphi(L)\omega(T_{set}) \geq \omega_{min}, \tag{6}$$

where $\omega(T_{set})$ represents the road segment variance degree of the dummy location set T_{set}.

The objective of EPED privacy protection is to maximize user benefits while considering the trade-off between service quality loss and limited segment diversity. This can be achieved through a series of linear programs. The final linear programming process is shown in Eq. 7:

$$\text{Maximize} \sum_{T_{set}} \chi. \tag{7}$$

Subject to the six constraints outlined below:

$$
\begin{cases}
\sum_{L} f(T_{set}|L)\varphi(L)d(L,\hat{L}) \geq \chi, (case\ 1) \\
\sum_{L,L^{\dagger}} \varphi(L)f(L^{\dagger}|L)d(L,L^{\dagger}) \leq Q_{loss}^{max}, (case\ 2) \\
\sum_{L,\hat{L},T_{set}} f(T_{set}|L)\varphi(L)h(\hat{L}|T_{set})d(L,\hat{L}) \leq \rho_{min}, (case\ 3) \\
\sum_{L,T_{set}} f(T_{set}|L)\varphi(L)\omega(T_{set}) \geq \omega_{min}, (case\ 4) \\
\sum_{T_{set}} f(T_{set}|L) = 1, (case\ 5) \\
\sum_{T_{set}} f(T_{set}|L) \geq 0, \forall L, T_{set}, (case\ 6).
\end{cases}
\tag{8}
$$

In summary, the proposed scheme aims to enhance user privacy by optimizing coverage area construction while considering potential attacks and balancing service quality. The optimization process is formulated using linear programming techniques and a game-theoretic approach.

4 Performance Evaluation and Security Analysis

In this section, we comprehensively evaluate the proposed EPED scheme's performance and security using a custom simulator developed with Java and Matlab. The analysis is conducted on a system with a 3.3 GHz Intel CPU, 12 GB of RAM, and Windows 11 OS. The simulation employs synthetic data in a real road network from Odinburg, Germany, covering 102.96 square kilometers, divided into a 16×18 grid. With 500 mobile users generating location-based service requests as they move along roads, we assess the scheme based on parameters like 100,000 historical queries, Voronoi unit of 4, 6105 vertices, 7035 road segments, and a 600-meter grid edge length.

4.1 Performance Analysis

A. Performance Analysis from the Dummy Location Set Generation: We evaluated the performance of the proposed EPED scheme for generating dummy location sets, considering location entropy and dispersion metrics. Comparing it with DLS [6], EDLS [6], HCLS [7], and a random generation scheme, we found the EPED scheme to slightly outperform others. As shown in Fig. 3(a), DLS occasionally chose zero-query probability dummy locations, while EDLS improved on DLS but overlooked location entropy. HCLS addressed zero-query probabilities but lacked further optimization. In contrast, the random generation method disregarded location entropy and might choose zero-query probability

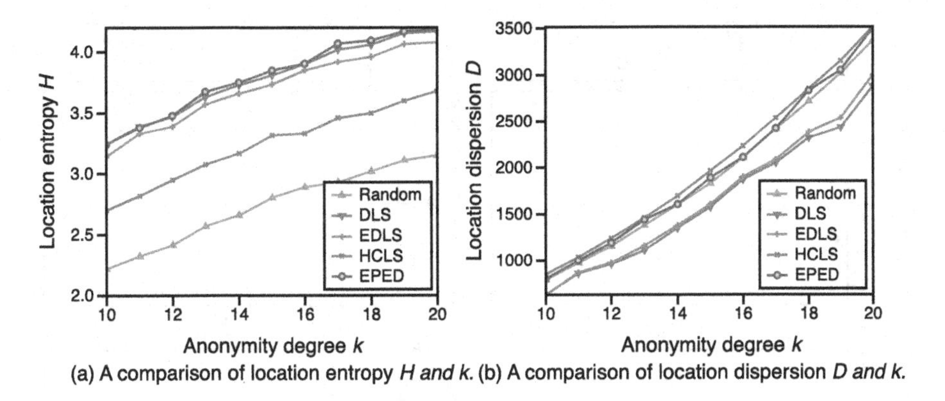

(a) A comparison of location entropy *H and k*. (b) A comparison of location dispersion *D and k*.

Fig. 3. A comparison of the different schemes.

dummy locations. Moreover, we compared these approaches in terms of location dispersion D and anonymity degree k, as presented in Fig. 3(b). DLS ignored location dispersion, relying on query probabilities for dummy position selection. EDLS improved DLS but lacked guaranteed maximum dispersion. HCLS aimed for maximum dispersion but disregarded location entropy. Conversely, our EPED scheme sequentially selected the k grids with the highest dispersion from the pre-selected $2k$ locations, ensuring both maximum dispersion and high location entropy. Based on the above analysis, it's evident that the EPED scheme, integrating location entropy and maximum dispersion, outperforms existing schemes [6,7]. The utilization of location offsetting for anonymity ensures the protection of genuine locations even in the presence of filtered false ones.

B. Performance Analysis of Covering Area Selection: The performance evaluation of the cloaking area is based on three key metrics: privacy protection level (ρ), road segment variance (ω), and service quality loss (Q_{loss}).

Figure 4(a) compares ρ and Q_{loss}^{max} among different schemes: HCLS-SG ([7]), our pre-optimization scheme (Base), and the proposed EPED optimization scheme. As Q_{max} increases, privacy protection level ρ improves for all schemes. Notably, our optimized EPED scheme outperforms pre-optimization scheme (Base) when the minimum road segment variance degree ω_{min} is set to 24, 25, or 26. Particularly, at $\omega_{min} = 26$, our EPED scheme surpasses the HCLS_SG scheme in privacy protection level ρ. In Fig. 4(b), the relationship between actual road segment variance ω and maximum service quality loss Q_{loss}^{max} is evident. EPED enhances road segment disparity compared to our pre-optimization scheme Base, while HCLS_SG prioritizes privacy over ω. Figure 4(c) and 4(d) illustrate how, under a specific privacy protection level (e.g., $\rho = 10$), both the privacy protection level ρ and the actual service quality loss Q_{loss}^{max} vary with changes in the minimum road segment variance ω_{min}. In Fig. 4(c), we observe a decrease in privacy protection level ρ for both the pre-optimization scheme (Base) and the post-optimization scheme (EPED) as the minimum road seg-

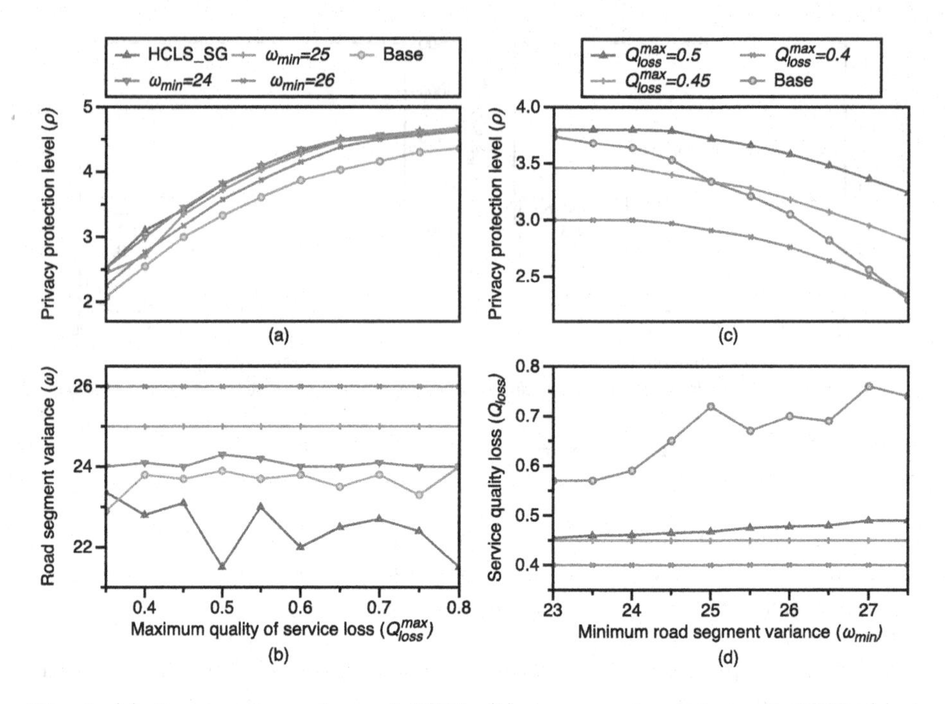

Fig. 4. (a) A comparison of ρ and Q_{loss}^{max}. (b) A comparison of ω and Q_{loss}^{max}. (c) A comparison of ρ and ω_{min}. (d) A comparison of Q_{loss} and ω_{min}.

ment variance (ω_{min}) increases. More restrictive constraints on Q_{loss}^{max} generally result in a lower privacy protection level ρ, potentially even below that of the pre-optimization scheme (Base). Notably, when the maximum quality of service loss Q_{max} is set to 0.5, the EPED scheme consistently outperforms the pre-optimization scheme (Base) in terms of privacy protection level ρ. Figure 4(d) demonstrates that the pre-optimization scheme (Base) exhibits a higher quality of service loss compared to the post-optimization scheme (EPED). This indicates minimal impact of ω_{min} on the actual quality of service loss Q_{loss}. With a Q_{loss}^{max} limit of 0.5, EPED consistently offers a superior privacy protection level ρ compared to the pre-optimization scheme (Base). These findings emphasize the adaptability of EPED's parameters, optimizing user benefits by achieving a balance between service quality loss, road segment variance, and privacy protection level.

4.2 Security Analysis

A. Security Analysis Based on Network Architecture. The network architecture we propose relies on semi-trusted third-party entities. These entities are expected to fulfill their duties independently, but they also possess the ability to gather and analyze user data within their respective domains.

In the scenario of a single entity compromise, three potential scenarios arise: a compromised pseudonym server may unveil the true identity tied to the pseudonym, while secure communication channels prevent access to query content and real location data; a compromised location covering server could expose the real location linked to the pseudonym, without enabling identification of the real identity or access to query specifics; in the event of a compromised service provider, access to the user's covering area and query content becomes feasible. Importantly, the proposed EPED scheme effectively shields user location privacy, making it difficult for attackers to deduce the actual location from the cloaking area and to determine the true user behind the pseudonym.

In the scenario where two entities are compromised, three possibilities arise: if the attacker controls both the pseudonym server and the location covering server, they can obtain the actual identity and real location, while query content remains inaccessible; if the attacker controls the pseudonym server and the service provider, they can access the real identity, query content, and the covering area, but deducing the actual location is not possible; in the case of the attacker controlling the service provider and the location covering server, they can access the real location and query content of the pseudonym user, yet determining the true identity behind the pseudonym remains unachievable. In summary, EPED effectively protects specific user private information, even when one or two entities are compromised.

B. Security Analysis Based on Dummy Location Construction. The security analysis of the dummy location construction scheme demonstrates its effectiveness against both weak and strong attackers, preventing their efforts to deduce actual locations from the constructed dummy locations. For weak attackers, the offset location construction addresses *center-of-cloaked-area* privacy attacks and mitigates the impact of real user distribution, rendering location distribution attacks ineffective. The selection of highly dispersed grids during dummy location set creation also thwarts location homogeneity attacks. Strong attackers find nearby offset positions with similar query probabilities, and the presence of intersections in the dummy location set introduces uncertainty into attempts at location homogeneity attacks. Probability distribution attacks face difficulties due to the close grid query probabilities within the dummy location set, while road segment diversity constraints ensure semantic information diversity. Through a game theory approach, the construction of the covering area scheme is optimized against inference attacks, guaranteeing maximum user privacy even against sophisticated attacker strategies. Furthermore, during the offset location selection and dummy location set construction process, we enhance resistance against various inference attacks by integrating location trajectory likelihood and three false location obfuscation techniques. This ensures user information privacy in real-world scenarios.

5 Related Work

In recent years, location privacy protection research has focused on various techniques, including pseudonym [4,11,12], spatial obfuscation [13–15], encryption [16,17], and dummy location [3,6,18] methods.

The pseudonym technique replaces a user's identity with a temporary pseudonym during queries [4,11]. To enhance privacy, Memon et al. proposed pseudonym replacement with mix zones, where multiple users exchange pseudonyms at a designated zone [4]. Additionally, Memon and Zhao introduced a multi-hybrid zone model to balance service quality and privacy protection [12].

Spatial obfuscation methods perturb queried location information to hinder attackers from identifying the user's real location [13–15]. For example, Agir et al. utilized adaptive privacy level adjustment based on user mobile locations for continuous query privacy protection [13]. Some studies employed differential privacy models to preserve user privacy while controlling disturbance [14,15].

The dummy location technique generates multiple dummy locations mixed with the user's real location for privacy protection [3]. Service providers cannot distinguish real from dummy locations, ensuring privacy. However, the quality of dummy locations and queries can affect security. Niu et al. proposed constructing dummy locations with similar query frequencies [18]. They also used dummy locations to improve the cache hit rate in group collaboration [6]. Shockley et al. introduced a protection strategy based on the Stackelberg game for privacy optimization [9].

Our EPED scheme integrates pseudonymity, spatial obfuscation, game theory, and virtual location techniques to enhance user privacy [3,6,9,12,13]. It hides the actual location and protects the user's real identity with controlled perturbation. Considering historical query distribution and virtual location coverage, it minimizes the risk of virtual location filtering. The EPED scheme leverages game theory and the Voronoi diagram model for optimized privacy and service quality control [9].

6 Conclusion

In conclusion, this paper introduces the EPED scheme, a novel privacy-preserving and dummy location technique designed to enhance the security of Location-Based Services (LBS). By shifting the load of the location covering server and segregating user identity, query, and location privacy, the EPED scheme reduces reliance on third-party servers and offers improved trustworthiness. Compared to conventional location preservation methods in LBS, our proposed scheme not only ensures high reliability of LBS transmissions but also significantly minimizes user privacy exposure and effectively combats potential collusion among various servers. Extensive performance evaluation further confirms the EPED scheme's ability to generate high-quality dummy location sets, making it highly resilient against speculative attacks based on extensive background knowledge.

References

1. Wang, X., Ma, J., Miao, Y., Liu, X., Zhu, D., Deng, R.H.: Fast and secure location-based services in smart cities on outsourced data. IEEE Internet Things J. **8**(24), 17639–17654 (2021)
2. Feng, J., Wang, Y., Wang, J., Ren, F.: Blockchain-based data management and edge-assisted trusted cloaking area construction for location privacy protection in vehicular networks. IEEE Internet Things J. **8**(4), 2087–2101 (2021)
3. Wu, Z., et al.: A location privacy-preserving system based on query range cover-up or location-based services. IEEE Trans. Veh. Technol. **69**(5), 5244–5254 (2020)
4. Jiang, H., Li, J., Zhao, P., Zeng, F., Xiao, Z., Iyengar, A.: Location privacy-preserving mechanisms in location-based services: a comprehensive survey. ACM Comput. Surv. (CSUR) **54**(1), 1–36 (2021)
5. Gao, H., Huang, W., Liu, T., Yin, Y., Li, Y.: PPO2: location privacy-oriented task offloading to edge computing using reinforcement learning for intelligent autonomous transport systems. IEEE Trans. Intell. Transp. Syst. **24**(7), 7599–7612 (2023)
6. Zhang, S., Choo, K.R., Liu, Q., Wang, G.: Enhancing privacy through uniform grid and caching in location-based services. Future Gener. Comput. Syst. **86**, 881–892 (2018)
7. Saravanan, P.S., Balasundaram, S.R.: Protecting privacy in location-based services through location anonymization using cloaking algorithms based on connected components. Wirel. Pers. Commun. **102**(1), 449–471 (2018)
8. Wen, R., Zhang, R., Peng, K., Wang, C.: Protecting locations with differential privacy against location-dependent attacks in continuous LBS queries. In: 20th IEEE International Conference on Trust, Security and Privacy in Computing and Communications, TrustCom 2021, Shenyang, China, 20-22 October 2021, pp. 379–386. IEEE (2021)
9. Yang, D., Ye, B., Zhang, W., Zhou, H., Qian, X.: Klpps: a k-anonymous location privacy protection scheme via dummies and stackelberg game. Sec. Commun. Netw. **2021**, 1–15 (2021)
10. Qi, L., Wang, X., Xu, X., Dou, W., Li, S.: Privacy-aware cross-platform service recommendation based on enhanced locality-sensitive hashing. IEEE Trans. Netw. Sci. Eng. **8**(2), 1145–1153 (2020)
11. Shao, Z., Wang, H., Zou, Y., Gao, Z., Lv, H.: From centralized protection to distributed edge collaboration: a location difference-based privacy-preserving framework for mobile crowdsensing. Secur. Commun. Netw. **2021**, 5855745:1–5855745:18 (2021)
12. Zhao, D., Jin, Y., Zhang, K., Wang, X., Hung, P.C.K., Ji, W.: EPLA: efficient personal location anonymity. GeoInformatica **22**(1), 29–47 (2018)
13. Lai, H., Xu, L., Zeng, Y.: An efficient location privacy-preserving authentication scheme for cooperative spectrum sensing. IEEE Access **8**, 163472–163482 (2020)
14. Yin, C., Xi, J., Sun, R., Wang, J.: Location privacy protection based on differential privacy strategy for big data in industrial internet of things. IEEE Trans. Industr. Inf. **14**(8), 3628–3636 (2018)
15. Mendes, R., Cunha, M., Vilela, J.P.: Impact of frequency of location reports on the privacy level of geo-indistinguishability. Proc. Priv. Enhancing Technol. **2020**(2), 379–396 (2020)

16. Wang, H., Qin, Y., Liu, Z., Li, Y.: Processing private queries based on distributed storage in location-based services. In: Proceedings of the ACM Turing Celebration Conference - China, ACM TUR-C 2019, Chengdu, China, 17-19 May 2019, pp. 2:1–2:6. ACM (2019)
17. Zhang, J., Li, C., Wang, B.: A performance tunable cpir-based privacy protection method for location based service. Inf. Sci. **589**, 440–458 (2022)
18. Wu, Z., Shen, S., Lian, X., Su, X., Chen, E.: A dummy-based user privacy protection approach for text information retrieval. Knowl. Based Syst. **195**, 105679 (2020)

W-MRI: A Multi-output Residual Integration Model for Global Weather Forecasting

Lihao Gan[1], Xin Man[1], Changyu Li[1], Lei She[2], and Jie Shao[1,2(✉)]

[1] University of Electronic Science and Technology of China, Chengdu 611731, China
{ganlihao,manxin,changyulve}@std.uestc.edu.cn, shaojie@uestc.edu.cn
[2] Sichuan Artificial Intelligence Research Institute, Yibin 644000, China

Abstract. Weather forecasting refers to the process in which science and technology are applied to predict the conditions of the atmosphere for a given time. In this paper, we present W-MRI, a multi-output residual integration model for global weather forecasting. We introduce residual mechanism into weather prediction, which can simulate changes in weather conditions, and elaborately design a residual network to integrate and constrain multi-output residuals. W-MRI can effectively extract the features of meteorological data, capture their internal relations, and make fast and accurate forecasts of multiple meteorological variables, such as surface wind speed and precipitation. We use the fifth-generation ECMWF Re-Analysis (ERA5) data to train W-MRI, with samples selected every six hours. Importantly, our proposed W-MRI outperforms FourCastNet in multi-variable weather forecasting under the same experimental conditions. Moreover, experiments show that our model has a stable and significant advantage in short-to-medium-range forecasting, and the longer the forecasting time-step, the more obvious the performance advantage of W-MRI, showing that the residual network has great advantages in weather forecasting.

Keywords: Weather forecasting · Residual network · Deep learning · Data-driven prediction

1 Introduction

Weather forecasting is the analysis of past and present weather observations, as well as the use of modern science and technology, to predict the state of the Earth's atmosphere at a specific location in the future. It plays a crucial role in key sectors such as transportation, logistics, agriculture, and energy production [2]. Numerical Weather Prediction (NWP) utilizes mathematical models [4,14] of the atmosphere and oceans to forecast the weather based on current weather conditions. It involves applying the laws of physics [1] to weather forecasting. Over the decades, with the increasing number of observations and the continuous improvement of modeling techniques and computing power, weather

X. Song et al. (Eds.): APWeb-WAIM 2023, LNCS 14332, pp. 209–222, 2024.
https://doi.org/10.1007/978-981-97-2390-4_15

forecasts have become more accurate, and the real-world applications of NWP [15,25] have grown. Although modern numerical weather prediction models have achieved satisfactory results, the underlying powerhouses of these models still rely on the physical principles governing the atmosphere and oceans, which may not fully meet the demands of complex climate forecasting [22]. Moreover, numerical weather prediction faces challenges in meeting the diverse needs of weather forecasting due to its high computational cost, the difficulty of solving nonlinear physical processes, and model deviations [26].

In recent years, there has been a growing interest in utilizing data-driven deep learning models for weather forecasting, as they require significantly lower computational resources [11] compared with traditional NWP models. By employing deep learning techniques to train observational and reanalysis data, data-driven models can bypass the need for solving complex nonlinear physical processes [21] and mitigate the biases [11,21] inherent in NWP models. Once trained, data-driven models can generate multi-step weather predictions by using autoregressive inference [6], which is considerably faster than traditional NWP models. Furthermore, meteorological data involved in weather forecasting exhibit characteristics such as a vast amount of data, diverse types [18] and interconnectedness, all of which can be effectively handled by data-driven models.

In weather forecasting, the weather conditions at the next time largely depend on the weather conditions at the previous time [11,18], which can be seen as a slight change that occurred in the previous moment. This observation aligns with the characteristics of residuals. Consequently, we introduce the residual mechanism into weather forecasting tasks. In mathematical statistics, a residual refers to the difference between the actual observed value and the estimated value [24]. In weather forecasting, the residual is regarded as the difference between the ground-truth and the prediction result. By modeling and fitting the difference between the ground-truth and the prediction result, we can utilize residuals to effectively characterize weather characteristics that exhibit variations between different time periods. We represent these weather variations in the form of residuals. This representation not only significantly improves computational efficiency [12], but also aligns with the intrinsic characteristics of weather data, thereby enabling better extraction of weather data features.

In this work, we propose a data-driven Multi-output Residual Integration model called W-MRI for global weather predictions. First, W-MRI utilizes the Naive method [16] as the base prediction, which represents the weather conditions at the previous moment. Then, we introduce the residual mechanism as the foundation of our model, and design residual modules to generate residuals. For each residual module, we employ three Fully Connected (FC) layers to simulate small changes in weather conditions. Next, all the residual blocks are passed through the integration constraint module, allowing them to be integrated. Finally, the model output is computed by adding the base prediction and the integrated residual. Through training on the high-resolution ERA5 dataset with two years of data, we compare the performance of our W-MRI model to FourCastNet [19], a recent AI-based data-driven weather forecasting model.

Our W-MRI demonstrates significant performance advantages in predicting short-to-medium-range global weather conditions.

Our technical contributions can be summarized as follows:

- We introduce the residual mechanism into weather forecasting, which enables weather forecasting models to explore the internal relationships within weather data and effectively handle the integration of heterogeneous data.
- We propose an efficient multi-output residual integration architecture to combine multiple residual blocks, allowing the model to effectively characterize various weather conditions.
- Extensive experiments demonstrate that our W-MRI model exhibits strong stability and superior performance in short-to-medium-range weather prediction.

2 Related Work

2.1 Numerical Weather Prediction

Numerical Weather Prediction (NWP) is a method used to forecast atmospheric movement and weather phenomena within a certain time period by solving fluid dynamics and thermodynamics equations that describe the evolution of weather under specific initial and boundary conditions, based on the current atmospheric conditions [2,14]. More specifically, given the initial conditions [27], the NWP model conducts simulations of atmospheric processes. By numerically solving the coupled system of partial differential equations, known as the Navier-Stokes equations, which describe the atmosphere in terms of momentum, mass, and energy [23], the model obtains the future atmospheric state for each grid cell. Some researchers have proposed grid refinement techniques [3] to increase the resolution and improve the NWP model's performance. Others have suggested fine-tuning physical parameterizations [17] to further enhance the accuracy of weather forecasts.

2.2 Deep Learning Weather Forecasting Methods

In recent years, researchers have shown increasing interest in the application of deep learning methods to weather forecasting. Deep learning based weather predictions offer several advantages over traditional Numerical Weather Prediction (NWP) methods, including low computational cost and fast processing speed [11,21]. These methods have demonstrated satisfactory results in predicting extreme weather events such as hurricanes [25]. Denby [5] employed a Convolutional Neural Network (CNN) for the classification of weather satellite images, while Xu et al. [28] utilized a combination of Generative Adversarial Network (GAN) and Long Short-Term Memory (LSTM) for cloud image prediction. Additionally, data-driven weather forecasting has gained attention, as exemplified by Dueben and Bauer [8], who employed a Multi-layer Perceptron

(MLP) model. More recently, FourCastNet [19], a data-driven weather forecasting model, has utilized the Vision Transformer (ViT) architecture [7] and Adaptive Fourier Neural Operators (AFNO) for high-resolution weather forecasting. FourCastNet achieves reliable and rapid forecasts with efficient computation. However, it requires a substantial amount of data for training and fine-tuning.

2.3 Residual Network

The residual network utilizes the residual unit as the fundamental building block of the network. The classic residual network, ResNet [9], constructs the network by creating a structure composed of residual units consisting of tens to hundreds of layers, thereby addressing the issue of network degradation. The residual mechanism has garnered increasing interest among researchers. iResNet [13] enhances the residual network model by addressing the problem of deep information loss in ResNet. ResNeSt [29] further advances ResNet by introducing the channel attention mechanism and the split-attention network. For the specific task of weather prediction, we design a novel multi-output residual integration network named W-MRI, which enables accurate and efficient global weather forecasting.

3 Preliminaries

3.1 Dataset

ERA5 [10] is a publicly available atmospheric reanalysis dataset produced by the European Centre for Medium-Range Weather Forecasts (ECMWF). The ERA5 dataset is generated through an optimal combination of observations from various measurement sources and is outputted using a numerical model that employs a Bayesian estimation process called data assimilation. It represents the best reconstruction and assessment of meteorological elements observed, stored as gridded data, providing a wealth of information about the Earth's climate and weather. The dataset covers the period from 1940 to the present and includes meteorological variables such as wind speed, precipitation, temperature, relative humidity, terrain height, mean sea level pressure, and many others. The spatial resolution of the data is 0.25° latitude and longitude, with hourly intervals, and 37 vertical pressure levels ranging from 1000 hPa to 1 hPa.

In this study, we use the ERA5 dataset to train W-MRI model. Specifically, we select six-hourly sampled data points (t0, t6, t12, t18), with each sample consisting of twenty atmospheric variables across five vertical levels (see Table 1 for more details). In addition, we allocate two years of data for training (2015 and 2016), one year for validation (2017), and one year for testing (2018).

3.2 Multi-variable Forecasting Problems

In this work, we focus on forecasting two important and challenging atmospheric variables, namely, (1) the wind velocities at a distance of 10m from the Earth's

Table 1. The abbreviations of atmospheric variables are as follows: U_{10} and V_{10} represent the zonal and meridional wind velocity from the surface, respectively, at a height of 2 m; T_2 m represents the temperature at 2 m from the surface; T, U, V, Z and RH represent the temperature, zonal velocity, meridional velocity, geopotential and relative humidity at specified vertical level; $TCWV$ represents the total column water vapor.

Vertical Level	Variables
Surface	U_{10}, V_{10}, T_{2m}, sp, $mslp$
10000 hPa	U, V, Z
850 hPa	T, U, V, Z, RH
500 hPa	T, U, V, Z, RH
50 hPa	Z
Integrated	$TCWV$

surface and (2) the 6-hourly total precipitation. These variables have been chosen for comparison purposes, aligning with the work conducted by FourCastNet [19]. The reasons for choosing these two variables are as follows: (1) Surface wind velocities and precipitation encompass numerous small-scale features and are influenced by them. Therefore, models employed for their forecasting should possess the capability to efficiently handle high-resolution data and provide accurate forecasts. (2) Near-surface wind speed forecasts are highly valuable as they play a crucial role in the planning of energy storage, grid transmission, and other operational considerations for both onshore and offshore wind farms. Furthermore, accurate wind speed forecasts assist in monitoring and tracking extreme wind events. (3) Total precipitation data exhibits strong random and nonlinear characteristics. Techniques based on artificial intelligence offer powerful capabilities in parametric inference for high-resolution discrete data. Hence, utilizing these techniques allows for effective prediction of total precipitation. Moreover, our model also provides forecasts for some other variables, including geopotential height, temperature, wind velocity, and relative humidity at different vertical levels.

Our W-MRI model is a data-driven, high-resolution short-to-medium-range global weather forecast system, which predicts the future atmosphere states based on the current atmosphere conditions. We collectively denote the modeled variables by the tensor X_t, where t denotes the time index. Each of the 20 variables is represented as a 2D field with a shape of 721×1440 pixels and a resolution of 0.25° latitude-longitude. Therefore, the tensor X has a shape of $20 \times 721 \times 1440$. W-MRI aims to generate forecasts up to 8 days in advance with a time interval of six hours in an autoregressive manner (Eq. 1):

$$\mathrm{W-MIR}(X_t) = \left\{\hat{X}_{t+1}, \hat{X}_{t+2}, \cdots, \hat{X}_{t+32}\right\}, \tag{1}$$

where X_t represents the initial weather condition and $\hat{X}_{t+\Delta t}$ is the prediction of the weather condition at time-step $t + \Delta t$.

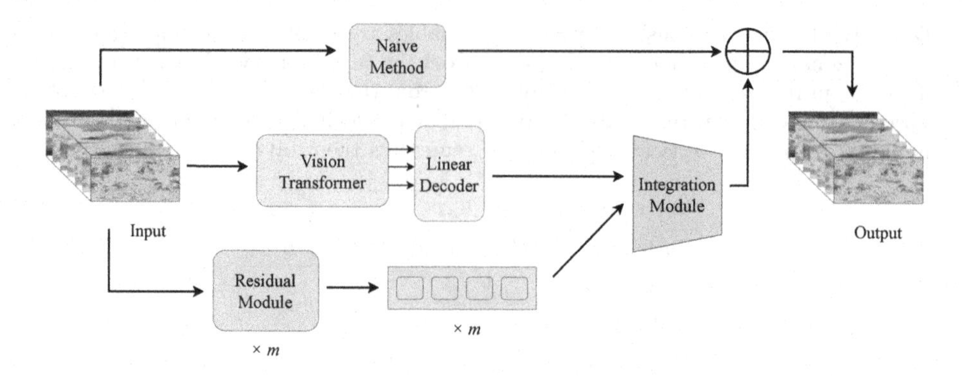

Fig. 1. The structure of W-MRI. First, the input image is fed into the ViT and several residual modules to generate multi-output residual blocks, which represent the change in the original input. Each residual module consists of three fully connected layers, with the output of the former residual module serving as the input of the latter. Next, all the residual blocks are passed through the integration constraint module, where they can be integrated. Finally, the output is computed by adding the base prediction and the integrated residuals.

4 Method

In this section, we give a detailed introduction to the proposed W-MRI model. As shown in Fig. 1, our W-MRI can be regarded as a multi-output residual network with an integration constraint module. The key component of W-MRI is the utilization of residual mechanism, along with the design of the residual module. This allows the network to effectively handle multi-variable data and extract intricate weather condition features with higher granularity.

4.1 ViT and Residual Model

Vision Transformer (ViT) [7] has emerged as a popular neural architecture due to its simplicity, computational efficiency, and scalability, making it well-suited for image-related tasks. We denote the input image as a high-dimensional tensor $X \in \mathbb{R}^{C \times H \times W}$, where C represents the number of atmosphere variables or channels. H and W are the height and width. ViT divides the input image into a sequence of 2D patches x_p, where the size of each patch is $p \times p$. This results in $N = HW/P^2$ patches in total. Then, the patches are flattened and compressed to a D-dimensional space using a linear projection (Eq. 2). After the patch and position embedding, the output of linear projection is send to the Transformer encoder, which performs feature extraction. The encoder consists of Multi-head Self-attention (MSA) architecture and Multi-Layer Perceptron (MLP) blocks (Eq. 3 and Eq. 4). Each MLP block contains two Fully Connected (FC) layer, and the activation function is ReLU, where $ReLU = \max(0, x)$. Before each block, Layer Normalization (LN) is applied. This process is repeated for L layers, and a linear decoder reconstructs the patches and generates the residual block based on the final encoder output.

$$z_0 = [x_{class}; x_p^1 E; x_p^2 E \cdots ; x_p^N E] + E_{pos}, \quad (2)$$

$$z'_\ell = MSA(LN(z_\ell - 1)) + z_{\ell-1}, \quad (3)$$

$$z_\ell = MLP(LN(z'_\ell)) + z'_\ell, \quad (4)$$

where $E \in \mathbb{R}^{(p^2 \cdot C) \times D}$, $E_{pos} \in \mathbb{R}^{(N+1) \times D}$ and $\ell = 1 \ldots L$.

The residual module consists of three FC layers, and we apply ReLU as the activation function in the first two layers. The last FC layer does not use any activation function, so we call it linear layer. Let x_{in} denote the input of residual model, and by applying residual module, we can represent:

$$x_{re} = x_{in} - R(x_{in}), \quad (5)$$

where x_{re} is the output of residual module and R denotes the residual module. Specifically, the operation of each residual module can be described by the following equations:

$$\begin{aligned} z_1 &= FC_1(x_{in}) = ReLU(L_1(x_{in})), \\ z_2 &= FC_2(z_1) = ReLU(L_2(z_1)), \\ R(x_{in}) &= L_3(z_2), \\ L_i(x) &= W_i x + b_i (i = 1, 2, 3), \end{aligned} \quad (6)$$

where W_i represents the matrix of linear transform and b_i is the bias. The output of each residual module serves as the input for the next residual module in a sequential manner. The input of the first module is the original image. As the input passes through the residual models, it undergoes transformations that extract subtle and dense features compared with the original input. Consequently, the residual x_{re} has a similar shape with the original input but its average is close to 0. This characteristic allows it to effectively simulate small perturbations in the initial weather conditions. By utilizing the residuals generated by the ViT and residual modules, our W-MRI model can capture the changes in each meteorological variable at the initial time and subsequently make accurate predictions of the weather conditions at future time-steps.

4.2 Integration and Constraint of Residual

In contrast to ResNet, which only utilizes the last residual in the deep network, our approach introduces a multi-output residual integration structure. This design aims to enhance the capacity for residual reuse and more accurately simulate the subtle variations in meteorological variables. The proposed structure not only integrates the output residuals from each component, but more importantly, it incorporates a constraint mechanism to regulate the integrated residuals. Let x_v denote the residual generated by ViT, R_i denote the i-th residual model, $x_{in,i}$ denote the input of the i-th residual model, $x_{out,i}$ denote the output of the i-th residual model. Then, we can write the residual integration as follows:

$$\begin{aligned} x_{rei} &= x_v + \sum_{i=1}^{m} R_i(x_{in,i}) \\ &= x_v + \sum_{i=1}^{m} x_{out,i}. \end{aligned} \tag{7}$$

To enhance the effectiveness of residual integration and capture the interdependencies between different parts, a learnable tensor with the same shape as the original input is trained by the ViT. This tensor is utilized to better represent and constrain the integration of residuals, reducing errors compared with the ground-truth and improving prediction accuracy. Specifically, the tensor is multiplied element-wise with the corresponding pixel in the integrated residual. By comparing the result with the ground-truth and calculating the loss, the model can be optimized. This architecture improves the efficiency of residual reuse in W-MRI and enhances the model's generalization capability. The final output is obtained by adding the constrained residual integration and the Naive prediction. The prediction results of W-MRI can be expressed as follows:

$$x_{pred} = x_{na} + constr(x_{rei}), \tag{8}$$

where x_{na} denotes the naive input and *constr* denotes the process of constraint.

5 Experiments

In this section, we introduce two important evaluation metrics in weather forecasting. We also show the results of W-MRI compared with FourCastNet on predicting multiple meteorological variables and provide visualization examples to demonstrate the superiority of W-MRI in weather prediction. Additionally, we conduct ablation study to analyze the effect of integration constraint module.

5.1 Evaluation Metrics

In order to be consistent with the work done by FourCastNet, for each forecast, we evaluate the latitude-weighted Anomaly Correlation Coefficient (ACC) and Root Mean Squared Error (RMSE) [20] for all of the variables included in the forecast. The latitude weighted ACC for a forecast variable v at forecast time-step l is defined as follows:

$$ACC(v,l) = \frac{\sum_{m,n} L(m)\tilde{X}_{pred}(l)[v,m,n]\tilde{X}_{true}(l)[v,m,n]}{\sqrt{\sum_{m,n} L(m)(\tilde{X}_{pred}(l)[v,m,n])^2 \sum_{m,n} L(m)(\tilde{X}_{true}(l)[v,m,n])^2}}, \tag{9}$$

where $\tilde{X}_{pred/true}(l)[v,m,n]$ represents the long-term-mean-subtracted value of predicted or true variable v at the location denoted by the grid co-ordinates

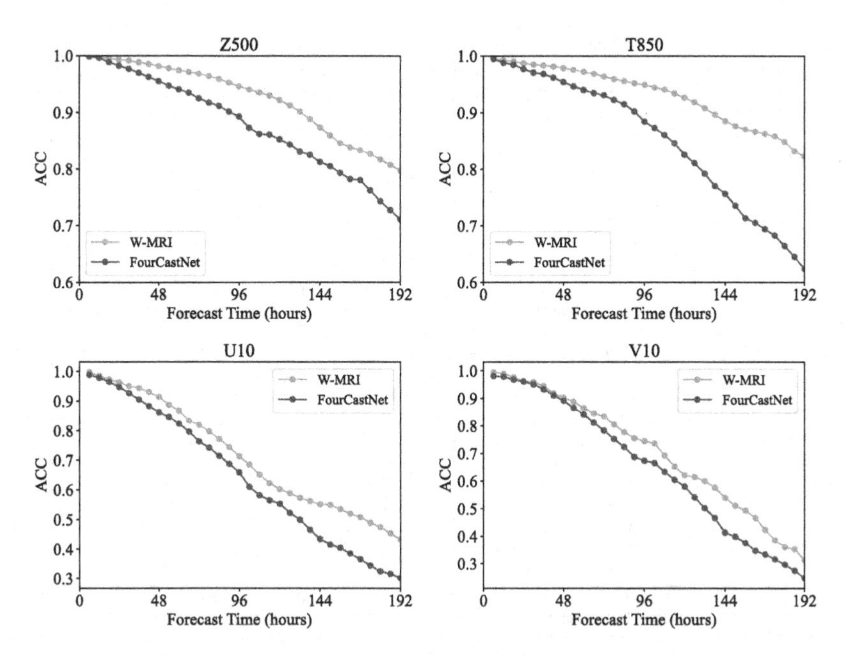

Fig. 2. Latitude weighted ACC for our W-MRI and the FourCastNet forecasts, at a 6-h interval over a 8-day lead time in the testing dataset corresponding to the calendar year 2018. (Color figure online)

(m, n) at the forecast time-step l. The long-term mean of a variable is just the mean value of it over a large number of historical samples in the training dataset. The long-term mean-subtracted variables $\tilde{X}_{pred/true}$ represent the anomalies of those variables that are not captured by the long term values. $L(m)$ is the latitude weighting factor at the co-ordinate m. The latitude weighting is defined as:

$$L(j) = \frac{\cos(lat(m))}{\frac{1}{N_{lat}}\sum_{j}^{N_{lat}} \cos(lat(m))}. \tag{10}$$

The latitude-weighted RMSE for a forecast variable v at forecast time-step l is defined by the following equation:

$$RMSE(v, l) = \sqrt{\frac{1}{NM}\sum_{m=1}^{M}\sum_{n=1}^{N} L(m)(X_{pred}(l)[v, m, n] - X_{true}(l)[v, m, n])^2}, \tag{11}$$

where $X_{pred/true}(l)[v, m, n]$ represents the value of predicted variable v at the location denoted by the grid co-ordinates (m, n) at the forecast time-step l.

5.2 Quantitative Forecasting Performance of W-MRI

We compare the ACC and RMSE metrics for each variable at each time-step (up to 8 days) with the corresponding FourCastNet forecasts under the same

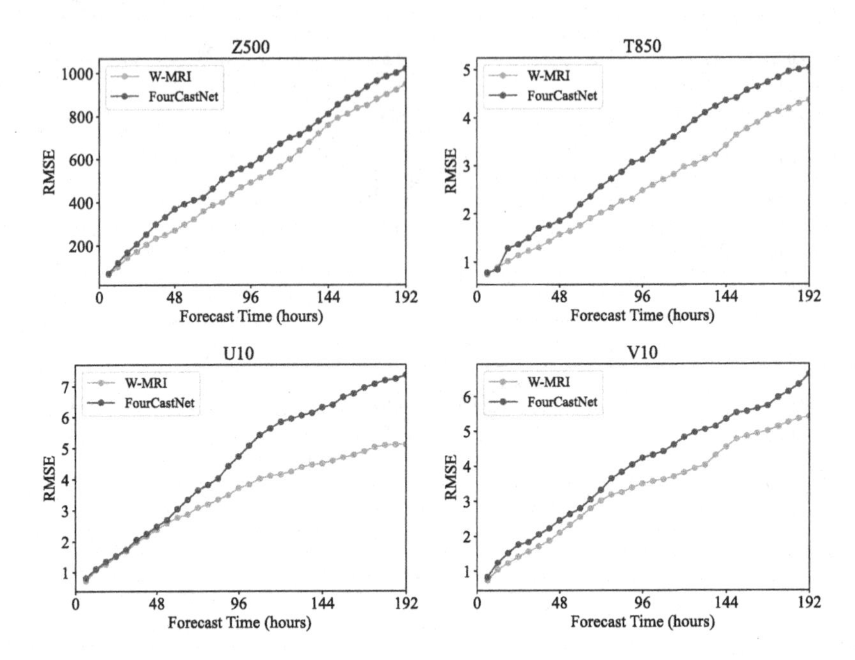

Fig. 3. Latitude weighted RMSE for our W-MRI and the FourCastNet forecasts, at a 6-h interval over a 8-day lead time in the testing dataset corresponding to the calendar year 2018. (Color figure online)

initial conditions. Figures 2 and 3 show the latitude-weighted ACC and RMSE for our W-MRI model (red line with markers) and the corresponding matched FourCastNet forecasts (blue line with markers). It can be seen that W-MRI has both higher ACC and lower RMSE than FourCastNet. Specifically, our proposed W-MRI consistently delivers stable and satisfactory results across a range of meteorological variables. In contrast, FourCastNet exhibits slightly lower prediction accuracy and stability. The advantages become more significant as the forecast time increases. This shows that our proposed W-MRI model has better performance than FourCastNet. We also provide some visual forecast results in Fig. 4. Compared with the ground-truth, our model has high accuracy in short-to-medium-range weather forecasting.

In weather and climate models, Total Precipitation (TP) represents the amount of water (rain or snow) that reaches the land surface. Forecasting TP is a crucial task in weather forecasting. Unlike other meteorological variables, TP exhibits nonlinear and sparse spatial characteristics. Since TP is influenced by multiple meteorological variables, it requires the utilization of meteorological variables from the previous time step to make accurate predictions. As shown in Fig. 5, we use the ACC and RMSE metrics to compare the performance of W-MRI and FourCastNet in predicting total precipitation. The results clearly demonstrate that our W-MRI model outperforms FourCastNet in accurately forecasting 6-h total precipitation. This indicates that W-MRI effectively captures the complex relationships between meteorological variables and provides more accurate predictions for TP.

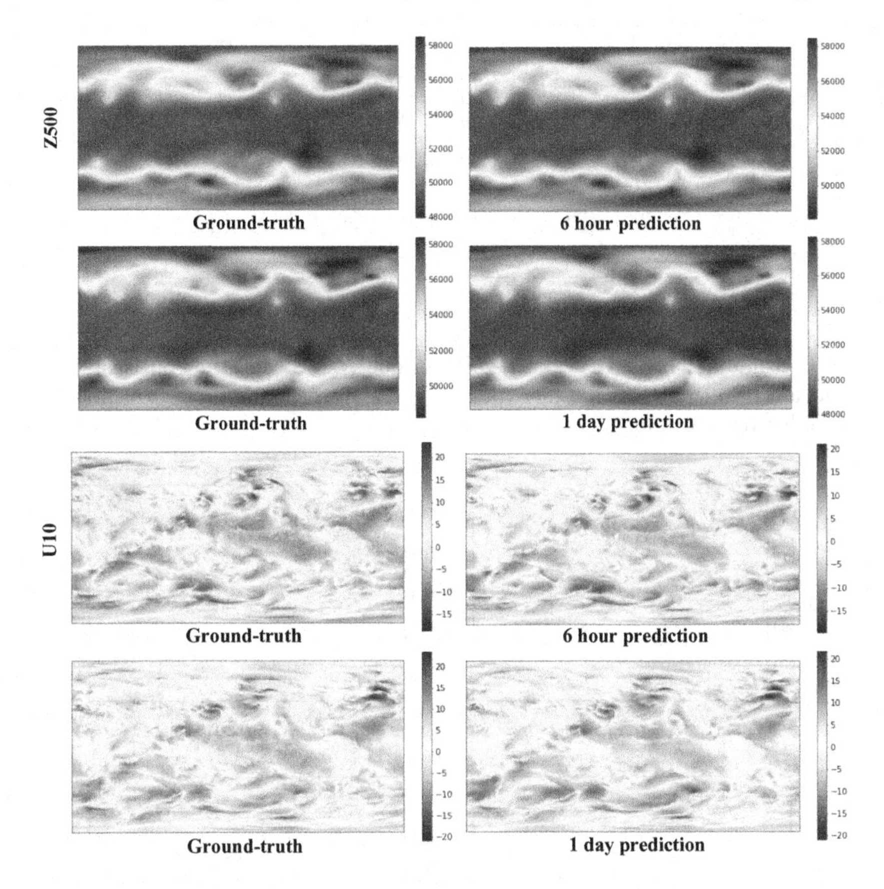

Fig. 4. Visualization examples of future state prediction for meteorological variables compared with ground-truth, including Z_{500} and U_{10}.

5.3 Effect of Integration Constraint Module

To fully leverage the predictive power of multi-output residuals and enhance their ability to capture subtle changes, we introduce an integration constraint module. This module is designed to integrate and constrain the multiple residuals generated by W-MRI. We conduct experiments to compare the performance of W-MRI with and without the integration constraint module, in order to evaluate the effectiveness of this module in enhancing weather forecasting skills. Our ablation results, as shown in Fig. 6, indicate that it is a crucial component in improving the accuracy of weather predictions.

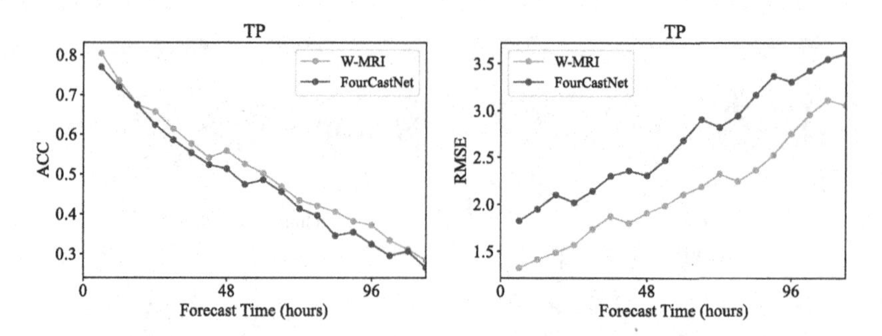

Fig. 5. Latitude weighted ACC and RMSE for our W-MRI and the FourCastNet forecasts, at a 6-h interval over a 3-day lead time in the testing dataset corresponding to the calendar year 2018 for TP.

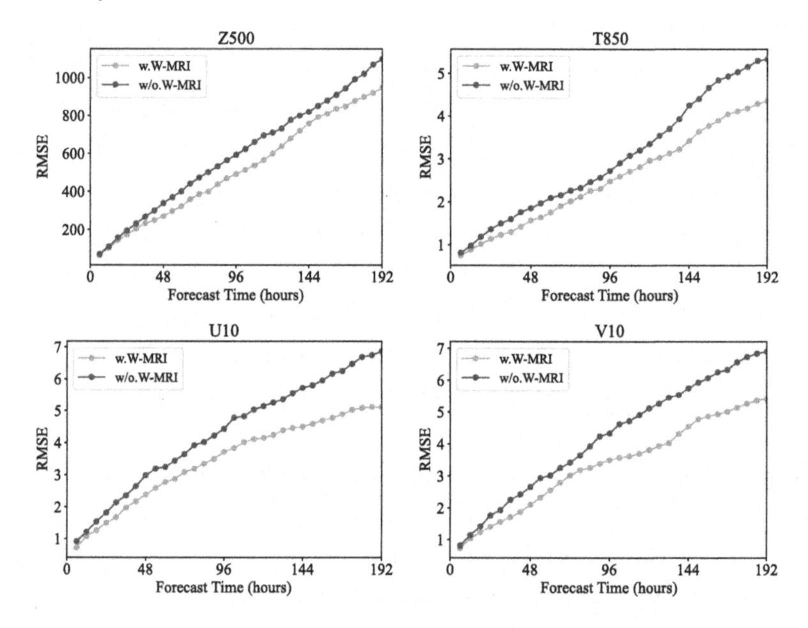

Fig. 6. Latitude weighted RMSE for W-MRI with or without the integration constraint module, at a 6-h interval over a 8-day lead time in the testing dataset corresponding to the calendar year 2018.

6 Conclusion

In this paper, we introduce the residual mechanism to the weather forecasting task and propose a multi-output residual integration model, W-MRI. This model effectively utilizes residuals to achieve short-to-medium-range global weather forecasting. Compared with the baseline model FourCastNet, our W-MRI exhibits greater stability and superior performance. These advantages are particularly prominent when the prediction time-step is increased. Furthermore, the residual network, optimized for meteorological tasks, can better capture the characteristics and internal relationships of meteorological data, thereby improving the model's generalization ability. For the future work, we plan to extend W-MRI by incorporating a more complicated residual network and applying it to a wide range of weather and climate prediction tasks.

Acknowledgments. This work was supported by the National Natural Science Foundation of China (No. 62276047 and No. 61832001) and Sichuan Science and Technology Program (No. 2023JDGD0016). The authors would like to thank Chenghong Zhang and Yuanchang Dong from Institute of Plateau Meteorology, China Meteorological Administration, and Chao Lu and Jian Chen from Chengdu Intelligent Computing Center for their useful insights and fruitful discussions in this research.

References

1. Abbe, C.: The physical basis of long-range weather forecasts. Mon. Weather Rev. **29**(12), 551–561 (1901)
2. Bauer, P., Thorpe, A., Brunet, G.: The quiet revolution of numerical weather prediction. Nature **525**(7567), 47–55 (2015)
3. Best, M.J.: Representing urban areas within operational numerical weather prediction models. Bound.-Layer Meteorol. **114**, 91–109 (2005)
4. Bjerknes, V.: The problem of weather prediction, considered from the viewpoints of mechanics and physics. Meteorol. Z. **18**(6), 663–667 (2009)
5. Denby, L.: Discovering the importance of mesoscale cloud organization through unsupervised classification. Geophys. Res. Lett. **47**(1), e2019GL085190 (2020)
6. Dewitte, S., Cornelis, J., Müller, R., Munteanu, A.: Artificial intelligence revolutionises weather forecast, climate monitoring and decadal prediction. Remote Sens. **13**(16), 3209 (2021)
7. Dosovitskiy, A., et al.: An image is worth 16 × 16 words: transformers for image recognition at scale. In: 9th International Conference on Learning Representations, ICLR 2021, Virtual Event, Austria, 3–7 May 2021 (2021)
8. Dueben, P.D., Bauer, P.: Challenges and design choices for global weather and climate models based on machine learning. Geosci. Model Dev. **11**(10), 3999–4009 (2018)
9. He, K., Zhang, X., Ren, S., Sun, J.: Deep residual learning for image recognition. In: 2016 IEEE Conference on Computer Vision and Pattern Recognition, CVPR 2016, Las Vegas, NV, USA, 27–30 June 2016, pp. 770–778 (2016)
10. Hersbach, H., Bell, B., Berrisford, P., Hirahara, S., et al.: The ERA5 global reanalysis. Q. J. R. Meteorol. Soc. **146**(730), 1999–2049 (2020)
11. Irrgang, C., et al.: Towards neural earth system modelling by integrating artificial intelligence in earth system science. Nat. Mach. Intell. **3**(8), 667–674 (2021)
12. Li, H., Tang, M., Liao, K., Shao, J.: A multi-output integration residual network for predicting time series data with diverse scales. In: PRICAI 2022: Trends in Artificial Intelligence - 19th Pacific Rim International Conference on Artificial Intelligence, PRICAI 2022, Shanghai, China, 10–13 November 2022, Proceedings, Part I, pp. 380–393 (2022)
13. Liang, Z., et al.: Learning for disparity estimation through feature constancy. In: 2018 IEEE Conference on Computer Vision and Pattern Recognition, CVPR 2018, Salt Lake City, UT, USA, 18–22 June 2018, pp. 2811–2820 (2018)

14. Lorenc, A.C.: Analysis methods for numerical weather prediction. Q. J. R. Meteorol. Soc. **112**(474), 1177–1194 (1986)
15. Lorenz, E.N.: Energy and numerical weather prediction. Tellus **12**(4), 364–373 (1960)
16. Makridakis, S., Spiliotis, E., Assimakopoulos, V.: The M4 competition: 100,000 time series and 61 forecasting methods. Int. J. Forecast. **36**(1), 54–74 (2020)
17. Navon, I.M.: Data assimilation for numerical weather prediction: a review, pp. 21–65 (2009)
18. Park, S.K., Xu, L. (eds.): Data Assimilation for Atmospheric, Oceanic and Hydrologic Applications (Vol. II). Springer, Heidelberg (2013). https://doi.org/10.1007/978-3-642-35088-7
19. Pathak, J., et al.: FourCastNet: a global data-driven high-resolution weather model using adaptive Fourier neural operators. CoRR abs/2202.11214 (2022)
20. Rasp, S., Dueben, P.D., Scher, S., Weyn, J.A., Mouatadid, S., Thuerey, N.: WeatherBench: a benchmark data set for data-driven weather forecasting. J. Adv. Model. Earth Syst. **12**(11), e2020MS002203 (2020)
21. Reichstein, M., et al.: Deep learning and process understanding for data-driven earth system science. Nature **566**(7743), 195–204 (2019)
22. Robert, A.: A semi-Lagrangian and semi-implicit numerical integration scheme for the primitive meteorological equations. J. Meteorol. Soc. Japan. Ser. II **60**(1), 319–325 (1982)
23. Rodwell, M.J., Richardson, D.S., Hewson, T.D., Haiden, T.: A new equitable score suitable for verifying precipitation in numerical weather prediction. Q. J. R. Meteorol. Soc. **136**(650), 1344–1363 (2010)
24. Rossini, N.S., Dassisti, M., Benyounis, K.Y., Olabi, A.G.: Methods of measuring residual stresses in components. Mater. Des. **35**, 572–588 (2012)
25. Schultz, M.G., et al.: Can deep learning beat numerical weather prediction? Phil. Trans. R. Soc. A **379**(2194), 20200097 (2021)
26. Simmons, A.J., Hollingsworth, A.: Some aspects of the improvement in skill of numerical weather prediction. Q. J. R. Meteorol. Soc. **128**(580), 647–677 (2002)
27. Stensrud, D.J.: Parameterization Schemes: Keys to Understanding Numerical Weather Prediction Models. Cambridge University Press, Cambridge (2007)
28. Xu, Z., Du, J., Wang, J., Jiang, C., Ren, Y.: Satellite image prediction relying on GAN and LSTM neural networks. In: 2019 IEEE International Conference on Communications, ICC 2019, Shanghai, China, 20–24 May 2019, pp. 1–6 (2019)
29. Zhang, H., et al.: ResNest: split-attention networks. In: IEEE/CVF Conference on Computer Vision and Pattern Recognition Workshops, CVPR Workshops 2022, New Orleans, LA, USA, 20–24 June 2022, pp. 2735–2745 (2022)

HV-Net: Coarse-to-Fine Feature Guidance for Object Detection in Rainy Weather

Kaiwen Zhang[1] and Xuefeng Yan[1,2](✉)

[1] Nanjing University of Aeronautics and Astronautics, Nanjing, China
{kaiwenzhang,yxf}@nuaa.edu.cn
[2] Collaborative Innovation Center of Novel Software Technology and Industrialization, Nanjing, China

Abstract. Object detection algorithms have been extensively researched in the field of computer vision, but they are still far from being perfect, especially in adverse weather conditions such as rainy weather. Traditional object detection models suffer from an inherent limitation when extracting features in adverse weather due to domain shift and weather noise, leading to feature contamination and a significant drop in model performance. In this paper, we propose a novel staged detection paradigm inspired by the human visual system, called Human Vision Network (HV-Net). HV-Net first extracts coarse-grained edge features and leverages their insensitivity to weather noise to reduce feature contamination and outline the edges of large and medium-sized objects. The subsequent network uses deep fine-grained features and edge-attentional features to generate clear images, enhancing the understanding of small objects that might be missed in edge detection and mitigating weather noise. The staged end-to-end pipeline design allows the clear features to be shared throughout the network. We validate the proposed method for rainy weather object detection on both real-world and synthetic datasets. Experimental results demonstrate significant improvements of our HV-Net compared to baselines and other object detection algorithms.

Keywords: Object detection · Edge detection · Adverse weather

1 Introduction

As traffic accidents continue to rise globally, research on intelligent vehicles (IVs) has become a critical area of focus within intelligent transportation systems. IVs rely heavily on fast and accurate visual perception, including object tracking [35], object detection [31], and lane detection [29], with object detection being the most critical component, allowing them to perceive their surroundings and make informed decisions.

Supported by organization the Basic Research for National Defense under Grant Nos. JCKY2020605C003.

X. Song et al. (Eds.): APWeb-WAIM 2023, LNCS 14332, pp. 223–238, 2024.
https://doi.org/10.1007/978-981-97-2390-4_16

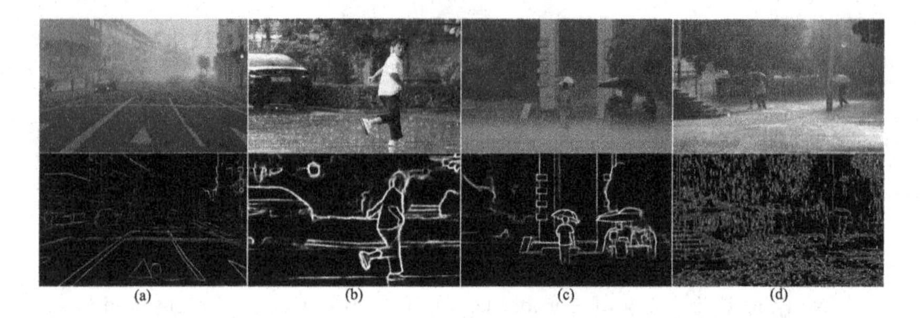

Fig. 1. A comparison of different edge maps generated from various rainy images. (a) depicts a synthetic rain image generated based on the model proposed in [12]. In (b), a real photograph showcases rain in the form of raindrops, while (c) showcases rain in the form of rain streaks. Notably, the deep learning-based method [19] successfully filters out the weather noise in the first three edge maps. In contrast, the method proposed by Canny [3] in (d) proves to be sensitive to weather noise.

Both CNN-based [18,24,26] and transformer-based [4] object detection methods operate smoothly on normal images. But they struggle to perform well in adverse weather conditions. The degradation in detector performance can be attributed to two main reasons. Firstly, the training data for most object detection methods are collected under sunny weather conditions, resulting in a significant domain shift when using images captured in adverse weather conditions. Secondly, when capturing rainy images, raindrops with irregular shapes and sizes may fall on vehicles, pedestrians, or other objects, leading to changes in the captured image. Additionally, raindrops can affect the refraction and propagation of light, while raindrop occlusion and increased fog caused by heavy rainfall make it difficult for detectors to extract favorable features as they do under normal weather conditions.

To enhance the performance of object detection in rainy conditions, previous methods can be roughly divided into two categories: single-image rain removal-based methods [11,21,25,33] and domain adaptation-based methods [6]. Deraining-based methods usually first input the image with rain noise into the net to obtain a high-quality clear image, and then input the clear image into the object detection model. However, this method has two main drawbacks. Firstly, while the deraining model smooths the edges of objects in the image to achieve better visual quality, it also causes feature loss to a certain extent, which plays a crucial role in the positioning and classification of the detection network. Secondly, non-end-to-end detection methods will inevitably incur additional time overhead in the deraining process. Therefore, these methods cannot guarantee detection accuracy. Domain Adaptation Faster R-CNN proposes a model that mitigates domain shift and proposes a consistency regularization to ensure domain invariance in the RPN network. However, in the case of significant differences between the target domain data and the source domain data, it may not achieve good adaptation effects, resulting in ambiguity about whether a certain region contains rainwater. Moreover, such methods have poor interpretability.

In order to improve the performance of object detection in rainy conditions and address some of the shortcomings of previous methods, a novel object detection model called Human Vision Network (HV-Net) is proposed in this paper. Inspired by the human visual system, even in challenging conditions like a rainy environment, humans prioritize and process information in a sequential manner. So we design our network, which consists of three stages: edge searching, image restoration, and object detection. To balance detection accuracy and speed, we employ one of the state-of-the-art one-stage detector YOLOv7-Tiny as our object detector and feature extractor for the entire model. As depicted in Fig. 1, edge detection can filter out a large amount of raindrop noise and outline medium to large objects. Therefore, in the edge searching stage, we use relatively shallow features to predict edge maps and use a lightweight decoder to generate edge-attentional features, enhancing the network's ability to locate medium to large objects and reducing interference from weather noise. In the image restoration stage, HV-Net uses deep multiscale features and edge-attentional features to generate clear images and enable the network to detect small objects missed in the edge detection stage. The object detection stage forms an end-to-end pipeline with the first two stages and shares some features guided from coarse to fine, thereby improving the performance of detection on rainy-day.

Our contributions are three-fold:

1. By simulating how humans detect objects in rainy conditions, we propose a novel object detection network that enhances the performance of the cutting-edge model by reducing feature contamination through coarse-to-fine feature guidance.
2. We split the detection process into three stages: edge detection, image restoration, and object detection, and propose an edge-guided attention mechanism to combine edge information with fine-grained features. Three stages share clear features.
3. HV-Net is validated on both synthetic and real-world datasets and achieves the most outstanding performance among another methods.

2 Related Work

2.1 Object Detection

Object detection is a fundamental and active research area within computer vision, and deep convolutional neural networks have revolutionized the field. Deep learning-based object detection methods can be broadly classified into two major categories.

The first category comprises two-stage region proposal-based approaches, which include pioneering methods such as R-CNN [10], Fast R-CNN [9], and Faster R-CNN [26], among others. These methods typically employ a region proposal network (RPN) to generate potential object bounding box proposals. These proposals are then fed into a separate classifier for object classification and a regressor for precise bounding box refinement. These methods achieve remarkable detection accuracy.

The second category is the one-stage methods, which approach object detection as a regression problem, simultaneously addressing object classification and localization. Representative methods in this category include the influential YOLO series [2,8,22–24] and SSD [18]. One-stage methods are renowned for their real-time processing speed as they directly predict object bounding boxes and class probabilities using CNNss, without the need for an explicit region proposal step. These methods offer high efficiency in terms of computation and memory usage, making them well-suited for applications requiring real-time processing, such as autonomous driving and video surveillance.

Recent studies have investigated methods based on the Transformer architecture, such as DETR [4]. DETR serves as a representative work in the field of object detection, being the first to adopt the Transformer for this task. DETR has demonstrated superior detection performance compared to Faster R-CNN. Notably, DETR does not require predefined candidate regions and eliminates the need for Non-Maximum Suppression (NMS) to remove redundant bounding boxes, resulting in a more elegant algorithmic process.

2.2 Single Image Deraining

In the early stages, manual priors based on low-level image statistics [1,13,27] were designed for rain pattern removal in images. With the impressive learning capabilities of CNNs, researchers increasingly started applying CNNs to image rain removal. Fu et al. proposed DerainNet [7], the first depth CNN-based approach for rain removal. They divided input images into a high-frequency detail layer and a low-frequency base layer, which provided a direct nonlinear mapping between rain-contaminated and rain-free images. Additionally, image enhancement techniques were employed to improve the visual quality of the images.

Qian et al. [21] used generative adversarial networks (GANs) for rain removal and re-model rainy images. Based on this model, the background image could be separated from the input image, and an attention distribution map was established. Attention mechanisms were integrated into the generator and discriminator of the GAN to learn the rain areas and their surrounding regions. The generator's attention mechanism emphasized the rain area and its neighboring regions, while the discriminator's attention mechanism evaluated the local consistency of the restored areas. Ren et al. proposed another approach named PReNet [25], which recursively removes rain from a single image using long short-term memory (LSTM) layers and ResNet blocks. This progressive method gradually eliminates rain from different parts of the image in each iteration, resulting in a rain-free image.

3 Proposed Method

Motivation. Biological studies [15,30] have shown that since the environment contains an enormous amount of information evolution has developed a step-by-step intake of this information by allocating the gaze and attention on particular

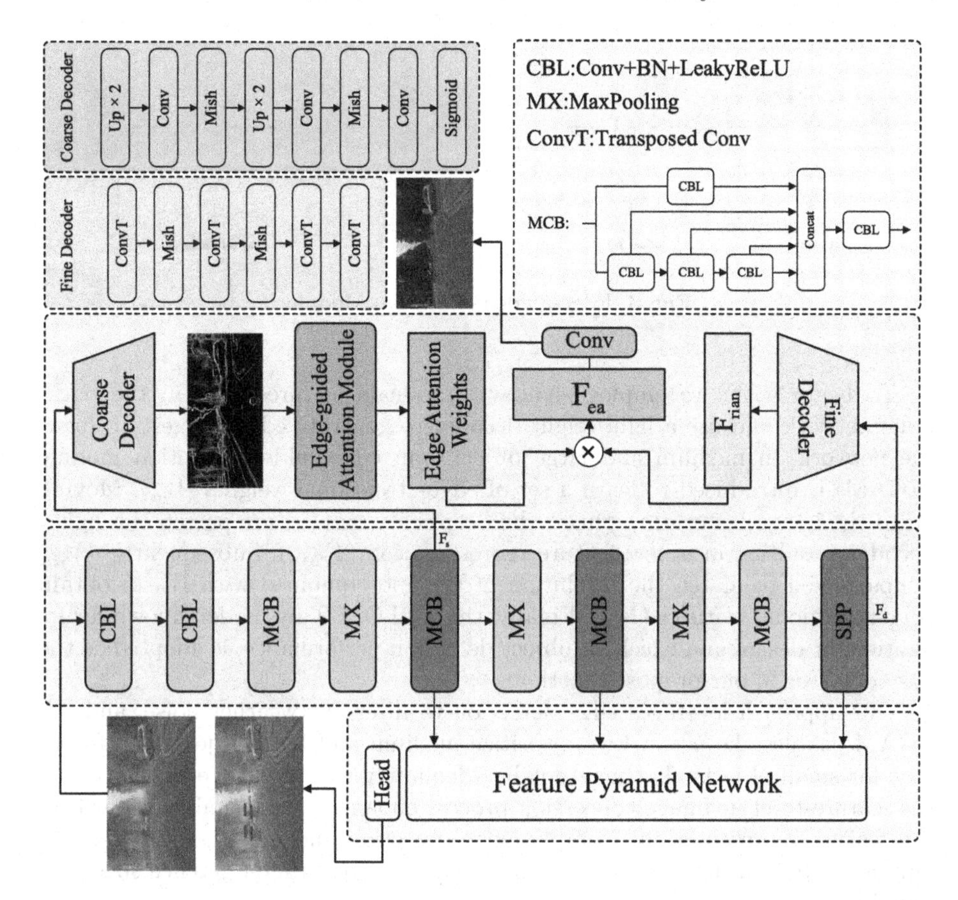

Fig. 2. Architecture of the proposed HV-Net.

parts of interest. For example, in a rainy environment, let's assume there is a black car of a certain brand within our field of view. Due to the presence of rain, we may not initially be concerned with the fine details of the car. However, even on a rainy day, the differences between the object's edges and the background are still apparent. Therefore, we can locate the specific position of the object and then identify it as a car based on prior knowledge. Finally, if necessary, we would be interested in its color and brand. For humans, this process is very rapid, but we suggest that this process can be divided into multiple stages, and each stage can be optimized independently.

Overview. The proposed HV-Net draws inspiration from the three stages involved in object detection performed by humans. It consists of three key stages: edge detection, image restoration, and object detection. The object detection stage shares a subset of rain-free features with the preceding two stages, thereby establishing an end-to-end network that maintains high performance and efficiency in both rainy and normal weather conditions.

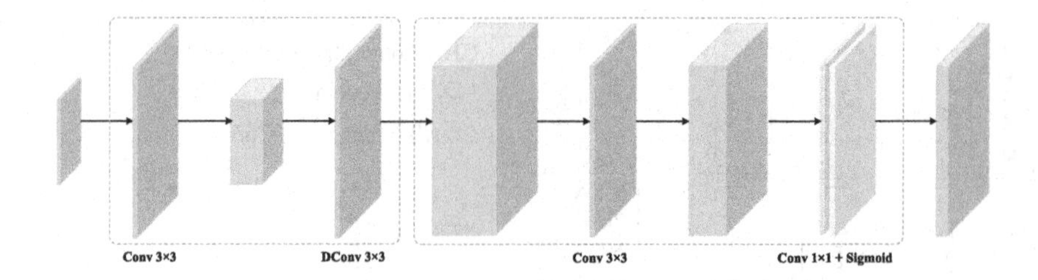

Fig. 3. Edge-guided Attention Module.

To begin with, we employ shallow large-scale features (F_s) in the backbone network and use a lightweight decoder to generate edge images. To focus the network on medium and large objects, an edge-guided attention module (EGAM) is introduced to learn a set of edge-attentional weights (W_a). Moving on to the image restoration phase, the deep features (F_d) are passed through a decoder, resulting in a new feature representation (F_{rain}). Subsequently, F_{rain} is upsampled to match the resolution of W_a and combined with W_a to obtain edge-attentional features (F_{ea}). Finally, the YOLOv7-Tiny model, known for its lightweight design and excellent object detection performance, is adopted as the base detector in our proposed method.

Our approach leverages edge detection to filter out weather noise, and the EGAM enables the network to prioritize medium and large objects. Additionally, for small objects that may not be adequately captured by edge detection, the feature-to-clear image conversion process enhances their significance within the feature representation. Notably, edge detection and image restoration are only activated during the training phase, ensuring real-time inference speed for individual images.

3.1 Generate the Edge Map

The first stage focuses on object discovery and localization. Edge information plays a crucial role in this process as it represents the transitional region between an object and the background. It aids the network in determining the approximate location of the object and provides valuable feature information such as shape, texture, and structure. This feature information contributes to distinguishing between different object classes to some extent. Extracting edges from an image incurs less computational overhead compared to more advanced computer vision tasks like object detection and semantic segmentation. Therefore, our approach uses edge information as a coarse-grained feature to guide the network's focus on medium and large objects in adverse weather conditions (Fig. 3).

Traditional methods based on image intensity and color gradients were initially considered for edge detection. Although these methods offer speed and accurate edge localization, they are sensitive to noise and prone to edge fragmentation. As depicted in Fig. 1, such traditional methods, like Canny [3], demonstrate poor robustness in complex scenes with high levels of weather noise, such as rainy images. In contrast, deep learning-based methods exhibit the capability to effectively filter out weather information and perform well in such challenging scenarios.

Therefore, we designed a lightweight decoder, named Coarse Decoder (CD). The blue region in Fig. 1 represents the overall structure of the decoder. Specifically, it obtains features of half the size of the original input image by using two sets of upsampling layers, 1×1 convolutional layers, and the Mish activation function. Finally, convolutional layers and the Sigmoid activation function are applied to generate the edge map. The use of 1×1 convolutional layers and the output of low-resolution features reduces the number of parameters in the entire decoder. The shallow, large-scale features (F_s) are fed into the CD, which directly generates the edge map supervised by the ground truth (GT). The generated edge map serves as a guiding reference for the network to learn attention weights in subsequent stages.

Loss Function. As mentioned in [32], the proportion of edge pixels in the overall pixel set is typically small in edge detection tasks. This imbalance between edge and non-edge pixels can significantly impact the learning process of the network and lead to a large number of non-edge pixels being detected in the results. To address this issue, a balancing weight β is introduced to automatically adjust the loss balance between positive and negative samples. The cross-entropy loss function is used to mitigate this problem. Specifically, if the number of edge pixels is small in the image, the value of β is reduced, resulting in a lighter weight for the first term in the loss function, which represents the loss for non-edge pixels. This helps to reduce the overall loss.

We compute the loss of each pixel with respect to its label as

$$L\left(X_i;W\right)=\begin{cases}\beta\cdot log\left(1-P\left(X_i;W\right)\right) & y_i=0\\ \left(1-\beta\right)\cdot log\left(X_i;W\right) & y_i=1\end{cases} \tag{1}$$

in which

$$\beta=\frac{\left|Y^+\right|}{\left|Y^+\right|+\left|Y^-\right|} \tag{2}$$

Y^+ and Y^- represent the numbers of edge and non-edge pixels, respectively, while W represents the set of network parameters. $P\left(x\right)$ is the sigmoid function.

3.2 From Edge-Attentional Features to Image

As mentioned in the previous section on motivation, visual information collected by the eyeballs is transmitted to the brain cortex, and the reconstructed images based on this visual information are continuously compared with our learned prior knowledge. We abstract this process as an image restoration task. We combine the coarse-grained edge features obtained in the initial stage with the subsequently acquired fine-grained features. Guided by a set of edge-attentional weights, we perform feature fusion and directly restore a clear image using this combined feature representation.

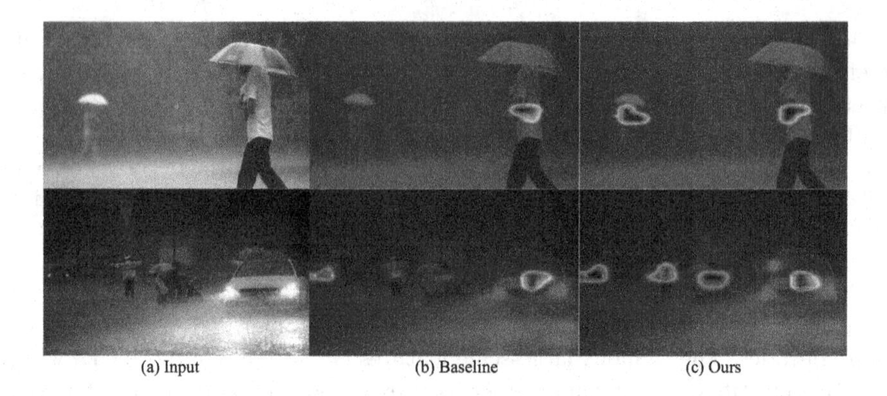

Fig. 4. Heatmap of the detection results of baseline and HV-Net.

Edge-Attentional Feature. As explained in Subsect. 3.1, edge detection plays a crucial role in eliminating weather noise and enhancing the details and contours of medium to large objects in images. To leverage this, the generated edge map in the edge detection stage guides the learning process of attention weights through the Edge-guided Attention Module (EGAM) in our network. The detailed structure of EGAM is illustrated in Fig. 4.

Initially, the edge map undergoes a 1×1 convolutional layer to increase its dimension. It then passes through a 3×3 dilated convolutional layer, preserving the feature map's details and spatial distribution while matching the input image's resolution. Subsequently, a 3×3 convolutional layer reduces the dimensionality, followed by a 1×1 convolutional layer and an activation function that generates the edge-attentional weight (W_a).

The features F_s, which carry semantic information, are fed into the FD (as depicted in the green part of Fig. 2). Through three deconvolutional layers and the Mish activation function, F_s is processed to obtain F_{rain} with the same resolution as the input image. W_a is then element-wise multiplied with F_{rain} on a channel-wise basis, resulting in F_{ea}. It is important to note that F_{ea} represents fine-grained features. Finally, a 1×1 convolutional layer is used to linearly combine the features from different channels in F_{ea}, thereby fusing information from various levels, enhancing the feature expressiveness, and producing a clear image with the incorporation of supervised signals.

Loss Function. In the image restoration stage we use the simple Mean Square Error (MSE) loss function to train the network, which can be expressed as:

$$L_{MSE} = \frac{1}{n} \sum_{i=1}^{n} (\bar{Y_i} - Y_i)^2 \tag{3}$$

where n is the batch size, $\bar{Y_i}$ is the ground truth, and Y_i is the restored clean image.

3.3 Object Detection Stage

As discussed in Sect. 2, a plethora of CNN-based object detection models have been proposed. Among them, the YOLO series detectors have gained significant recognition as prominent single-stage detection models. YOLOv7, an improved iteration of its predecessors, introduces a coarse-to-fine lead head guided label assigner in data processing. Additionally, it incorporates an increased number of skip connection structures within the backbone network to enhance feature extraction. Moreover, YOLOv7 integrates a specialized SPP structure that combines the CSP structure within the SPP module, resulting in a significant residual edge that aids in optimizing feature extraction.

To further enhance the detection performance, YOLOv7-tiny leverages FPN [17], using the three valid feature layers from the backbone for feature fusion. The purpose of feature fusion is to combine information from features of different scales. Notes that the F_s used in the initial two stages and F_d represent two of the three effective feature layers. In other words, these three stages share some common clear features. This approach facilitates the generation of higher-resolution layers by leveraging rich semantic layers, ultimately reconstructing sharper images. Extensive experiments have demonstrated the success of the YOLOv7-tiny across various scenarios, striking a balance between speed and detection accuracy. Therefore, we chose it as the base detector for our proposed multi-scale object detection framework.

4 Experimental Results

In this section, we will discuss the experimental results of HV-Net, as well as the results of other detection methods for adverse weather. The purpose is to evaluate their performance and compare their generalization ability using different datasets. For our experiments, we use two datasets: a synthetic rainy dataset called RainCityscapes [12] and a real-world rain dataset named RID [16].

We comprehensively compare the proposed HV-Net with object detection methods that use transfer learning and deraining-based methods. Firstly, we compare our approach with three state-of-the-art object detection methods: FasterRCNN [26], DETR [4], SFA-Net [14], and YOLOv7-Tiny [31]. Subsequently, we combine each of the two deraining algorithms (MPRNet [33], DRS-former [5]) with baseline and compare them with our approach.

All experiments were implemented using PyTorch. The hardware configuration included two NVIDIA RTX 3090Ti GPUs, an Intel Core i9-12900KF CPU, and 32 GB RAM. For images with a size of 640×640, the average inference time was 0.0642 s.

4.1 Datasets

Computer vision tasks, such as object detection, heavily rely on extensive data. However, real rain datasets are limited, and obtaining annotated pairs of images with and without rain in the same scene is practically infeasible. To overcome this challenge, we use the RainCityscapes dataset, which synthesizes rainy-day images from 285 real images. The rain generation model in RainCityscapes leverages depth information to synthesize rain and fog with various parameters, generating 36 images with different degrees of rain based on a single clear image. Furthermore, the images in RainCityscapes are annotated, making them suitable for the object detection task. We partitioned the dataset into training, validation, and test sets with an 8:2:1 ratio.

To assess the model's detection performance in real rainy conditions, we also employ the RID (Rain in Driving Set) dataset [16] as an additional test set. RID comprises 2495 images depicting real-world rain scenes, and the labeled object types in its subset are consistent with those in our training set. Most images have a resolution of 1920×990, with a few exceptions of 4023×3024. The high resolution and weather noise present in these images make the dataset particularly challenging. Although this is the first task-driven evaluation scheme for deraining algorithms. But this dataset can also be effectively used for object detection in adverse weather conditions (Table 1).

Table 1. Details of the training and testing datasets used in our experiment.

Dataset	Type	Num	Train/Test	Per	Bus	Car	Mot	Bic
RainCityscapes	Synthetic	8488	Train	30394	799	93826	1458	8601
RainCityscapes	Synthetic	944	Test	3518	65	10538	162	1047
RID	Real	2495	Test	1135	613	7332	968	268

Due to the unavailability of manually annotated labels in the Cityscapes dataset for the network's edge detection stage, a solution was devised. We used the RCF [19] network to generate edge maps, which were then used as labels for the edge detection stage. RCF is a highly effective network for edge detection, leveraging convolutional features at various scales. Given that our main focus is object detection, the edge maps generated by RCF adequately fulfill our requirements.

4.2 Implementation Details

Our HV-Net consists of three stages, and each stage requires supervision. For training, we use the RainCityscapes dataset, which provides both clear images and corresponding rainy images of the same scene. In the first stage, we generate edge map labels using RCF, while the labels in the second stage are the corresponding clear images from Cityscapes. The object detection labels are JSON files also obtained from Cityscapes.

Throughout our experiments, we fix the image size at 640×640. During training, we employ an SGD optimizer [34] with a batch size of 16. The initial learning rate is set to 1×10^{-4}, and we use the Cosine annealing scheduler [20] to adjust the learning rate during training. To enhance the network's generalization ability and learning speed, we apply label smoothing techniques [28].

We empirically set the total number of training epochs to 220. For the last 20 epochs, we reduce the batch size to 8 and further decrease the learning rate to 1×10^{-5}. It is important to note that we employ only simple data augmentation methods, such as rotation and color shifting. This choice is made to ensure that complex augmentation methods (e.g., CutMix and Mosaic used by YOLOv7) do not hinder the image recovery task and indirectly impact the performance of the object detection task when features are shared among the three tasks.

Beside, in order to compare the performance of each method comprehensively and fairly, we report mean average precision (mAP) as a criterion with a threshold of 0.5 for evaluation.

4.3 Qualitative and Quantitative Results

To validate the effectiveness of HV-Net, we conducted a comparative analysis with six other object detection algorithms on both synthetic and real-world datasets. We used a pre-trained model from the VOC dataset and fine-tuned it using the RainCityscapes dataset (Fig. 5).

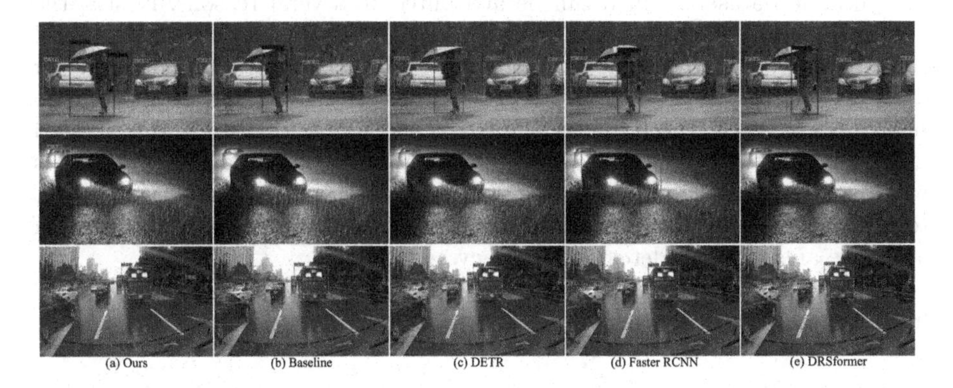

Fig. 5. Detection results by different methods on both synthetic and real-world rain datasets with a confidence threshold of 0.5. From left to right columns: the detection results by our HV-Net, YOLOv7-Tiny, DETR, Faster R-CNN, and DRS-fomer+baseline. We keep only the prediction bounding boxes with confidence greater than 0.5. Obviously, our method is able to detect more objects with higher confidence.

Table 2 presents a quantitative comparison between HV-Net and the other six methods in terms of mAP (mean average precision) on the real dataset. It is evident that the performance enhancement achieved by the combination of image deraining and object detection is limited compared to the baseline. Notably, even the two advanced deraining algorithms exhibit a decline in performance for certain categories. In contrast, our method surpasses the baseline by 4.15% and achieves the highest mAP, outperforming the state-of-the-art approach. Our approach significantly improves the AP for the car, bus, and person categories. These specific classes represent large-sized objects in the images, and our method effectively enhances the local semantic information gradients for these categories.

Table 2. Object detection results on the RID. Red and blue colors are used to indicate the 1st and 2nd metrics, respectively.

Method	Publication	Person	Car	Bicycle	Bus	Motorbike	*mAP*
YOLOv7-Tiny [31]	arXiv'22	28.15	64.13	18.12	40.95	9.69	32.21
Base+MPRNet [33]	CVPR'21	29.03	62.79	17.45	41.77	9.82	32.17
Base+DRSformer [5]	CVPR'23	27.74	66.02	16.78	42.17	8.31	32.20
FasterRCNN [26]	CVPR'15	26.08	48.56	13.90	33.94	8.97	26.29
DETR [4]	ECCV'20	35.08	62.81	16.75	47.24	9.95	34.52
SFA-Net [14]	TNNLS'22	–	–	–	–	–	32.68
HV-Net	Ours	35.85	69.44	19.67	46.44	10.33	36.36

As shown in Table 3, deraining-based methods also exhibit poor performance on synthetic datasets, which can be attributed to several reasons. Firstly, the preprocessing of images results in the loss of potential information that could contribute to improved detection accuracy. Secondly, the training datasets for deraining models are often generated by synthetically adding 2D rain streaks to photos. In contrast, the RainCityScape dataset used in this study is generated using a novel rain model that considers scene depth, rain streaks, and fog, thus making it more representative of real-world rainy outdoor scenes. It is important to note that there is no direct causal relationship between the visual quality of an image and the final detection accuracy.

In conclusion, our method surpasses other approaches in terms of performance on both test datasets.

Table 3. Object detection results on the RainCityscapes.

Method	Per	Car	Bic	Bus	Mot	*mAP*
YOLOv7-Tiny	59.69	62.38	49.47	45.93	54.19	54.33
Base+MPRNet	58.64	61.58	48.62	47.87	52.39	53.82
Base+DRSformer	60.79	65.54	46.51	43.29	46.08	52.44
FasterRCNN	42.07	68.56	43.18	44.86	55.57	50.65
DETR	56.85	77.68	49.24	67.24	29.95	56.19
HV-Net	56.33	73.28	56.58	47.96	54.72	57.77

4.4 Ablation Study

In this section, we conducted ablation experiments on the RID dataset to further evaluate the effectiveness of HV-Net. The quantitative results presented in Table 4 demonstrate the effectiveness of each stage, including the edge detection stage and the image restoration stage. Model a consists of the backbone network and the edge detection sub-network, while model b consists of the backbone network and the image restoration sub-network. Model c represents our final model.

Table 4. Ablation study

Components	Baseline	a	b	d
ED	w/o	✓	w/o	✓
IR	w/o	w/o	✓	✓
mAP	32.21	34.17	33.91	36.36

Effectiveness of Edge Detection and Image Restoration. By comparing models a, b, and the baseline, we observed that the mAP metric is superior to the baseline. This indicates that multi-stage processing of features can improve the model's performance. In conventional object detection methods, detecting large objects on rainy days is also challenging due to domain shift. The addition of the ED reduces feature contamination throughout the network, enabling the network to focus on large objects. On the other hand, IR uses deep and fine-grained features, which helps extract latent information from images that are often difficult to process. In conclusion, both stages contribute to improving the accuracy of rainy-day object detection.

5 Conclusion

We propose a novel staged detection paradigm called Human Vision Network (HV-Net) for object detection in adverse weather conditions, particularly rainy weather. HV-Net consists of three stages: edge detection, image restoration, and object detection. In the edge detection stage, coarse-grained edge features are extracted to outline large and medium-sized objects and reduce feature contamination from weather noise. The subsequent image restoration stage uses deep fine-grained features and edge-attentional features to generate clear images, enhancing the understanding of small objects and mitigating weather noise. The object detection stage employs a one-stage detector and shares clear features throughout the network. Experimental results on real-world and synthetic datasets demonstrate the superiority of HV-Net compared to baselines and other object detection algorithms. HV-Net achieves significant improvements in rainy weather object detection and overcomes the limitations of traditional methods.

Future work can focus on the application of HV-Net in other computer vision tasks, such as semantic segmentation or instance segmentation, which can be investigated to leverage its potential in broader contexts.

References

1. Barnum, P.C., Narasimhan, S., Kanade, T.: Analysis of rain and snow in frequency space. Int. J. Comput. Vision **86**, 256–274 (2010)
2. Bochkovskiy, A., Wang, C.Y., Liao, H.Y.M.: YOLOv4: optimal speed and accuracy of object detection. arXiv preprint arXiv:2004.10934 (2020)
3. Canny, J.: A computational approach to edge detection. IEEE Trans. Pattern Anal. Mach. Intell. **6**, 679–698 (1986)
4. Carion, N., Massa, F., Synnaeve, G., Usunier, N., Kirillov, A., Zagoruyko, S.: End-to-end object detection with transformers. In: Vedaldi, A., Bischof, H., Brox, T., Frahm, J.M. (eds.) Computer Vision–ECCV 2020: 16th European Conference, Glasgow, UK, 23–28 August 2020, Proceedings, Part I 16, pp. 213–229. Springer, Cham (2020). https://doi.org/10.1007/978-3-030-58452-8_13
5. Chen, X., Li, H., Li, M., Pan, J.: Learning a sparse transformer network for effective image deraining. In: Proceedings of the IEEE/CVF Conference on Computer Vision and Pattern Recognition, pp. 5896–5905 (2023)
6. Chen, Y., Li, W., Sakaridis, C., Dai, D., Van Gool, L.: Domain adaptive faster R-CNN for object detection in the wild. In: Proceedings of the IEEE Conference on Computer Vision and Pattern Recognition, pp. 3339–3348 (2018)
7. Fu, X., Huang, J., Ding, X., Liao, Y., Paisley, J.: Clearing the skies: a deep network architecture for single-image rain removal. IEEE Trans. Image Process. **26**(6), 2944–2956 (2017)
8. Ge, Z., Liu, S., Wang, F., Li, Z., Sun, J.: YOLOX: exceeding YOLO series in 2021. arXiv preprint arXiv:2107.08430 (2021)
9. Girshick, R.: Fast R-CNN. In: Proceedings of the IEEE International Conference on Computer Vision, pp. 1440–1448 (2015)
10. Girshick, R., Donahue, J., Darrell, T., Malik, J.: Rich feature hierarchies for accurate object detection and semantic segmentation. In: Proceedings of the IEEE Conference on Computer Vision and Pattern Recognition, pp. 580–587 (2014)
11. Hnewa, M., Radha, H.: Object detection under rainy conditions for autonomous vehicles: a review of state-of-the-art and emerging techniques. IEEE Signal Process. Mag. **38**(1), 53–67 (2020)
12. Hu, X., Fu, C.W., Zhu, L., Heng, P.A.: Depth-attentional features for single-image rain removal. In: Proceedings of the IEEE/CVF Conference on Computer Vision and Pattern Recognition, pp. 8022–8031 (2019)
13. Huang, D.A., Kang, L.W., Wang, Y.C.F., Lin, C.W.: Self-learning based image decomposition with applications to single image denoising. IEEE Trans. Multimedia **16**(1), 83–93 (2013)
14. Huang, S.C., Hoang, Q.V., Le, T.H.: SFA-Net: a selective features absorption network for object detection in rainy weather conditions. IEEE Trans. Neural Networks Learn. Syst. **34**(8), 5122–5132 (2022)
15. Konstantopoulos, P., Crundall, D.: The driver prioritisation questionnaire: exploring drivers' self-report visual priorities in a range of driving scenarios. Accid. Anal. Prev. **40**(6), 1925–1936 (2008)

16. Li, S., et al.: Single image deraining: a comprehensive benchmark analysis. In: Proceedings of the IEEE/CVF Conference on Computer Vision and Pattern Recognition, pp. 3838–3847 (2019)
17. Lin, T.Y., Dollár, P., Girshick, R., He, K., Hariharan, B., Belongie, S.: Feature pyramid networks for object detection. In: Proceedings of the IEEE Conference on Computer Vision and Pattern Recognition, pp. 2117–2125 (2017)
18. Liu, W., et al.: SSD: single shot multibox detector. In: Leibe, B., Matas, J., Sebe, N., Welling, M. (eds.) Computer Vision–ECCV 2016: 14th European Conference, Amsterdam, The Netherlands, 11–14 October 2016, Proceedings, Part I 14, pp. 21–37. Springer, Cham (2016). https://doi.org/10.1007/978-3-319-46448-0_2
19. Liu, Y., Cheng, M.M., Hu, X., Wang, K., Bai, X.: Richer convolutional features for edge detection. In: Proceedings of the IEEE Conference on Computer Vision and Pattern Recognition, pp. 3000–3009 (2017)
20. Loshchilov, I., Hutter, F.: SGDR: stochastic gradient descent with warm restarts. arXiv preprint arXiv:1608.03983 (2016)
21. Qian, R., Tan, R.T., Yang, W., Su, J., Liu, J.: Attentive generative adversarial network for raindrop removal from a single image. In: Proceedings of the IEEE Conference on Computer Vision and Pattern Recognition, pp. 2482–2491 (2018)
22. Redmon, J., Divvala, S., Girshick, R., Farhadi, A.: You only look once: unified, real-time object detection. In: Proceedings of the IEEE Conference on Computer Vision and Pattern Recognition, pp. 779–788 (2016)
23. Redmon, J., Farhadi, A.: YOLO9000: better, faster, stronger. In: Proceedings of the IEEE Conference on Computer Vision and Pattern Recognition, pp. 7263–7271 (2017)
24. Redmon, J., Farhadi, A.: YOLOv3: an incremental improvement. arXiv preprint arXiv:1804.02767 (2018)
25. Ren, D., Zuo, W., Hu, Q., Zhu, P., Meng, D.: Progressive image deraining networks: a better and simpler baseline. In: Proceedings of the IEEE/CVF Conference on Computer Vision and Pattern Recognition, pp. 3937–3946 (2019)
26. Ren, S., He, K., Girshick, R., Sun, J.: Faster R-CNN: towards real-time object detection with region proposal networks. In: Advances in Neural Information Processing Systems, vol. 28 (2015)
27. Sun, S.H., Fan, S.P., Wang, Y.C.F.: Exploiting image structural similarity for single image rain removal. In: 2014 IEEE International Conference on Image Processing (ICIP), pp. 4482–4486. IEEE (2014)
28. Szegedy, C., Vanhoucke, V., Ioffe, S., Shlens, J., Wojna, Z.: Rethinking the inception architecture for computer vision. In: Proceedings of the IEEE Conference on Computer Vision and Pattern Recognition, pp. 2818–2826 (2016)
29. Tabelini, L., Berriel, R., Paixao, T.M., Badue, C., De Souza, A.F., Oliveira-Santos, T.: Keep your eyes on the lane: real-time attention-guided lane detection. In: Proceedings of the IEEE/CVF Conference on Computer Vision and Pattern Recognition, pp. 294–302 (2021)
30. Treue, S.: Neural correlates of attention in primate visual cortex. Trends Neurosci. **24**(5), 295–300 (2001)
31. Wang, C.Y., Bochkovskiy, A., Liao, H.Y.M.: YOLOv7: trainable bag-of-freebies sets new state-of-the-art for real-time object detectors. In: Proceedings of the IEEE/CVF Conference on Computer Vision and Pattern Recognition, pp. 7464–7475 (2023)
32. Xie, S., Tu, Z.: Holistically-nested edge detection. In: Proceedings of the IEEE International Conference on Computer Vision, pp. 1395–1403 (2015)

33. Zamir, S.W., et al.: Multi-stage progressive image restoration. In: Proceedings of the IEEE/CVF Conference on Computer Vision and Pattern Recognition, pp. 14821–14831 (2021)
34. Zhang, S., Choromanska, A.E., LeCun, Y.: Deep learning with elastic averaging SGD. In: Advances in Neural Information Processing Systems, vol. 28 (2015)
35. Zhang, Y., et al.: ByteTrack: multi-object tracking by associating every detection box. In: Avidan, S., Brostow, G., Cissé, M., Farinella, G.M., Hassner, T. (eds.) Computer Vision–ECCV 2022: 17th European Conference, Tel Aviv, Israel, 23–27 October 2022, Proceedings, Part XXII, pp. 1–21. Springer, Cham (2022). https://doi.org/10.1007/978-3-031-20047-2_1

Vehicle Collision Warning System for Blind Zone in Curved Roads Based on the Spatial-Temporal Correlation of Coordinate

Qiao Meng(✉), Xinli Li, Yu'an Zhang, and Junyi Huangfu

Computer Technology and Application Department, Qinghai University, Xining 810016, China
250345481@qq.com

Abstract. Traffic safety has been an important research topic in intelligent transportation, especially the special terrain of mountainous areas, which increases the traffic accident rate. The main contribution of this paper is to propose a warning system for vehicles meeting in blind zones of mountain roads with low-cost, stable and reliable communication, and high accuracy data prediction. Among them, a corner-matched tracking algorithm based on special blocks and a bidirectional traffic estimation strategy based on coordinate correlation in spatiotemporal space were designed for the first time, providing reliable judgment information for the warning system. Moreover, communication methods without network environment is applied to the proposed system, solving the problem of weak network infrastructure in mountainous areas. Finally, the application performance shows that our system and its algorithm have sufficient robustness under complex weather conditions.

Keywords: Traffic Safety · Curved Road · Target Extraction · Communication Mode

1 Introduction

Over the past several years, with the rapid development of the automobile industry and the continuous improvement of the road transport infrastructure, the convenience of residents' travel has been greatly improved. However, while enjoying the convenience of transportation, people also face various transportation problems, especially traffic safety issues. Since 1978, the number of road traffic fatalities in China has always been the highest in the world. In 2022, there were a total of 2.123 million traffic accidents in China, of which more than 60% were serious accidents on roads in mountainous areas. The reason for this phenomenon is due to the terrain. Many provinces and regions in China are mountainous, and roads in mountainous areas have complex linear conditions and diverse driving environments, such as uneven terrain, steep slopes, cliffs near mountains, narrow road width, small turning radius and large blind area of sight, et al.[1–3] Therefore, if the driver is slightly careless or mishandled, it can lead to an accident on a mountain bend when the two cars meeting, and even cause a car to overturn or fall, with a very high fatality rate. It can be seen that if the occurrence of major traffic accidents on mountainous

X. Song et al. (Eds.): APWeb-WAIM 2023, LNCS 14332, pp. 239–253, 2024.
https://doi.org/10.1007/978-981-97-2390-4_17

roads cannot be effectively controlled, it will inevitably affect the economic development of mountainous areas and surrounding areas. So, how to adopt the concept of intelligent transportation systems and advanced technologies for real-time dynamic monitoring and warning is an important research topic to achieve traffic safety goals.

Several studies have been published concerning on a warning for vehicles meeting on mountain curves. The most commonly method in China is to use spherical mirrors to determine the situation of incoming cars on both sides of the curve, which has the advantages of low cost, easy installation, no energy consumption, and convenient maintenance [4]. In other novel studies, sensors and microwave radars are used for bend warning, which can only detect and recognize the single road vehicles. When cars meet, the difficulty of information processing increases with the shortening of the distance between the two vehicles [5]. In [6], a design of traffic intelligence early warning system was proposed, in which the wireless vehicle detector and LED warning screen are installed on the road side to realize an early warning of vehicles meeting on the bend. In [7], a safety warning system for vehicle running on curves, where sensing coils are buried to monitor the running condition of vehicles. Other relatively mature intelligent transportation products are mostly applied to traffic guidance and information prompts on highways or urban roads in cities, while mountain bends have almost no related applications due to their unique geographical environment and diverse bend characteristics [8–10].

Our main work is to propose a low-cost and high-precision early warning system in blind areas for the current traffic safety requirements of mountain curves. Our main research ideas are derived from the current situation of mountain curves in China, with a focus on solving the driving conditions hazardous problems caused by complex geographic environment and poor network quality in mountain areas. We investigated the possibility of installing cameras on a large number of curves in different mountain areas, tested some computer vision recognition methods, and selected the best strategy for the final wireless communication mode, ensuring the reliability and stability of the system in terms of accuracy and speed.

The primary contributions of this paper are as follows:

1. We proposed a coordinate correlation algorithm in spatiotemporal space based on target tracking to estimate traffic status, which can achieve simultaneous warning of oncoming vehicles on both sides of the curve.
2. We carry out a communication mode that does not require networking conditions.
3. The warning system we developed just needs to be installed on the roadside. Compared with existing methods, it does not damage the road surface and existing facilities, and the installation and debugging process does not affect normal traffic, achieving low cost and high efficiency.

2 Materials and Methods

Figure 2 shows the overview architecture of the proposed system. The system is composed of three modules, which are video acquisition module, computer vision analysis module and communication module. First, video capture is used to obtain video images on both sides of the bend. Second, the computer vision processing module tracks and analyzes the collected video sequence to provide reliable data information for the system early warning. Finally, the communication module will transmit the identification information to the variable message signs, which will alert the driver. In this section, we will introduce the three-part approach.

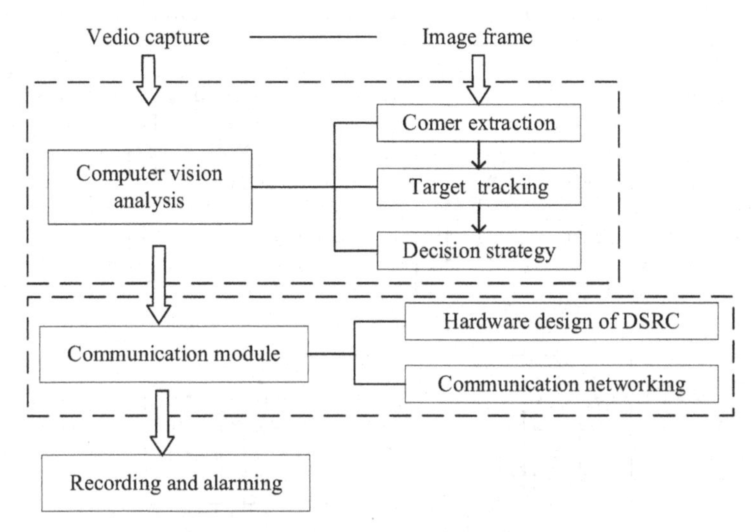

Fig. 1. System overview of the proposed approach.

2.1 Target Tracking Method

- *Block-based corner extraction*

Target tracking is the basis of target analysis, and tracking results will have a great impact on objective understanding and analysis in the following-up processing [11–14]. Since the real-time and accuracy requirements of the system, we propose a block-based corner matching method. Firstly, the three-frame difference method [15–17] is used to separate the target from the background. The specific calculation formula is as follows:

$$\begin{cases} I_1(x, y) = |f_i(x, y) - f_{i-1}(x, y)| > t \\ I_2(x, y) = |f_i(x, y) - f_{i+1}(x, y)| > t \\ I_i(x, y) = I_1(x, y)|I_2(x, y) \end{cases} \tag{1}$$

In the formula, $f_i(x, y)$, $f_{i-1}(x, y)$ and $f_{i+1}(x, y)$ represent the current frame, the previous frame and the following frame respectively, and t is the threshold used for segmentation. In addition, $I_i(x, y)$ is the result of image logical operations.

Extracting features from detected targets is the main task for tracking. The Moravec [18] algorithm takes a neighborhood around a pixel as a patch, and detects the correlation between this patch and other surrounding patches to achieve corner extraction. However, this corner extraction method based on window movement is sensitive to edge points, and it is prone to mistakenly recognizing the edges outside the window as corners [19, 20]. Therefore, this paper proposes a block-based corner extraction strategy based on the Moravec algorithm. In this method, firstly, selecting corner points based on a block basis, and take the block where the corner points are located as the center. Then, selecting a certain size area around the block as the matching template, and searching the matching position of the corner point in the time series to achieve target tracking.

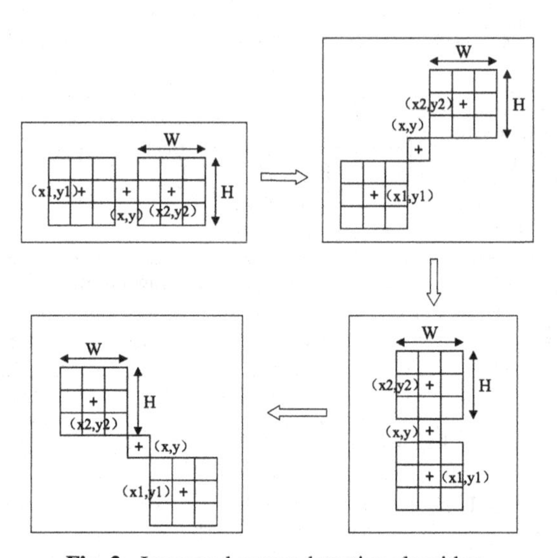

Fig. 2. Improved corner detection algorithm.

As shown in Fig. 2, in the four directions of horizontal, vertical and two diagonal lines. The center points of the two blocks involved in the calculation differ by one pixel from the detection point, where the center point (x, y) is the detection point, and $(x1, y1)$ and $(x2, y2)$ are the center points of the blocks involved in the calculation. The new corner detection algorithm can be calculated according to the following formula.

$$v_1 = \sum_{a=-w}^{w} \sum_{b=-h}^{h} |g(x - i + a, y + b) - g(x + i + a, y + b)| \tag{2}$$

$$v_2 = \sum_{a=-w}^{w} \sum_{b=-h}^{h} |g(x - i + a, y - j + b) - g(x + i + a, y + j + b)| \tag{3}$$

$$v_3 = \sum_{a=-w}^{w} \sum_{b=h}^{h} |g(x + a, y - j + b) - g(x + a, y + j + b)| \tag{4}$$

$$v_4 = \sum_{a=-w}^{w} \sum_{b=-h}^{h} |g(x + i + a, y - j + b) - g(x - i + a, y + j + b)| \tag{5}$$

$$PointValue(x, y) = \min\{v_1, v_2, v_3, v_4\} \tag{6}$$

In the above formula, we define some measurements for corner points, including $w = int(W/2), h = int(H/2)$ and *PointValue*. In addition, *i* and *j* respectively represent the horizontal and vertical pixel distances between the detected point and the center point of the block, which can be adjusted according to actual situations.

- *Research on tracking strategy*

The overall strategy of corner tracking is to extract corners and use them as tracking keys to search for the positions of corner in each subsequent frames images. Unlike point matching methods such as pixel recursive search based on spatial domain, object matching method and phase correlation search based on transform domain, this research adopts a target tracking method based on block matching. It needs to determine the tracking position by calculating the similarity between blocks in subsequent frames that have the same size as the blocks containing corners in the previous frame image, so as to achieve vehicle tracking.

The tracking algorithm based on block matching consists of three steps, as shown in Fig. 3. Firstly, select a block centered around the corner points with a size of $m \times n$ as the template, and assume that the motion of each pixel within the block remains consistent. Secondly, it is necessary to set the search range in the image to be searched, and use a certain strategy to traverse all positions in the search area, calculating the similarity between blocks with a size of $m \times n$ and the templates. Thirdly, select the block that is most similar to the template as the matching block.

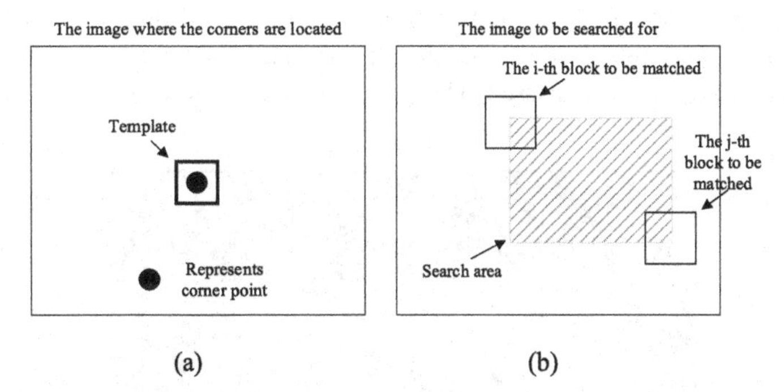

Fig. 3. Principle of Block Matching Algorithm. (a) is the image where the corners are located. (b) is the image to be searched for matching points.

Accurate matching is the key to obtaining effective trajectories. Common matching criteria include methods such as mean absolute deviation (MAD), mean squared error (MSE), and normalized cross correlation function (NCCF), which involve a lot of multiplication operations and have high time complexity. To reduce computational complexity, our research selects Sum of Absolute Difference (SAD) as the optimal matching point selection criterion, and the calculation formula is as follows:

$$SAD(x, y) = \sum_{m=1}^{H} \sum_{n=1}^{W} |G_N(m, n) - G_{N-1}(m + x, n + y)| \tag{7}$$

After determining the matching criteria, it is necessary to find a correct matching search method to estimate the moving target and continuously determine its motion trajectory. Therefore, this paper adopts the Full Search Method (FS) to matching all sub-blocks within the search window and calculating their SAD values, taking the block with the smallest value of SAD as the optimal matching block [21, 22]. In addition, the size of the search window can be selected based on image characteristics, so it is necessary to select the appropriate area to reduce the computational burden during the entire search process. For example, the search window size can be appropriately reduced for the video image with a small target motion range. Conversely, when the motion range of the target in the video image is large, a larger search window can be selected.

In order to obtain reliable matching results, the size of the template and search range are dynamically determined based on the distance between the target and the camera. At close-range, a large template is used for search and matching within a larger search range, while at far range, a small template is used for search and matching within a smaller range. The specific approach is to divide the image into several levels, with the bottom level having the largest target size and search range. Meanwhile, from bottom to top, the template and search scope of each layer are reduced in a certain proportion, which should be consistent with the proportion of the shape changed from near to far.

2.2 Traffic Condition Analysis

Based on the above tracking method, detecting the oncoming vehicle based on the vehicle's trajectory on the curve can be divided into two steps. Firstly, the ROI area for vehicles detection can be selected. After that, if a trajectory appears in the detection area, it indicates that a vehicle has passed. The detection area is shown in Fig. 4.

Fig. 4. Example of ROI area for vehicle detection.

It is important to determine the driving direction based on the change of the vehicle position. As shown in Fig. 5(a), assuming that the position of the tracking point is represented by the coordinate (x, y) in the image, so the tracking trajectory called line 1 of the current vehicle moves from point A to point B, and the values of its tracking point coordinates (x, y) changes. It is particularly noteworthy that the value of the x coordinate in the image changes from larger to smaller. Moreover, as shown in Fig. 5(b), the trajectory called line 3 on the current vehicle in the opposite direction moves from point C to point D, and the position of the tracking points (x, y) changes, where the value of x coordinate changes from small to large in the image.

(a)

(b)

Fig. 5. Location changes of tracing points. (a) The vehicle moved from point A to point B on the side of the curve. (b) The vehicle moved from point C to point D on other side of the curve.

2.3 Module of Communication

In wireless communication technology, DSRC (Dedicated Short Range Communication) can be well used in the field of intelligent transportation, transmitting the information of real-time navigation, road condition, traffic flow, danger warning and emergency notification, and even providing feedback on video and image information about traffic accident scenes [23].

DSRC is commonly used in data collection systems, automated monitoring systems, remote network control, and medical monitoring in the field of intelligent transportation [24]. We adopt the 490 MHz DSRC communication subsystem, which is mainly used for the transmission of short-distance wireless information between video detection module and display module. Therefore, the detailed technical parameters for designing DSRC are shown in Table 1.

The hardware design of the DSRC communication module mainly includes two parts: power supply and interface. For the reason of complex environment of mountain curves, there are many interference factors on the power supply, which can easily cause problems such as surges and voltage jumps, leading to the burning of communication

Table 1. Technical parameters of DSRC communication module.

Number	Technical index	Parameter			Remarks
1	Modulation Mode	FSK, GFSK, OOK			
2	Frequency	490 MHz–510 MHz			
3	Power	≧18 dBm	≧27 dBm	≧33 dBm	
4	Transmission range	500 m	1500 m	3000 m	No occlusion
5	Receiving sensitivity	−118 dBm			
6	Emission Current	≤100 mA	≤600 mA	≤1400 mA	
7	Receiving Current	≤46 mA	≤46 mA	≤46 mA	
8	Transfer rate	0.123 kbps–250 kbps			
9	Interface type	RS485			
10	Baud rate	1200–115200 bps			User configurable
11	Data bits	8, 9			User configurable
12	Check bits	Odd check; even check; none check			User configurable
13	Stop bit	1, 2			User configurable
14	Antenna interface	SMA, 50Ω			
15	Power supply	DC24 V/200 mA			
16	Operation temperature	−40 °C–85 °C			
17	Storage temperature	−40 °C–125 °C			
18	Working humidity	10%–90% relative humidity			

modules. Therefore, in terms of power supply, we have adopted Transient Suppression Diodes (TVS) to suppress the generation of transient voltage. Correspondingly, in terms of communication interface, we select RS485 or RS232 interface for corresponding communication based on the specific conditions of mountain bends, and the specific method is to connect it to the wireless receiving terminal, as shown in Fig. 6 and Fig. 7.

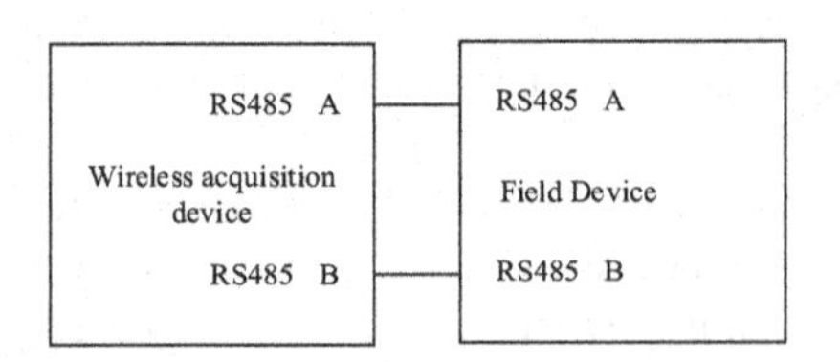

Fig. 6. RS485 communication connection.

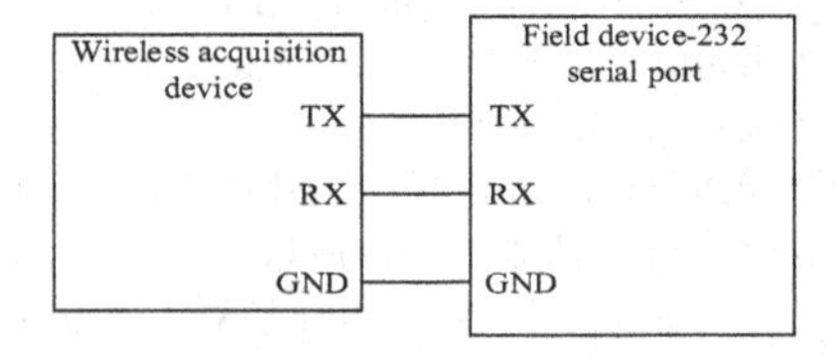

Fig. 7. RS232 communication connection.

3 Results

3.1 Software Testing

Vehicle detection based on computer vision is the core algorithm of warning system for meeting vehicles on mountain curves. This part of the experiment mainly applies the collected video to test the warning system of mountain bend, mainly verifying the performance of the software module. The experimental process can be divided into three steps, as shown in Fig. 8. First, input the video and obtain a sequence of images, where each image has a size of 720 × 288. Then, set the ROI area for detection, specify the driving direction, and use the block based corner tracking method to obtain the tracking trajectory. Finally, determine whether 10 points can be tracked stably. If so, that is prove that a vehicle has entered from the bend.

Fig. 8. Vehicles detection in the Bend. (a) Selected ROI Area. (b) Tracking method testing.

In the experiment, different colors were set to represent the output information. For the vehicles on both sides of the curve, setting the color of display box to red indicates that a vehicle is approaching from the other side of the curve and it needs to be driven carefully, while green indicates that no vehicle is approaching from the other side of the curve and it can be passed safely. Figure 9 shows the method of getting different software programs to work together and the results of the software tests. We found that the software can analyze the situation of oncoming vehicles in different directions on the curve and provide timely warning information.

To further verify the effectiveness of tracking, we conducted tests on different roads in cities of Qinghai and Xi'an to see whether our method can accurately detect the vehicles in the bends with various environments, including four scenarios: mountains, cliffs, highways, and urban roads. The improved algorithms are used to analyze the video input to the system, determining the direction of the incoming vehicles, and transmitting the results to the display equipment. The results of the experiment are divided into three parts. Conditions for the test scenario are addressed in the first part. The second part shows the detection and tracking accuracy and error rate of the warning system in the bend. The final part compares the performance of the improved method and the original method.

Fig. 9. Software collaborative work.

As shown in Table 2, "Height" refers to the installation height of the camera, and "Angle" refers to the viewing angle of the camera. In addition, "Detection" represents the number of frames correctly detected, and "Inspection" represents the number of frames to be tested, while "Accuracy rate" and "Fps" respectively represent the detection accuracy and frames per second. The experiment result in Table 2, the accuracy of the original corner tracking method for testing data in various scenarios is only about 64%, while the improved block-based corner tracking algorithm can achieve an accuracy of about 88%. The reason for this situation is that the original corner tracking method does not have rotation invariance and has a strong response to edge points. However, the improved block-based corner tracking algorithm has good adaptability to rotation, especially block matching criteria added in improved method has stronger stability than point matching, so its accuracy has been greatly improved. On the other hand, due to the reduction of matching times in block matching, the average processing time for every image with a resolution of 720×288 in various environments is about 3ms. It is thus clear that the improved method has good accuracy and real-time performance.

Table 2. Software function analysis.

Part III	Part I				Part II		
	Scenarios	place	Height	Angle	Detection/Inspection	Accuracy rate (%)	Fps
Original method	mountains	In Qinghai	10	15	532/855	62.2	18
	cliffs	In Qinghai	10	15	297/549	54.1	17
	highways	In Xi'an	11	15	651/898	72.5	19
	urban roads	In Xi'an	11	15	515/761	67.7	19
Our method	mountains	In Qinghai	10	15	736/855	86.1	21
	cliffs	In Qinghai	10	15	498/549	90.7	20
	highways	In Xi'an	11	15	795/898	88.5	23
	urban roads	In Xi'an	11	15	688/761	90.4	23

3.2 Field Application

To ensure the stability of the system, we use RS232 serial communication as the communication channel of this system, and design the Printed Circuit Board (PCB) shown in Fig. 11 based on DSRC communication according to the circuit schematic shown in Fig. 10. In addition, set parameters of coordinator and collector for the communication in our system, as shown in Table 3.

On the basis of completing the above work, all software algorithms and hardware circuits were integrated to obtain the final test prototype. After that, the prototype was applied to conduct one-way warning experiment for eight weeks on a mountain bend in Qinghai, with a sampling frequency of 26 frames/s. In the period of the first four weeks, we collected data under different conditions, which can be used to calibrate the system parameters. In the next four weeks, we conducted practical applications on the verified system to evaluate its performance.

Some experimental results of the system during the actual testing phase are shown in Fig. 12. As shown in the first image of Fig. 12, we have noticed that the testing equipment installed in the mountain bends, which is mainly responsible for video acquisition and processing, and sending the results to the information board. In addition, the other two images in Fig. 12 respectively show the recognition results of the system for vehicles coming from a certain direction on the mountain bend in real scenarios.

Table 3. Parameter Configuration List for ZigBee Network.

	Baud rate	Data bit	PAN_ID	Local address	Destination address	Channel	Data transfer rate
Coordinator	COM1	8	12	6	253	8	3
Collector	COM1	8	12	212	6	8	3

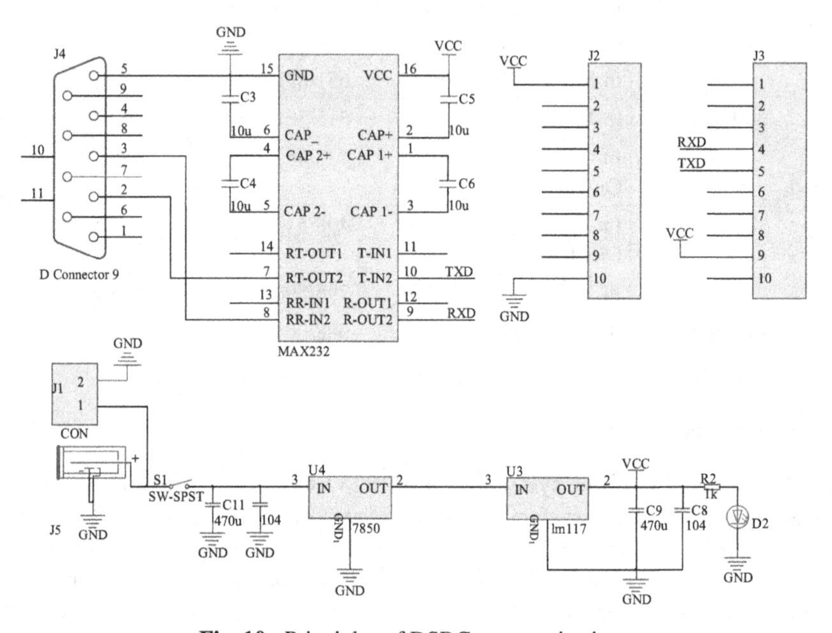

Fig. 10. Principles of DSRC communication.

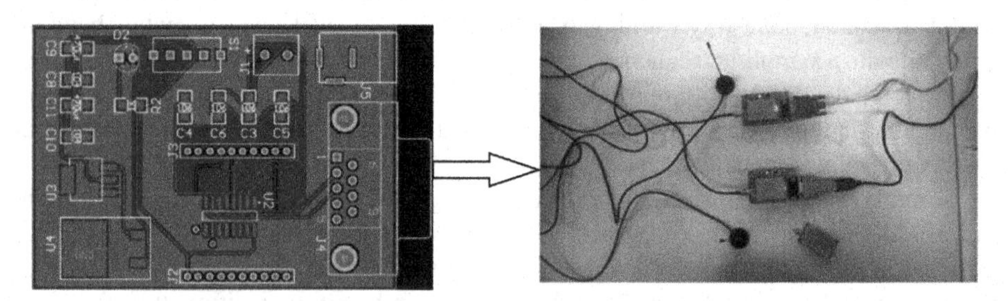

Fig. 11. Circuit board and physical map of DSRC communication.

To test the performance of the system in real environments, we compared the experimental result in four scenarios: cloudy, sunny, rainy and nighttime. The method of comparison is to compare the results automatically determined by the statistical system with the results determined by the human eyes. The comparative results of the experiment are shown in Table 4.

The evaluation indicators in Table 4 need to be explained. 'Error detection' represents the number of frames in which an error judgment occurred. 'No detection' indicates the number of frames in which an incoming vehicle was not detected. From the results in Table 4, it can be seen that the error rate and missing rate (the coming vehicle is not found) are relatively high in nighttime video and rainy video. It is worth noting that the performance of image acquisition and detection algorithm are affected to some extent by low-level lighting in night time or obstructed vision in rainy days, which directly affects the judgment of incoming vehicles. In addition, the communication latency of the system is relatively delayed during rainy days and night time. Overall, our system can adapt to various climate conditions, with an average accuracy of 87%, and the communication delay is also within a permissible range. Therefore, it can be seen that the system we designed has good portability and robustness.

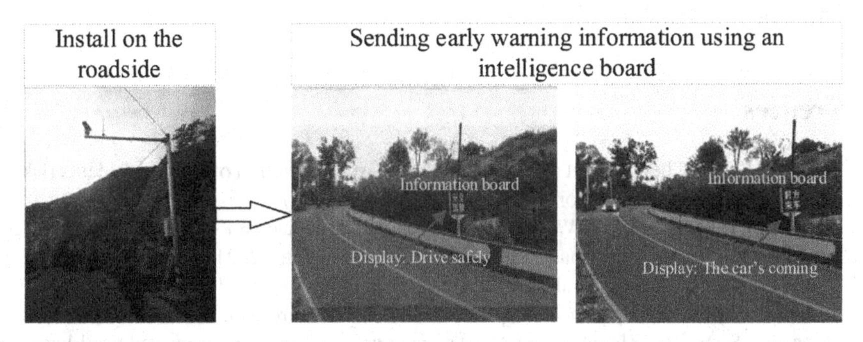

Fig. 12. Example of actual application. The first picture shows the road testing equipment, and the other pictures shows the information board installed in the bend.

Table 4. Parameter Configuration List for ZigBee Network.

	Scenarios	Error detection/Inspection	No detection/Inspection	Accuracy rate (%)	Communication Latency (ms)
Video-1	Sunny	51/843	28/843	90.6	350
Video-2	Rainy	38/630	61/630	84.2	530
Video-3	Cloudy	46/867	25/867	91.8	350
Video-4	Nighttime	61/593	42/593	82.6	460

4 Conclusions

The automatic warning system for mountain bends proposed in this paper is based on the technology of computer vision and applies video analysis methods to detect oncoming vehicles on both sides of the bend. At the same time, artificial intelligence inference methods are used to determine whether oncoming vehicles are about to meet in the bend. After that, the information board (warning light) of the system provides warning information to the driver in advance, enabling them to control the speed as soon as possible, thereby avoiding traffic accidents. The advantages of the system we designed are as follows. Firstly, the system can operate independently and control autonomously, without the need for networking. Secondly, solar energy can provide electricity to our system without laying cables. Finally, the system is installed on the roadside without damaging existing facilities on the road, and the installation and shakedown test of hardware do not stop the traffic in the road. Since the installation position, height and bend curvature of the system will affect the performance of the corner tracking algorithm, in future work, we will further improve the target recognition method to improve the performance of the system and achieve the goals of industrialization and marketization.

Acknowledgments. This research was funded by Natural Science Foundation of Qinghai Province, grant number: 2023-ZJ-989Q.

References

1. Zhao, H., Li, X., Cheng, H., et al.: Deep learning-based prediction of traffic accidents risk for internet of vehicles. China Commun. **19**(2), 214–224 (2022)
2. Wang, W., Yang, S., Zhang, W.: Risk prediction on traffic accidents using a compact neural model for multimodal information fusion over urban big data (2021)
3. Han, F.C., Cao, J., Xu, T., et al.: Threshold determination of lane departure warning system for mountainous freeways. In: Transportation Research Board Annual Meeting (2011)
4. Wang, X., Song, W., Zhang, B., et al.: An early warning system for curved road based on OV7670 image acquisition and STM32. Comput. Mater. Continua (2019)
5. Gao, C., Xie, Y., Ouyang, D., et al.: Design of a warning system for pedestrians and non-motorized vehicles. In: International Conference on Civil Engineering, Architecture and Building Materials (2014)
6. Vijayalakshmi, L., Pradeepa, P., Sairam, S.: Driver warning system in hill bends. In: 2018 International Conference on Intelligent Computing and Communication for Smart World (I2C2SW). Kongu Engineering College Department of ECE, Perundurai, India (2018)
7. Xie, R., Dong, X., Deng, K., et al.: Design of traffic intelligence early warning system for blind zones in mountain road. Highway **65**(7), 5 (2020)
8. Zhao, X., Chen, Y., Li, H., et al.: A study of the compliance level of connected vehicle warning information in a fog warning system based on a driving simulation. Transport. Res. Part Traffic Psychol. Behav. **76**, 215–237 (2021)
9. Vishwakarma, L., Das, D.: SmartCoin: a novel incentive mechanism for vehicles in intelligent transportation system based on consortium blockchain. Veh. Commun. **33** (2022)
10. Bykov, N.V.: Impact of counteracting vehicles on the characteristics of a smart city transport system (2022)
11. Arun, K., Prabhakar, C.J.: Detection and tracking of lane crossing vehicles in traffic video for abnormality analysis. Int. J. Eng. Adv. Technol. **10**(4), 1–9 (2021)
12. Kong, X., Chen, Q., Gu, G., et al.: Particle filter-based vehicle tracking via HOG features after image stabilisation in intelligent drive system. IET Intel. Transport Syst. **6**, 13 (2019)
13. Zhang, W., Jiao, L., Liu, F., Li, L., Liu, X., Liu, J.: MBLT: learning motion and background for vehicle tracking in satellite videos. IEEE Trans. Geosci. Remote Sens. **60**, 1–15 (2021)
14. Song, S., Li, Y., Huang, Q., et al.: A new real-time detection and tracking method in videos for small target traffic signs. Appl. Sci. **11**(7), 3061 (2021)

15. Lei, M., Geng, J.: Fusion of three-frame difference method and background difference method to achieve infrared human target detection. In: 2019 IEEE 1st International Conference on Civil Aviation Safety and Information Technology (ICCASIT). IEEE (2019)
16. Shang, M., Zeng, S., Jiang, L.: A foreground detection algorithm based on improved three-frame difference method and improved Gaussian mixed model. In: International Conference on Graphics and Image Processing (2019)
17. Simsek, E., Ozyer, B.: Selected three frame difference method for moving object detection. Int. J. Intell. Syst. Appl. Eng. **9**(2), 48–54 (2021)
18. Zhang, K., Yang, Z., Zhang, X., et al.: Comparative study on feature point extraction operators based on low altitude photogrammetric images (3), 30–35 (2021)
19. Ji, B.: Improvement of moravec point feature algorithm and research on low altitude oblique image matching. Fujian Qual. Manag. **000**(010), 274 (2020)
20. Feng, Y., Zhu, D., Hu, L., et al.: Research on algorithm for extracting feature points from aerial photographic images. Softw. Guide **18**(8), 5 (2019)
21. Sardari, F., Moghaddam, M.E.: A hybrid occlusion free object tracking method using particle filter and modified galaxy based search meta-heuristic algorithm. Appl. Soft Comput. **50**, 280–299 (2016)
22. Zhang, Q., Fan, G., et al.: Real-time automatic obstacle detection method for traffic surveillance in urban traffic. J. Signal Process. Syst. Signal Image Video Technol. **82**, 357–371 (2016)
23. Jin, Y., Liu, X., Zhu, Q.: DSRC & C-V2X comparison for connected and automated vehicles in different traffic scenarios (2022)
24. Abboud, K., Omar, H., Zhuang, W.: Interworking of DSRC and cellular network technologies for V2X communications: a survey. IEEE Trans. Veh. Technol. **65**, 9457–9470 (2016)

Local-Global Cross-Fusion Transformer Network for Facial Expression Recognition

Yicheng Liu, Zecheng Li, Yanbo Zhang, and Jie Wen(✉)

School of Computer Science and Technology, Harbin Institute of Technology, Shenzhen, China
jiewen_pr@126.com

Abstract. Facial Expression Recognition (FER) has received increasing attention in the computer vision community. For FER, there are two challenging issues among the facial images: large inter-class similarity and small intra-class discrepancy. To address these challenges and obtain a better performance, we propose a Local-Global Cross-Fusion Transformer network in this paper. Specifically, the method seeks to obtain a more discriminative facial representation by sufficiently considering the local features of multiple local regions of the face and global face features. In order to extract the critical local area features of the face, a local feature decomposition module based on facial landmarks is designed. In addition, a local-global cross-fusion Transformer is designed to enhance the synergistic correlation between local features and global features using the cross-attention mechanism, which can maximize the focus on key regions while considering the connection information among local regions. Extensive experiments conducted on three mainstream expression recognition datasets, RAF-DB, FERPlus, and AffectNet, show that the method outperforms many existing expression recognition methods and can significantly improve the accuracy of expression recognition.

Keywords: facial expression recognition · facial landmark · cross-attention mechanism · local and global facial features

1 Introduction

Facial expression is a form of human non-verbal communication, which can convey rich emotional information. The research of expression recognition can promote the better development of computer science in the direction of intelligence and humanization, which has crucial practical significance.

However, FER in complex scenes still faces a great challenge due to two reasons: (1) Large inter-class similarity: Similar images with subtle changes between them are often part of different expression categories. As illustrated in Fig. 1, two expression images of surprise and anger both show an open form at the mouth, and two sub-images of the mouth are very similar; (2) Small intra-class discrepancy: Due to various factors such as age, gender, and race, similar expressions show greater variability on different faces, for example, the two images of sadness have very significant differences in the eye area.

X. Song et al. (Eds.): APWeb-WAIM 2023, LNCS 14332, pp. 254–269, 2024.
https://doi.org/10.1007/978-981-97-2390-4_18

To address the above two challenges, in this paper, we propose a new deep network, called the local-global cross-fusion Transformer network (LGCFTN), based on the following observations. (1) People usually pay more attention to some salient local area features when classifying expressions such as eyes, nose, mouth, etc. (2) Changes in expression are the result of the joint action of multiple facial regions. Therefore, individual local regions cannot be analyzed in isolation. (3) Global features can provide information about the association among local regions.

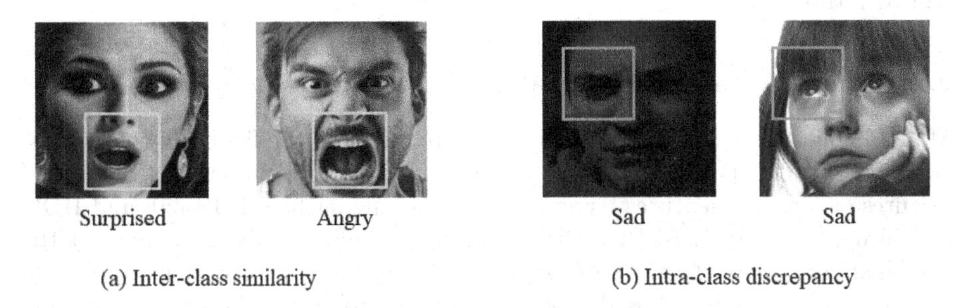

Fig. 1. Inter-class similarity and intra-class discrepancy. Images are selected from the AffectNet [17] dataset.

To this end, our LGCFTN seeks to sufficiently fuse the global facial features and several local features corresponding to the key facial areas to obtain a discriminative FER network. To achieve this goal, firstly, a local feature decomposition module is proposed to automatically extract local regional features based on facial landmarks. These feature blocks can accurately cover facial regions that are highly correlated with expressions. Secondly, in order to extract the connectivity between these regional blocks and highlight the important regional blocks while reducing the attention to irrelevant regional blocks, a Transformer-based cross-fusion network is designed to explore the relationship between different regional blocks in the global context. Due to the cross-attention mechanism, the features of key regions can be highlighted, such that a more discriminative facial representation can be obtained for FER. To validate the effectiveness, extensive experiments are conducted on three popular benchmarks. The experimental results show that the proposed method can significantly improve the accuracy of FER. The ablation experiments verify the effectiveness of fusing the local and global facial features for FER. Overall, our contributions are summarized as follows:

- We propose a new local-global cross-fusion Transformer network to alleviate the challenging issues caused by the large inter-class similarity and small intra-class discrepancy properties of faces in FER tasks.
- We design a simple local feature decomposition module that can obtain discriminative local feature representations based on the location of facial landmarks.

- A cross-fusion transformer module is introduced to integrate the global and local facial features. The module enables the network to focus more on the salient local facial features so as to obtain a more discriminative facial representation for FER.

2 Related Work

In this section, facial expression recognition and the related work of Transformer will be introduced.

2.1 Facial Expression Recognition

Many scholars have conducted a lot of research work in the field of facial expression recognition. Traditional methods generally require manual extraction of features or learning features using shallow learners, such as LBP [20] and HOG [6]. [5] used unsupervised Principal Component Analysis (PCA) to extract the principal component features of faces that retain more information for expression recognition. [22] proposed a geometrical descriptor based on the calculation of distances from coordinates of facial fiducial points, which are used as features for training support vector machines (SVM) to classify emotions. However, these features lack the ability to generalize in some challenging scenarios, such as poor lighting conditions, or the presence of occlusions. Deep learning approaches prefer to design and build deep neural networks to perform expression recognition tasks in an end-to-end form.

With the rapid development and advancement of computer technology, the development of deep learning has been greatly facilitated and the performance of deep learning methods has substantially outperformed traditional methods. [18] proposed a multi-scale Contractive Convolutional Network (CCNET) to obtain the local invariant features of the face. This method achieved a high accuracy rate on the then-latest facial expression recognition dataset, i.e., the Toronto facial expression dataset. [9] proposed a Deep Attentive Center loss (DACL) to adaptively select a subset of important feature elements to enhance the discriminative power of the model. [26] addressed the problems of occlusion and pose change in expression recognition. The method divided face images into multiple regions and designs a regional attention network for weighted analysis of multiple expression regions. [32] proposed a global multiscale and local attention network (MA-Net) to cope with the facial expression recognition problem in the field environment. MA-Net can acquire discriminative global and local features, so it can deal with the problems of occlusion and pose change well. [31] proposed an attention mechanism-based approach to enhance sensitivity to label noise and to learn stronger feature representations adaptively from large amounts of noisy data. [30] proposed the Meta-Face2Exp framework to extract deviated knowledge information from auxiliary face recognition data through a meta-optimization framework.

2.2 Transformer

Transformer [24] has made a huge breakthrough in the field of natural language processing. For computer vision tasks, Vision Transformer (ViT) [8] treats images as multiple patches. It achieved very good results in several vision downstream tasks. The task of expression recognition is essentially a classification task of images, and naturally, there are many research works applying ViT to expression recognition. [29] first used ViT for facial expression recognition, which exploits the self-attention mechanism of Transformer to explore the global relationship between multiple local patches. [16] proposed a Vision Transformer model with feature fusion, which integrates LBP features and CNN features through global-local attention and global self-attention mechanisms to improve expression recognition accuracy. [11] designed a teacher-student model for modeling the probability distribution of frontal and multi-pose facial expressions to solve the problems of facial pose change and occlusion in expression recognition. Unlike existing Transformer-based expression recognition works that directly utilized the standard Transformer structure, we propose an improved cross-attention mechanism in this paper, which enables the Transformer to correlate and fuse local region features and global features belonging to two dimensions.

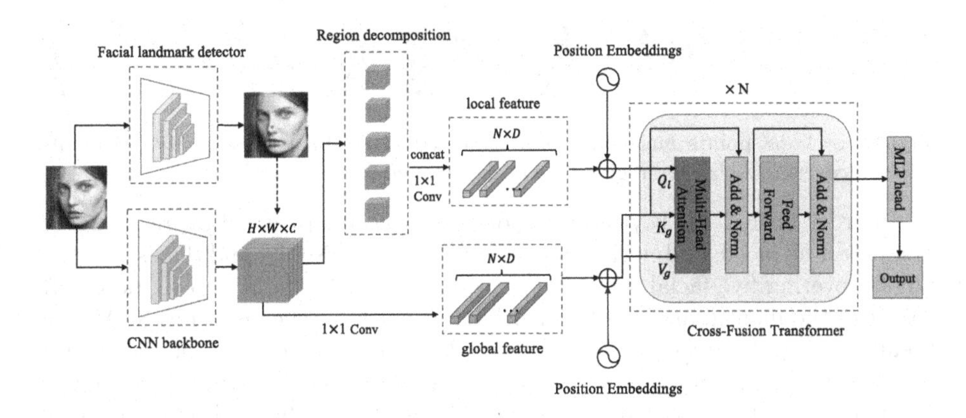

Fig. 2. The overall architecture of our proposed LGCFTN.

3 Method

In this section, the overall framework of our method will be introduced first, and then the specific technical details will be presented separately.

3.1 Overall Framework

The overall architecture of our method is shown in Fig. 2, which mainly consists of a convolutional neural network (CNN) backbone, facial landmark detector, local feature decomposition (LFD), and cross-fusion Transformer. The network takes an RGB facial image as input. The feature map of the image is extracted by CNN backbone IR50 [7], while the facial landmark of the image is extracted by the facial landmark detector MobileFaceNet [3]. Then, LFD decomposes the feature maps of the whole face into multiple local feature maps based on the coordinates of the locations of facial areas. Each local feature map and global feature map are transformed into sequence form. Finally, local and global features are fused using the cross-fusion Transformer for FER.

3.2 Local Feature Decomposition (LFD)

Local regional features of faces are important manifestations of facial expressions, so extracting this local regional information is beneficial to improve the performance of the FER task. Inspired by this, LFD is designed to extract the features of local discriminative regions that are highly related to facial expressions.

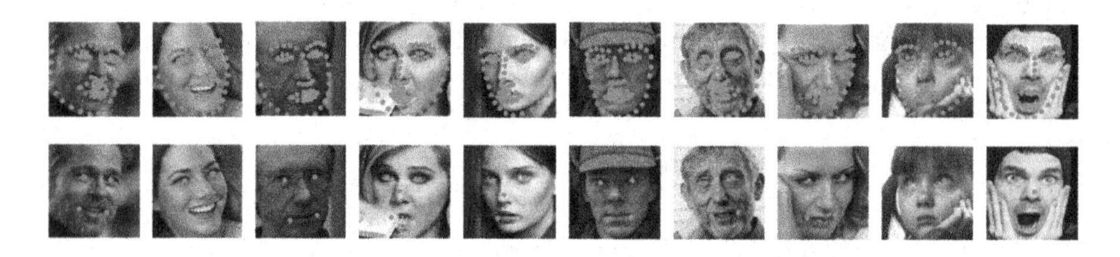

Fig. 3. row1: 68 points facial landmark detected by MobileFaceNet; row2: 5 points selected covering facial key regions about expression.

In order to find the key facial regions associated with facial expressions, we first detect 68 facial landmark points using MobileFaceNet [3] and then select five points covering five facial information regions, i.e., left eye, right eye, nose, left corner of the mouth, and right corner of the mouth, as shown in Fig. 3. Among them, two points are used to locate the left and right corners of the mouth respectively to determine the position of the mouth, and other regions use one point to locate the center of the region. In the commonly used 68-point facial landmark labeling method, each eye of a human face is localized by 6 points. In order to accurately locate the center of the eye as much as possible, we calculate the average of the horizontal and vertical coordinates of each of these six points to obtain an average point to locate an eye. Then, a square area is truncated with a set radius based on the center of the facial landmark point of each image to obtain the local feature map.

For the input image of the face, the image is first processed using the backbone network, and then the feature map generated by the last convolutional layer is decomposed to obtain the local features of the regions of the face. Compared with the direct decomposition of the original image, this approach is beneficial to reduce the total number of parameters of the model because all regions of the image are processed by the same convolutional neural network, instead of setting up a separate convolutional network for each region block. The size of the feature map obtained from IR50 is 14×14, and the number of channels of the feature map is 256. After decomposing the feature map, five local feature maps are obtained, and the size of each local feature map is 7 × 7. The local feature maps obtained after decomposition are shown in Fig. 4. In the division, the value of the area points beyond the original feature map range is set to 0.

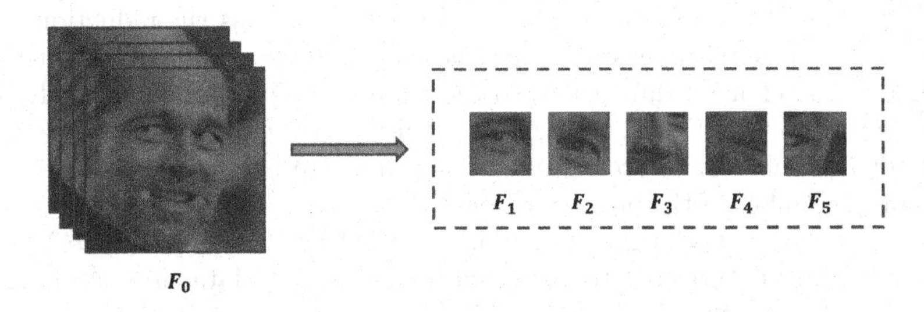

Fig. 4. The local feature maps obtained after decomposition

After the above process, the complete set of features of the facial image is extracted, where the CNN backbone is represented as $r(\cdot;\theta)$, and the global features of the image F_0 can be represented as follows:

$$F_0 = r(I;\theta), \tag{1}$$

where θ denotes the parameters of the CNN backbone.

Let $F_1, F_2, ..., F_k$ be the local features produced by LFD, the set of features X can be expressed as follows:

$$X = [F_0, F_1, F_2, ..., F_k], \tag{2}$$

3.3 Cross-Fusion Transformer

Local facial features generated by LFD can provide rich information on facial regions associated with expressions, while global features contain information about the connection among several local regions. After obtaining the above global and local features, a cross-fusion Transformer network is used to fuse both two features to obtain a fused feature representation for FER.

The following are the details about this module:

Feature Map Preprocessing. Transformer was originally proposed in the field of natural language processing and was used to process sequence-shaped inputs. In order to use Transformer for feature fusion, the global feature map and the five local feature maps are first converted into the sequence.

Let $F_0 \in R^{h_g \times w_g \times c_g}$ be the global feature map extracted by the CNN backbone, where h_g and w_g denote the height and width of the feature map, respectively, and c_g denotes the number of channels of the feature map. In order to keep the same size as the local feature map when input to the Transformer, the global feature map is first dimensionally reduced to its initial 1/2 height and width. Then the global feature map is sliced along the channel dimension, and they are rearranged into a sequence of feature vectors $F_g \in R^{c_g \times (\frac{h_g}{2} \times \frac{w_g}{2})}$, which can be succinctly expressed as $F_g \in R^{N \times D}$, where $N = c_g$ denotes the input number of patches of the Transformer. $D = \frac{h_g}{2} \times \frac{w_g}{2}$ denotes the dimension of a single patch feature. For local feature maps, each feature map is first processed using a convolution kernel of 1×1, and the number of channels becomes c/r after each feature map is convolved with 1×1 kernel, where r is the reduction factor to reduce the feature dimension. Then, the five feature maps are concatenated together in the channel dimension, so the channel number of the concatenated feature map should be $5c/r$. Therefore, when the reduction factor r is set to 5, the channel number of the concatenated feature map is c, which is the same as the channel number of the original global feature map, and then the same shape change operation is performed to obtain the final local feature $F_l \in R^{N \times D}$. At this point, both features are transformed into a form of data that can be input to the Transformer encoder.

Finally, following [8], a learnable class token is appended to the sequence of input vectors. And a learnable 1D position embedding is added to the input sequence to inject inter-sequence position information.

Cross-Fusion Transformer Encoder. The self-attention mechanism in Transformer can handle the relationship within a single-dimensional sequence well. Based on the self-attention mechanism, many research works [2,15,28] have improved the cross-attention mechanism to learn the complementary relationship between two-dimensional features. To be able to fully exploit the connection information between local and global features, we use the cross-attention mechanism to model the bidirectional correspondence between local and global features.

For the self-attention operation in Transformer, the input sequence is first mapped into three matrices by linear transformation: queries Q, keys K, and values V. Then the attention weights are computed, which are defined as follows:

$$Attention(Q, K, V) = softmax(\frac{QK^T}{\sqrt{d_k}})V, \tag{3}$$

where Q, K, and V represent the query, key, and value matrices, respectively, and $\sqrt{d_k}$ is the dimension of the key.

In order to achieve the fusion association between the two features and complement each other, the query matrix Q is generated by the local feature F_l, while the key matrix K and the value matrix V are generated by the global feature F_g:

$$Q_l = F_l W_Q, K_g = F_g W_K, V_g = F_g W_V, \tag{4}$$

where W_Q, W_K, and $W_V \in R^{D \times D}$ are linear transformation. After obtaining Q_l, F_g, and V_g, the cross-attention is calculated as follows:

$$Attention(Q_l, K_g, V_g) = softmax(\frac{Q_l K_g^T}{\sqrt{d_k}})V_g, \tag{5}$$

By doing so, the global feature is joined with the prior information brought by the local features. At the same time, the local features get more complementary information under the guidance of the global feature.

The cross-fusion multi-head cross-attention (CFMCA) performs a parallel cross-attention operation and concatenates the outputs of multiple attention heads into a final output. Adding residual connections after the CFMCA and then adding the multi-layer perception (MLP) to form the complete cross-fusion Transformer encoder. The output of the cross-fusion Transformer encoder F_{out} is computed as follows:

$$F^{'} = CFMCA(Q_l, K_g, V_g) + F_g, \quad (6)$$

$$F_{out} = MLP(Norm(F^{'})) + F^{'}, \quad (7)$$

where $CFMCA(\cdot)$ represents the cross-fusion multi-head cross-attention block, $Norm(\cdot)$ is the layer normalization, and $MLP(\cdot)$ is the multi-layer perception.

3.4 Loss Function

For the loss function, we use the standard label smoothing cross-entropy loss [23]. It can reduce overfitting and mitigate the effects of noisy labels in FER datasets. A comprehensive description of the training process is presented in Algorithm 1.

Algorithm 1. Training process of LGCFTN

Input: Training images $\mathcal{X}$ with expression labels $\mathcal{Y}$; Batch size B; Number of batches N; Training epochs E; Optimizer Adam.
Output: Parameters of the trained model.
1: **for** $e = 0 \rightarrow E$ **do**
2: **for** $k = 0 \rightarrow N$ **do**
3: Sample a batch $(\mathcal{X}_{batch}, \mathcal{Y}_{batch})$ from $(\mathcal{X}, \mathcal{Y})$.
4: **for** each sample $\in (\mathcal{X}_{batch}, \mathcal{Y}_{batch})$ **do**
5: Extract the feature map F_0 of using (1).
6: Extract the local feature map of according to LFD.
7: Transform local feature maps and global feature map into sequence F_l and F_g.
8: Compute the fused feature F_{out} using (4), (5), (6), (7).
9: Generate the predicted labels $\mathcal{P}$ according to the F_{out} via MLP.
10: **end for**
11: Compute the label smoothing cross-entropy $\mathcal{L}$ of $\mathcal{X}_{batch}$ according to $\mathcal{P}$ and $\mathcal{Y}_{batch}$.
12: Use the optimizer to update the network parameters.
13: **end for**
14: **end for**
15: return trained model

4 Experiment

4.1 Experiment Setup

Dataset. RAF-DB [14] is a real-world large-scale FER dataset, which contains 29,672 real-world facial images collected by Flickr's image search API and independently annotated by about 40 human annotators. In our experiments, a single-label subset provided by RAF-DB is used as the benchmark dataset for algorithm evaluation, which has 15,339 expression images containing 12,271 training images and 3,068 test images covering seven expression categories (happy, surprised, sad, angry, disgusted, fearful, and neutral). AffectNet [17] is the largest publicly available dataset in the field of FER, which contains about 1M facial images. It mainly contains 8 expression categories (neutral, happy, angry, sad, fear, surprise, disgust, and contempt). We conduct our experiments on AffectNet based on two subsets of AffectNet. AffectNet-7class consists of 280,000 training images and 3,500 validation images. AffectNet-8class consists of 283,000 training images and 4,000 validation images. FERPlus [1] is an extended version from FER2013, a large-scale face expression dataset collected by Google Image Search API, where all images are aligned and cropped to 48 $\times$ 48 grayscale images. It contains 28,709 training images, 3,589 validation images, and 3,589 test images covering eight emotion categories (happy, surprised, sad, angry, disgusted, fearful, neutral, and contemptuous) and two special labels for unknown and non-human faces. The specific dataset configuration is shown in Table 1.

Table 1. Detailed size of the experimental dataset

Dataset	Train size	Test size	Classes
RAF-DB	12271	3068	7
AffectNet-7class	280401	3500	7
AffectNet-8class	283501	4000	8
FERPlus	28236	3137	8

Implementation Details. We adopted the ir50 [7] network pre-trained on the Ms-Celeb-1M [10] dataset as the image backbone due to its excellent performance in face recognition. And MobileFaceNet [3] is used as the facial landmark detector. The feature dimension D was set to 512. For the cross fusion Transformer encoder, the number of cross-attention heads is 8 and each level of encoders consists of 8 blocks. Our LGCFTN is trained with the Adam [12] optimizer. The initial learning rate was set as 4e−4 and was reduced with a gamma of 0.1. The batch size was set to 256.

4.2 Comparison with the State-of-the-Art Methods

The experimental results of the proposed method and some state-of-the-art FER methods are listed in Table 2. On the RAF-DB dataset, our proposed method achieves a test accuracy of more than 92%, which exceeds the best-performing TransFER method among the previous methods by 1.2%. On AffectNet, the largest dataset in the current FER task, our proposed method achieves an accuracy of 67.49% on the 7-class dataset and an accuracy of 63.47% on the 8-class dataset, which are both superior to the existing methods. It indicates that our proposed method can have good generalization ability on large-scale datasets. On the FERPlus dataset, our method achieves 92.06% test accuracy, which is 1.23% higher than the TransFER method. Overall, the results above clearly demonstrate that the proposed method is highly competitive.

Table 2. Comparison with the state-of-the-art methods

method	RAF-DB	FERPlus	AffectNet-7class	AffectNet-8class
SCN [25]	87.03	89.39	–	60.23
LDL-ALSG [4]	85.53	–	59.35	–
RAN [26]	86.90	89.16	–	–
DACL [9]	87.78	–	65.20	–
KTN [13]	88.07	90.49	63.97	–
FDRL [19]	89.47	–	–	–
TransFER [29]	90.91	90.83	66.23	–
DAN [27]	89.70	–	65.69	62.09
EfficientFace [33]	88.36	–	63.70	60.23
MA-Net [32]	88.42	–	64.53	60.29
Meta-Face2Exp [30]	88.54	–	64.23	–
EAC [31]	90.35	–	65.32	–
Ours	92.11	92.06	67.49	63.47

In order to better understand the performance of our model on each facial expression category, Fig. 5 presents the confusion matrices on each dataset we evaluated. More specifically, it is obvious that the 'Happy' class is the easiest on both Affect-7class and Affect-8class, followed by 'Sad', 'Surprise', and 'Fear'. On RAF-DB, our LGCFTN model has high accuracy on the 'Surprise', 'Happy', 'Sad', 'Neutral', and 'Anger' classes. On FERPlus, our method performs well on the 'Neutral', 'Happy', 'Surprise', and 'Angry' classes, while the 'Contempt' and 'Fear' classes are very challenging for recognition. Possible reasons for the large gaps in the performance include the appearance similarities among facial expression categories as well as the skewed class distribution in the training dataset.

4.3 Param and FLOPs Comparison

The total number of parameters (Param) and floating-point operations (FLOPs) are two evaluation metrics that measure the size and complexity of the model. Here we list the Param and FLOPs of our LGCFTN model in Table 3.

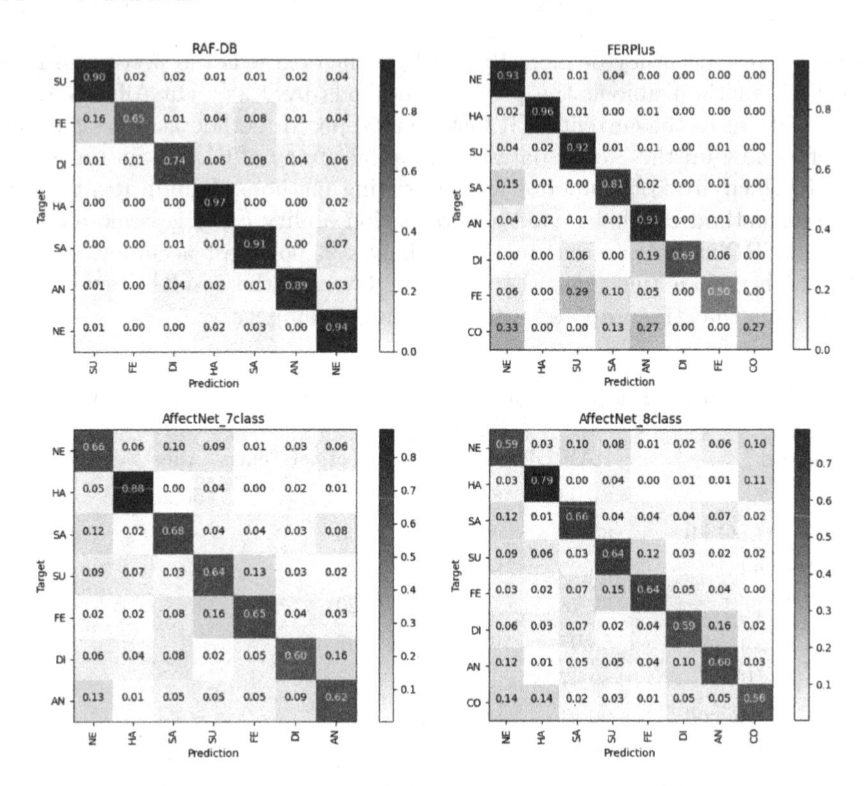

Fig. 5. The confusion matrices of our method on RAF-DB, FERPlus, and AffectNet test sets.

Compared with DMUE and TransFER, our method achieves better test accuracy on RAF-DB and AffectNet datasets with smaller Param and FLOPs. Specifically, compared with DMUE, our method gets 2.69% higher test accuracy on RAF-DB and 4.38% higher test accuracy on AffectNet-7class while reducing 31.5M Param and 5.1G FLOPs. Compared with TransFER, our method gets 1.2% higher test accuracy on RAF-DB and 1.26% higher test accuracy on AffectNet-7class while reducing 18.3M Param and 7.0G FLOPs.

Table 3. Comparison of Param and FLOPs

Method	Param	FLOPs	RAF-DB	AffectNet-7class
DMUE [21]	78.4M	13.4G	89.42	63.11
TransFER [29]	65.2M	15.3G	90.91	66.23
LGCFTN	46.9M	8.3G	92.11	67.49

4.4 Ablation Study

Effectiveness of the Proposed Modules. In order to verify the effectiveness of the proposed modules, an ablation study is designed to investigate the effects of these two modules on the expression recognition task. The experimental results are shown in Table 4. Among them, (a) the baseline method means directly using the features extracted by IR50 for expression classification, which is the simplest way of using CNN for the FER task. (b) Based on the baseline method, the local feature decomposition module is added. Then the two features are concatenated to obtain the fusion feature for classification. The performance of the model is better than simply using a global feature of the image. (c) On the basis of (b), the Transformer-based cross-fusion module is further used to fuse the local features with the global features through the cross-attention mechanism, which is the complete method we proposed. We can observe that the performance of the model improved further, which verifies the effectiveness of the cross-fusion Transformer.

Table 4. Effectiveness of the proposed modules

method	RAF-DB	AffectNet-7class	AffectNet-8class
(a) baseline	88.93	63.86	58.86
(b)+ local feature decomposition	90.67	66.50	62.36
(c)+ cross fusion Transformer	92.11	67.49	63.47

Effectiveness of Cross-Attention. To further verify the effectiveness of the cross-attention mechanism, a comparison experiment between the standard Transformer and the cross-attention Transformer is set up. The experimental results of the comparison are shown in Table 5. The experimental results show that the test accuracy of the model decreases when using the self-attention mechanism. This is because the self-attention mechanism is unable to discern the difference between local and global features. Cross-attention, on the other hand, can obtain information about the differences between local and global features. Information exchange between local and global features can be achieved in the process of cross-fusion, thus obtaining a stronger fused feature.

Table 5. Effectiveness of cross-attention mechanism.

method	RAF-DB	AffectNet-7class	AffectNet-8class
self-attention	91.23	66.87	62.87
cross-attention	92.11	67.49	63.47

Evaluation of Region Generation Strategies. We evaluate the fixed cropping, random cropping, and facial landmark-based cropping methods on RAF-DB with the default setting for other modules, the results are shown in Table 6. The length and width of the local blocks in the random cropping method are set to 7. And the number of local blocks is set to 3, 5, 7, and 9 for the comparison of the experiment. The fixed cropping is to set 5 coordinate points and crop out 5 local blocks with 5 coordinates as the center, the distribution of 5 points are (0.3, 0.3), (0.5, 0.5), (0.6, 0.6), (0.3, 0.6), (0.6, 0.3). This setting is to cover the eye, nose, and mouth position of the face as much as possible. Facial landmark-based cropping is specified in the Sect. 3.2.

Table 6. Evaluation of region generation strategies on RAF-DB.

method	Acc. (%)
random(3)	90.64
random(5)	91.12
random(7)	91.03
random(9)	90.67
fixed	91.76
facial landmark-based	92.11

The results show that facial landmark-based cropping achieves the best performance. The accuracy of the fixed cropping is slightly lower than the facial landmark-based cropping. Since the face images may contain occlusion and tilt poses, the fixed position cannot accurately locate the expression-related regions of the face. Random cropping obtains the worst overall performance. The performance decreases when the number of random cropping blocks increases to 7. This is mainly because increasing the number of blocks leads to too much redundant and irrelevant information.

5 Conclusions

In this paper, we have proposed a new local-global cross-fusion transformer network for the FER task. Firstly, a local feature decomposition module has been proposed to extract local regional features based on facial landmarks, which can provide diverse local information on multiple local areas of the face. Secondly, the Transformer-based cross-fusion module is designed to fuse the local and global features. With the cross-attention mechanism, our network can sufficiently explore the information of local features and global features, such that a discriminative facial representation yet good FER performance can be obtained. Extensive experiments on three public FER datasets verify that our method outperforms the state-of-the-art methods.

Acknowledgements. This work is supported by the Higher Education Stability Support Program Project (Grant No. GXWD20220811173317002) and Shenzhen Science and Technology Program (Grant No. RCBS20210609103709020).

References

1. Barsoum, E., Zhang, C., Ferrer, C.C., Zhang, Z.: Training deep networks for facial expression recognition with crowd-sourced label distribution. In: Proceedings of the 18th ACM International Conference on Multimodal Interaction, pp. 279–283 (2016)
2. Chen, C.F.R., Fan, Q., Panda, R.: CrossViT: cross-attention multi-scale vision transformer for image classification. In: Proceedings of the IEEE/CVF International Conference on Computer Vision, pp. 357–366 (2021)
3. Chen, S., Liu, Y., Gao, X., Han, Z.: MobileFaceNets: efficient CNNs for accurate real-time face verification on mobile devices. In: Zhou, J., et al. (eds.) Biometric Recognition: 13th Chinese Conference, CCBR 2018, Urumqi, China, 11–12 August 2018, Proceedings 13, pp. 428–438. Springer, Cham (2018). https://doi.org/10.1007/978-3-319-97909-0_46
4. Chen, S., Wang, J., Chen, Y., Shi, Z., Geng, X., Rui, Y.: Label distribution learning on auxiliary label space graphs for facial expression recognition. In: Proceedings of the IEEE/CVF Conference on Computer Vision and Pattern Recognition, pp. 13984–13993 (2020)
5. Cotter, S.F.: Sparse representation for accurate classification of corrupted and occluded facial expressions. In: 2010 IEEE International Conference on Acoustics, Speech and Signal Processing, pp. 838–841. IEEE (2010)
6. Dalal, N., Triggs, B.: Histograms of oriented gradients for human detection. In: 2005 IEEE Computer Society Conference on Computer Vision and Pattern Recognition (CVPR 2005), vol. 1, pp. 886–893. IEEE (2005)
7. Deng, J., Guo, J., Xue, N., Zafeiriou, S.: ArcFace: additive angular margin loss for deep face recognition. In: Proceedings of the IEEE/CVF Conference on Computer Vision and Pattern Recognition, pp. 4690–4699 (2019)
8. Dosovitskiy, A., et al.: An image is worth 16 × 16 words: transformers for image recognition at scale. arXiv preprint arXiv:2010.11929 (2020)
9. Farzaneh, A.H., Qi, X.: Facial expression recognition in the wild via deep attentive center loss. In: Proceedings of the IEEE/CVF Winter Conference on Applications of Computer Vision, pp. 2402–2411 (2021)
10. Guo, Y., Zhang, L., Hu, Y., He, X., Gao, J.: MS-Celeb-1M: a dataset and benchmark for large-scale face recognition. In: Leibe, B., Matas, J., Sebe, N., Welling, M. (eds.) Computer Vision–ECCV 2016: 14th European Conference, Amsterdam, The Netherlands, 11–14 October 2016, Proceedings, Part III 14, pp. 87–102. Springer, Cham (2016). https://doi.org/10.1007/978-3-319-46487-9_6
11. Huang, Y.F., Tsai, C.H.: PIDViT: pose-invariant distilled vision transformer for facial expression recognition in the wild. IEEE Trans. Affect. Comput. **14**(4), 3281–3293 (2022)
12. Kingma, D.P., Ba, J.: Adam: a method for stochastic optimization. arXiv preprint arXiv:1412.6980 (2014)
13. Li, H., Wang, N., Ding, X., Yang, X., Gao, X.: Adaptively learning facial expression representation via CF labels and distillation. IEEE Trans. Image Process. **30**, 2016–2028 (2021)

14. Li, S., Deng, W., Du, J.: Reliable crowdsourcing and deep locality-preserving learning for expression recognition in the wild. In: Proceedings of the IEEE Conference on Computer Vision and Pattern Recognition, pp. 2852–2861 (2017)
15. Lin, W., et al.: CAT: cross-attention transformer for one-shot object detection. arXiv preprint arXiv:2104.14984 (2021)
16. Ma, F., Sun, B., Li, S.: Facial expression recognition with visual transformers and attentional selective fusion. IEEE Trans. Affect. Comput. **14**(2), 1236–1248 (2021)
17. Mollahosseini, A., Hasani, B., Mahoor, M.H.: AffectNet: a database for facial expression, valence, and arousal computing in the wild. IEEE Trans. Affect. Comput. **10**(1), 18–31 (2017)
18. Rifai, S., Bengio, Y., Courville, A., Vincent, P., Mirza, M.: Disentangling factors of variation for facial expression recognition. In: Fitzgibbon, A., Lazebnik, S., Perona, P., Sato, Y., Schmid, C. (eds.) Computer Vision–ECCV 2012: 12th European Conference on Computer Vision, Florence, Italy, 7–13 October 2012, Proceedings, Part VI 12, pp. 808–822. Springer, Cham (2012). https://doi.org/10.1007/978-3-642-33783-3_58
19. Ruan, D., Yan, Y., Lai, S., Chai, Z., Shen, C., Wang, H.: Feature decomposition and reconstruction learning for effective facial expression recognition. In: Proceedings of the IEEE/CVF Conference on Computer Vision and Pattern Recognition, pp. 7660–7669 (2021)
20. Shan, C., Gong, S., McOwan, P.W.: Facial expression recognition based on local binary patterns: a comprehensive study. Image Vis. Comput. **27**(6), 803–816 (2009)
21. She, J., Hu, Y., Shi, H., Wang, J., Shen, Q., Mei, T.: Dive into ambiguity: latent distribution mining and pairwise uncertainty estimation for facial expression recognition. In: Proceedings of the IEEE/CVF Conference on Computer Vision and Pattern Recognition, pp. 6248–6257 (2021)
22. Maximiano da Silva, F.A., Pedrini, H.: Geometrical features and active appearance model applied to facial expression recognition. Int. J. Image Graph. **16**(04), 1650019 (2016)
23. Szegedy, C., Vanhoucke, V., Ioffe, S., Shlens, J., Wojna, Z.: Rethinking the inception architecture for computer vision. In: Proceedings of the IEEE Conference on Computer Vision and Pattern Recognition, pp. 2818–2826 (2016)
24. Vaswani, A., et al.: Attention is all you need. In: Proceedings of the 31st International Conference on Neural Information Processing Systems, pp. 6000–6010 (2017)
25. Wang, K., Peng, X., Yang, J., Lu, S., Qiao, Y.: Suppressing uncertainties for large-scale facial expression recognition. In: Proceedings of the IEEE/CVF Conference on Computer Vision and Pattern Recognition, pp. 6897–6906 (2020)
26. Wang, K., Peng, X., Yang, J., Meng, D., Qiao, Y.: Region attention networks for pose and occlusion robust facial expression recognition. IEEE Trans. Image Process. **29**, 4057–4069 (2020)
27. Wen, Z., Lin, W., Wang, T., Xu, G.: Distract your attention: multi-head cross attention network for facial expression recognition. Biomimetics **8**(2), 199 (2023)
28. Xu, X., Wang, T., Yang, Y., Zuo, L., Shen, F., Shen, H.T.: Cross-modal attention with semantic consistence for image-text matching. IEEE Trans. Neural Networks Learn. Syst. **31**(12), 5412–5425 (2020)
29. Xue, F., Wang, Q., Guo, G.: Transfer: learning relation-aware facial expression representations with transformers. In: Proceedings of the IEEE/CVF International Conference on Computer Vision, pp. 3601–3610 (2021)

30. Zeng, D., Lin, Z., Yan, X., Liu, Y., Wang, F., Tang, B.: Face2Exp: combating data biases for facial expression recognition. In: Proceedings of the IEEE/CVF Conference on Computer Vision and Pattern Recognition, pp. 20291–20300 (2022)
31. Zhang, Y., Wang, C., Ling, X., Deng, W.: Learn from all: erasing attention consistency for noisy label facial expression recognition. In: Avidan, S., Brostow, G., Cissé, M., Farinella, G.M., Hassner, T. (eds.) Computer Vision–ECCV 2022: 17th European Conference, Tel Aviv, Israel, 23–27 October 2022, Proceedings, Part XXVI, pp. 418–434. Springer, Cham (2022). https://doi.org/10.1007/978-3-031-19809-0_24
32. Zhao, Z., Liu, Q., Wang, S.: Learning deep global multi-scale and local attention features for facial expression recognition in the wild. IEEE Trans. Image Process. **30**, 6544–6556 (2021)
33. Zhao, Z., Liu, Q., Zhou, F.: Robust lightweight facial expression recognition network with label distribution training. In: Proceedings of the AAAI Conference on Artificial Intelligence, vol. 35, pp. 3510–3519 (2021)

Answering Spatial Commonsense Questions by Learning Domain-Invariant Generalization Knowledge

Miaopei Lin[1], Jianxing Yu[2,3](✉), Shiqi Wang[2], Hanjiang Lai[1,3], Wei Liu[2,3], and Jian Yin[2,3]

[1] School of Computer Science and Engineering, Sun Yat-sen University, Guangzhou, China
linmp3@mail2.sysu.edu.cn,
[2] School of Artificial Intelligence, Sun Yat-sen University, Zhuhai, China
wangshq25@mail2.sysu.edu.cn,
[3] Guangdong Key Laboratory of Big Data Analysis and Processing, Guangzhou, China
{yujx26,laihanj3,liuw259,issjyin}@mail.sysu.edu.cn

Abstract. Existing spatial commonsense reading comprehension (SC-RC) systems struggle to answer questions from unknown domains or any out-of-domain distributions, which prevents them from being deployed in real applications. Unsupervised domain adaptation (UDA) in QA has emerged as a major approach to address this challenge. However, existing methods mainly rely on generating synthetic data and pseudo-labeling target domains, which not only consumes extra computational resources but also places high demands on noise filtering of the generated data. To tackle these problems, we propose a UDA framework, called LEGRN-DIG, for spatial commonsense question answering using unlabeled target domain data. This framework avoids the use of labeled or pseudo-labeled target instances and noisy synthetic data. We apply domain-invariant generalization learning to integrate features of the target domain into the source domain and still use the source domain for supervised training. Extensive experiments are conducted to illustrate the effectiveness and robustness of our model.

Keywords: spatial commonsense reasoning · question answering · domain-invariant generalization

1 Introduction

In the process of human reading, spatial commonsense about object properties such as shape, size, distance, and position can help humans understand and construct the described scene in their minds. This holds similar significance for intelligent agents in comprehending and reasoning about natural language text [12]. For example, when someone wants to go to a wider body of water, we can infer that they are more likely to go to the ocean than to the lake (see Q2

X. Song et al. (Eds.): APWeb-WAIM 2023, LNCS 14332, pp. 270–285, 2024.
https://doi.org/10.1007/978-981-97-2390-4_19

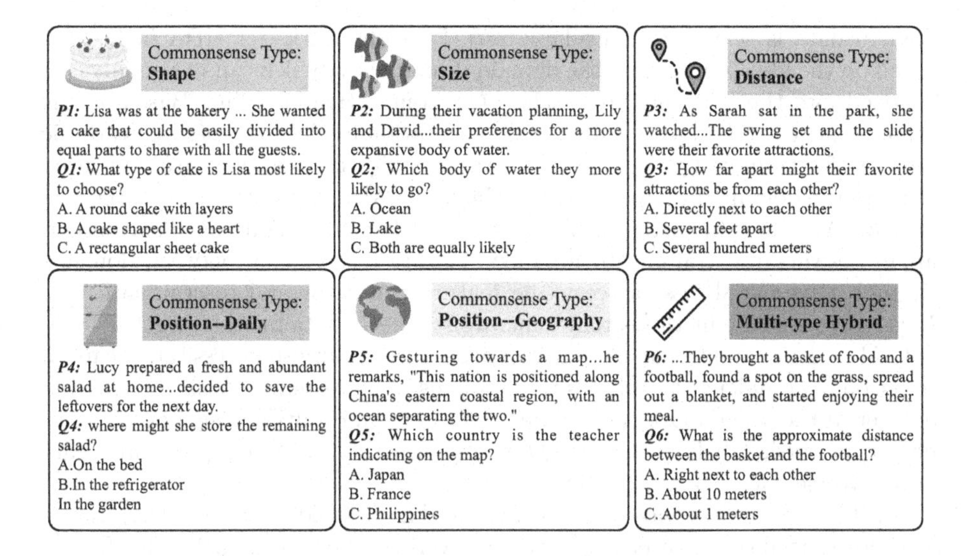

Fig. 1. Examples with various spatial reasoning types in CRCSpatial

in Fig. 1). Existing works have paid attention to applying spatial common sense in machine reading comprehension (MRC). Some researchers have developed a large-scale knowledge graph covering a variety of spatial commonsense [20], and benchmark datasets [12,19], thus propelling research in spatial commonsense reading comprehension (SCRC).

It is crucial to acquire sufficient commonsense knowledge for this task. Given that commonsense knowledge Graphs (KG), such as ConceptNet [17] and WebChild 2.0 [20], have high-quality commonsense triples but insufficient coverage, while pre-trained language models (PLMs) capture rich but potentially noisy knowledge, many works consider fusing knowledge from these two sources. Some works design pre-training tasks based on KG to implicitly learn structured knowledge. Conversely, various efforts explicitly encode KG information to facilitate structured reasoning. These methods involve encoding text and KG using PLMs and Graph Neural Networks (GNNs), followed by strategies such as direct concatenation, as seen in [15,29], enhancing the representation of one modality with features from another modality, as in [3–5,11], or introduce simple interaction layers to facilitate mutual augmentation, as in [24,28]. Additionally, for spatial commonsense, LEGRN [12] proposes cross-attention layers and edge-augmented graph encoder to enhance the model's semantic comprehension, enabling it to perform reasoning over complex spatial relationships.

However, existing MRC models struggle to answer questions from unknown domains or any out-of-domain distributions, which makes them less reliable when deployed to real scenarios. Unsupervised domain adaptation (UDA) in QA has emerged as a promising avenue to address this challenge, which aims to enhance the robustness of MRC models in cross-domain scenarios where labeled target domain data is scarce or unavailable [27]. Previous works for UDA in QA are based on the idea to minimize the domain discrepancy, which can be roughly categorized into the following approaches: (1) Contrastive Learning. Based on the target contexts, they introduce question generation models to generate synthetic datasets for training QA systems [25]. Additionally, contrastive learning is used to minimize domain discrepancy using Maximum Mean Discrepancy (MMD) [26] or Sliced Wasserstein Distance (SWD) [8]; (2) Adversarial Training. This method applies adversarial training to minimize feature discrepancy between domains through adversarial loss between the feature generator and discriminator [9]. Nevertheless, these methods that generate synthetic data for the target domain not only require additional computational resources but also necessitate a careful selection of confidence thresholds to distinguish noisy samples from the training dataset. If the pseudo-labeling for the target domain results in a significant number of incorrect labels exceeding the confidence threshold, it could potentially have adverse effects on adaptation.

In this paper, we proposed an unsupervised domain adaptation (UDA) framework for spatial commonsense question answering (shown in Fig. 2). This framework avoids the need to explicitly employ noisy pseudo-labeled target domain instances to train the answer predictor; instead, it remains supervised by source domain data. We introduced a method for learning domain-invariant features, utilizing domain-invariant generalization learning to identify and optimize target instances that deviate from the source domain. The main contributions of this paper include:

- We reveal the issue of domain distribution shift in spatial reading comprehension and address the issue from the perspective of invariant learning, which is new for this task.
- We propose an unsupervised domain adaptation framework that utilizes invariant generalization to alleviate the impact of domain shift for the task of spatial commonsense reading comprehension and provides theoretical guarantees for mitigating domain bias.
- We conduct extensive experiments to show the superiority and robustness of our methods.

2 Approach

In this section, we formalize the task of unsupervised domain adaptation(UDA) for multiple choice spatial commonsense reading comprehension (MC-SCRC) and then introduce the overview of our framework.

2.1 Setup

In UDA, we consider the labeled source domain $\mathcal{D}_s$ and unlabeled target domain $\mathcal{D}_t$. Our goal is to improve QA performance on the target domain by training with labeled source domain data and unlabeled target domain data [27].

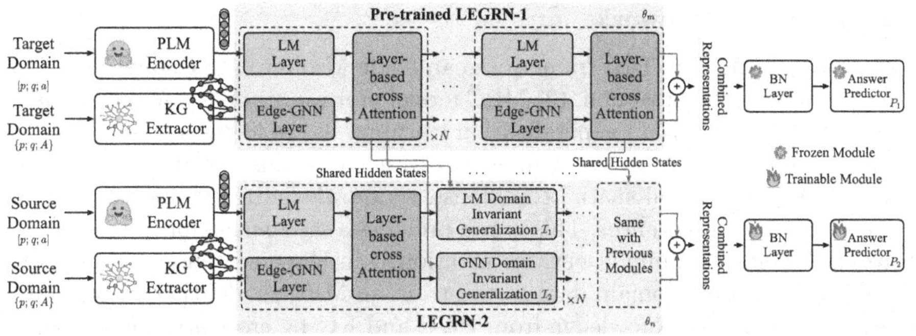

Fig. 2. Architecture of proposed UDA framework for spatial commonsense reading comprehension. It constructs a reasoning graph of sentences for each given QA context p, q, A. Our method first pre-trains a LEGRN encoder on the source domain and respectively applies LM/GNN domain invariant generalization after LM and Edge-GNN of another LEGRN encoder.

MC-SCRC Task: Given a context paragraph p, a question q, a set of candidate answers A and an external knowledge graph (KG) providing spatial commonsense knowledge related to the p, q, A, the objective is to predict the correct answer $a_i \in A$. We transform this problem into measuring the coherence scores between p, q, and each option $a_i \in A$, and predict the most plausible answer as $argmax_{a_i \in A}\ p(a_i|p, q)$.

Data: We define a sample in the labeled source data $\boldsymbol{X}_s$ from $\mathcal{D}_s$ by a quad, $\left\{p_s^{(i)}, q_s^{(i)}, A_s^{(i)}, a_s^{(i)}\right\} \in \boldsymbol{X}_s$, consisting of a context paragraph $p_s^{(i)}$, a question $q_s^{(i)}$, a set of candidate answers $A_s^{(i)}$ and a groud truth answer $a_s^{(i)}$, where $a_s^{(i)} \in A_s^{(i)}$. Similarly, A sample in the unlabeled target data $\boldsymbol{X}_t$ from $\mathcal{D}_t$ is defined by a triplet, $\left\{p_t^{(i)}, q_t^{(i)}, A_t^{(i)}\right\} \in \boldsymbol{X}_t$, consisting of only a context paragraph $p_t^{(i)}$, a question $q_t^{(i)}$ and a set of candidate answers $A_t^{(i)}$. Furthermore, it is required to input an external knowledge graph, KG, that provides spatial commonsense knowledge for the model.

Model: We denote the MRC model as function $\boldsymbol{f}$ which predicts an answer $a_t^{(i)}$ given the context paragraph $p_t^{(i)}$, question $q_t^{(i)}$ and candidate answers set $A_t^{(i)}$, namely $a_t^{(i)} = \boldsymbol{f}\left(p_t^{(i)}, q_t^{(i)}, A_t^{(i)}, KG\right)$. Our goal is to optimize the function $\boldsymbol{f}$ to maximize the performance of answering questions in the target domain $\mathcal{D}_t$, that is, minimize the loss for the target domain distribution $\boldsymbol{X}_t$:

$$\boldsymbol{f}^* = \min_{\boldsymbol{f}} \mathcal{L}\left(\boldsymbol{f}, \mathbf{X}_t\right) \tag{1}$$

2.2 Overall Framework

The proposed LEGRN-DIG framework can be divided into three stages: (1) Spatial Commonsense Extraction, (2) MRC model pretraining on the source domain, and (3) Domain-Invariant generalization learning. In the first stage, for each QA context p, q, A from both the source and target domains, we retrieve relevant triples from an external spatial commonsense knowledge graph and construct a reasoning graph of sentences. This graph provides a mechanism for inter-sentence traversal based on common entities mentioned in these sentences. In the next stage, we use source domain data to pre-train an MRC model that incorporates spatial commonsense knowledge from PLM and KG by cross-attention. Specifically, we apply LEGRN as the encoder and introduce an additional batch norm layer and answer predictor P_1. This model is utilized to learn features of target domain instances. Lastly, we conduct domain-invariant generalization learning on another MRC model. We augment the source domain with pseudo labels and feature information to obtain another target domain-aware answer predictor P_2. The overview of our framework is shown in Fig. 2.

2.3 Spatial Commonsense Knowledge Extraction from KG

Given a QA context $\{p, q, A\}$ and a spatial commonsense knowledge KG, we construct a directed reasoning graph of sentences, G, with edges indicating textual coherency. First, co-referent links are constructed for each sentence $s_i \in [p; q]$. We add a link between two sentences if they share a resolved entity during coreference resolution. The direction of this link is based on the order of occurrence, with only forward links added. These links form disjoint and highly coreferent subgraphs.

Then we link the entities mentioned in the paragraph p, question q, and candidate answer $a_i \in A$ to the given KG. We search for the paths (less than 3 hops) from entities in p and q to entities in a_i and merge the covered triples into a graph where nodes are triples and edges are the relation between triples. If triples tri_i and tri_j contain the same entity, we will add an edge from the previous triple tri_i to the next triple tri_j. To acquire contextual word representations for KG triples, we convert the triples into natural language sentences using the relation template. For instance, we transfer *(book, on, bookshelf)* into *"book is on bookshelf"*. Here we get a spatial commonsense knowledge subgraph G_{kg}.

Then we join these separate subgraphs by entity linking applied to all sentences $s_i \in [p; q]$ and $tri_i \in G_{kg}$ to obtain the final reasoning graph of sentences G (shown in Fig. 3). It provides a mechanism for inter-sentence traversal based on the common entities mentioned in these sentences. Furthermore, all traversals in G are highly coherent, forming logically connected chains of sentences that can be comprehended independently.

2.4 MRC Model Pretraining with Spatial Commonsense

In this section, we will describe our MRC model pre-trained on $\boldsymbol{X}_s$ from source domain $\mathcal{D}_s$. Specifically, we apply the N layers LM Edge-GNN Reasoner Net-

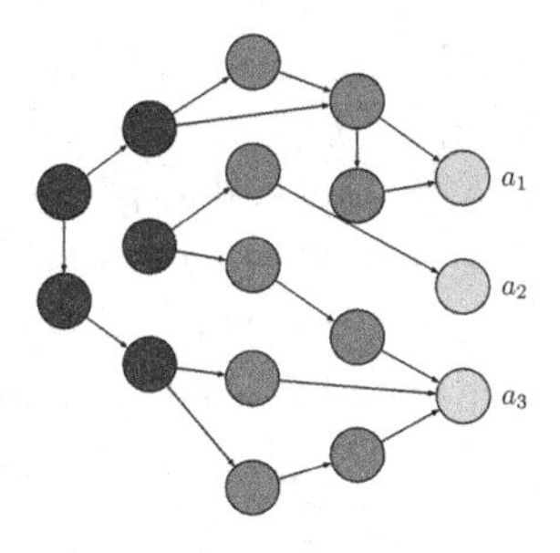

Fig. 3. Example of a reasoning graph of sentences G given a QA context $\{p, q, A\}$, where red nodes, green nodes, and yellow nodes represent sentences from [paragraph p; question q], the external spatial commonsense knowledge graph KG, and the candidate answer set A, respectively. Note that we set a constraint that the distance from red nodes to yellow nodes does not exceed 3 hops.

works (LEGRN) [12] as the MRC encoder, which connects N layers LM and Edge-GNN at each layer by the Layer-based Cross Attention, enabling a more deep fusion and mutual enhancement of spatial commonsense knowledge from pre-trained language models (PLM) and knowledge graphs(KG). This empowers it to answer intricate spatial commonsense questions. LEGRN takes two kinds of sources as inputs: one is the concatenated text $[p, q, a_i]$ of the context paragraph$p$, question$q$ and candidate answer $a_i \in A$, and the other is the reasoning sentence graph G extracted from QA context $\{p, q, A\}$ and KG described in §2.3.

Text Representation. We apply a PLM (e.g.RoBERTa) on the input $[p, q, a_i]$ to get the initial text representation $H_T^{(0)}$ and process it by an MLP to unify its hidden size. Then we employ N layers transformer encoder as the language model (LM) to encode the text. For the *l-th* LM layer, we get the text representation by: $H_T^{(l)} = Transformer\left(H_T^{(l-1)}\right)$.

Graph Representation. For the input reasoning sentence graph G, we initialize sentence node embeddings $H_G^{(0)}$ by a PLM and edge embeddings $H_E^{(0)}$ by random initialization, and we also process them by an MLP to unify their hidden size. Subsequently, we feed them into N layers Edge-GNN [12], which is capable of simultaneously encoding nodes and edges of a graph and enhancing node representations with edge embeddings. As a result, node representations can better encapsulate the information of the graph. For the *l-th* Edge-GNN layer, we get the graph representation by: $H_G^{(l)} = Edge\text{-}GNN\left(H_G^{(l-1)}\right)$.

Integration of Textual and Graph Representations. To integrate the representations of text and graph deeply, we apply two transformers [21] based on cross-attention after each LM Layer and Edge-GNN Layer separately. In detail, for the outputs of the *l-th* LM Layer $H_T^{(l)}$ and the *l-th* Edge-GNN Layer $H_G^{(l)}$, we map them to query $Q_T^{(l)}$, $Q_G^{(l)}$, key $K_T^{(l)}$, $K_G^{(l)}$ and value $V_T^{(l)}$, $V_G^{(l)}$ matrices. We then interchange the key-value pairs in multi-head attention to compute correlation between the two sources, where $Q_T^{(l)}$, $K_G^{(l)}$, $V_G^{(l)}$ for text transformer and $Q_G^{(l)}$, $K_T^{(l)}$,$V_T^{(l)}$ for graph transformer. We get the text representation that is grounded in structured spatial commonsense knowledge $H_T^{(l)} = \textit{Cross-Attention Transformer}\left(H_T^{(l)}\right)$, and graph representation improved by linguistic nuances in QA context $H_G^{(l)} = \textit{Cross-Attention Transformer}\left(H_G^{(l)}\right)$.

Answer Predictor. We obtain the final fused representation by concatenating the output text and graph representations from LEGRN. We apply batch normalization(BN) on it and finally compute the probability of a_i being the correct answer as

$$p(a_i|p, q) = MLP(BN(Pool(H_T^{(N)}) \oplus Pool(H_G^{(N)}))) \tag{2}$$

.

Then we optimize the whole MRC model end-to-end using cross-entropy loss only on $\boldsymbol{X}_s$ from source domain $\mathcal{D}_s$.

2.5 Domain Invariant Generalization

In this section, we present the process of our domain-invariant generalization learning. In §2.4, we have pre-trained an MRC model with N layers of LM Edge-GNN Reasoner Networks (LEGRN-1) on the source domain. We denote its answer predictor as P_1 and LEGRN parameters as θ_m of the MRC model. In domain-invariant generalization learning, we introduce another LEGRN framework, LEGRN-2. We propose to inject the feature information of the target domain $\mathcal{D}_t$ into the source domain $\mathcal{D}_s$ by hidden state sharing at each layer $l \in N$ between the two LEGRN (shown in Fig. 2). This enables the features of the target domain to be updated in the LEGRN-2 encoder supervised on the source domain. Let $\delta_1(x, x')$ and $\delta_2(y, y')$ be the learnable domain shift vectors the input text and reasoning graph between the source domain instance and target domain instance. We propose a learnable LM domain invariant generalization layer $\mathcal{I}_1(x, \delta_1(x, x'))$ after each LM layer and a learnable GNN domain invariant generalization layer $\mathcal{I}_2(y, \delta_1(y, y'))$after each Edge-GNN layer of MRC model LEGRN-2. They are designed to transfer the parameters of source domain predictor P_1 to the target domain-aware predictor P_2 and update LEGRN-2 parameters θ_n with features of the target domain.

Domain Invariant Generalization Layer $\mathcal{I}_1$ and $\mathcal{I}_2$. Given the domain shift between the source domain and the target domain, the proposed domain invariant generalization layer operates to fuse the features, namely hidden states, of the target domain with the domain shift vector at each layer of LM and Edge-GNN.

The domain shift vectors δ_1 and δ_2 solely capture features in target domain that differ from source domain. Let $H_{T-s}^{(l)}$, $H_{T-t}^{(l)}$ obtained from *l-th* LM decoder and $H_{G-s}^{(l)}$, $H_{G-t}^{(l)}$ obtained from *l-th* Edge-GNN decoder be the hidden states for the source domain and target domain respectively. We define the domain shift vector at *l-th* layer of LM and Edge-GNN as the difference between the average pooled vector of hidden states:

$$\delta_1^{(l)}\left(H_{T-t}^{(l)}, H_{T-s}^{(l)}\right) = \text{avg}\left(W_1 H_{T-t}^{(l)}\right) - \text{avg}\left(W_1 H_{T-s}^{(l)}\right), \tag{3}$$

$$\delta_2^{(l)}\left(H_{G-t}^{(l)}, H_{G-s}^{(l)}\right) = \text{avg}\left(W_2 H_{G-t}^{(l)}\right) - \text{avg}\left(W_2 H_{G-s}^{(l)}\right), \tag{4}$$

where $W_1, W_2 \in \mathbf{R}^{k \times d}$ and are the linear transform parameters, k is a hyper-parameter that indicates the domain shift information resides in a k-dimensional subspace. Then we apply the LM domain invariant generalization layer as:

$$\mathcal{I}_1\left(H_{T-s}^{(l)}, \delta_1^{(l)}\left(H_{T-t}^{(l)}, H_{T-s}^{(l)}\right)\right) = H_{T-s}^{(l)} + W_1^T \delta_1^{(l)}\left(H_{T-t}^{(l)}, H_{T-s}^{(l)}\right), \tag{5}$$

and GNN domain invariant generalization layer follows a similar definition:

$$\mathcal{I}_2\left(H_{G-s}^{(l)}, \delta_2^{(l)}\left(H_{G-t}^{(l)}, H_{G-s}^{(l)}\right)\right) = H_{G-s}^{(l)} + W_2^T \delta_2^{(l)}\left(H_{G-t}^{(l)}, H_{G-s}^{(l)}\right), \tag{6}$$

In this way, the domain shift vectors adhere to the identity property, $\delta_1(x, x) = 0, \delta_2(x, x) = 0$. This implies we can plug in the domain invariant generalization layer only during training, and during inference on the target domain, $\delta_1\left(H_{T-t}^{(l)}), H_{T-t}^{(l)})\right) = 0$, $\delta_2\left(H_{G-t}^{(l)}), H_{G-t}^{(l)})\right) = 0$. To minimize the feature difference between the two domains, it is crucial to choose an appropriate subspace dimension k for domain shift vectors.

Training and Inference. Considering that the instance from the target domain consists of only paragraph, question, and candidate answer set, the ground truth answer is the pseudo-answer from the predictor P_1 of the pre-trained MRC model. During training, we randomly sample target domain instances with the same batch size as the source domain and we enforce alignment between them, i.e. belong to the same question types (denoted as $\boldsymbol{X}_s \approx \boldsymbol{X}_t$). Specifically, we identify the question types in a fine-grained manner using classification of interrogative words, dependency parse, semantic role labeling, and NER described in [6]. In the training stage, the weights of the batch normalization layer and predictor P_1 are frozen and a subset of LEGRN-1 parameters θ_m are shared with LEGRN-2. And as shown in Fig. 2, the parameters of LEGRN-2 and predictor P_2 are updated under target domain awareness and source domain supervision. Predictor P_2 initialized with Predictor P_1. We optimize the model with cross-entropy loss $\mathcal{L}_{ce}$:

$$\min_{\boldsymbol{f}} \mathcal{L}_{ce}\left(\boldsymbol{f}, \boldsymbol{X}_s \approx \boldsymbol{X}_t\right) \tag{7}$$

where $\boldsymbol{f}$ consists of parameters θ_n and parameters of predictor P_2 and domain invariant generalization layer $\mathcal{I}_1$ and $\mathcal{I}_2$. Finally, we obtain the target-aware domain adapted MRC model $\boldsymbol{f}^*$.

3 Experiments

In this session, we comprehensively evaluated the efficacy of our approach in various aspects.

3.1 Dataset and Knowledge Graph

We conduct experiments on CRCSpatial (Spatial Commonsense Reading Comprehension) dataset [12] with $4k$ paragraphs and $40k$ spatial commonsense questions to evaluate the improvement of our proposed method. CRCSpatial contains six types of spatial commonsense reasoning questions, which are *Shape*(12%), *Size*(10%), *Distance*(14%), *Position*(42%: including 37% for daily scenes and 5% for geographic scenes) and *Hybrid reasoning*(22%), with several examples shown in Fig. 1. Considering CRCSpatial split the dataset into training, development, and test sets at an 8:1:1 ratio, we take the training set as the source domain $\mathcal{D}_s$, development and test sets as the target domain $\mathcal{D}_t$.

Regarding the external spatial commonsense knowledge base, we use WebChild2.0 [20], with over 2 million disambiguated concepts connected by over 18 million assertions. WebChild2.0 contains a large number of spatial commonsense triples, including the shape of objects (e.g. *hasShape*), size of objects (e.g. *hasSize*), relationships between objects (e.g. *largerThan, nextTo, above, on, around*). As we transform the retrieved relevant triplet into sentence nodes when constructing the reasoning graph, we initialize node embeddings by PLM, i.e. RoBERTa, and edge embeddings by random assignment.

3.2 Baseline Models

For the compared methods, we select models and classify them into 2 groups: fine-tuned PLMs which do not use the KG, and KG-based models which leverage structured external knowledge for reasoning. Furthermore, we also compared our model with the baseline encoder we used, LEGRN, human performance as obtained in [12] to validate the effectiveness of our domain adaptation method.

Group1 are Fine-tuned PLMs. To study the role of using KG as an external knowledge source, we conduct a comparison between our approach and the baseline fine-tuned LMs on the CRCSpatial. These models include GPT-2 (2019) [16], BERT-large (2019) [7], RoBERTa-large (2019) [14], ERNIE2.0-large (2020) [18]. ERNIE2.0 stands out as a semantic knowledge-enhanced pre-trained model built upon BERT, designed to facilitate continuous learning.

Group2 are KG-based models. To assess the capability of our LEGRN-DIG in amalgamating representations of both modalities, we also conducted comparisons with alternative state-of-the-art methods that combine LM and KG. These models include: (1) GconAttn (2019) [22], which applies the Match-LSTM model in the field of text matching to align knowledge concepts; (2) KagNet (2019) [11], which extracts QA-related subgraphs from the KG and encodes the relational paths by GCN and LSTM; (3) MHGRN (2020) [3], which introduces a multi-hop relational reasoning module to acquire graph representation at a path-level;

(4) QA-GNN (2021) [23], which introduces a QA content node into the subgraph, allowing combined reasoning over both QA context and KG, that is, the one we utilize to construct subgraph; and (5) GreaseLM (2022) [28], which fuse embeddings of LM and GNN by a two-layers MLP.

3.3 Implementation Details

We use RoBERTa-large [14] as the text encoder to initialize QA context representation. We set the dimension (D = 100) and the number of layers (L = 4) of the LM Edge-GNN Reasoner Networks, with a dropout rate of 0.2 applied to each layer. We train the model with the Adam optimizer using two GPUs (GeForce RTX3090). The batch size is 64, and the learning rate for the LM module and GNN module are set to 10^{-5} and 10^{-3}, respectively. Under the aforementioned configuration, we pre-trained an MRC model on the training set (i.e., source domain), with a LEGRN encoder, an additional batch normalization layer, and MLP predictor P_1. During training the domain adaptation model, we utilized the frozen predictor P_1 to generate labels for the target domain. In this stage, the only additional hyper-parameter is the dimension of domain shift vectors (i.e., k), which we set to 256. This k value was determined through experiments from the range [64, 128, 256, 512, 768]. The remaining settings are the same as the above in the pre-training stage.

3.4 Performance Comparison Result

Figure 4 shows the results of varying approaches and human performance. We observe consistent improvements compared to fine-tuned PLMs and KG-based models, specifically, 14.03% dev accuracy, 14.47% test accuracy over RoBERTa, and 10.77% dev accuracy, 10.38% test accuracy over the prior best KG-based system, GreaseLM. The improvement over PLMs indicates that the incorporation of external spatial commonsense knowledge bolsters the model's comprehension of spatial information within the text. Moreover, the improvement over the existing KG-based methods underscores the superior utilization of knowledge graphs for collaborative reasoning by the LEGRN. Our domain-invariant generalization model (LEGRN-DIG) outperformed LEGRN by 3.47% on the dev set and 3.10% on the test set, which suggests the capacity of our method to address the training-test domain shift and enhance model generalization.

3.5 Ablation Study

The ablation study on each of our model components is shown in Fig. 5 using the CRCSpatial dev set. Domian-Invariant Generalization (DIG), is the key component of LEGRN, its removal would result in a performance drop of $\sim$ 2.56%. We also test replacing the reasoning graph of sentence (GoS) by entity graph in the spatial commonsense knowledge extraction (§2.3), resulting in a $\sim$ 1.02% drop in performance, which suggests that sentence graph can more accurately

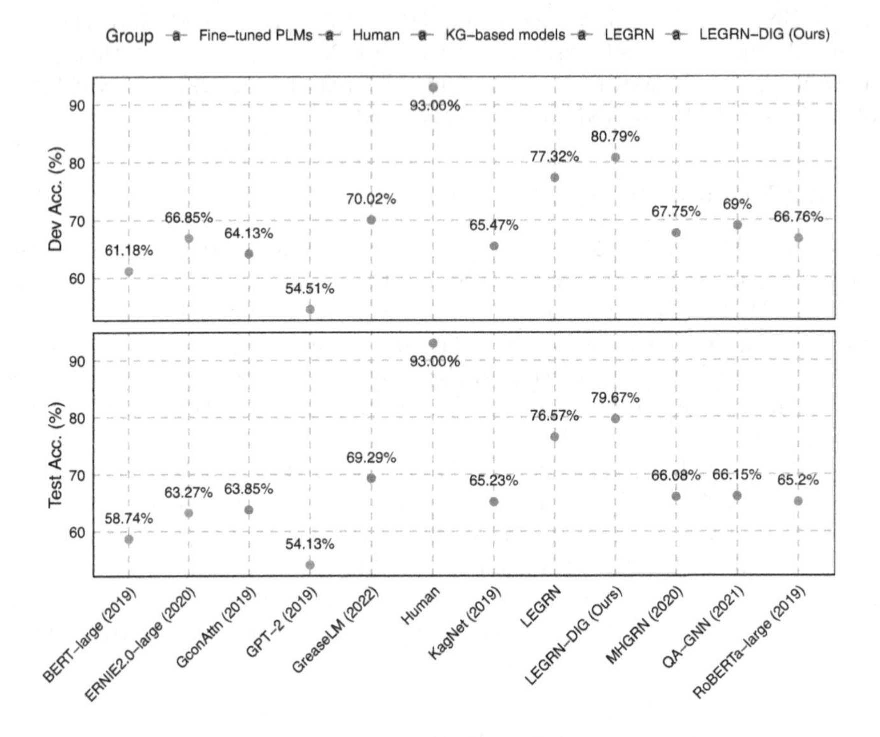

Fig. 4. Dev /Test accuracy on CRCSpatial.

capture semantic relationships and contextual information of the text, thereby enhancing the model's comprehension and reasoning capabilities towards textual content. Furthermore, if we remove Domian-Invariant Generalization and replace the reasoning graph of sentences simultaneously, i.e. w/o DIG & GoS (it becomes LEGRN), performance will drop by $\sim$ 3.47%. We also verify the effectiveness of external spatial commonsense knowledge graph (KG). Removing the extraction of knowledge from KG, the performance drop significantly by $\sim$ 8.28%, which proves KG supplements high-quality commonsense knowledge to enhance semantic understanding ability of our model. Finally, when we remove both the DIG module and the use of KG, the performance further declines to $\sim$ 69.14%, falling below some of the baseline models. For the dimension of domain shift vectors, we find k = 256 works best on the dev set (see Fig. 6).

3.6 Case Study

Benefiting from a large amount of commonsense knowledge about the spatial locations in WebChild2.0, our model performs very well on the questions about spatial position reasoning, compared to baseline models (see Q1.1 in Table 1). And, by performing multiple layer Layer-based Attention to connect LM and GNN modules and domain invariant generalization, our model performs well on

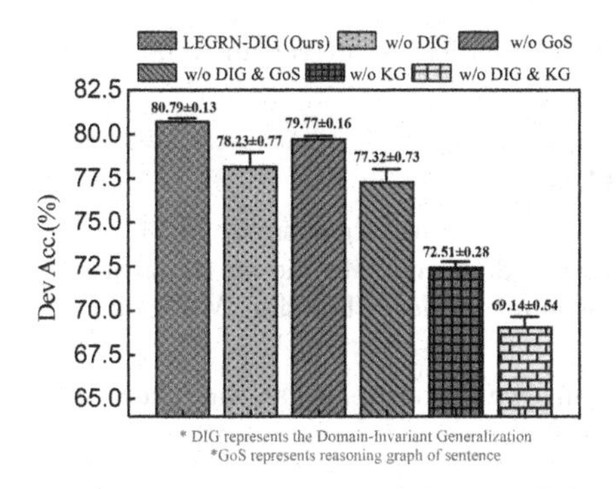

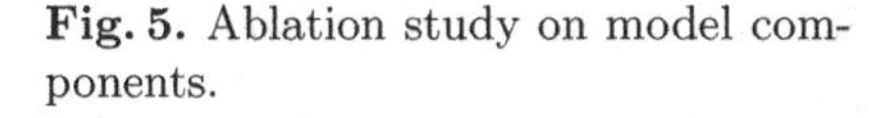

Fig. 5. Ablation study on model components.

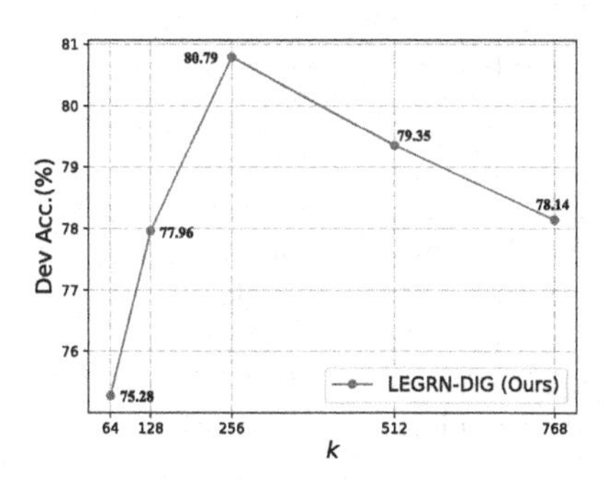

Fig. 6. Effect of the dimension of domain shift vectors (parameter k)

multi-hop spatial commonsense reasoning. For example, the multi-hop reasoning of Q1.2 in Table 1, the model goes through these steps: *I was drinking my water → My friend Sara had also taken out her water bottle → Sara had put her backpack on the ground near her → Sara was likely drinking from her water bottle close by while I was drinking my water.* By chaining these reasoning steps about the spatial locations of the backpacks and water bottles, the model can determine that Sara was likely drinking from her water bottle near me. The multi-hop attention connects the information across sentences to perform this commonsense reasoning.

3.7 Error Analysis

We randomly select 100 errors made by our approach from the dev set, unveiling 4 phenomena: (1) Insufficiency of evidence (38%). For instance, the Q2 in Table 1 requires commonsense knowledge about distance. However, our model struggled to extract relevant knowledge from the external KG. This problem can be alleviated by deploying more advanced extraction strategies and incorporating more knowledge sources. (2) Discrepancy with Human Commonsense (28%). In cases where questions have two valid candidate answers, humans discern the plausible choice grounded in commonsense, while our model erroneously selects an option inconsistent with human commonsense. (3) Relatively weak in multi-type hybrid reasoning (20%). For example, in Q3 in Table 1, the model failed to make the multi-hop connection that since I was chopping on the cutting board, which was on the counter, I must have been standing in front of the kitchen counter. It seems the model lacked the spatial commonsense reasoning to link the chopping action to the location of the cutting board on the counter. (4) Unanswerable Questions (14%). The model struggles to appropriately handle "None of the above", as it lacks direct entailment from the given paragraph or question.

Table 1. Examples of predictions of our model. ✔ indicates correct answers and ✗ shows prediction errors.

Correct Case
P1: My friends and I went on a hike in the woods. When we got tired, we stopped to take a break. I took my backpack off and set it down next to me on the ground. My friend Sara also put her backpack on the ground on her other side. We took out our water bottles and snacks.
Q1.1: When I reached into my backpack for my water bottle, what was likely the closest thing to me?
A. A tree B. Sara's backpack ✔C. My backpack
Q1.2: As I was drinking my water, what was Sara likely doing that was close by?
A. Climbing a tree B. Eating a snack ✔C. Drinking from her water bottle
Error Cases
P2: Yesterday, my friend James and I were at the park. We sat down on a bench to eat the sandwiches we had brought for lunch...
Q2: How far is it from where James and I are sitting?
A. A few dozen centimeters ✗B. Adjacent ✔C. Not enough evidence
P3: I was cooking dinner in the kitchen. I took some vegetables from the refrigerator and started chopping them on the cutting board on the counter...
Q3: Where was I most likely standing while chopping the vegetables?
✗A. At the kitchen table B. Next to the refrigerator ✔C. At the kitchen counter

4 Related Work

Spatial Commonsense: Spatial commonsense has garnered attention in prior research. Some studies focus on extracting and structurally storing spatial commonsense knowledge from visual or textual sources [2,13]. One of the most prominent is WebChild 2.0 [20], which comprises 18 million triplet-form assertions covering various types of spatial commonsense knowledge. For MRC, Commonsense QA [19] considers the position relationship of "location at". Another recent dataset, CRCSpatial [12], focuses on a broader range of spatial commonsense types, including shape, size, position, and distance.

Commonsense Reading Comprehension: Previous works attempt to enhance the semantic understanding ability of models by integrating knowledge from both the PLM and structured knowledge bases. Some of these works design new pre-training tasks to implicitly acquire structured knowledge [10,18]. Conversely, various efforts explicitly encode KG information to facilitate structured reasoning. One straightforward method involves the direct concatenation of KG representation and QA context representation from two independent modules [15,29]. Another strategy enhances the representation of one modality using knowledge from another modality –augment text representations with KG subgraph embeddings [5,11], or boost graph-based reasoning ability with text representations [3,4,23]. Recent efforts have also been dedicated to exploring bidirectional interactions between these modalities. GreaseLM [24,28] proposes to jointly update the LM and GNN representations by fusing embeddings of LM and GNN with a simple MLP. To achieve a more profound fusion of the two modalities, [12] introduced a layer-based cross-attention mechanism to fuse representations from both LM and GNN.

Unsupervised Domain Adaptation in QA: Unsupervised domain adaptation (UDA) in Question Answering (QA) has garnered significant attention due to its potential to enhance QA systems' performance across diverse domains without the need for labeled target domain data [27]. In previous works, methods for unsupervised domain adaptation can be roughly categorized into the following approaches: (1) Contrastive Learning. Based on target context, they introduce question generation models to generate proxy datasets for training QA systems [25]. Additionally, contrastive learning is used to minimize domain discrepancy using Maximum Mean Discrepancy(MMD) [26] or Sliced Wasserstein Distance (SWD) [8]; (2) Adversarial Training. This method applies adversarial training to minimize feature discrepancy between domains through adversarial loss between the feature generator and discriminator [9]. In collaborative training, utilization of pseudo labeling and iterative refinement of these labels can lead to improvements [1].

5 Conclusion

In this paper, we proposed a novel unsupervised domain adaptation (UDA) framework, called LEGRN-DIG, for spatial commonsense question answering. This framework avoids the use of labeled or pseudo-labeled target instances and noisy synthetic data. We propose domain-invariant generalization learning to integrate features of the target domain into the source domain and still use the source domain for supervised training. Experimental results illustrated the effectiveness of our approach.

Acknowlendgement. This work is supported by the National Natural Science Foundation of China (62276279, 62002396), the Key-Area Research and Development Program of Guangdong Province (2020B0101100001), the Tencent WeChat Rhino-Bird Focused Research Program (WXG-FR-2023-06), and Zhuhai Industry-University-Research Cooperation Project (2220004002549).

References

1. Cao, Y., Fang, M., Yu, B., Zhou, J.T.: Unsupervised domain adaptation on reading comprehension. Proc. AAAI **34**, 7480–7487 (2020)
2. Elazar, Y., Mahabal, A., Ramachandran, D., Bedrax-Weiss, T., Roth, D.: How large are lions? inducing distributions over quantitative attributes. In: Proceedings of the 57th ACL, pp. 3973–3983 (2019)
3. Feng, Y., Chen, X., Lin, B.Y., Wang, P., Yan, J., Ren, X.: Scalable multi-hop relational reasoning for knowledge-aware question answering. In: Proceedings of the EMNLP, pp. 1295–1309 (2020)
4. Foolad, S., Kiani, K.: Luke-graph: a transformer-based approach with gated relational graph attention for cloze-style reading comprehension. arXiv preprint arXiv:2303.06675 (2023)
5. Han, Z., Feng, Y., Sun, M.: A graph-guided reasoning approach for open-ended commonsense question answering. arXiv preprint arXiv:2303.10395 (2023)
6. Keklik, O.: Automatic question generation using natural language processing techniques. Ph.D. thesis, Izmir Institute of Technology (Turkey) (2018)
7. Kenton, J.D.M.W.C., Toutanova, L.K.: Bert: pre-training of deep bidirectional transformers for language understanding. In: Proceedings of NAACL-HLT, pp. 4171–4186 (2019)
8. Kolouri, S., Nadjahi, K., Simsekli, U., Badeau, R., Rohde, G.: Generalized sliced wasserstein distances. Adv. Neural Inf. Process. Syst. **32** (2019)
9. Lee, S., Kim, D., Park, J.: Domain-agnostic question-answering with adversarial training. In: Proceedings of the 2nd Workshop on MRQA, pp. 196–202 (2019)
10. Levine, Y., et al.: Sensebert: Driving some sense into bert. In: Proceedings of the 58th ACL, pp. 4656–4667 (2020)
11. Lin, B.Y., Chen, X., Chen, J., Ren, X.: Kagnet: knowledge-aware graph networks for commonsense reasoning. In: Proceedings of the 9th EMNLP-IJCNLP, pp. 2829–2839 (2019)
12. Lin, M., Wang, M.-x., Yu, J., Wang, S., Lai, H., Liu, W., Yin, J.: Spatial commonsense reasoning for machine reading comprehension. In: Yang, X., et al. (eds.) Advanced Data Mining and Applications, ADMA 2023, Part II. LNCS, vol. 14177, pp. 347–361. Springer, Cham (2023). https://doi.org/10.1007/978-3-031-46664-9_24
13. Liu, X., Yin, D., Feng, Y., Zhao, D.: Things not written in text: exploring spatial commonsense from visual signals. In: Proceedings of the 60th ACL, pp. 2365–2376 (2022)
14. Liu, Y., et al.: Roberta: a robustly optimized bert pretraining approach. arXiv preprint arXiv:1907.11692 (2019)
15. Park, J., et al.: Relation-aware language-graph transformer for question answering. Proc. AAAI Conf. Artif. Intell. **37**(11), 13457–13464 (2023)
16. Radford, A., Wu, J., Child, R., Luan, D., Amodei, D., Sutskever, I., et al.: Language models are unsupervised multitask learners **1**(8), 9 (2019)
17. Speer, R., Chin, J., Havasi, C.: Conceptnet 5.5: an open multilingual graph of general knowledge. Proc. AAAI **31** (2017)
18. Sun, Y., et al.: Ernie 2.0: a continual pre-training framework for language understanding. Proc. AAAI **34**, 8968–8975 (2020)
19. Talmor, A., Herzig, J., Lourie, N., Berant, J.: Commonsenseqa: a question answering challenge targeting commonsense knowledge. In: Proceedings of NAACL-HLT, pp. 4149–4158 (2019)

20. Tandon, N., De Melo, G., Weikum, G.: Webchild 2.0: fine-grained commonsense knowledge distillation. In: Proceedings of ACL 2017, System Demonstrations, pp. 115–120 (2017)
21. Vaswani, A., et al.: Attention is all you need. Adv. Neural Inf. Process. Syst. **30** (2017)
22. Wang, X., et al.: Improving natural language inference using external knowledge in the science questions domain. Proc. AAAI **33**, 7208–7215 (2019)
23. Yasunaga, M., Ren, H., Bosselut, A., Liang, P., Leskovec, J.: Qa-gnn: reasoning with language models and knowledge graphs for question answering. In: Proceedings of the NAACL 2021, pp. 535–546 (2021)
24. Ye, Q., Cao, B., Chen, N., Xu, W., Zou, Y.: Fits: fine-grained two-stage training for knowledge-aware question answering. arXiv preprint arXiv:2302.11799 (2023)
25. Yue, X., Yao, Z., Sun, H.: Synthetic question value estimation for domain adaptation of question answering. In: Proceedings of the 60th ACL (2022)
26. Yue, Z., Zeng, H., Kou, Z., Shang, L., Wang, D.: Contrastive domain adaptation for early misinformation detection: a case study on covid-19. In: Proceedings of the 31st CIKM, pp. 2423–2433 (2022)
27. Yue, Z., Zeng, H., Kou, Z., Shang, L., Wang, D.: Domain adaptation for question answering via question classification. In: Proceedings of the 29th COLING, pp. 1776–1790 (2022)
28. Zhang, X., et al.: Greaselm: graph reasoning enhanced language models for question answering. In: Proceedings of ICLR (2022)
29. Zheng, C., Kordjamshidi, P.: Relevant commonsense subgraphs for "what if..." procedural reasoning. In: Findings of ACL 2022, pp. 1927–1933 (2022)

Global and Local Structure Discrimination for Effective and Robust Outlier Detection

Canmei Huang, Li Cheng(✉), Feng Yao, and Renjie He

College of System Engineering, National University of Defense Technology, Changsha, China
{chengli09,yaofeng,herenjie}@nudt.edu.cn

Abstract. Deep outlier detection on high dimensional data is an important research problem with critical applications in many areas. Though promising performance has been demonstrated, we observe that existing methods characterized outliers only from a single perspective, which leads to reducing the distinction of inliers/outliers with the growth of training epochs. This in turn hurts the robustness and effectiveness of outlier detection since the optimal training epoch on a special dataset is unknown in unsupervised scenarios. In this paper, we propose a DNN based framework with both global and local structure discrimination for effective and robust Outlier Detection, named GOOD. The global module compacts the data (mainly inliers) since the majority of data are inliers, while the local module scatters the data (mainly outliers) based on that outliers reside in low-probability density areas. These two modules are cleverly united by a self-adaptive weighting strategy that trades off the degree of complementary and competitive cooperation. The complementary views can help effectively detect outliers with diverse characteristics, and such competitive learning can prevent a single module from learning the entire data too well and ensure robust detection performance. Comprehensive experimental studies on datasets from diverse domains show that GOOD significantly outperforms state-of-the-art methods by up to 30% improvement of AUC while performing much more robustly with the growth of training epochs.

Keywords: Deep outlier detection · Global structure discrimination · Local structure discrimination · Self-adaptive weighting

1 Introduction

Unsupervised outlier detection is a fundamental problem that has broad application domains, such as medical diagnosis [8], fraud detection [1], and information security [32], to name just a few. Although fruitful progress has been made in the last several years, conducting robust and effective unsupervised outlier detection on high-dimensional data remains a challenging task. Fortunately, deep neural networks (DNN) [13] presents a way to learn relevant features automatically, with exceptional success in learning useful representations on high-dimensional data, such as images and texts. Therefore, DNN based outlier detection (deep outlier detection) has recently attracted considerable attention and demonstrated promising performance.

X. Song et al. (Eds.): APWeb-WAIM 2023, LNCS 14332, pp. 286–300, 2024.
https://doi.org/10.1007/978-981-97-2390-4_20

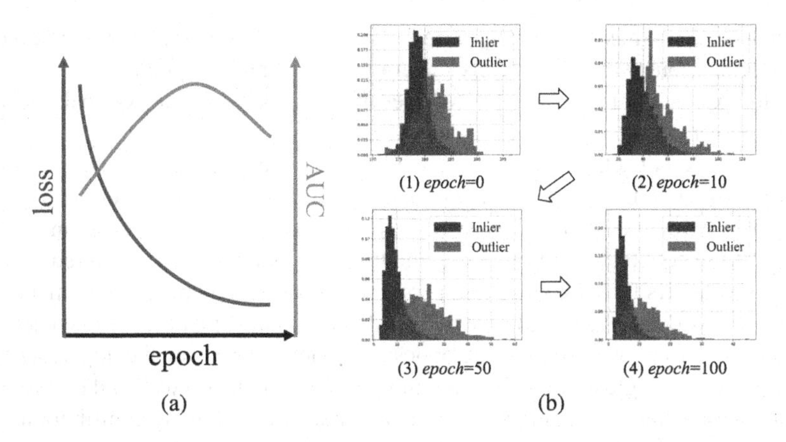

Fig. 1. (a) The curve of AUC and loss values of different deep outlier detection methods with the growth of training epochs. There is an interesting observation: after learning for some epochs, the training loss continues to decrease and may become (near-)zero, but the performance is decreasing. (b) The loss values' distribution of inliers and outliers. It is not a surprise to see that the gap between the loss distributions of inliers and outliers gets smaller after some epochs.

Feeding DNN's learned representations into a separated outlier detection method is an intuitive solution that has been first studied [7]. These approaches could easily lead to suboptimal performance because representation learning in the first step is unaware of the subsequent outlier detection task. Recently, several end-to-end solutions are proposed to combine the force of representation learning and outlier detection. They try to fit the entire data into a DNN based model such as autoencoder (AE) [9], AE based Gaussian mixture model [34] and deep one-class classification [20]. Deviations from this global model are then deemed to be outlier scores. Although it seems reasonable based on the intrinsic class imbalance of inliers/outliers, we observe that such methods usually demonstrate unstable performance across multiple real-world datasets. As shown in Fig. 1(a), the performance may decrease with the growth of training epochs while the training loss is still decreasing.

To explore the underlying reason, we further trace the loss values' distribution of inliers and outliers. As shown in Fig. 1(b), the inliers dominate the model training with the randomly initialized network, but the gap between the loss distributions of inliers and outliers gets smaller after learning for some epochs. Accordingly, the explanation could be: although the modeling on outliers is different from that on inliers and some of them do demonstrate unusually high modeling deviations, a significant amount of outliers could also lurk with a normal level of deviation, which usually happens when the global model with high complexity trained with large training epochs. We call such a phenomenon "overfitting" (similar to the "overfitting" in supervised learning) since both inliers and outliers fit the global model well. Since it is impossible to get the optimal training epochs in unsupervised scenarios with no labels, "early stop" with a given fixed training epoch is widely adopted in the aforementioned methods. Unfortunately, the optimal training epochs on each dataset with different characteristics may be quite different, specifying the training epochs in advance may lead to unsatisfying performance. A challenge naturally raised is to design a deep outlier detection method that performs robustly and effectively under uncertain training epochs.

In this paper, we propose a DNN based framework with both **G**lobal and l**O**cal structure discrimination for effective and robust **O**utlier **D**etection, named GOOD. The global module compacts the data (mainly inliers) since the majority of data are inliers, while the local module scatters the data (mainly outliers) based on that outliers reside in low-probability density areas. Moreover, we propose a self-adaptive weighting strategy to trade off the cooperation of these two modules. On one hand, the two modules characterize the outliers from two different perspectives, such a complementary manner can help effectively detect outliers that may have diverse characteristics. On the other hand, the objectives of the two modules are opposite to some degree, such competitive learning can prevent a single module from learning the entire data too well and ensure robust detection performance. In summary, we mainly make the following contributions:

- We propose an effective and robust deep outlier detection algorithm with global and local structure discrimination. Moreover, we theoretically show the foundation of the global module and local module.
- We propose a self-adaptive weighting strategy that can dynamically adjust the weights of each module. With such a strategy, the global module and local module can cooperate in a complementary and competitive cooperation to ensure the effectiveness and robustness of GOOD.
- The extensive experiments on diverse datasets confirm the robustness and effectiveness of the proposed GOOD.

2 Related Work

The strong modeling capacity of DNN makes it essential and natural to introduce DNN to outlier detection [7]. In early attempts, representations are learned separately in a preceding step before these representations are then fed into classical outlier detection methods. Deep AE (of various types) is the predominant approach used for deep outlier detection [7]. Several works plug the learned embeddings by AE into classical outlier detection methods, such as OC-SVM, iForest, and so on [11,21,30]. Such naive two-stage solutions may lead to unsatisfying performance since the learned representations may not be optimally served for the subsequent outlier detection method.

In contrast, recent works try to employ the representation learning objective directly for detecting outliers. Deep AE also plays a fundamental role in such methods. Simply employing the reconstruction errors as outlier scores is the very first attempt [2,9,22]. To enrich the network architecture, some variants of AE are proposed for the purpose of outlier detection, including denoising AE [27], sparse AE [17], and variational AE [12]. However, the performance of reconstruction based methods is limited by the fact that they only conduct outlier analysis from a single aspect, that is, reconstruction error. Moreover, AE has the objective of dimensionality reduction and does not target outlier detection directly.

In order to combine AE and outlier detection objective, several works are proposed to learn an AE based outlier detection model. Zhou et al. [33] propose a decoupled solution that combines a deep AE with robust PCA, which decomposes the inputs into a low-rank part from inliers and a sparse part from outliers; Xia et al. [29] use deep AE directly and propose a variant that estimates inliers by seeking a threshold that maximizes the inter-class variance of AE's reconstruction loss. A loss function is designed to encourage the separation of estimated inliers/outliers; Zong et al. [34] jointly optimize a deep AE and an estimation network to perform simultaneous representation learning and density estimation; Inspired by kernel-based one-class classification, Ruff et al. [20] learns a neural network transformation that attempts to map most of the data network representations into a hypersphere of minimum volume. In such models, both the inliers and outliers may be learned well due to the strong modeling capacity of DNN, leading to unsatisfying detection performance To avoid this, early-stop with given training epochs is widely used. However, it is impossible to set optimal epochs since different datasets may have different characteristics. Thus such methods usually demonstrate unstable performance on different datasets.

Apart from AE, Schlegl et al. [23] have recently proposed a novel deep outlier detection method based on Generative Adversarial Networks (GANs) called AnoGAN. Unlike AnoGAN which uses a standard GAN, ALAD builds upon bi-directional GANs and also includes an encoder network that maps data samples to latent variables, to avoid the computationally expensive inference procedure [31]. Similar to AE, this generative approach also uses reconstruction errors to determine if a data sample is anomalous and does not have the objective of outlier detection. It is also worth noting that there are some works designed for specific applications, like using CNN in the image or video data [28], and LSTM in NLP [25]. Since they are customized for given applications, we do not detail them here.

3 The Proposed Method

3.1 Framework

We consider outlier detection problems defined over a set of n data objects $\mathbf{X} = \{\mathbf{x}_1, \mathbf{x}_2, \ldots, \mathbf{x}_n\}$, where each data object $\mathbf{x}_i \in \mathbb{R}^d$ is described as a d-dimensional real-valued vector. There is an unknown partition that divides X into a set of inliers $\mathbf{X}^+ = \{\mathbf{x}_1^+, \mathbf{x}_2^+, \ldots, \mathbf{x}_{n^+}^+\}$ and a set of outliers $\mathbf{X}^- = \{\mathbf{x}_1^-, \mathbf{x}_2^-, \ldots, \mathbf{x}_{n^-}^-\}$, so that $\mathbf{X} = \mathbf{X}^+ \cup \mathbf{X}^-$. n^+ and n^- are the number of inliers and outliers, respectively. $\rho = n^-/n$ is the outlier ratio of the data set. Our goal is to obtain an outlier detector $s(\cdot) : \mathbf{x}_i \rightarrow \mathbb{R}$ assigns outlier scores to objects in $\mathbf{x}$. In practice, a larger $s(\mathbf{x})$ indicates a higher likelihood of $\mathbf{x}$ to be an outlier.

We aim to find a good neural network $f_\theta(\cdot)$ which makes embedded points of $\mathbf{X}$ more suitable and stable for the outlier detection tasks. To this end, two components, i.e., global and local structure discrimination, are adopted from different perspectives as illustrated in Fig. 2. They are united by a self-adaptive weighting strategy and work in a complementary and competitive manner. We present detailed descriptions in the following.

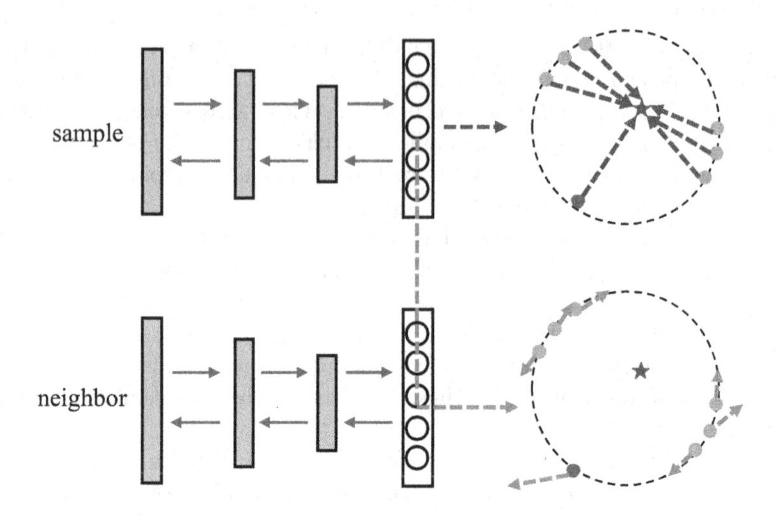

Fig. 2. The overview of GOOD. It tries to learn a latent representation that can differentiate inliers/outliers by a deep neural network. The last layer of the network is a normalization layer which units the outputs to avoid divergence. There are two key modules, i.e., global structure discrimination and local structure discrimination. The global module attempts to map the data representations into a hypersphere (the purple arrow), and the local module tries to push away the data representations from its neighbors (the orange arrow). The two modules are integrated by a self-adaptive weighting strategy. (Color figure online)

3.2 Global Structure Discrimination

Inspired by OC-SVM [24] and its deep version [20], the global module tries to compact all the objects in the embedded space. It uses a neural network transformation $f(\cdot)$ rather than the traditional kernel function. The aim is to jointly learn the network parameters θ together with minimizing the volume of a data-enclosing hypersphere in embedded space characterized by center $\mathbf{c} \in \mathcal{Z}$:

$$L_g(\theta, \mathbf{X}) = \frac{1}{n} \sum\nolimits_{i=1}^{n} ||\mathbf{z}_i - \mathbf{c}||^2 \tag{1}$$

where $\mathbf{z}_i = f_\theta(\mathbf{x}_i)$ is learned representation of $\mathbf{x}_i$.

The global module contracts the sphere by minimizing the mean distance of all data representations to the center. Stated less formally, to map the data (on average) as close to center $\mathbf{c}$ as possible, the neural network must extract the common factors of variation. Since the majority of the dataset are inliers, it is reasonable to obtain that inliers are closely mapped to center $\mathbf{c}$, whereas outliers are mapped further away from the center or outside of the hypersphere. To be more rigorous, we can derive the following theorem:

Theorem 1. *At the early training stage of the global module, with the randomly initialized network weights θ^0, giving an inlier $\mathbf{x}^+$ and an outlier $\mathbf{x}^-$, let $E(\Delta L_g(\theta^t, \mathbf{x}^+))$, $E(\Delta L_g(\theta^t, \mathbf{x}^+))$ be the exception values of the inlier's and outlier's loss reduction in the t-th epoch, respectively. For a small value of $t \geq 1$, we have:*

$$E(\Delta L_g(\theta^t, \mathbf{x}^+)) > E(\Delta L_g(\theta^t, \mathbf{x}^-)) \tag{2}$$

Proof. Consider an object $\mathbf{x}_i$, its negative gradient is the fastest network updating direction to reduce $\mathbf{x}_i$'s loss:

$$-\nabla_\theta L_g(\theta, \mathbf{x}_i) = -\frac{2}{n} * \frac{\partial f}{\partial \theta}(\mathbf{x}_i) \tag{3}$$

However, the network weights are actually updated by the negative gradient of the all objects:

$$\begin{aligned}-\nabla_\theta L_g(\theta, \mathbf{X}) &= -\frac{1}{n}\sum\nolimits_{i=1}^{n} \nabla_\theta L_g(\theta, \mathbf{x}_i) \\ &= -\frac{2}{n^2}\sum\nolimits_{i=1}^{n} \frac{\partial f}{\partial \theta}(\mathbf{x}_i)\end{aligned} \tag{4}$$

It is different from the best updating direction for each individual object. Then the network weights is updated as:

$$\theta^{(t)} = \theta^{t-1} - \delta \nabla_\theta L_g(\theta, \mathbf{X}) \tag{5}$$

where δ is the step size.

We compute the reduced loss of $\mathbf{x}_i$ in the t-th iteration is:

$$\begin{aligned}\Delta L_g(\theta^t, \mathbf{x}_i) &= L_g(\theta^{t-1}, \mathbf{x}_i) - L_g(\theta^t, \mathbf{x}_i) \\ &\approx \delta \nabla_\theta L_g(\theta^{t-1}, \mathbf{X}) \cdot \nabla_\theta L_g(\theta^{t-1}, \mathbf{x_i}) \\ &= \delta |\nabla_\theta L_g(\theta^{t-1}, \mathbf{X})||\nabla_\theta L_g(\theta^{t-1}, \mathbf{x}_i)| \cos \angle_i\end{aligned} \tag{6}$$

where $\angle_i$ is the angel between $\nabla_\theta L_g(\theta^{t-1}, \mathbf{X})$ and $\nabla_\theta L_g(\theta^{t-1}, \mathbf{x}_i)$.

It has been shown that the angle of inlier is smaller than the angle of a outlier based on the intrinsic class imbalance between inliers/outliers [28,29], that is:

$$\cos \angle^{+} > \cos \angle^{-} \tag{7}$$

Since θ is randomly initialized and t is small, we can assume that:

$$E(|\nabla_\theta L_g(\theta^{t-1}, \mathbf{x}^{+})|) \approx E(|\nabla_\theta L_g(\theta^{t-1}, \mathbf{x}^{-})|) \tag{8}$$

thus we have

$$E(\Delta L_g(\theta^t, \mathbf{x}^{+})) > E(\Delta L_g(\theta^t, \mathbf{x}^{-})) \tag{9}$$

Theorem 1 confirms that the inliers dominate the training at the early stage. Thus the distance between embedded point $\mathbf{z}$ and $\mathbf{c}$, $s_g(\mathbf{x}) = ||\mathbf{z} - \mathbf{c}||^2$, can be a outlier score measure. The higher $s_g(\mathbf{x})$ is, the more likely $\mathbf{x}$ to be an outlier.

3.3 Local Structure Discrimination

Inspired by the distance based outlier detection method: outliers are the data points with the highest average distance to their respective neighbors [6,19], local structure discrimination tries to push objects far away from its neighbors in the embedded space:

$$\max \frac{1}{n}\sum\nolimits_{i=1}^{n} ||\mathbf{z}_i - \hat{\mathbf{z}}_i||^2 \tag{10}$$

where $\mathbf{z}_i = f_\theta(\mathbf{x}_i)$, $\hat{\mathbf{z}}_i = f_\theta(\hat{\mathbf{x}}_i)$ are respectively the learned representations of $\mathbf{x}_i$ and $\hat{\mathbf{x}}_i$, and $\hat{\mathbf{x}}_i$ is a random sample from the k nearest neighbors of $\mathbf{x}_i$. Thus the local loss function is:

$$L_l = -\frac{1}{n}\sum_{i=1}^{n} ||\mathbf{z}_i - \hat{\mathbf{z}}_i||^2 \tag{11}$$

Put differently, at the core of outlier detection is density estimation: given a lot of input samples, outliers are those ones residing in low probability density areas [34]. Considering x_i as an outlier candidate, its mapping point in $\mathcal{Z}$ should be as far as possible from its neighbors: if x_i is truly an outlier, it should be well discriminated from its neighboring objects. In contrast, it is difficult to put an inlier far away from its neighbors in the embedded space since they reside in high-density areas. As a result, inliers are closely mapped to their neighbors, whereas outliers and their neighbors are scattered in the embedded space. We also provide a more formal conclusion:

Theorem 2. *At the early training stage of the local module, with the randomly initialized network weights θ^0, giving inlier $\mathbf{x}^+$ and outlier $\mathbf{x}^+$, and $E(\Delta L_l(\theta^t, \mathbf{x}^+))$, $E(\Delta L_l(\theta^t, \mathbf{x}^+))$ are the exception values of the inlier's and outlier's loss reduction in the t-th epoch, respectively. For a small value of t, we have:*

$$E(\Delta L_l(\theta^t, \mathbf{x}^-)) > E(\Delta L_l(\theta^t, \mathbf{x}^+)) \tag{12}$$

Proof. For the local loss, the negative gradient of a given object $\mathbf{x}_i$ is:

$$-\nabla_\theta L_l(\theta, \mathbf{x}_i) = -\frac{2}{n} * (\frac{\partial f}{\partial \theta}(\mathbf{x}_i) - \frac{\partial f}{\partial \theta}(\hat{\mathbf{x}}_i)) \tag{13}$$

Similarly, we can obtain the reduced loss of $\mathbf{x}_i$ in the t-th epoch is:

$$\Delta L_l(\theta^t, \mathbf{x}_i) = \delta|\nabla_\theta L_l(\theta^{t-1}, \mathbf{X})||\nabla_\theta L_l(\theta^{t-1}, \mathbf{x}_i)| \cos \angle_i \tag{14}$$

Since outliers reside in low probability area, the gap between inlier $\mathbf{x}^+$ and its neighbor $\hat{\mathbf{x}}^+$ is smaller than that between $\mathbf{x}^-$ and $\hat{\mathbf{x}}^-$, we assume that:

$$E(|\nabla_\theta L_l(\theta^{t-1}, \mathbf{x}^-)|) > E(|\nabla_\theta L_l(\theta^{t-1}, \mathbf{x}^+)|) \tag{15}$$

The distribution of data is unknown and θ is randomly initialized, it is reasonable to assume $\cos \angle^+ \approx \cos \angle^-$ here, thus we have:

$$E(\Delta L_l(\theta^0, \mathbf{x}^-)) > E(\Delta L_l(\theta^0, \mathbf{x}^+)) \tag{16}$$

Theorem 2 validates that the outliers scatter further than inliers at the early training stage. Thus the distance between embedded point $\mathbf{z}$ and its neighbors $\hat{\mathbf{z}}_j$, i.e., $s_l(\mathbf{x}) = \frac{1}{k}\sum_{\hat{\mathbf{z}} \in knn(\mathbf{z})} ||\mathbf{z} - \hat{\mathbf{z}}||^2$, can also be an outlier score measure. The higher $s_l(\mathbf{x})$ is, the more likely $\mathbf{x}$ to be an outlier.

3.4 Complementary and Competitive Cooperation

The proposed GOOD is a joint consideration of the above modules, the objective is defined as:

$$\mathcal{L} = \alpha_g L_g + \alpha_l L_l \tag{17}$$

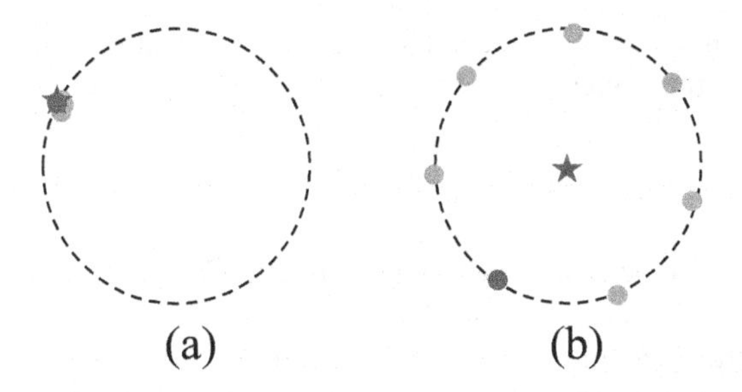

Fig. 3. The illustrations of overfitting.

where $\alpha_g > 0$ and $\alpha_l > 0$ are coefficients that controls the degree of cooperation. Accordingly the outlier score of a data object $\mathbf{x}$ can be calculated as:

$$s(\mathbf{x}) = s_g(\mathbf{x}) + s_l(\mathbf{x}) \tag{18}$$

We show below that the cooperation of the two modules can identify outliers in an **effective** and **robust** manner.

The two modules are based on the different characteristics of outliers. Specifically, the global module obtains the deviation from the learned global model, and the local module captures the outliers with the regional density. Such a complementary manner helps to **effectively** detect outliers with different behaviors.

Due to the strong modeling capacity of DNN, either the aforementioned global module or local module may yield trivial solutions caused by "overfitting" if one of them is adopted alone. Both inliers and outliers are learned well, which could seriously hurt the detection performance. To be more specific, suppose that the capacity of the DNN is large enough, we can obtain the following propositions:

Proposition 1. *($\alpha_g = 1$ and $\alpha_l = 0$) Consider the following trivial solution: Let θ_0 be the set of all-zero network weights (as shown in Fig. 3(a)). For this choice of parameters, the network maps any input to the same output, i.e., $f_\theta(\mathbf{x}) = \mathbf{c}_0$. And the center of the hypersphere is given by $\mathbf{c} = \mathbf{c}_0$, all errors in the empirical sums of the objectives become 0. Thus $\theta = \theta_0$ and $\mathbf{c} = \mathbf{c}_0$ are optimal solutions.*

Proposition 2. *($\alpha_g = 0$ and $\alpha_l = 1$) Consider the following trivial solution: all the embedded points are uniformly scattered on the hypersphere (as shown in Fig. 3(b)). In this case, the distances between all the point pairs are equal. This choice of parameters is an optimal solution for L_l.*

Fortunately, it can be solved by the competitive fashion with the two modules. As shown in Fig. 3, when one of these two modules achieves the ideal optimal solution, the other one falls into the worst-case. Specifically, the distance between each object and its neighbors is 0 when all the objects are mapped to the center, and the objects are quite far from the center when each object and its neighbors are scattered everywhere. Taking a deeper insight into the loss function, we can see that the objectives of the two modules are opposite to some extent: the compacting in global structure modeling makes the objects denser in embedded space, and the scattering in local structure prevents objects getting too close to the center:

$$||\mathbf{z}_i - \mathbf{c}||^2 + ||\hat{\mathbf{z}}_i - \mathbf{c}||^2 \geq \frac{||\mathbf{z}_i - \hat{\mathbf{z}}_i||^2}{2} \tag{19}$$

the above equation can be easily derived according to Cauchy-Schwarz inequality.

The above propositions also show that the values of α_g and α_l should be set carefully. Improper settings would also lead to "overfitting", the module with a large enough weight dominates the training process and achieves the trivial solution. Since different datasets have different characteristics, it is impossible to obtain a proper setting that is optimal for all the datasets. To address this issue, we propose **a self-adaptive weighting strategy**, which dynamically trades off the weights of the two modules. With the aim to make the two modules learn at a consistent pace, the core idea is to give larger weight to the module that gets less loss reduction. Specifically, for the $(t+1)$-th epoch, α_g and α_l are set according to the different conditions of loss reduction:

- $\Delta L_g > 0$ and $\Delta L_l > 0$. Both the global loss and local loss are decreasing, the weight of the module should be inversely proportional to its loss reduction: $\alpha_g = 1$, $\alpha_l = \frac{\Delta L_g}{\Delta L_l}$.
- $\Delta L_g > 0$ and $\Delta L_l < 0$. The global loss is decreasing and the local loss is increasing, which indicates that the global module dominates this training epoch. To alleviate this problem, we ignore the global module in the next epoch: $\alpha_g = 0$, $\alpha_l = 1$.
- $\Delta L_g < 0$ and $\Delta L_l > 0$. Similar to the above condition, we ignore the local module in the next epoch: $\alpha_g = 1$, $\alpha_l = 0$.
- $\Delta L_g < 0$ and $\Delta L_l < 0$. Both the global loss and local loss are increasing, which indicates that convergence is achieved.

where $\Delta L_g = \Delta L_g(\theta^t, \mathbf{X})$, $\Delta L_l = \Delta L_l(\theta^t, \mathbf{X})$.

In summary, with this self-adaptive weighting strategy, the module with less learning profits would get more weight in the next iteration. Such a balanced training process ensures that the trivial solution would never be achieved. Then the proposed GOOD can **robustly** detect outliers under uncertain training epochs.

3.5 Optimization

We use Adam to optimize the parameters of the neural network in our objectives using backpropagation. Training is carried out until convergence or it reaches given training epochs. α_g and α_l are initialized as $\alpha_g = 1, \alpha_l = 1$, and then they are adjusted by the self-adaptive weighting strategy. Given α_g and α_l, the gradients of L with respect to latent representation $\mathbf{z}_i$ and cluster center $\mathbf{c}$ can be computed as:

$$\frac{\partial L}{\partial \mathbf{z}_i} = \frac{2}{n}(\alpha_g \mathbf{z}_i - \alpha_l \mathbf{z}_i - \alpha_g \mathbf{c} + \alpha_l \hat{\mathbf{z}}_i) \tag{20}$$

$$\frac{\partial L}{\partial \mathbf{c}} = \frac{2}{n}\sum\nolimits_{i=1}^{n} \alpha_g(\mathbf{c} - \mathbf{z}_i) \tag{21}$$

4 Experiments

4.1 Evaluation Setup

Datasets and Metrics. We follow the standard procedure from previous literature [20, 28] to construct a series of datasets: Given a standard benchmark, all objects from a class with one common semantic concept are retrieved as inliers, while outliers are randomly sampled from the rest of classes by an outlier ratio ρ. We vary ρ from 0.05 to 0.30 by a step of 0.05. The assigned inlier/outlier labels are strictly unknown to outlier detection methods and only used for evaluation. Each class of a benchmark is used as inliers in turn and the performance on all classes is averaged as the overall UOD performance. The experiments are repeated 10 times to report the average results. Five widely used benchmarks, including two image datasets (MNIST [14], USPS [4]), one text dataset (REUTERS10K [15]), one network security dataset (KDDCUP99 [26]) and one human activity recognition dataset (HAR [3]), are used for experiments. [1] In total, 186 datasets are generated based on these benchmarks and only part of the results are reported in the following due to space limit. The full results are provided in the supplementary. As for evaluation, we adopt two commonly-used metrics: AUC as a threshold-independent one and F1-score as a threshold-based one [5, 10]. In particular, the threshold to identify outliers is set as ρ.

Table 1. The network architectures for used datasets.

Dataset	Network Architecture
MNIST	784-256-64-16-64-256-784
USPS	256-64-16-64-256
KDDCUP99	118-64-16-64-118
REUTERS10K	2000-1024-256-64-16-64-256-1024-2000
HAR	561-256-64-16-64-256-561

Implementation Details and Compared Methods. For GOOD, the network architectures are set according to the dimensions of the datasets, which are shown in Table 1. The batch size is set as 64. We compare GOOD with the baselines and existing state-of-the-art deep outlier detection methods: (1) IF [16]: It is a classical outlier detection method used as baseline; (2) AE [18]: It directly uses AE's reconstruction loss to perform outlier detection; (3) AE-IF [30]: It feeds AE's learned representations into isolation forest (IF); (4) Deep one-class classification (DEEPOC) [20]; (5) Deep autoencoding gaussian mixture model (DAGMM) [34]; (6) Adversarially Learned Anomaly Detection (ALAD) [31]. For all deep outlier detection methods above, the hyperparameters of the compared methods are set to recommended values (if provided) or the same values of our proposed GOOD. The source codes are provided in the supplementary.

[1] KDDCUP99 contains semantically real inliers and outliers, so no further inlier sampling is adopted.

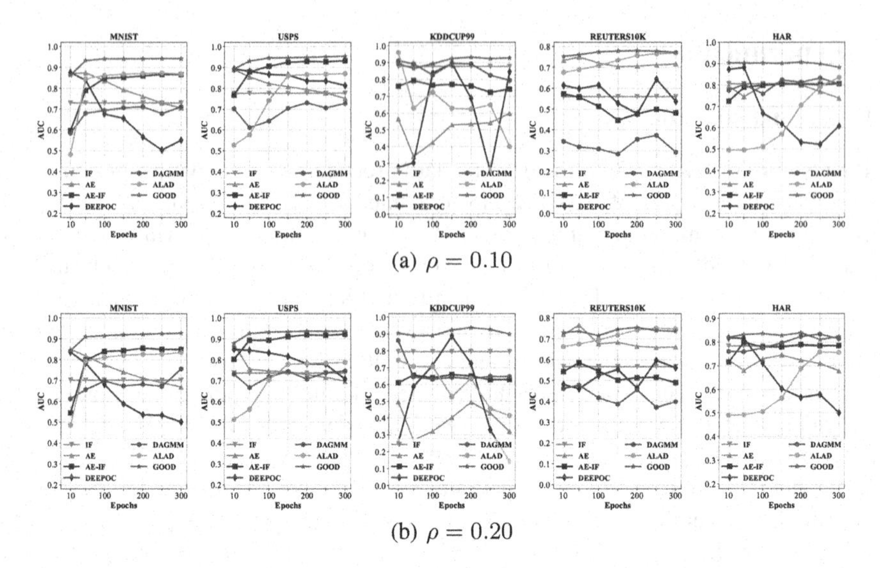

(a) $\rho = 0.10$

(b) $\rho = 0.20$

Fig. 4. Outlier detection performance with the growth of training epochs with outlier ratio (a) $\rho = 0.10$ and (b) $\rho = 0.20$.

4.2 Robustness Test w.r.t. Training Epochs

Settings. To show the robustness of the proposed GOOD, we report the performance of all the methods with the growth of training epochs. The training epochs are set as [10, 50, 100, 150, 200, 250, 300].

Results. Due to space limit, here we only report the AUC score with $\rho = 0.1$ and $\rho = 0.2$ as shown in Fig. 4. We draw the following observations from those results: Above all, GOOD is consistently more robust than the other deep outlier detection methods. In particular, the AUC score of GOOD increases significantly in the initialization and varies in a small margin after that. Thus we can easily set proper training epochs for GOOD (e.g., larger than 100) to obtain a promising performance in practical applications. In contrast, other deep methods have sharply changed AUC score with the growth of training epochs. And they have a quite different variation tendency on different datasets. It is difficult to set optimal training epochs for such deep methods. This demonstrates the strong robustness of GOOD in tapping the characteristics of outliers/inliers to well optimize the outlier scores.

4.3 Effectiveness Test on Diverse Datasets

Settings. To make a fair comparison, we obtain the optimal training epochs for each competitor (except iForest) according to the overall performance on the 5 datasets. Specifically, we average the AUC scores performed on the 5 datasets and select the training epochs with the highest value. Then we get the optimal training epochs 10,

Table 2. AUC/F1-score of GOOD and its competitors. The number in the brackets along with the method's name is the optimal epochs for the corresponding method. The best performance is in bold.

Dataset	ρ	IF(*)	AE(10)	AE-IF(250)	DEEPOC(50)	DAGMM(10)	ALAD(200)	GOOD(150)
MNIST	0.10	0.731/0.428	0.870/0.512	0.866/0.560	0.839/0.431	0.583/0.163	0.872/0.632	**0.941/0.662**
	0.20	0.702/0.508	0.849/0.588	0.849/0.651	0.783/0.514	0.613/0.323	0.825/0.659	**0.919/0.690**
USPS	0.10	0.777/0.480	0.891/0.483	0.927/0.626	0.882/0.496	0.703/0.329	0.866/0.568	**0.947/0.687**
	0.20	0.736/0.539	0.872/0.580	0.916/0.689	0.845/0.571	0.731/0.491	0.781/0.595	**0.934/0.716**
KDDCUP99	0.10	0.877/0.329	0.565/0.035	0.724/0.333	0.304/0.157	0.911/0.729	0.625/0.385	**0.924/0.781**
	0.20	0.795/0.351	0.496/0.089	0.628/0.296	0.588/0.116	0.861/0.694	0.637/0.316	**0.923/0.713**
REUTERS10K	0.10	0.561/0.178	0.734/0.394	0.500/0.102	0.598/0.205	0.344/0.132	0.756/0.456	**0.778/0.478**
	0.20	0.566/0.281	0.720/0.466	0.515/0.210	0.460/0.166	0.461/0.151	0.742/0.520	**0.745/0.532**
HAR	0.10	0.806/0.627	0.795/0.413	0.801/0.645	0.882/0.624	0.776/0.533	0.707/0.373	**0.902/0.646**
	0.20	0.784/0.631	0.724/0.468	0.785/**0.680**	0.816/0.623	0.761/0.544	0.688/0.445	**0.828**/0.660

250, 50, 10, 200, and 150 for AE, AE-IF, DEEPOC, DAGMM, ALAD, and GOOD, respectively.[2]

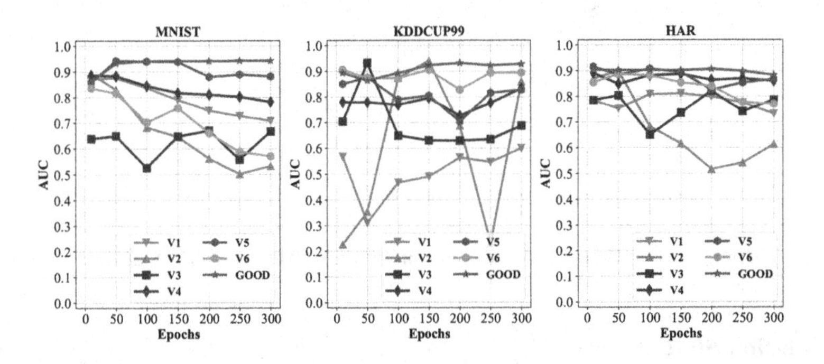

Fig. 5. Outlier detection performance of GOOD and its variants.

Results. We report the AUC/F1-score on each dataset under $\rho = 0.10$ and $\rho = 0.20$ in Table 2. We draw the following observations from those results: Most importantly, GOOD overwhelmingly outperforms existing deep methods by a large margin. As Table 2 shows, GOOD usually improves AUC/F1-Score by 10% to 60% when compared with state-of-the-art deep methods. GOOD performs best on all 10 setups in AUC and 9 of 10 setups in F1-score performance. In particular, GOOD obtains substantially better average improvement than IF (20.5%), AE(19.3%), AE-IF(13.5%), DEEPOC(21.7%), DAGMM(31.1%), ALAD(17.9%) in terms of AUC; in terms of F1-Score, the improvement GOOD achieves is much more substantial than IF (51.3%), AE(69.9%), AE-IF(36.3%), DEEPOC(66.8%), DAGMM(59.2%), ALAD(32.2%). These results are due to the reason that GOOD effectively captures both local structure and global structure to well optimize the outlier scores, resulting in high-quality outlier rankings; while the

[2] Note that this is different from the default settings of training epochs in the corresponding papers since different datasets are adopted here, thus the reported results may also be different.

competitors only have a single structure modeling, resulting in weak capability of discriminating some intricate outliers from inliers. It is also worth noting that deep methods obtain better performance than the traditional method IF, especially on relatively higher-dimensional datasets, which confirms the essentiality of deep outlier detection. Hence, the observations above have justified GOOD as a highly effective unsupervised outlier detection solution.

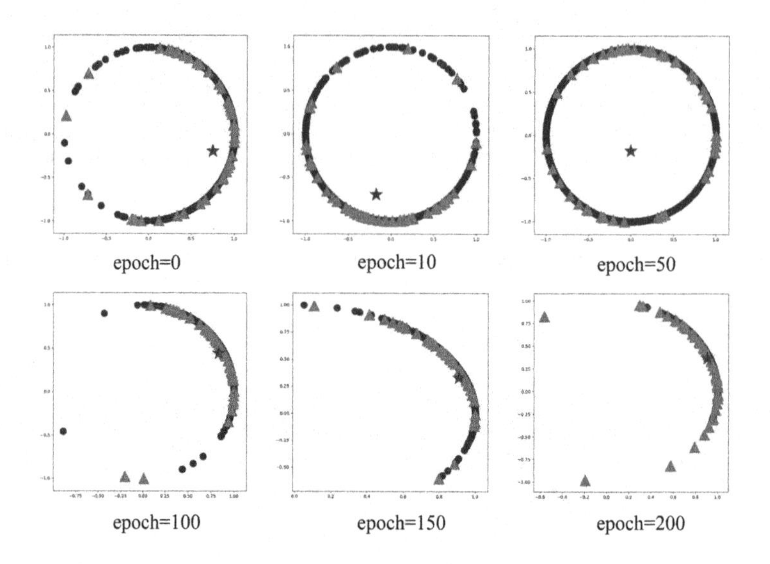

Fig. 6. Visualization of learned representation on the subset of MNIST during training.

4.4 Ablation Study

Settings. The proposed GOOD consists of two modules: global structure and local structure discrimination. We perform an ablation study to analyze the contributions of two modules. Moreover, to validate the effectiveness of the self-adaptive weighting strategy, we test the performance of GOOD with fixed α_g and α_l during the training. All in all, we test the detection performance on six variants of GOOD, which are denoted as V1(only local module, i.e., $\alpha_g = 0$ and $\alpha_l = 1$), V2(only global module, i.e., $\alpha_g = 1$ and $\alpha_l = 0$), V3($\alpha_g = 1$ and $\alpha_l = 0.25$), V4($\alpha_g = 1$ and $\alpha_l = 0.5$), V5($\alpha_g = 1$ and $\alpha_l = 1$), and V6($\alpha_g = 1$ and $\alpha_l = 2$). The above experiments are conducted on the datasets with outlier ratio $\rho = 0.1$.

Results. The performance of GOOD and its six variants are shown in Fig. 5. Unlike the variants with fluctuating performances, a union of global module and local module with a self-adaptive weighting strategy produces the most effective and robust performance, which reflects the necessity of the three parts.

4.5 Visualization

Settings. To give a visual effect of the cooperation of the global and local modules, we temporarily set the dimensionality of learned representations as 2 and plot the distribution of inliers/outliers under different epochs. The experiments are conducted on a subset of MNIST.

Results. As shown in Fig. 6, we can see that after the initial learning stage, the inliers(green points) become denser than outliers(red triangle), which helps us to effectively pick outliers out.

5 Conclusion

In this paper, we propose a deep outlier detection algorithm that simultaneously performs global and local structure discrimination. The two modules are united by a self-adaptive weighting strategy, which trades off the complementary and competitive cooperation. The extensive experiments confirm the robustness and effectiveness of the proposed framework.

References

1. Ahmed, M., Mahmood, A.N., Islam, M.R.: A survey of anomaly detection techniques in financial domain. FGCS **55**, 278–288 (2016)
2. An, J., Cho, S.: Variational autoencoder based anomaly detection using reconstruction probability. Spec. Lect. IE **2**(1) (2015)
3. Anguita, D., Ghio, A., Oneto, L., Parra, X., Reyes-Ortiz, J.L.: A public domain dataset for human activity recognition using smartphones. In: ESANN (2013)
4. Cai, D., He, X., Han, J., Huang, T.S.: Graph regularized nonnegative matrix factorization for data representation. TPAMI **33**(8), 1548–1560 (2010)
5. Campos, G.O., et al.: On the evaluation of unsupervised outlier detection: measures, datasets, and an empirical study. DMKD **30**(4), 891–927 (2016)
6. Cao, L., Yang, D., Wang, Q., Yu, Y., Wang, J., Rundensteiner, E.A.: Scalable distance-based outlier detection over high-volume data streams. In: ICDE, pp. 76–87. IEEE (2014)
7. Chalapathy, R., Chawla, S.: Deep learning for anomaly detection: a survey. arXiv preprint arXiv:1901.03407 (2019)
8. Chandola, V., Banerjee, A., Kumar, V.: Anomaly detection: a survey. CSUR **41**(3), 15 (2009)
9. Chen, J., Sathe, S., Aggarwal, C., Turaga, D.: Outlier detection with autoencoder ensembles. In: SDM, pp. 90–98. SIAM (2017)
10. Cheng, L., Wang, Y., Liu, X., Li, B.: Outlier detection ensemble with embedded feature selection. In: AAAI, pp. 3503–3512 (2020)
11. Erfani, S.M., Rajasegarar, S., Karunasekera, S., Leckie, C.: High-dimensional and large-scale anomaly detection using a linear one-class svm with deep learning. Pattern Recogn. **58**, 121–134 (2016)
12. Kingma, D.P., Welling, M.: Auto-encoding variational bayes. arXiv preprint arXiv:1312.6114 (2013)
13. LeCun, Y., Bengio, Y., Hinton, G.: Deep learning. Nature **521**(7553), 436–444 (2015)
14. LeCun, Y., Bottou, L., Bengio, Y., Haffner, P., et al.: Gradient-based learning applied to document recognition. Proc. IEEE **86**(11), 2278–2324 (1998)
15. Lewis, D.D., Yang, Y., Rose, T.G., Li, F.: Rcv1: a new benchmark collection for text categorization research. JMLR **5**, 361–397 (2004)
16. Liu, F.T., Ting, K.M., Zhou, Z.H.: Isolation forest. In: ICDM, pp. 413–422. IEEE (2008)

17. Makhzani, A., Frey, B.: K-sparse autoencoders. arXiv preprint arXiv:1312.5663 (2013)
18. Masci, J., Meier, U., Cireşan, D., Schmidhuber, J.: Stacked convolutional auto-encoders for hierarchical feature extraction. In: Honkela, T., Duch, W., Girolami, M., Kaski, S. (eds.) Artificial Neural Networks and Machine Learning. ICANN 2011, pp. 52–59. Springer, Heidelberg (2011). https://doi.org/10.1007/978-3-642-21735-7_7
19. Pang, G., Ting, K.M., Albrecht, D.: Lesinn: detecting anomalies by identifying least similar nearest neighbours. In: ICDMW, pp. 623–630. IEEE (2015)
20. Ruff, L., et al.: Deep one-class classification. In: ICML, pp. 4393–4402 (2018)
21. Sabokrou, M., Fayyaz, M., Fathy, M., Klette, R.: Fully convolutional neural network for fast anomaly detection in crowded scenes. arXiv preprint arXiv:1609.00866 (2016)
22. Sakurada, M., Yairi, T.: Anomaly detection using autoencoders with nonlinear dimensionality reduction. In: MLSDA Workshop, p. 4. ACM (2014)
23. Schlegl, T., Seeböck, P., Waldstein, S.M., Schmidt-Erfurth, U., Langs, G.: Unsupervised anomaly detection with generative adversarial networks to guide marker discovery. In: Niethammer, M., et al. (eds.) Information Processing in Medical Imaging. IPMI 2017, pp. 146–157. Springer, Cham (2017). https://doi.org/10.1007/978-3-319-59050-9_12
24. Schölkopf, B., Platt, J.C., Shawe-Taylor, J., Smola, A.J., Williamson, R.C.: Estimating the support of a high-dimensional distribution. Neural Comput. **13**(7), 1443–1471 (2001)
25. Singh, A.: Anomaly detection for temporal data using long short-term memory (LSTM) (2017)
26. Tavallaee, M., Bagheri, E., Lu, W., Ghorbani, A.A.: A detailed analysis of the KDD cup 99 data set. In: CISDA, pp. 1–6. IEEE (2009)
27. Vincent, P., Larochelle, H., Bengio, Y., Manzagol, P.A.: Extracting and composing robust features with denoising autoencoders. In: ICML, pp. 1096–1103. ACM (2008)
28. Wang, S., et al.: Effective end-to-end unsupervised outlier detection via inlier priority of discriminative network. In: NeurIPS, pp. 5960–5973 (2019)
29. Xia, Y., Cao, X., Wen, F., Hua, G., Sun, J.: Learning discriminative reconstructions for unsupervised outlier removal. In: ICCV, pp. 1511–1519 (2015)
30. Xu, D., Ricci, E., Yan, Y., Song, J., Sebe, N.: Learning Deep Representations of Appearance and Motion for Anomalous Event Detection, pp. 8.1–8.12 (2015)
31. Zenati, H., Romain, M., Foo, C.S., Lecouat, B., Chandrasekhar, V.: Adversarially learned anomaly detection. In: ICDM, pp. 727–736. IEEE (2018)
32. Zhang, J., Zulkernine, M.: Anomaly based network intrusion detection with unsupervised outlier detection. In: ICC, vol. 5, pp. 2388–2393. IEEE (2006)
33. Zhou, C., Paffenroth, R.C.: Anomaly detection with robust deep autoencoders. In: SIGKDD, pp. 665–674. ACM (2017)
34. Zong, B., et al.: Deep autoencoding gaussian mixture model for unsupervised anomaly detection (2018)

A Situation Knowledge Graph Construction Mechanism with Context-Aware Services for Smart Cockpit

Xinyi Sheng[1], Jinguang Gu[1,2,3](✉), and Xiaoyu Yang[1]

[1] College of Computer Science and Technology, Wuhan University of Science and Technology, Wuhan 430065, China
{tudouxia,yangxy1216}@wust.edu.cn

[2] Institute of Big Data Science and Engineering, Wuhan University of Science and Technology, Wuhan 430065, China
simon@ontoweb.wust.edu.cn

[3] Key Laboratory of Rich-media Knowledge Organization and Service of Digital Publishing Content, National Press and Publication Administration, Beijing 100038, China

Abstract. With the continuous development of intelligence and network connectivity, the smart cockpit gradually transforms into a multifunctional value space. Smart devices are heterogeneous, massive, complex, and contextually dynamic, which makes the services provided by the system inaccurate. Introducing knowledge graphs in smart cockpit situations can meet users' needs in specific scenarios while delivering experiences that exceed expectations. This paper constructs a smart cockpit situation model with context, service, and user as the core elements, not only refining the context dimension but also incorporating context into the definition of service. Firstly, we analyze the elements that constitute the smart cockpit situation model and explore the connection between them. Secondly, a top-down approach is used to construct the smart cockpit situation ontology using the smart cockpit situation model as a guide. Finally, the smart cockpit situation model is instantiated to build a knowledge graph for fitness scenarios. The research results show that the coverage relationships between scenarios are inferred based on the coverage relationships between contexts. Furthermore, we verify the context can improve the accuracy of the service with a family travel scenario example. The situation knowledge graph constructed in this paper cannot only comprehensively describe the smart cockpit scene data, but also the service can adapt to the dynamic changes of contextual data.

Keywords: Smart cockpit · Knowledge graph · Context-aware service

1 Introduction

With the development of intelligent and connected cars, high-end car brands have upgraded features such as intelligent car control, voice assistants, and cock-

X. Song et al. (Eds.): APWeb-WAIM 2023, LNCS 14332, pp. 301–315, 2024.
https://doi.org/10.1007/978-981-97-2390-4_21

pit safety [1,2]. They Rely on integrating artificial intelligence, big data, cloud computing, and other technologies to provide an intelligent base platform for smart cockpits. The smart cockpit has become the "third space," in addition to the workspace and living space, and has become a must-have for major car brands [3]. ICVTank predicts that China's smart cockpit market will reach 1030 billion by 2025. China's smart cockpit market will steadily rise and may become the world's largest smart cockpit market.

The smart cockpit uses sensors and devices to understand the driver's behavior, needs, or emotions, providing friendly human-machine interaction and a more comfortable driving experience [4]. In the interaction between the user and the Smart Cockpit, relevant data needs to be collected. The user's request for the Smart Cockpit needs to be understood to provide a precise and personalized service. The data involved in the smart cockpit scenario is too large, messy, and sparse [5]. How to use this data and extract valuable information is crucial.

As far as we know, there are no articles related to building the knowledge graph of the smart cockpit in the existing research, and modeling the smart cockpit scenes faces the following problems. First, the scene data is multi-source heterogeneous: various information is obtained from smart device interfaces in different formats. The scenario data generated while the car is driving is massive; second, the flexibility of the service: the need to establish the association relationship between contextual knowledge and vehicle services brings complexity to the description of the service.

This paper defines the value of the smart cockpit from a scenario perspective and uses a "top-down" approach to design the smart cockpit model. The data contained in the scenes are defined from multiple perspectives to provide a more comprehensive and detailed picture of the scenes. The user experience is analyzed through scenario-specific user requests, considering the relationship between the adaptation of the service to the scenario.

2 Related Work

2.1 Ontology Model

Knowledge graphs describe the knowledge framework with ontologies that can organize heterogeneous data into structured knowledge. Therefore, we refer to knowledge graphs in smart cities and smart homes, which are also in the field of smart fields. Chen Xing et al. [6] proposed a conceptual model of a knowledge graph for context-aware services for smart homes. The model contains five entities: location, user, context, device, and service. The conceptual model also defines the relationships between the above concepts, such as located, sensed, provided, monitored, raised, lowered, and assigned. The model covers all the base entities in the scenario and refines the four relationships of the service to context. It can better describe the state metrics of the environment or service configuration parameters. Ma Yazhong [7] uses people, things, and events as the core elements and integrates the IoT ontology to extend a knowledge ontology that can cover the basic concepts of the city. The applicability of the knowledge graph is enhanced by using the urban IoT ontology. Saba [8] has designed smart home ontologies for home electrical energy management. The model contains the concepts of Occupant, Environment, and Building. The model will determine whether energy consumption is abnormal for the same activity under similar or dissimilar conditions, depending on different conditions (environment, user preferences, etc.). Zaoui et al. [9] construct a smart city architecture, building an ontology of over 200 concepts represented by humans, devices, places, environment, services, etc. The model improves solutions for ontologies in big data environments due to the frequent changes in parameters and states of appliances. Swenja et al. [10] construct contextual ontologies, and the upper ontology defines RealEntity and VirtualEntity; RealEntity subclasses contain Machine, User, and Product, and VirtualEntity subclasses are State and VirtualNode.

In existing research, some ontologies focus on users, services, and devices without considering contextual information and are unable to improve accurate services based on contextual information. Some studies consider contextual information, but contextual content is redundant (e.g., location can be a dimension of context), or context is portrayed singularly (e.g., context is focused on people, location). Some ontologies contain too many concepts with varying parameter descriptions. The comparison of existing models is shown in Table 1. Based on the above problems, this paper focuses on users, services, and contexts in this paper's model of the smart cockpit, refines the contextual classification, and unifies the contextual data format.

Table 1. Comparison of ontological models.

Survey	Goal	Pros	Cons	Context
Smart Home [6]	Service recommendation	Rule-based reasoning	Relatively complex inference rules	Light, temperature
Smart City [7]	Management Monitoring	Big Data	Complex ontology	Spatiotemporal
Smart Home [8]	Energy saving	Multi-user management	Frequent scene similarity judgments	User information
Smart City [9]	Web APIs recommendation	Contextual hierarchy	Poor scalability	No
Smart Manufacturing [10]	Item recommendation	Adaptive production processes	No inference can be made	No

2.2 Contextual Definitions

Currently, the definition of "context" is not unique. Pradeep et al. [11] define "context as any information used to describe the state of an entity, either as a physical entity, a virtual entity, or a concept." Yavari et al. [12] define context as information about all entities (i.e., people, places or things) associated with a given IoT service. Chen Kuan et al. [13] define context as "the task knowledge or basic information that characterizes the important relationships between the surrounding application (users, objects, interactions) and the application itself." In different research areas, contextual classification will categorize contexts according to specific systems or user needs. Augusto et al. [14] emphasize that context impacts beneficiaries and services, considering users' preferences at multiple levels. They define contexts as base contexts and higher-order contexts based on the way they obtain contextual information. Bang et al. [15] propose four important types of contextual information, namely location, identity, time, and activity, and refine the four contexts by combining the concepts of primary and secondary contexts. Dinh et al. [16] defined contextual categories as historical information, network, system, user, etc., based on the source of contextual information. Arezou et al. [17] define a data context and an application context. The data context is the context associated with the collected data. The application context is the information about the entity issuing the query context.

In summary, the above context definitions describe the basic information of entities, where the entities have the user, environment, and program as the core elements. The context classification model focuses on access to the context without considering the relationship between the context and the service. The classification approach in this paper classifies contexts according to the sources of context information and divides contexts into four categories of contexts: physical, software, user, and service, based on the relationship between contexts and services.

2.3 Contextual Services

In the field of recommender systems, ChaoWu [18] proposed an automatic rule discovery algorithm based on user-item interaction, which leverages user feedback information (e.g., comments, clicks, etc.) in constructing the knowledge graph. Mezni et al. [19] proposed a deep context-aware recommendation model. It considers three entities - user, service, and context - and creates five relationships between the three entities. Considering the similarity between the three entities and themselves, it can combine the current context with the historical context services to predict the services by the user in the current context. Yavari et al. [12] assign an identifier to each context in an existing rule and then calculate the identifier of the application context based on the multiplication of the prime numbers associated with the application context.

Some studies focus only on the association between 'user items' and less on their contextual information, which would not be conducive to generating suitable recommendations. And the model is based on rule-based recommendations, but in reality, situations are complex and variable, and rules cannot cover them all. Some models consider the preference relationship of the user to the context and the adaptation relationship of the service to the context. However, the invocation relationship between the user and the service is missing. In this paper, the smart cockpit model is divided into a service layer, a user layer, and a context layer, taking into account the five relationships of the above paper and the new relationship between the user's invocation of the service. Using a combined IOPE-based approach [21] can solve the problem of service homogeneity.

3 Smart Cockpit Situation Model

Smart cockpits are intelligent bodies with autonomous characteristics. Xu Wei [22] and others pointed out that AI systems can have a certain degree of human-like cognition, learning, self-adaptation, independent operation, and other capabilities. It can autonomously perform specific tasks in specific scenarios and can autonomously perform tasks in unanticipated scenarios that previous automation technologies could not function. The smart cockpit can provide services to meet users' needs according to specific scenes. Thus the services are autonomous. Smart Cockpit Knowledge Graph can link the similarity of entities. Through the similarity between entities, the system can provide services in scenarios that have not occurred before. The smart cockpit has a human-machine interaction relationship, and the situation model is analyzed according to the interaction logic.

3.1 Situational Components

From a smart cockpit perspective, a scene is a situation in a specific time and space, which can be defined as a set of contexts. Contexts are perceptual data about the entities in a scene and can describe the current environment. The situation is the totality of behavior patterns in a given scenario, and it contains scenario data, behaviors, and relationships between these entities. The situation includes the scene at the time of the user's request, which is the basis for contextual execution. It describes the environment in which the user's request arises and explains the interrelationship of the various objects in the situation. This paper abstracts and defines the Smart Cockpit situation. It involves three abstractions: user, service, and context, and four interaction relations: "invocation," "preference," "adaptation," and "similarity," four kinds of interaction relations.

User. The user, as the subject of the smart cockpit situation, has a cognitive level, thinking ability, educational background, physical characteristics, regional characteristics, etc. The user is the situation's core element and initiator of the request. There are two roles for the user in the smart cockpit situation: the driver and the passenger. The driver has control of the car and is the necessary element of the smart cockpit. The driver's mood, fatigue level, and driving age determine the safety of the vehicle driving. Therefore, it is essential to monitor the physical characteristics of the driver. The driver of the same car is not fixed. The vehicle is able to target the user's identity based on face detection, thus providing services to different users. (e.g., seat adjustment, customized news feeds).

Context. As the environment in which the smart cockpit is located is dynamic and changing, the services provided will be influenced by multi-dimensional contextual factors such as the user domain, the information space domain, and the physical domain. There is no uniform standard for contextual model construction. The selection of contextual elements is highly subjective and needs to be analyzed in the context of specific domains and specific application scenarios. Context classification or fine granularity can affect the accuracy of recommendations.

Service. Service gives users a more convenient experience by considering scenarios and psychological requirements. In the vehicle system, the service gives more freedom to the user. It allows the user to control the vehicle in various ways, such as through voice control and manual control, for example, by voice control, manual control, etc. Also, in some scenarios, the vehicle enjoys control, e.g., automatic parking and entry. Physical services (e.g., assisted driving, fault detection, body control, seat heating) and software services (e.g., online entertainment functions, voice navigation, telephone services) exist depending on the vehicle service provider.

3.2 Situational Element Relationships

The relationship between situational elements is a relationship between two entities. There are relationships between users, contexts, and services. There is a preference relationship between users and contexts, an adaptation relationship between services and contexts, and an invocation relationship between users and services. It also includes a similar relationship between user, context, and service, as shown in Table 2. If a user invokes a service in a specific context, this fact is linked through entities (user-services, service-contexts), which can record the services initiated by the user. Based on the user similarity and the user's invocation relationship to the service, the system will autonomously ask whether to invoke the service. By actively recommending services to users that meet their interests and preferences, we reduce the difficulty of service selection and achieve personalized service delivery. Similarity relationships (user-user, service-service, context-context) help to infer hidden relations among them, allowing recommending the same service to similar users and recognizing that similar contexts can also link to the same service.

Table 2. Relationship table of situational elements.

Relationships	Subject	Context
Preference	User	Context
Adaptation	Services	Context
Invocation	User	Services
U-Similar	User	User
S-Similar	Service	Service
C-Similar	Context	Context

3.3 Situational Model

In this paper, after analyzing the components of the situation and the relationship between them, the Smart Cockpit Situation Model (SCSM) is constructed, namely

$$SCSM = (CE(u, s, c), R(ce)) \tag{1}$$

where SCSM identifies the Smart Cockpit Situation Model; CE (u, s, c) identifies the situational components, i.e., user u, service s, context c; and R(ce) identifies the relationship between the situational components. The smart cockpit situation model is a framework for the panoramic description of the smart cockpit, as shown in Fig. 1.

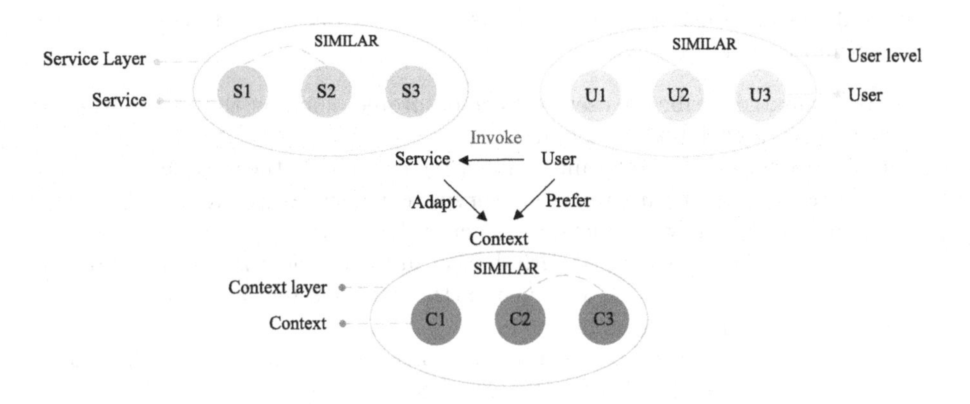

Fig. 1. Smart Cockpit Situation Model.

4 Smart Cockpit Situation Ontology

As seen from the Smart Cockpit situation model, the user, context, and service are the three elements of the model. This paper uses a top-down approach to design the class hierarchy.

4.1 User Class Design

Depending on the role, the user can invoke different services. For example, the driver can control the status lock of the windows and doors, while the passenger can only open and close the windows if the driver unlocks them. Therefore, the users in the smart cockpit are divided into two categories: driver and passenger.

1. Driver: has the right to control the car, controlling its direction, speed, braking, and other operations to ensure the safe operation of the car. The driver has the right to choose the route and destination to travel. And they also can adjust the temperature, stereo, and seats in the car to improve comfort.

2. Passengers: Passengers are users other than the driver. The needs of passengers vary by age. A safety seat is required for babies and young children. Adult passengers usually need more privacy or freedom, such as adjusting the lights, etc. Elderly passengers are sensitive to driving times and need to stop for breaks when driving for more than two hours.

4.2 Contextual Class Design

Smart Cockpit scenarios include not only snapshots of the environment (e.g., location, time, temperature, etc.) but also descriptions of entities in the environment (e.g., user's emotional state, attention, device status, etc.). The variety of contexts is comprehensive. So contexts are divided into physical and application contexts in terms of data access, user, and service contexts regarding the relationship between contexts and entities.

1. UserContext describes a user request or profile, such as the location, user preferences, user interests, moods, etc.
2. ServiceContext is the relevant context for evaluating the service, the characteristics of which vary from one service category to another, such as the response time, trustworthiness, satisfaction level, etc.
3. PhysicalContext is associated with devices in the physical world that describe the state of a physical device, such as the vehicle's fuel level, the speed of travel, the status of the windows, etc.
4. SoftwareContext obtains data from the software system and describes environmental data, such as weather, road conditions, etc.

4.3 Service Class Design

The automotive system is a complex system with integrated software and hardware. It is necessary to achieve integrated collaboration between software and hardware components, such as unifying data formats, communication protocols, and interface definitions. Therefore, the functions provided by various software and hardware of the automotive system are unified and abstracted as services. Considering the characteristics of services in automotive systems, there are two main categories:

1. Physical services: provided by physical devices such as sensors and actuators, responsible for sensing information about the physical world and controlling the physical world devices act on the physical world. Physical services are atomic services that cannot be partitioned into more minor granular services.
2. Software services: provided by the software on board the vehicle system, responsible for the business processing part. It can provide business functions such as data analysis and decision-making. The provision of software services is mainly affected by network resources.

4.4 Ontology Construction

The smart cockpit situation model with user, service, and context as the core extends the knowledge ontology. The ontologies consist of eight types: scenario, context, context value, context type, user, user request, service, and service provider, as shown in Fig. 2.

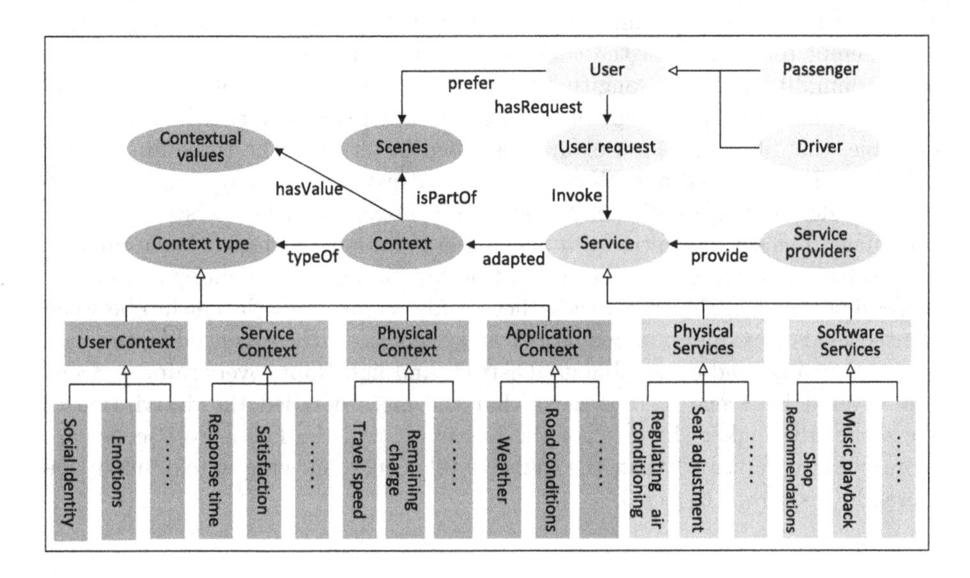

Fig. 2. Smart Cockpit Situational Knowledge Graph Ontology.

A scenario consists of a set of contexts, representing the set of all contextual data over a time horizon {C1, C2,......, C3}. It has attributes such as scene type, start time, end time, and scene description. Contexts have a context type and a context value; the context value represents a specific value, and the context type has four subclasses. The user entity has attributes such as name, occupation, driving age, and city.

This paper reuses OWL-S [23] to describe service ontologies. By analyzing various service combination approaches, this paper adopts the AI-based approach and treats web services as described by IOPE. Not only is I/O matching considered, but also the antecedents and effects (P/E) of the service. A smart cockpit service is defined as a five-tuple scs<Is, Os, Cp, Ce>, where cs are the service name; Is denotes the input parameters of the service, Os indicates the output parameters of the service, Cp is the contextual precondition, and Ce is the contextual execution effect.

From the service perspective, to achieve more accurate service recommendations, contextual factors' influence on service recommendations must be considered. By incorporating context into the service definition, services should be able to adapt to dynamic changes in context.

5 Case Study

5.1 Scenario Reasoning

The scenario is defined above as consisting of a set of contexts, which have a context type and a context value. By defining context types, scenes can be described by different dimensions. Each dimension has its own structure, which distinguishes it from the other dimensions of contextual information. Context values for different dimensions of the context have different data types, e.g., temperature, humidity, latitude, longitude, state, etc. Thus, it is possible to explicitly represent and reason through one context more specifically than another. For example, the different scenarios are described in terms of the three dimensions DayOfWeek, Location, and WeatherType. Saturday, in Scenario 1, is covered by the wider context of Weekend, and both contexts describe the scenario from the same dimension. Intuitively, if one context covers the other, it has a broader perspective. If there is a covering relationship between the contexts in one or more dimensions of the scene, then there is also a covering relationship between the scenes. For example, in Scenario 1 and Scenario 2, the Weekend covers Saturday, SanyangRoad covers JianganDistrict, and lightRain covers rain, so Scene 1 is covered by Scene 2. It is clear that the more detailed the description of the scene, the higher the level of the hierarchy, and the lower-level scenes are more broadly described than the higher-level scenes, showing an extended tree hierarchy, as shown in Fig. 3.

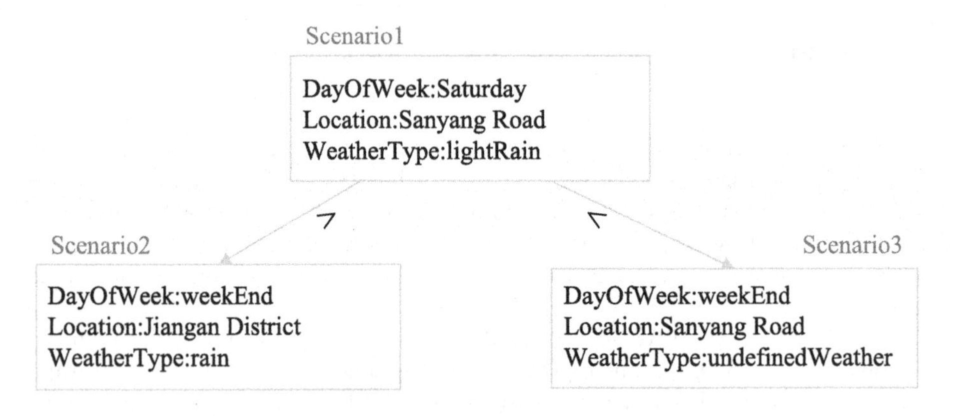

Fig. 3. Overlay relationships between scenes.

5.2 Service Effectiveness Analysis

A family trip scenario is used as an example to show that referring to contextual information when making service recommendations can improve service recommendation accuracy. Validate the usability of the methods in this paper. The users of the services, in this case, include drivers, children, and passengers. The

Table 3. Relationship table of situational elements.

Service Name	API	Contextual elements
Account Login	POST /smartCockpit/login	Facial features
Turn on the air conditioning	POST /smartCockpit/start_aircon	Temperature, season, room temperature
Adjusting the seat	POST /smartCockpit/adjust_seat	UserId
Search for animated films	GET /babyBus/search_cartoon? q=SpongBob	Age of user
Navigating destinations	POST /amap/navigation	/
Estimated range	GET /smartCockpit/estimate _range?battery_level=20	Remaining battery, remaining battery life
Recommended Charging Stations	GET /smartCockpit/recommend _charging_stations	Scoring, remaining openings, distance

name, interface, service function, and main context definitions for each service are shown in Table 3 below:

1. The driver is a user of the service, and when the driver sits in the cab, the in-car camera identifies the driver's facial features to determine the user's identity. The contextual information to be considered includes the time of day and facial features.
2. Passengers invoke to turn on the air conditioning service, which is provided by the physical device. The service is started considering the current temperature in the vehicle, the season, and the age of the passenger, which are set to 32°C, summer, and nine years old.
3. The car seat is adjusted according to the driver's physical characteristics and driving needs. For example, adjust the seat angle, front and rear position, backrest angle, etc.
4. The cartoon playback service is provided by third-party software carried in the car and is a software service. Children in the car are shown the cartoon SpongeBob SquarePants on the rear seat monitor; the context of the service is the age of the passenger.
5. Navigation services are provided by in-car navigation software, such as Gaode Maps, and are software services. The service call refers to the destination entered by the user.
6. During vehicle travel, it is detected that only 20% of the remaining battery is left. Based on the distance and temperature context, it is estimated that the destination cannot be reached, and the service alerts the driver whether to proceed to the charging port.
7. The recommended charging post service searches for three charging posts nearby based on geographical location. The provision of this service is influenced by the contextual factors of rating, remaining availability, and location.

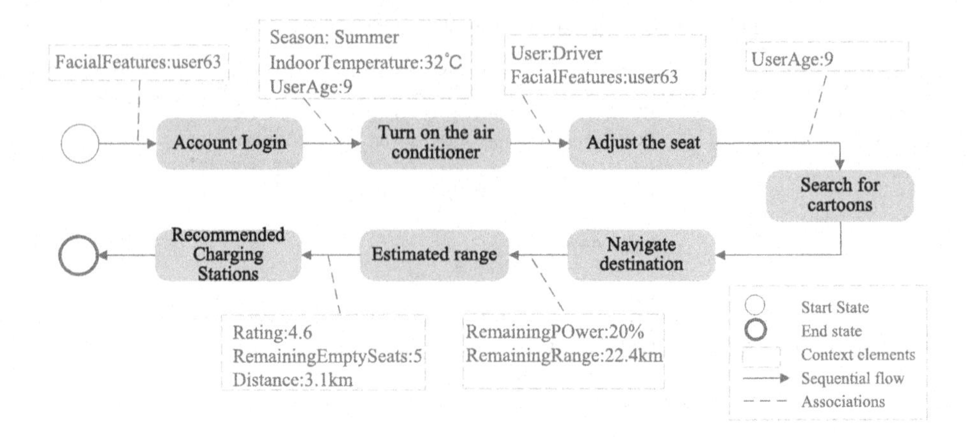

Fig. 4. Service flow of family outing scenario.

The above example provides an analysis of how context can improve the effectiveness of service recommendations. By analyzing the flowchart of the family outing scenario (shown in Fig. 4), it can be seen that context affects the availability of services and the selection of services. If the contextual features are ignored when invoking a service and the demand is met solely from the perspective of the service functionality, this may lead to problems such as the unavailability of the service or failure to meet the user's expectations.

5.3 Situation Knowledge Graph Example

In this paper, we build an example of an off-duty fitness scenario for young people living in urban life. The above example is shown in Fig. 5. When the initial user registers, basic information about the user is collected, such as gender, age, city, and interests. Based on the number of times the user searches for content, the system records the user's preference to visit food stores, tourist attractions, hospitals, etc. For example, the Smart Cockpit recorded that Allen would visit the gym at least four times a week. So when Ellen gets off work, the system recognizes that it's Tuesday and recommends his preferred content. Ellen chooses to navigate to the gym service. The system invokes a series of services, starting with the Gaode Map to navigate to the gym. The display shows multiple routes for the user to choose from, and after selecting a route, the system enables voice navigation. At the same time, the system asks where to park near the gym and displays information about the car park, such as the number of spaces left and the number of available spaces. While driving on Renaissance Road, Ellen detects that it is in an evening rush hour scenario based on the speed, road conditions, and other contexts. As a result, the system asks if music will be played, and the service refers to the user's mood, weather, and other information. At this point, Ellen is depressed, and it is raining lightly, so the system plays delighted music. Finally, when Ellen arrives at his destination and is ready to get out of the car, the system plays a voice reminder, "Please take an umbrella with you when you get out of the car," based on the contextual premise. The Smart Cockpit Situation Knowledge Graph can also facilitate some downstream tasks such as multi-round conversations [24].

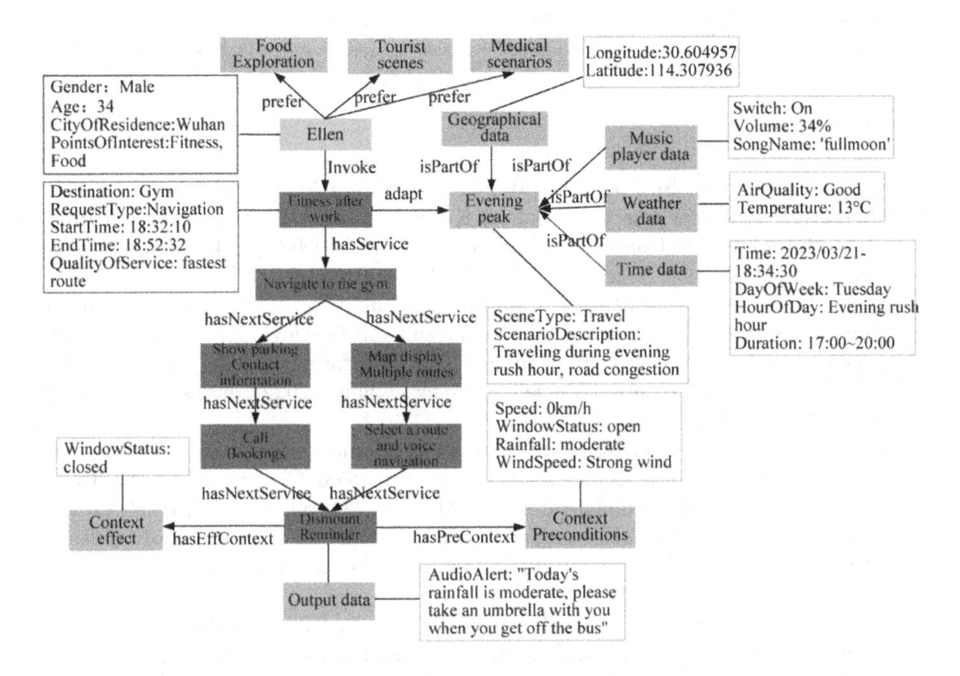

Fig. 5. Smart Cockpit Situation Knowledge Graph for Evening Peak Travel.

6 Conclusions and Future Work

This paper proposes a smart cockpit situation model with user, service, and context as core elements. We first analyze the elements that make up the smart cockpit situation model and explore their connections. The Smart Cockpit Context Model takes user, service, and context as the core elements and contains six relationships between the elements. Second, we use the smart cockpit situation model as a guide to building the ontology using a top-down approach. At the context layer, the context dimension is refined. The software and hardware functions provided by the car are abstracted into service entities. And then reasoning about the coverage relationship between scenes based on the coverage relationship between contexts, using the family travel scenario example to validate the service accuracy provided by the model. Finally, we constructed a fitness scenario smart cockpit situation graph to instantiate the model. The research results show that the situation graph constructed in this paper cannot only comprehensively describe the smart cockpit scene but also can adapt to the dynamic changes of contextual data. In the future, we will collect more smart cockpit situation data to further improve ontology. We also plan to explore more refined methods for context-aware service recommendations.

Acknowledgements. This work was funded by the National Natural Science Foundation of China grant number U1836118, Key Laboratory of Rich Media Digital Publishing, Content Organization and Knowledge Service grant number ZD2022-10/05, and National key research and development program grant number 2020AAA0108500.

References

1. Sun, X., Chen, H., Shi, J., Guo, W., Li, J.: From HMI to HRI: human-vehicle interaction design for smart cockpit. In: Kurosu, M. (ed.) Human-Computer Interaction. Interaction in Context. HCI 2018. LNCS, vol. 10902, pp. 441–454. Springer, Cham (2018). https://doi.org/10.1007/978-3-319-91244-8_35
2. Liang, B., Tang, Z.: Multi-modal information analysis of automobile intelligent human-computer interaction. In: Sugumaran, V., Sreedevi, A.G., Xu, Z. (eds.) Application of Intelligent Systems in Multi-modal Information Analytics. ICMMIA 2022. LNCS, vol 136, pp. 658–666. Springer, Cham (2022). https://doi.org/10.1007/978-3-031-05237-8_81
3. Meng, F., Zhu, X., et al.: Application and development of AI technology in automobile intelligent cockpit. In: 2022 3rd International Conference on Electronic Communication and Artificial Intelligence (IWECAI), pp. 274–280. IEEE (2022)
4. Li, W., et al.: Cogemonet: a cognitive-feature-augmented driver emotion recognition model for smart cockpit. IEEE Trans. Comput. Soc. Syst. **9**(3), 667–678 (2021)
5. Lin, S., Zou, J., Zhang, C., Lai, X., Mao, N., Fu, H.: Understanding user requirements for smart cockpit of new energy vehicles: a natural language process approach. Tech. rep, SAE Technical Paper (2022)
6. Xing, C., Zhiming, H., Xinshu, Y., Yun, M., Yiyan, C., Wenzhong, G.: Approach to modeling and executing context-aware services of smart home at runtime. J. Softw. **30**(11), 3297–3312 (2019)
7. Yazhong, M., et al.: Construction and application of city brain knowledge graph. J. Chin. Inf. Process. **36**(04), 48–56 (2022)
8. Saba, D., Sahli, Y., Hadidi, A.: An ontology based energy management for smart home. Sustain. Comput. Informatics Syst. **31**, 100591 (2021)
9. Sayah, Z., Kazar, O., Lejdel, B., Laouid, A., Ghenabzia, A.: An intelligent system for energy management in smart cities based on big data and ontology. Smart Sustain. Built Environ. **10**(2), 169–192 (2021)
10. Swenja, S., Gerst, N., Keller, C., Thomas, S.: Defining a context model for smart manufacturing. Procedia Comput. Sci. **204**, 22–29 (2022)
11. Pradeep, P., Krishnamoorthy, S.: The mom of context-aware systems: a survey. Comput. Commun. **137**, 44–69 (2019)
12. Yavari, A., Jayaraman, P.P., Georgakopoulos, D.: Contextualised service delivery in the internet of things: parking recommender for smart cities. In: 2016 IEEE 3rd World Forum on Internet of Things (WF-IoT), pp. 454–459. IEEE (2016)
13. Chen, K.T., Chen, H.Y.W., Bisantz, A.: Adding visual contextual information to continuous sonification feedback about low-reliability situations in conditionally automated driving: a driving simulator study. Transport. Res. F: Traffic Psychol. Behav. **94**, 25–41 (2023)

14. Augusto, J.C.: Contexts and context-awareness revisited from an intelligent environments perspective. Appl. Artif. Intell. **36**(1), 694–725 (2022)
15. Bang, A.O., Rao, U.P.: Context-aware computing for IoT: history, applications and research challenges. In: Goyal, D., Chaturvedi, P., Nagar, A.K., Purohit, S.D. (eds.) Proceedings of Second International Conference on Smart Energy and Communication: ICSEC 2020, pp. 719–726. Springer, Singapore (2021). https://doi.org/10.1007/978-981-15-6707-0_70
16. Dinh, L.T.N., Karmakar, G., Kamruzzaman, J.: A survey on context awareness in big data analytics for business applications. Knowl. Inf. Syst. **62**, 3387–3415 (2020)
17. Panah, A.S., Yavari, A., van Schyndel, R., Georgakopoulos, D., Yi, X.: Context-driven granular disclosure control for internet of things applications. IEEE Trans. Big Data **5**(3), 408–422 (2017)
18. Wu, C., et al.: Knowledge graph-based multi-context-aware recommendation algorithm. Inf. Sci. **595**, 179–194 (2022)
19. Mezni, H., Benslimane, D., Bellatreche, L.: Context-aware service recommendation based on knowledge graph embedding. IEEE Trans. Knowl. Data Eng. **34**(11), 5225–5238 (2021)
20. Yavari, A., Jayaraman, P.P., Georgakopoulos, D., Nepal, S.: Contaas: an approach to internet-scale contextualisation for developing efficient internet of things applications. In: Bui, T. (ed.) 50th Hawaii International Conference on System Sciences, HICSS 2017, Hilton Waikoloa Village, Hawaii, 4–7 January 2017, pp. 1–9. ScholarSpace/AIS Electronic Library (AISeL) (2017)
21. Wang, X., Feng, Z.: Semantic web service composition considering iope matching. J. Tianjin Univ. **50**(9), 984–996 (2017)
22. Xu, W., Ge, L., Gao, Z.: Human-AI interaction: an emerging interdisciplinary domain for enabling human-centered AI. CAAI Trans. Intell. Syst. **16**(4), 605–621 (2021)
23. Bastani, F., Zhu, W., Moeini, H., Hwang, S.Y., Zhang, Y., et al.: Service-oriented IoT modeling and its deviation from software services. In: 2018 IEEE Symposium on Service-Oriented System Engineering (SOSE), pp. 40–47. IEEE (2018)
24. A task-oriented multi-turn dialogue mechanism for the intelligent cockpit (accepted at APWeb-WAIM 2023)

A Task-Oriented Multi-turn Dialogue Mechanism for the Smart Cockpit

Xiaoyu Yang[1], Xinyi Sheng[1], and Jinguang Gu[1,2,3](✉)

[1] College of Computer Science and Technology, Wuhan University of Science and Technology, Wuhan 430065, China
yangxy1216@wust.edu.cn, sxywust@126.com

[2] Institute of Big Data Science and Engineering, Wuhan University of Science and Technology, Wuhan 430065, China
simon@ontoweb.wust.edu.cn

[3] Key Laboratory of Rich-media Knowledge Organization and Service of Digital Publishing Content, National Press and Publication Administration, Beijing 100038, China

Abstract. As an important carrier and platform for AI applications, the dialogue system for smart cockpits will become an important scenario for AI applications. However, there are some problems with the existing smart cockpit dialogue system, such as a lack of relevant corpus and knowledge annotations. These problems lead to the low accuracy of response representation generated by task-oriented dialogue systems for smart cockpits, which cannot meet the needs of smart human-computer interaction. To this end, this paper designs cockpitWOZ, a multi-round Chinese conversation dataset with knowledge annotation in the smart cockpit domain, which contains 2.9k conversations in four domains, including restaurants, attractions, music, and itineraries. On this basis, a new knowledge-driven task-based dialogue model is designed in this paper based on multiple baseline models. Firstly, a knowledge graph named cockpitKG based on the smart cockpit scenario is constructed to enhance the responsiveness of the model. Secondly, the smart cockpit system is built and a label replacement method is used to meet the requirements for real-time interaction during vehicle movement. Finally, the experimental results show that this model outperforms the baseline model and demonstrates that the introduction of background knowledge and label replacement can lead to higher-quality responses in the smart cockpit dialogue system.

Keywords: Smart Cockpit · Task-oriented Dialogue Systems · Knowledge Graph

1 Introduction

With the rapid development of the smart car industry, the smart cockpit has gradually become an important platform for artificial intelligence applications [1]. The essence of a smart cockpit is to integrate information like human face and

X. Song et al. (Eds.): APWeb-WAIM 2023, LNCS 14332, pp. 316–330, 2024.
https://doi.org/10.1007/978-981-97-2390-4_22

voice patterns through various forms of interaction modes such as touch, voice control and gestures with in car voice assistants. They then form a comprehensive user input, and communicate and express with the user like a human through sound effects, augmented implementation and other technologies to complete the user's instructions. As a product of human computer interaction, the quality task-oriented dialogue system [2] has great potential and commercial values in the field of smart cockpit.

A quality smart cockpit task-oriented dialogue system aims to provide real-time assistance to users in human-machine dialogue. Nowadays, dialogue systems have become a hot research topic in the field of Natural Language Processing (NLP), while introducing background knowledge is particularly important [3], as it can provide crucial information for the dialogue.

However, it is a big challenging for current task-oriented dialogue systems to use background knowledge to generate effective interaction. Background knowledge can be represented as a knowledge graph [4] or unstructured text [5].

Recently, there have been many studies introducing background knowledge to dialogue systems. Zhou et al. [6] published a Chinese multi-domain dialogue dataset and verified through extensive experiments that introducing background knowledge can improve the performance of natural language generation. Xu et al. [7] proposed an open-domain task-based dialogue model (KnowHRL) based on knowledge graphs and incorporated session goal prediction. KnowHRL was shown to outperform state-of-the art baseline models in terms of user interest consistency, conversation coherence, and knowledge accuracy; Yang et al. [8] put forward a conversation model that integrates knowledge graphs into a task-based conversation system, and they incorporated a new recurrent unit structure to effectively utilize structural information in the conversation histories.

Conversation	Label Replace Conversation
User：最近的公园在哪？ where's the nearest parking garage	User：最近的<category>在哪？ where's the nearest <destinations>
Smart cockpit：检测到您目前位于北京西郊，距离您最近的公园是北京香山公园。您想知道怎么去吗？ It is detected that you are currently located in the western suburbs of Beijing. The nearest park to you is Beijing Xiangshan Park. Would you like directions there?	Smart cockpit：检测到您目前位于<nowLocation>，距离您最近的<category>是<destinations>。您想知道怎么去吗？ It is detected that you are currently located in the <nowLocation>. The nearest <category> to you is <destinations>. Would you like directions there?
User：好呀，请帮我选择一条避开交通拥堵的路线。 Ok, please help me choose a route to avoid the traffic jam	User：好呀，请帮我选择一条避开交通拥堵的路线。 Ok, please help me choose a route to avoid the traffic jam
Smart cockpit: 好的，正在为您导航。 Ok, navigating for you	Smart cockpit: 好的，正在为您导航。 Ok, navigating for you

Fig. 1. An example in cockpitWOZ. On the right side is the dialog data set that introduces the label, where the content of < > is the replacement label.

Although previous works have achieved great success, there has been little research on the dynamic and real-time considerations of the smart cockpit task-oriented dialogue system during the driving process because of lacking public benchmarks. Meanwhile, a lack of high-quality and annotated dialogue datasets limited studies in task-oriented dialogue systems. The gap between existing dialogue corpora and real-life human-car dialogue data hindered their applications as well. Most in-car voice assistants are based on question-and-answer dialogue, where they search and execute instructions in the existing knowledge base. Human-car dialogue systems are usually single-turn, template-driven. And for robots, each answer is independent and context-free. Real-life applications are much more complex than previous research, so how to consider the dynamic nature of the driving process and introduce background knowledge into task oriented dialogue systems in the smart cockpit domain to make human-machine interaction more natural has become a challenge.

To address these issues, this paper proposes a method for introducing background knowledge into dialogue systems in the smart cockpit domain to improve the quality of human-computer dialogue.

The main contributions of this paper include three aspects:

1) We construct a multi-round conversation dataset cockpitWOZ for smart cockpit, containing 40K discourses and 2.9K conversations in 4 domains (restaurant, attraction, music, and itinerary). We build a knowledge annotation tool to obtain triple knowledge for enhancing task-oriented conversation systems in the smart cockpit domain. Also, as shown in Fig. 1, we introduce the label replacement mechanism, which integrates labels into the dataset to meet the real-time needs of vehicle movement.
2) We construct smart cockpit knowledge graphs named cockpitKG. Firstly, we build the smart cockpit ontology model based on contexts, services and users. And then we integrate entity classes such as human, vehicle and trigger action in the smart cockpit knowledge graph. To solve the problems of semantic difficulties and the inability to handle the differences between the knowledge graph and natural language, we embed it into our model.
3) The experimental results demonstrate that incorporating knowledge into the smart cockpit task-based dialogue system improves accuracy, response diversity, and perplexity of conversation generation compared to the previous model.

2 Data Construction

In order to deeply study the task-based dialogue system in the smart cockpit domain and simulate knowledge interaction in the human-vehicle dialogue, multi-round human-vehicle dialogue in the smart cockpit domain is collected and constructed, and is used to build a knowledge graph in the smart cockpit domain.

2.1 Knowledge Graphs Construction

Smart cockpit can collect various types of information through sensors and internet connected devices in the car, such as weather, traffic, driver's facial features, and system status, analyze the driver's or passenger's emotions and behaviors, and provide personalized and accurate services or recommendations to ensure a comfortable and safe riding experience for users. We integrate multiple scenarios during the user's driving process. By utilizing the user's basic information and historical data of daily interaction, understanding their core demands, and analyzing the core compositional elements of smart cockpit scenario, we improve the knowledge graph of smart cockpit scenario and apply it to the smart cockpit dialogue system in this study. By integrating the knowledge graph that integrates into users' daily life scenes into the dialogue system in this study, it provides users with a new experience of travel, entertainment, driving advice, and safety travel recommendations.

The smart cockpit scenario describes a snapshot of the environment and a specific situation in a specific time and space. This article constructs the ontology model of the field through the top-down approach, and the smart cockpit ontology is shown in Fig. 6 in Appendix A. The user class refers to the driver or passenger, who uses the user request to call for services. Service providers provide corresponding services, and the context has context types and context values. The context types are divided into user context, service context, application context, and physical context. For the service context, we reuse OWL-S to describe web services [9] and divide them into physical services and software services. The other classes involved in the ontology model of smart cockpit are introduced below: 1) User: defined as the initiator of the request and having attributes such as name, age, gender, and city of residence. 2) User request: describing the details of the request, such as the start time, end time, and content of the request. 3) Service: defined as physical or software services, where physical services are provided by physical devices such as car sensors and actuators, such as air conditioning services; software services are provided by software carried in the onboard system, such as navigation and music playback functions. 4) Context type: defined as user context to describe data related to user preferences, such as interests and emotions; service context is used to evaluate services, such as response time and satisfaction; physical context includes context factors related to physical devices, such as car window status and remaining power; and application context is used to obtain related types of data from the software system, such as describing weather, traffic, and geography.

We instantiate the ontology model to transform these abstract concepts into specific instance objects. Based on these instantiated objects, we construct the knowledge graph of the smart cockpit.

2.2 Dialogue Collection and Construction

We recruited four annotators to simulate multi-turn conversations in a smart cockpit scenario, referring to speech dialogue sets about smart cockpit found online, in the absence of specific goals and constraints, and built the conversation dataset cockpitWOZ in the smart cockpit domain. During the conversation, two speakers both had access to the knowledge graph, and the annotators were required to record the related knowledge triples used to generate the dialogue as knowledge annotations at the turn level. In order to ensure the naturalness of the generated conversations, we filtered out low-quality dialogues, which contain grammatical errors, inconsistencies in knowledge facts, etc. To make the dialogue

Table 1. Dialogue statistical indicators.

Domain	train	dev	test	Total
# dialogues	2069	294	596	2989
# turn	14068.5	1914.5	4106.5	20270.5
# utterances	28137	3829	8214	40541
# token	370098	48729	105595	529627
Avg.# utterances per dialogues	13.15	12.73	13.86	13.56

more realistic, labels were used to represent real-time locations, surrounding scenes, recommended types, and other relevant information in the construction process, for real-time replacement in actual use cases. In total, 2989 dialogues were collected and constructed, covering four domains of human-car dialogue: restaurants, attractions, music, and itinerary planning. The dataset was divided into training, development, and testing sets based on a 7:1:2 ratio. Dialogue data statistics are presented in Table 1.

2.3 Dialogue Annotation

As the obtained dataset did not include sentence-level triple knowledge annotations, we invited four annotators and provided them with training to give them a basic understanding of smart cockpit. Their primary task was to annotate the dialogue data by providing knowledge annotations for the sentence-level triple. To facilitate the annotation process, reduce costs, and save time, a corresponding annotation system was developed. The annotation system used a separate front-end and back end approach. The annotation tool presented a group of dialogues on the page to mark the triple knowledge mentioned in the dialogue. The subject, predicate, and object in the drop-down box all came from the constructed smart cockpit domain KG.

After obtaining a group of dialogues, the frontends request backends interfaces, and backends uses SPARQL statements to query all the subject, predicate, and objects in the KG, storing the query results and associated information in the frontends drop down box. When the subject, predicate, or object has been determined, the other two drop-down boxes filter out irrelevant data to improve the efficiency of the annotation. Each sentence can submit multiple triple annotations. The annotation tool is shown in Fig. 2.

3 Model Architecture

3.1 Model Design

Conventional task-oriented dialogue systems find appropriate response sentences based on the input text. Some models input the sentence to be replied to as text

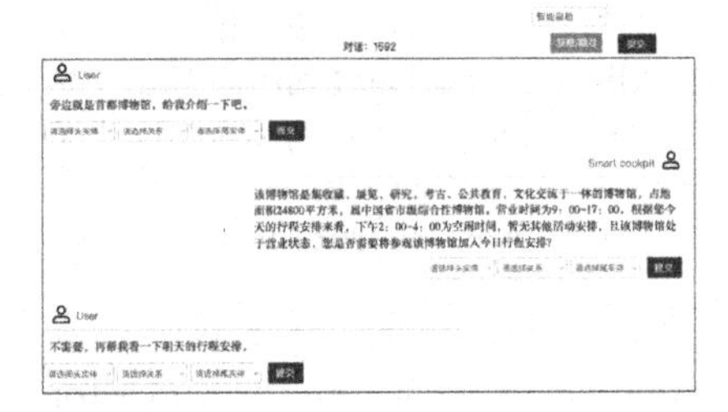

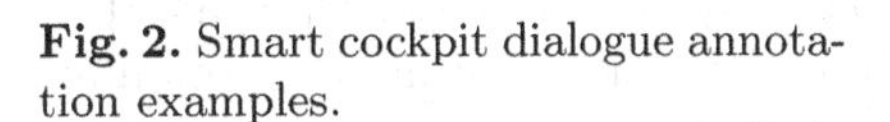

Fig. 2. Smart cockpit dialogue annotation examples.

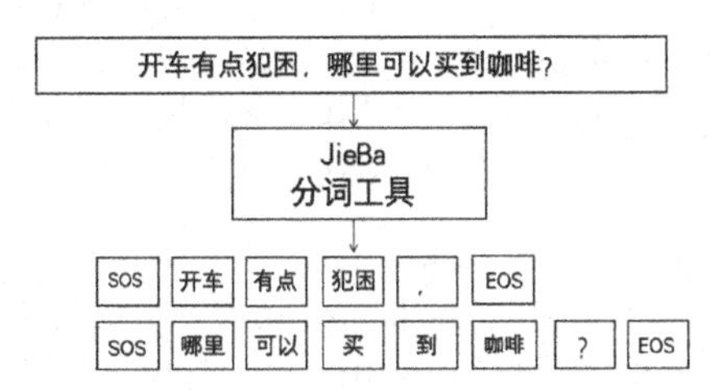

Fig. 3. Words segmentation examples.

information into the model, but this method can only capture the semantic information of this sentence and cannot obtain context information. In order to solve the above problems, our model concatenates all historical dialogue sentences and inputs them into the model as information text to capture context information, which generates higher responses. At the same time, in order to solve the semantic difficulties in capturing natural language problems in the smart cockpit domain and the differences between existing QA methods in processing knowledge graphs and natural language problems, the method of embedding smart cockpit knowledge graphs is adopted. The key idea is to better preserve the key information in KG, represent each entity and predicate in the smart cockpit knowledge graph as low dimensional vectors, and use the learned vectors for dialogue system design. This process is simple to model with few parameters, and low computational complexity, which can train a TransE model that specifies semantic relationships between entities and relationship vectors. In the TransE model, entities and relationships can be represented as RDF data triple forms (head entity, relationship, tail entity) or written as (h, r, t). From a semantic perspective, triples can represent both semantic relationships between concepts and instances in the ontology knowledge base, and can also represent attribute values for data items. The overall model algorithm mainly obtains natural language statements from the model based on knowledge triples $Triples_t$ and dialogues $H_{sentence}$, which can be achieved using generator-based BERT (Bidirectional Encoder Representation from Transformers) [10] models, retrieval based Seq2Seq (Sequence-to-Sequence) [11] models, and HRED (Hierarchical Recurrent Encoder-Decoder) [12] models. The specific steps are as follows:

First, preprocess the sentence, that is, tokenize the historical dialogue and add marker symbols "SOS" and "EOS" at the beginning and end of each sentence. For example, $H_{sentence}$= "开车有点犯困，哪里可以买到咖啡?" is tokenized as shown in Fig. 3.

Secondly, in order to enhance the key information in the dialogue and improve the accuracy of the response, we introduce a key-value memory module. We treated all knowledge triples mentioned in a dialogue as the knowledge information in the memory module. For a triple that is indexed by i, we represented the key memory and the value memory, respectively, as a key vector k_i and a value vector v_i, where k_i is the average word embedding of the head entity and the relation, and v_i is those of the tail entity. We used a query vector q to attend to the key vectors k_i $(i = 1, 2, ...)$: $a_i = \text{softmax}_i\ (q^T k_i)$, then the weighted sum of the value vectors v_i $(i = 1, 2, ...)$, $W_{tt} = \sum_i a_i v_i$, was incorporated into the decoding process.

Thirdly, the segment $H_{sentence}$ is vectorized using the large-scale word vector database open-sourced by Tencent AI Lab, obtaining an open-source word vector W_{hs} of 200 dimensions. The obtained W_{hs} and W_{tt} are fused and input into the encoder of the retrieval-based model or the generation-based model, obtaining the hidden state $Hidden_s$ of the current input text.

Fourthly, we pass the obtained hidden state $Hidden_s$ into the decoder, the retrieval based model retrieves the top 10 candidate response sentences $Sentence_t$ which match the text information, while the generation-based model decodes directly and maps the decoded vector to words to obtain a $Sentence_t$ composed of a vocabulary sequence.

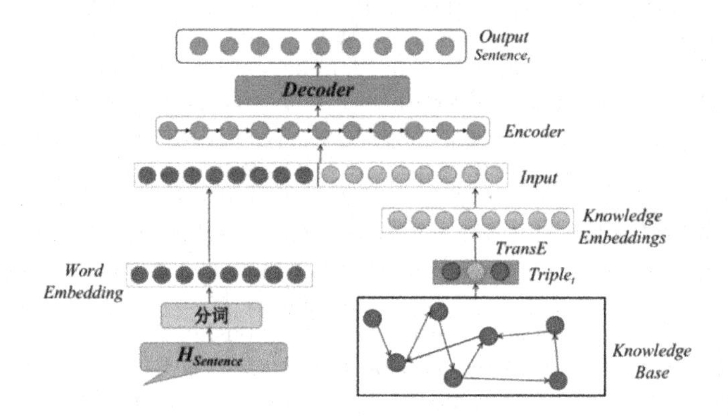

Fig. 4. Model construction.

Common generation-based models include Seq2Seq and HRED models, while the BERT model is a commonly used retrieval-based model. The specific model structure is shown in Fig. 4, and the formula is shown in (3.1).

$$Sentence_t = \max\left(soft\max\left(\exp\left(\left(I_t^{sp}\right)^T\right)\right)\right) \tag{3.1}$$

$I_t^{sp} = \left[W_{hs}; W_{tt}\right]$, I_t^{sp} contains the open-source word vector W_{hs} and the knowledge vector W_{tt} cockpit scene, All models were trained using the cross entropy loss function, as shown in formula (3.2), where $\{y_t^*\}$ represents the target sentence.

$$Loss = \frac{1}{T}\sum_{t=1}^{T} \log P(y_t^*) \tag{3.2}$$

4 Experiment and Evaluation

4.1 Implementation Details

All models in the experiment were run on a server with the following configuration: Tesla P100-PCIE graphics card, Intel(R)Xeon(R) CPU E5-2600 v4 @ 2.00 GHz CPU, 64G memory. At the same time, the experiment code was developed based on it. All word embedding dimensions and hidden layers were set to 200, the minimum sample size was 16, and the initial learning rate was set to 0.001. The training data accounted for 70% of the total data, the test data accounted for 20%, and the validation data accounted for 10%.

4.2 Baseline

This paper conducted experiments using the generation-based Seq2Seq and HRED models, as well as the retrieval-based Bert model. Among them, Seq2Seq is an encoder-decoder model with an enhanced attention mechanism. The output of the decoder is jointly determined by the output of the encoding side and the output of the previous time step on the decoding side. The formula of the Seq2Seq model is shown in (4.1).

$$p(y_1, \cdots, y_T \mid x_1, \cdots x_T) = \sum_{t=1}^{T} p(y_t \mid y_1, \cdots, y_{t-1}, \nu) \tag{4.1}$$

HRED is a hierarchically recursive encoder-decoder model consisting of three parts: 1) Encoder-side RNN: Similar to the encoder side of the Seq2Seq model, it encodes the input conversation into a fixed-length vector representation at the last hidden layer. 2) Context RNN: It takes the last hidden output of the encoder-side RNN and the output of the context RNN from the previous time step as input to effectively retain the historical dialogue information. 3) Decoder-side RNN: Unlike the Seq2Seq model, the decoder-side RNN not only depends on the hidden state output of the previous decoding step but also associates with context information, i.e., including the output of the context RNN. The RNN that manages the context is represented as a formula (4.2).

$$h_n^{context} = f(h_{n-1}^{context}, h_{n-1}^{encode}) \tag{4.2}$$

Where $h_n^{context}$ represents the hidden layer output of context management RNN; h_{n-1}^{encode} represents the output of the encoding end. After the context information is introduced into the decoding end, the decoding end is rewritten as (4.3).

$$h_n^{decode} = f(y_{n-1}, h_{n-1}^{decode}, h_n^{context}) \tag{4.3}$$

The retrieval-based Bert model encodes the input sentence and selects the top k highest matching candidates from the corpus as a reply. In this paper, the BM25 algorithm is used for query matching, and the best candidate with the highest probability is selected from the candidate set as a reply.

4.3 Automatic Evaluation

Metrics. Common evaluation metrics include BLEU-k (bilingual evaluation understudy, k-gram) [13], Distinct-k (k-gram) [14], and Perplexity (PPL) [15]. BLEU-k is a commonly used measure that calculates the k-gram overlap between the generated sentence and the reference words, used to evaluate the accuracy and fluency of sentences. Distinct-k is used to evaluate the diversity of generated responses. In the generation model, the ranking is determined by the PPL values of the candidate answers. Perplexity (PPL) is calculated to evaluate whether the generated results conform to grammar and fluency, as shown in formulas (4.4), (4.5), and (4.6).

$$BLEU-n=\frac{\sum_{c\in candidates}\sum_{n-gram\in c}Count_{clip}\left(n-gram\right)}{\sum_{c'\in candidates}\sum_{n-gram'\in c'}Count\left(n-gram'\right)} \tag{4.4}$$

$$Distinct-n=\frac{Different\left(n-gram\right)}{Total\left(n-gram\right)} \tag{4.5}$$

$$ppl(W)=P(w_1w_2\cdots w_n)^{-\frac{1}{N}}=\sqrt[N]{\frac{1}{P(w_1w_2\cdots w_n)}} \tag{4.6}$$

Experimental Results and Analysis. The main research focuses on introducing knowledge-aware BERT models, comparing them with knowledge-aware generation models Seq2Seq and HRED models, and also comparing them with models without introducing knowledge. Among them, the model enhanced by knowledge is represented as "+know". The experimental results are shown in Table 2. Since the sentences of the BERT model come from the human corpus, calculating Perplexity is meaningless, so it will not be shown anymore.

Table 2. Comparison of experimental results.

Model	BLEU - 1/2/3/4				Distinct - 1/2/3/4				PPL
HRED	35.76	27.24	23.08	20.60	3.55	9.46	15.82	21.98	17.28
Seq2Seq	37.08	28.69	24.50	21.95	4.12	11.04	17.94	24.32	16.82
BERT	84.79	81.65	79.58	78.04	9.51	31.22	46.02	53.94	–
HRED+ know	37.46	28.82	24.61	22.07	3.95	9.58	15.88	22.06	17.37
Seq2Seq+ know	37.94	29.33	25.04	22.42	4.25	11.16	17.98	24.39	16.73
BERT+ know	84.92	81.72	79.73	78.21	9.63	31.29	46.19	54.02	–

From the above experimental results, it can be seen that after introducing knowledge, all models have improved in all indicators except PPL. Among all the baselines, the knowledge-aware BERT model performs best in the smart cockpit domain and achieves the best performance because it retrieves gold truth

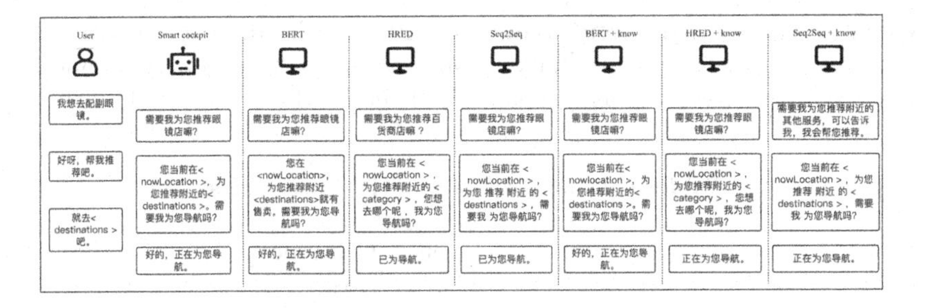

Fig. 5. Model example.

response at a relatively high ratio. The HRED model performs best among the generation-based models. However, after introducing knowledge triples, the performance improvement of the BERT model among all models is relatively small, because the feature differences between a single sentence and knowledge triples are not prominent in the shallow layer of BERT, so the improvement for the BERT model is limited. The knowledge enhanced Seq2Seq model improves by 0.64 points in BLEU-2 compared to Seq2Seq in the baseline and achieves better performance. In the generation-based dialogue model, the more utterances contained in the dialogue, the more difficult it is to model, so the performance of retrieval-based models is generally better than that of generation-based models. In addition, more diverse knowledge also requires models to use knowledge more flexibly. Figure 5 lists examples of sentences generated by different models. In which, <destinations> represents the destination label, <nowLocation> represents the current location label, and <category> represents the location type label. By replacing the labels with real-time location information through the navigation system, the real-time nature of the scene can be ensured in the dataset usage.

4.4 Manual Evaluation

Metrics. In addition to using automatic evaluation metrics for automatic evaluation during the research process, artificial evaluation metrics were also used to conduct the manual evaluation of the dialogue generation results. The evaluation metrics are defined as follows:

Fluency is defined as whether the response is fluent and natural. If the sentence has grammatical errors, is not fluent and difficult to understand, a score of 0 is given; if the sentence contains some grammatical errors, but is still understandable, a score of 1 is given; if the sentence is fluent and similar to human languages, a score of 2 is given.

Coherence is defined as whether the sentence is coherent with the context. If it is irrelevant to the context, a score of 0 is given; if it is related to the last sentence but not the entire history, a score of 1 is given; if it is related and reasonable in the context, a score of 2 is given.

Experimental Results and Analysis. 40% of the data generated by the models was randomly selected for manual evaluation. The models enhanced by knowledge are represented as "+know", and the results are shown in Table 3.

Table 3. Comparison of artificial tests results.

Model	Fluency	Coherence
HRED	1.35	0.94
Seq2Seq	1.39	1.02
BERT	1.86	1,.49
HRED+know	1.42	1.06
Seq2Seq+know	1.50	1.35
BERT+know	1.86	1.66

The experimental results above show that the BERT model performs better than other models in these two metrics because the candidate sentences come from a human corpus, so retrieval-based models score higher in fluency. For the original model, if only the historical dialogue data is considered, as the dialogue turns increases and the historical dialogue information grows, the context connectivity will be lost. However, after introducing knowledge triples, the model has improved in these two aspects.

4.5 Real Time Label

Evaluation Indices. In order to make the dialogue more in line with the dynamic nature of driving, this paper proposes a label replacement method and designs evaluation metrics to verify its effectiveness and accuracy. We conduct manual evaluations of the generated results, with and without the use of labels. The evaluation metrics are defined as follows: Real-time accuracy is defined as the metric evaluating the accuracy of the recommended content in the generated sentences. If the location in the recommended content is unrelated to the actual surrounding scene, zero points are given. If some of the recommended locations are related to the actual scene, one point is given. If the recommended content fully matches the real surrounding scene, two points are given.

Table 4. Comparison of label results.

Model	Real-time Accuracy
BERT+know	0.13
BERT+know+label	2

Experimental Results and Analysis. Randomly selecting 40% of the conversation datasets without labels and comparing them with the corresponding conversation datasets with labels, where the model with labels is denoted as "+label", the results are shown in Table 4.

From the above results, it can be seen that replacing location information in the dialogue dataset with labels improves the accuracy of the generated sentences and avoids unrelated recommendations. The recommended location is related to the current actual scene and can change with the change of the vehicle's position, which is in line with the actual application scenario. The model without labels can only recommend the locations mentioned in the dataset. When the vehicle's position changes, the recommended location cannot be changed in real-time to provide real help to users, greatly limiting the relevance and availability of the recommendations. As a result, the score did not exceed 1.00.

5 Related Work

Recently, there have been significant advancements in task-oriented dialogue systems due to the increase of publicly available dialogue datasets. However, due to the lack of background knowledge, dialogue models may generate dull and repetitive content, causing conversations to deviate from the main topic and produce useless responses in dialogue assistants. Currently, connecting dialogue with external knowledge has become an effective way to solve the degradation problem, and some external knowledge sources such as open-domain knowledge graphs [5], common-sense knowledge bases [4], or background documents [16] have become the main research objects in this field.

Based on knowledge, dialogue generation aims to generate information-rich and meaningful responses based on the dialogue context and external sources of knowledge. So far, researchers have collected knowledge-based dialogues using crowdsourcing platforms for various tasks such as open-domain dialogue [17] and conversational recommendation [18]. Workers are required to respond based on knowledge in structured knowledge bases [19] or unstructured documents [20]. Using recent advances in large-scale language models [21], researchers have also built knowledge-intensive dialogue systems by fine-tuning such language models in an end-to-end manner [22]. In this study, we build a dialogue with human annotators from scratch based on a structured knowledge graph. As annotating knowledge for dialogue is expensive and time-consuming, we designed a new knowledge annotation tool to efficiently and conveniently annotate the constructed dataset.

Previously, Eric et al. [23] released a multi-turn in-vehicle dialogue dataset that spans three different tasks in the automotive personal assistant space: calendar scheduling, weather information retrieval, and point-of-interest navigation. Although each dialogue turn has knowledge annotation, it does not consider the dynamics and real-time nature of the vehicle's movement, making it unsuitable for practical scenarios. Therefore, we introduced a label replacement mechanism in the dataset construction process to substitute locations and other data in the dialogue with labels for real-time conversation during vehicle travel, and we demonstrated that this label replacement method is effective.

6 Conclusion and Future Work

This article constructs a multi-turn Chinese dialogue dataset (cockpitWOZ) with knowledge annotation in the field of smart cockpit, and a knowledge graph based on the smart cockpit scenario, and applies them to improve the baseline model. The triple knowledge is introduced into the retrieval-based BERT model, the generation-based Seq2Seq model, and the HRED model to prove the efficiency of knowledge introduction. The experimental results show that introducing triple knowledge will improve the model in terms of perplexity, accuracy, and fluency. Among them, the BERT model with triple knowledge has the highest response accuracy. In addition, location, place name, and festival information in the dialogue are replaced with labels for calling interfaces in actual scenarios to help with real-life applications for the system.

Currently, our work still has some limitations including: 1) Due to the high cost of annotating dialogue datasets, dialogue datasets with knowledge annotation are few, which limits the research of dialogue systems in the field of smart cockpit. Future works can consider data augmentation or small samples and zero-shot learning. 2) There are currently no unified evaluation metrics for smart cockpit dialogue systems. Researchers can build more comprehensive evaluation metrics. 3) With the rapid development of large language models, future work will be fine-tuned based on large language models.

A Appendix: Figure from "A Situation Knowledge Graph Construction Mechanism with Context-Aware Services for Smart Cockpit (submit to APWEB-WAIM2023)"

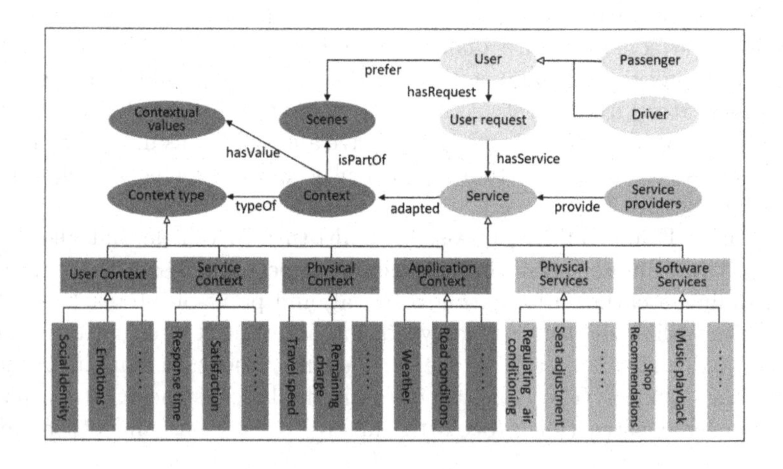

Fig. 6. Smart cockpit ontology model.

References

1. Sun, X., Chen, H., Shi, J., Guo, W., Li, J.: From HMI to HRI: human-vehicle interaction design for smart cockpit. In: Kurosu, M. (ed.) HCI 2018, Part II. LNCS, vol. 10902, pp. 440–454. Springer, Cham (2018). https://doi.org/10.1007/978-3-319-91244-8_35
2. Zhao, Y.J., Li, Y.L., Lin, M.: A review of the research on dialogue management of task-oriented systems. J. Phys. Conf. Ser. **1267**, 012025 (2019)
3. Zhou, K., Zhao, W.X., Bian, S., Zhou, Y., Wen, J.R., Yu, J.: Improving conversational recommender systems via knowledge graph based semantic fusion. In: Proceedings of the 26th ACM SIGKDD International Conference on Knowledge Discovery & Data Mining, pp. 1006–1014 (2020)
4. Zhou, H., Young, T., Huang, M., Zhao, H., Xu, J., Zhu, X.: Commonsense knowledge aware conversation generation with graph attention. In: IJCAI, pp. 4623–4629 (2018)
5. Ghazvininejad, M., et al.: A knowledge-grounded neural conversation model. In: Proceedings of the AAAI Conference on Artificial Intelligence, vol. 32 (2018)
6. Zhou, H., Zheng, C., Huang, K., Huang, M., Zhu, X.: KdConv: a Chinese multi-domain dialogue dataset towards multi-turn knowledge-driven conversation. arXiv preprint arXiv:2004.04100 (2020)
7. Xu, J., Wang, H., Niu, Z., Wu, H., Che, W.: Knowledge graph grounded goal planning for open-domain conversation generation. In: Proceedings of the AAAI Conference on Artificial Intelligence, vol. 34, pp. 9338–9345 (2020)
8. Yang, S., Zhang, R., Erfani, S.: GraphDialog: integrating graph knowledge into end-to-end task-oriented dialogue systems. arXiv preprint arXiv:2010.01447 (2020)
9. Martin, D., et al.: OWL-S: semantic markup for web services. W3C Member Submission **22**(4) (2004)
10. Devlin, J., Chang, M.W., Lee, K., Toutanova, K.: BERT: pre-training of deep bidirectional transformers for language understanding. arXiv preprint arXiv:1810.04805 (2018)
11. Sutskever, I., Vinyals, O., Le, Q.V.: Sequence to sequence learning with neural networks. Adv. Neural Inf. Process. Syst. **27** (2014)
12. Boutaba, R., et al.: A comprehensive survey on machine learning for networking: evolution, applications and research opportunities. J. Internet Serv. Appl. **9**(1), 1–99 (2018)
13. Papineni, K., Roukos, S., Ward, T., Zhu, W.J.: BLEU: a method for automatic evaluation of machine translation. In: Proceedings of the 40th Annual Meeting of the Association for Computational Linguistics, pp. 311–318 (2002)
14. Li, J., Galley, M., Brockett, C., Gao, J., Dolan, B.: A diversity-promoting objective function for neural conversation models. arXiv preprint arXiv:1510.03055 (2015)
15. Jelinek, F., Mercer, R.L., Bahl, L.R., Baker, J.K.: Perplexity-a measure of the difficulty of speech recognition tasks. J. Acoust. Soc. Am. **62**(S1), S63–S63 (1977)
16. Zhou, K., Prabhumoye, S., Black, A.W.: A dataset for document grounded conversations. arXiv preprint arXiv:1809.07358 (2018)
17. Dinan, E., Roller, S., Shuster, K., Fan, A., Auli, M., Weston, J.: Wizard of wikipedia: knowledge-powered conversational agents. arXiv preprint arXiv:1811.01241 (2018)
18. Hayati, S.A., Kang, D., Zhu, Q., Shi, W., Yu, Z.: Inspired: toward sociable recommendation dialog systems. arXiv preprint arXiv:2009.14306 (2020)

19. Moon, S., Shah, P., Kumar, A., Subba, R.: OpenDialKG: explainable conversational reasoning with attention-based walks over knowledge graphs. In: Proceedings of the 57th Annual Meeting of the Association for Computational Linguistics, pp. 845–854 (2019)
20. Feng, S., Fadnis, K., Liao, Q.V., Lastras, L.A.: DOC2DIAL: a framework for dialogue composition grounded in documents. In: Proceedings of the AAAI Conference on Artificial Intelligence, vol. 34, pp. 13604–13605 (2020)
21. Guu, K., Lee, K., Tung, Z., Pasupat, P., Chang, M.: Retrieval augmented language model pre-training. In: International Conference on Machine Learning, pp. 3929–3938. PMLR (2020)
22. Li, Y., Hayati, S.A., Shi, W., Yu, Z.: DEUX: an attribute-guided framework for sociable recommendation dialog systems. arXiv preprint arXiv:2105.00825 (2021)
23. Eric, M., Manning, C.D.: Key-value retrieval networks for task-oriented dialogue. arXiv preprint arXiv:1705.05414 (2017)

MEOM: Memory-Efficient Online Meta-recommender for Cold-Start Recommendation

Yan Luo and Ruoqian Zhang(✉)

School of Compute Science and Technology, Soochow University, Suzhou, China
20214227028@stu.suda.edu.cn, rqzhang@suda.edu.cn

Abstract. Online recommender systems aim to provide timely recommended results by constantly updating the model with new interactions. However, existing methods require sufficient personal data and fail to accurately support online recommendations for cold-start users. Although the state-of-the-art method adopts meta-learning to solve this problem, it requires to recall previously seen data for model update, which is impractical due to linearly increasing memory over time. In this paper, we propose a memory-efficient online meta-recommender MEOM that can avoid the explicit use of historical data while achieving high accuracy. The recommender adopts MAML as a meta-learner and particularly adapts it to online scenarios with effective regularization. Specifically, an online regularization method is designed to summarize the time-varying model and historical task gradients, such that overall model optimization direction can be acquired to parameterize a meaningful regularizer as a penalty for next round model update. The regularizer is then utilized to guide model updates with prior knowledge in a memory-efficient and accurate manner. Besides, to avoid task-overfitting, an adaptive learning rate strategy is adopted to control model adaptation by more suitable learning rates in dual levels. Experimental results on two real-world datasets show that our method can significantly reduce memory consumption while keeping accuracy performance.

Keywords: Online Recommendation · Meta Learning · Cold Start

1 Introduction

Recommender systems have been widely considered to be effective in overcoming information overload [5,12,23]. In real-world recommender systems, tremendous amounts of interaction data are sequentially generated at high velocity [2]. Since new interactions provide the latest evidence of user preferences [18,23], it is necessary to continuously update the model with new interactions, so as to provide timely personalized recommendations and avoid being stale.

The problem of online recommendation has aroused wide concern [2,20] recently, and several methods have been proposed to deploy recommender systems into online settings, such as SML [23] and SSGAN [12]. Despite of this,

X. Song et al. (Eds.): APWeb-WAIM 2023, LNCS 14332, pp. 331–346, 2024.
https://doi.org/10.1007/978-981-97-2390-4_23

these online recommendation methods always require sufficient interactions of a target user in modelling her complex preferences. Unfortunately, as indicated by [13,15,17], the majority of users in real-world recommendation systems have little historical data, preventing existing online recommendation methods [8,18,19] to provide precise recommendations for cold-start users. It thus calls for improved solutions that can well support cold-start recommendations with the ability of efficient model updates in online scenarios.

Recent years have witnessed extensive research on cold-start recommendations offline and online. They adopt meta-learning such as MAML [7] to successfully overcome the cold-start problem in recommendation systems. The meta-learned recommenders [5,9] can learn well-generalized global parameters, providing improved initialization for cold-start recommendation. Despite of this, the above meta-learned methods are offline models without continuous updates, and thus cannot support online recommendations. Recently, state-of-the-art method FORM [16] proposes an online meta-learner for recommendations, which supports continuous model updates under the meta-learning framework. Unfortunately, this method requires recalling historical data to form the training set for model updates, making it difficult to be long-termly employed in online scenarios. Consequently, to provide accurate recommendations and avoid explicitly recalling previously seen data for model updates, a memory-efficient meta-recommender is required.

However, we face several technical challenges when designing memory-efficient online meta-recommenders. Firstly, existing meta-recommenders [5,9,16] are weak in maintaining long-term knowledge, resulting in the explicit use of historical data for model updates. This requires highly expensive memory and I/O cost, as the data size increases over time. It thus requires to figure out the overall optimization direction by summarizing useful experiences of past updates, so that online models can be updated without recalling historical data while keeping recommendation accuracy. Secondly, existing meta-learning based methods [2,9,22] apply a fixed learning rate for model updates, which is hard to learn adequately from users (or interactions) with rich information. In fact, the model should pay more attention to data-rich users in global updates and interactions with high prediction loss in local updates. Thus, an adaptive learning rate strategy needs to be designed for model updates with more suitable learning rates. Besides, existing meta-recommenders [3] are unable to encode users' preferences included in interactions of cold-start users that are helpful for the recommendation. Whereas, existing methods cannot simultaneously solve the above challenges well.

To address the aforementioned challenges, we develop a memory-efficient online meta-recommender MEOM[1] which takes advantage of a meaningful regularizer to avoid buffering historical data while keeping recommendation accuracy. We first design an online regularization method to summarize the time-varying model and historical task gradients which are used to parameterize a meaningful regularizer as a penalty for the next round model update, thereby the overall

[1] Our code is available at: https://anonymous.4open.science/r/MEOM-master-4239,

model optimization direction can be acquired from the regularizer. Secondly, dual-level adaptive learning rates are introduced to avoid task overfitting. We adopt a local learning rate adaptation prediction losses in the local update and a global learning rate adaptation in the number of user interactions in the global update. As a result, the model can provide fast adaptation by more suitable learning rates. In the end, we adopt cluster-based preference modelling that relieves cold-start problems by transferring dynamic preferences among similar users. The main contributions of this work are summarized as follows:

- We propose a memory-efficient online meta-recommender (MEOM), which is the first to address cold-start problems without a buffer in online scenarios.
- We introduce an online regularization method to acquire the overall model optimization direction. It summarizes the time-varying model and historical task gradients to parameterize the regularizer as a penalty for next round model updates. The regularizer contains prior knowledge to carefully guide model updates without recalling historical data while maintaining accuracy.
- We adopt dual-level adaptive learning rate to control model adaptation by more suitable learning rates, and design cluster-based preference modelling to transfer dynamic preferences among similar users.
- We conduct experiments on two datasets and show that MEOM improves memory footprint and keeps performance compared with existing baselines.

2 Related Work

2.1 Online Recommendation

A common challenge for recommender systems is how to efficiently and effectively update the model in online scenarios where massive data arrives at a high speed. A simple way is to update the model on both all historical and new data [18] because it can capture both long-term and short-term user preferences to avoid forgetting and being stale, whereas they would increase massive training time and memory [2,6]. To overcome this drawback, some reservoir-based methods are proposed to store sampling historical data which helps the model maintain long-term preferences. For instance, SPMF [19] updates the model based on both randomly sampled historical data stored in the reservoir and new data, it is prone to overfitting and performs worse than updating with all historical data. To avoid forgetting and being stale, several methods have been proposed. [8,23] transfer the old model into the new model (update with new data only) so that the model recaptures short-term preferences while maintaining long-term preferences.

2.2 Cold-Start Recommendation

Alleviating the cold-start problem has been widely studied recently, some solutions have been proposed, and they can be categorized into two major types.

Extra-Information-Assisted Methods. The first type is extra-information-assisted methods, [17] regard content as auxiliary information to learn factors including preferences when addressing cold-start problems. Meanwhile, [11,25] introduce preferences in other domains into the target cold-start domain, but they all have privacy issues. To overcome the issues, [13] treat interactions of similar warm users as the data of target cold users to convert cold-start to warm-start.

Few-Shot Learning Methods. The other type is the few-shot learning method. [10] captures the knowledge of data-rich users (or items) from one domain to a related domain alleviating cold-start problems by transfer learning. Furthermore, some methods adopt meta-learning to alleviate the cold-start problem. [5,9,22] regard the recommendation for a user or item as one task and train static global parameters. Then, cold-start tasks are fine-tuned with a few interactions based on the initialization of global parameters. Besides, [14,16] transfer meta-based models into the online setting by introducing a task-gradient-based regularization to avoid overfitting, when training the global parameter. However, these meta-learning-based methods need to buffer all the data collected so far.

In summary, limited by offline training sets or buffering increasing historical data, these methods cannot simultaneously solve the cold-start problem and memory-efficiently update the model in the online setting directly.

3 Problem Definition

3.1 Problem Formulation

We formulate the cold-start recommendation problem and then extend it to the online setting. Given a user $u \in U$ with profile p_u and item $i \in I$ with profile p_i, the interaction between them is denoted as score $y_{u,i}$. Our goal is to predict the unknown score $y_{u,i}$ between cold-start user u and item i. Next, in online settings, the streaming user-item interactions are represented as $\{D_0, ..., D_{t-1}, D_t, D_{t+1}, ...\}$, where D_t denotes user-item interactions newly collected at time period t. We aim to update the current model $\boldsymbol{\theta}_t$ with new data D_t to $\boldsymbol{\theta}_{t+1}$ which serves for near feature recommendation. The model update procedure is detailed in Sect. 4.3.

3.2 Meta-learning Setting

Under the meta-learning framework, we consider the recommendation for a user as a learning task. We set cold-start users as test tasks U^{test} for meta-testing and other users as train tasks U^{train} for meta-training. For each user $u \in U^{train} \cup U^{test}$, the rated items I_u of u will be divided into support sets I_u^S and query sets I_u^Q, which are used to locally and globally update the model.

4 Our Proposed Method

4.1 Overview of MEOM

In this paper, we propose a memory-efficient online meta-recommender MEOM to support accurate online cold-start recommendations without buffering historical data. The framework of MEOM is shown in Fig. 1. We first introduce the base recommender model in Sect. 4.2, then design an online regularization method to update global parameters in a memory-efficient and accurate manner in Sect. 4.3. To serve more accurate recommendations of cold-start users, we further adopt dual-level adaptive learning rates to control model adaptation in Sect. 4.4, and cluster-based preference modelling is detailed in Sect. 4.5.

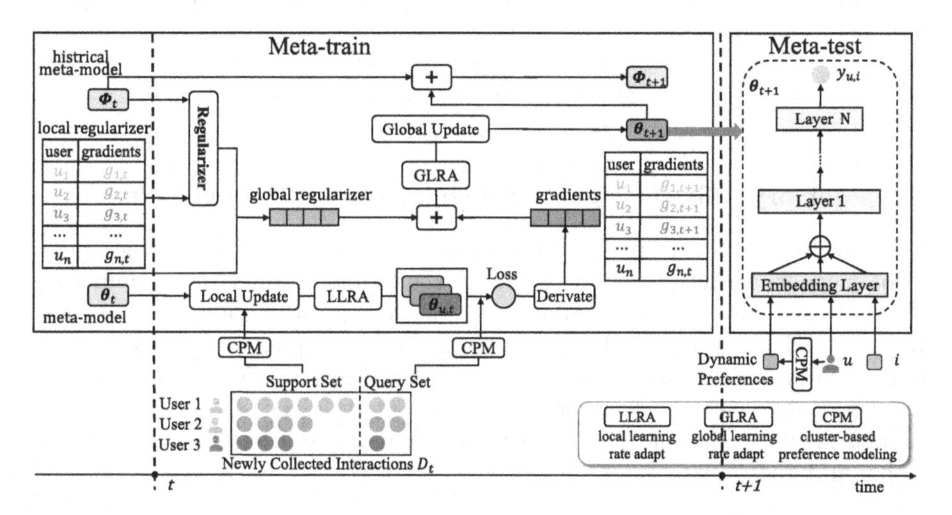

Fig. 1. The overview of the MEOM framework. During period t, MEOM conducts two processes. In meta-train, a new dataset D_t is used to update the meta-model, it consists of two parts: local update and regularizer-based global update. In meta-test, $\boldsymbol{\theta}_{t+1}$ is used as initial parameters of user models training to recommend.

4.2 Base Recommender Model

We use the classical meta-learner MAML [7] as our base recommender. The embedding vectors of users and items are learned based on their profile $p_u \in \mathbb{R}^{d_u}$ and $p_i \in \mathbb{R}^{d_i}$. In addition, we utilize the embedding of user's preferences $c_u \in \mathbb{R}^{d_i}$ from the preference modelling as auxiliary information. Furthermore, multi-layered perception (MLP) is adopted to learn embedding vectors of the user, item and preferences $e_u, e_i, e_{c_u} \in \mathbb{R}^{d_e}$:

$$e_u = f_{\omega_u}(p_u), e_i = f_{\omega_i}(p_i), e_{c_u} = f_{\omega_c}(c_u) \tag{1}$$

where f denotes MLP parameterized by $\boldsymbol{\omega}$; d_u and d_i represent the dimensions of the user profile vector and item profile vector; d_e is the embedding size. Our base recommender is a multilayered neural network, which is widely applied to recent recommendation researches [5,9,16,21]. The score between u and i is predicted as:

$$y_{u,i} = f_{\boldsymbol{\theta}}(e_u, e_i, e_{c_u}) = f_{\boldsymbol{\theta}}(e_u \oplus e_i \oplus e_{c_u}) \quad (2)$$

where f indicates the multilayered neural network parameterised by θ and $e_u \oplus e_i \oplus e_{c_u}$ is the concatenation of user embedding, item embedding and user's preferences embedding.

There are two processes to update the meta-based recommender: local update and global update. In the local update, the user model is trained on the support set I_u^S with the purpose of minimising the loss of predictions, as $\boldsymbol{\theta}_u = \arg\min_{\boldsymbol{\theta}_u} L_u(f_{\boldsymbol{\theta}_u}, I_u^S) = \boldsymbol{\theta}_u - \alpha \nabla L_u(f_{\boldsymbol{\theta}_u}, I_u^S)$, where the L is the loss function, and α is the local learning rate to u. In each training batch, the global parameter is updated as:

$$\widetilde{\boldsymbol{\theta}} = \boldsymbol{\theta} - \beta \sum_{u \in U} \nabla L_u(f_{\boldsymbol{\theta}}, I_u^Q) \quad (3)$$

4.3 Memory-Efficient Update for Online Meta-Recommender

Inspired by the idea in [1], to achieve a memory-efficient online meta-recommender, we introduce an online regularization method that summarizes the optimization direction on all data so far by explicitly storing both a time-varying model and historical task gradients. Based on prior knowledge, a regularizer is employed to guide model updates in a memory-efficient and accurate manner. Specifically, in each period, we conduct two processes to achieve continuous memory-efficient parameters update in online settings: local update and global update.

Online Local Update. Same as meta-learning based approaches [9,16], we take MAML [7] as a meta-learner to learn common knowledge of previous tasks.

During the period t, we use the newly collected dataset D_t to update the parameters. For each $u \in U^{train}$, we first query u's preferences c_u of u through CPM. The task model's parameters are initialized by the global parameters $\boldsymbol{\theta}_t$, and then trained on the support set I_u^S and c_u with some gradient descent steps to obtain task-specific model $\boldsymbol{\theta}_u$, which is optimized as:

$$\boldsymbol{\theta}_u = \boldsymbol{\theta}_u - \alpha_u \nabla L_u(f_{\boldsymbol{\theta}_u}, I_u^S) \quad (4)$$

where L is loss function calculated as $L_u(f_{\boldsymbol{\theta}_u}, I_u^S) = \frac{1}{|I_u^S|} \sum_{i \in I_u^S} (\hat{y}_{u,i} - y_{u,i})^2$. And α_u is the local learning rate of u, Sect. 4.4 is referred for details.

Online Global Update. To avoid stale and maintain long-term preferences, previous methods try to modify the model with all previously seen data which is impractical in terms of memory consumption. A simple way is to update the model with new data, but this can easily lead to overfitting. Hence, a memory-efficient method is required to update the meta-recommender online.

In greater detail, we update global parameters with locally updated task models on their query sets during the period $t < T$. Unfortunately, all losses have not been revealed and the gradients on only new data may be a local optimization direction for global parameters. As such we need to correct the direction to avoid overfitting by exploiting previously seen losses. The goal of updating process is to optimize the performance of global parameters $\boldsymbol{\theta}$ (initial parameters) across all tasks:

$$\boldsymbol{\theta}_{t+1} = \arg\min_{\boldsymbol{\theta}} \sum_{u \in U_t} L_u(f_{\boldsymbol{\theta}}, I_u^Q) + R_t \tag{5}$$

where R_t is the regular used to debias and correct the local optimal direction.

Inspired by the idea in [16,18], we design an online regularization method to summarize the historical meta-model and historical task gradients, which encode previous task experiences and parameterize a meaningful regularizer as a penalty for next round model update to avoid forgetting. The key of the method is that the model contains knowledge of training data and gradients indicate the optimization direction of data the model was trained on. Based on these views, the historical data can be implicitly leveraged by storing both a historical meta-model $\boldsymbol{\phi}$ and historical task gradients g, which are continuously refreshed after the global parameters are converged.

Parameters Update. To better correct the gradients g' obtained by the derivative of loss on new data to the model optimized direction on all data seen so far, we utilize a meaningful regularizer as a penalty to get involved in each round of model update, which is parameterized by both the time-varying model $\boldsymbol{\phi}$ and historical task gradients g as follows:

$$R_t(\boldsymbol{\theta}) = \gamma_1 \|\boldsymbol{\theta}\|_1 + \frac{\gamma_2}{2} \|\boldsymbol{\theta} - \boldsymbol{\phi}_t\|_2^2 - g_t \tag{6}$$

where γ_1 and γ_2 are the regular term coefficients. The first term makes the parameters more sparse which is conducive to fast online updates and the second term guides the parameter to optimize in the overall losses descent direction. The third negative term is used to reduce the learning of the parameters on new data.

Therefore, the global parameters are memory-efficiently updated with gradient corrections of the regularizer penalty on the query set. In this way, the global parameters can effectively avoid the overfitting problem caused by updating the model without explicitly recalling historical data. The regularized optimization direction across all training tasks $u \in U$ can be used to update $\boldsymbol{\theta}$:

$$\boldsymbol{\theta}_{t+1} = \boldsymbol{\theta}_t - \beta_{U_t} \{ \sum_{u \in U_t} \nabla L_t(f_{\boldsymbol{\theta}_t}, I_u^Q) + R_t(\boldsymbol{\theta}_t) \} \tag{7}$$

where U_t are tasks in newly collected dataset D_t, and β_{U_t} is the adaptive global learning rate detailed in Sect. 4.4. The overall model optimization direction can be acquired after penalising the loss function with the regularizer term $R_t(\boldsymbol{\theta})$.

Regularizer Update. After $\boldsymbol{\theta}_{t+1}$ converges, the time-varying model $\boldsymbol{\phi}$ and historical task gradients g are also updated with current loss, so as to capture all prior knowledge and get involved in the next round model update to ensure overall model optimization direction. To effectively adjust the direction of model optimization, we directly acquire the optimized direction of new data from the experiences learned by the parameters to refresh the historical task gradients g:

$$g_{t+1} = g_t - \gamma_2(\boldsymbol{\theta}_{t+1} - \boldsymbol{\phi}_t) \tag{8}$$

Besides, to ensure the model $\boldsymbol{\phi}$ is fresh and contains knowledge of all previously seen data, the new parameters $\boldsymbol{\theta}_{t+1}$ updated with the new historical task gradients g_{t+1} is utilized to modify $\boldsymbol{\phi}$, so as to ensure that $\boldsymbol{\phi}$ pays more attention to overall data experiences. $\boldsymbol{\phi}$ is modified by aggregation operation:

$$\boldsymbol{\phi}_{t+1} = \frac{1}{2}(\boldsymbol{\phi}_t + \boldsymbol{\theta}_{t+1} - \gamma_2 g_{t+1}) \tag{9}$$

where γ_2 is a learning rate for modifying historical task gradients g and $\boldsymbol{\phi}$.

During the meta-test phase, the new global parameters $\boldsymbol{\theta}_{t+1}$ will be used as initial parameters of task models. We recommend new items to cold-start users and collect their feedback. Then we fine-tune the model with such feedback by a small amount of gradient descent steps until the task-specific model converges.

4.4 Dual-Level Adaptive Learning Rate for Meta-recommender

Apart from the aforementioned methods, to provide more efficient recommendations for cold-start users, MEOM adopts dual-level adaptive learning rates.

User-Level Adaptive Learning Rate. To ensure adequate learning of the task, a user-level adaptive learning rate is used to control model adaptation in local update by a more suitable learning rate.

Interactions with high losses indicate that the model learned less of it. Thus, the model effectively captures preferences contained in interactions by attaching importance to them. However, the variance of the loss represents the dispersion of the interactions, hence the model should reduce the focus on these interactions. We consider both the mean and variance of the loss to design an adaptive learning rate in the local update that determines how much the task model should be updated in response to the interactions. The local learning rate of u is adapted as:

$$\alpha_u = LLRA(\alpha, \{\ell_i\}_{i=1}^N) = (1 - \frac{1}{1+\varepsilon_u})\alpha, \quad \varepsilon_u = \frac{E(\{\ell_u\}_{i=1}^N)}{Var(\{\ell_u\}_{i=1}^N)} \tag{10}$$

where α is a shared local learning rate; ℓ is the loss of task-specific model on its support set; E and Var are the mean and variance of batch losses. Thereby, the lower learning size is correlated with the task having lower loss.

Model-Level Adaptive Learning Rate. MEOM uses training batch tasks $U \in U^{train}$ to update the global parameters. Inspired by the idea in [2], we take the number of users' learning shots into consideration to update the global parameters, as the users' preferences are contained in their interactions.

The global parameters should pay more attention to users with rich interactions, since this represents more information about the user. Furthermore, the user with richer data means that it is more informative, thus the global parameters should increase the focus on these users. Consequently, we consider the amount of a user's interactions I_u to design an adaptive global learning rate that determines how much the global parameters should be updated in response to the user. And the global learning rate of a user u is adapted as:

$$\beta_u = GLRA(\beta, I_u) = (1 - \frac{1}{1 + |I_u|})\beta \tag{11}$$

where β is a shared global learning rate, and $|I_u|$ is the amount of it.

4.5 Cluster-Based Preference Modelling

To enrich the preferences of cold-start users, we adopt cluster-based preference modelling (CPM) to transfer the dynamic preferences of similar users.

Users' dynamic preferences are included in user-item interactions. Considering that similar users have shared preferences, we apply soft-clustering to cluster users based on their profile vector $p_u \in \mathbb{R}^{d_u}$ to query the dynamic preferences of similar users, taking it as auxiliary preferences of cold-start users as features for the recommendation. Here we follow the traditional soft-clustering, p_u is viewed as query vectors for the clusters. Next, the soft probability vector $\pi_u^k \in \mathbb{R}^{d_i}$ is computed by calculating the distance from the query vector to the learned centre of each cluster $\{\rho^k\}_{k=1}^K$ as $\pi_u^k = \text{softmax}(\langle q_u, \rho^k \rangle)$, where $\langle \cdot, \cdot \rangle$ denotes the distance of two vectors. To aggregate dynamic preferences $dp_u \in \mathbb{R}^{d_i}$ of similar users, we weigh user-interacted items by the user's score of it:

$$dp_u = \sum_{i=1}^{N} w_i p_i, w_i = \text{softmax}(y_{u,i}) = \frac{\exp(y_{u,i})}{\sum_{j=1}^{N} \exp(y_{u,j})} \tag{12}$$

where p_i denotes the profile vector of u interacted item i, and N is the number of u interacted items. The softmax function is used to normalize the score in $[0, 1]$, and $\sum_{i=1}^{N} w_i = 1$. After obtaining each user's dynamic preferences vector dp_u and probability vectors $\{\pi_u^k\}_{k=1}^K$, we compute the dynamic preferences center dp^k of cluster C_k by the weighting average as $dp^k = \sum_{u \in C_k} \pi_u^k \cdot dp_u$. Finally, the dynamic representation c_u of user u which contains shared preferences from similar users formulated with the learned dynamic preferences centres $\{dp^k\}_{k=1}^K$

$$c_u = \sum_{k=1}^{K} \pi_u^k \cdot dp^k \tag{13}$$

where c_u will be one of the inputs to the model, details are in Eq. 2. The CPM is updated with newly collected data after each round of parameter convergence.

5 Experiments

5.1 Datasets

To evaluate the effectiveness of MEOM, we conduct experiments with two real-world datasets: Movielens 1M[2] and Yelp[3]. (1) **Movielens 1M:** Movielens contains interactions between users and items over three years. (2) **Yelp:** Yelp collects interactions between users and items such as supermarkets and restaurants. For both datasets, to simulate online scenarios where the interactions arrive sequentially, we sort the interactions by timestamp. Then divide the MovieLens and Yelp dataset into 45 and 38 periods respectively, with each time period having an equal amount of interaction. We further split the periods of each dataset into offline/online training sets: for Movielens the ratio is 25/20 and for Yelp the ratio is 20/18. We partition cold-start users who have fewer than 5 interactions so far and used to test the model, the rest compose training sets. The interactions of a user during each period are divided into support sets and query sets with a ratio 80:20. The characteristics of datasets are summarized in Table 1.

Table 1. Basic statistics of two datasets.

Datasets	#User	#Item	#Interactions	Sparsity
Movielens	6,040	3,952	1,000,209	95.81%
Yelp	122,816	59082	3,014,421	94.67%

5.2 Evaluation Settings

To evaluate the performance of our proposed model, we show experimental results under two evaluations: Mean Absolute Error (MAE) and top-N Normalized Discounted Cumulative Gain ($NDCG@N$). N is set to 5 and 10. For fair comparisons, all baselines adopt the same base recommender. The embedding size of all vectors and the layers of the base model is set to 32 and 3 of all methods. We fine-tune in advance to get optimal hyper-parameters of baselines. In our proposed model, we set the batch size to 64, and the epoch is set to 28. The local and the global learning rate α and β are set to 1$e-$6 and 1$e-$5 for Movielens and to 1$e-$5 and 1$e-$4 for Yelp. The regular term coefficients γ_1, γ_2 is set as 1$e-$5, 1$e-$5. Furthermore, we tune the hyper-parameter: the clusters number K of preference modelling in $\{5, 6, ..., 20\}$.

5.3 Baseline Methods

We compare our proposed model with the following baselines, and they can be classified into three categories: online general methods (SPMF, SML, MeLON),

[2] https://grouplens.org/datasets/movielens/.
[3] https://www.yelp.com/dataset/.

offline cold-start methods (Wide&Deep, DIN, MeLU, MAMO) and online cold-start method (FORM). Particularly, we adapt all offline cold-start methods to online scenarios by periodical full-update.

- **SPMF** [19]**:** Reservoir-based methods. It updates the model with new and historical data sampled in the reservoir.
- **SML** [23]**:** This method introduces a convolutional neural network meta-model to combine the last recommender-model with the new recommender-model, which are trained on historical data and new data respectively.
- **MeLON** [8]**:** It learns a meta-model both on graph attention network and multi-layer perceptron to reweight the importance of each interaction to each parameter when updating the recommender model.
- **Wide & Deep** [4]**:** This method recommend for users by jointly training a wide linear model and deep neural networks to extract users' preferences.
- **DIN** [24]**:** Deep Interest Network. It provides recommendations by encoding users' preferences considering the relevance of historical behaviours.
- **MeLU** [9]**:** It studies a meta-model on all users, which can initialize the user-specific-model and fast adapt to provide personalized recommendations.
- **MAMO** [5]**:** Meta-recommender. It designs a memory matrix based on user features by personalizing user embedding to provide personal initialization.
- **FORM** [16]**:** This is a state-of-art meta-learning based methods for cold-start recommendation. It requires to buffer all historical data to form the training set and uses an online regularization to update the model.

5.4 Experimental Results

Performance Comparison. To validate the effectiveness of our method, we compare the performance of user cold-start recommendations over all baselines. The results are shown in Table 2.

We can see that our proposed model achieves the best performance compared to other baselines on two benchmark datasets, even without buffering historical data. This is due to the effectiveness of the regularizer that summarizes prior knowledge to participate in model updates and the cluster-based preference modelling enriching the cold-start users' preferences. Besides, the online general methods obtain lower performances than other methods due to the need for sufficient data for learning user preferences. All offline methods show superior performance than online general methods for the reason that they adopt few-shot learning strategy to alleviate the cold-start problem.

In particular, MEOM outperforms state-of-the-art FORM by 1.29% and 0.87% on average, which indicates that our method successfully summarizes previous tasks experiences in a regularizer for model update and provides more useful information to the recommender than sampling tasks. The best offline method MAMO performs lower than MEOM by 3.48% and 4.76% probably due to MAMO only using prediction losses to update the model. The best online method MeLON is worse than our method by 10.28% and 6.67%, because it can

Table 2. Experiment Results on Movielens and Yelp.

Model	Movielens-1M			Yelp		
	NDCG@5	NDCG@10	MAE	NDCG@5	NDCG@10	MAE
SPMF	0.4805	0.6610	0.9982	0.5865	0.6957	0.9663
SML	0.5649	0.7593	0.9317	0.6894	0.7509	0.9197
MeLON	0.6103	0.7746	0.9181	0.7341	0.7986	0.8920
Wide&Deep	0.6581	0.7914	0.9083	0.7609	0.8117	0.8698
DIN	0.6523	0.7851	0.9203	0.7681	0.8205	0.8633
MeLU	0.6607	0.8004	0.8976	0.7572	0.8078	0.8702
MAMO	0.6810	0.7996	0.8835	0.7596	0.8018	0.8760
FORM	0.6960	0.8214	0.8697	0.7887	0.8306	0.8385
MEOM	**0.7025**	**0.8411**	**0.8650**	**0.7941**	**0.8339**	**0.8257**

provide improved initialization for cold-start users by well-generalized parameters. The strong performance shows that MEOM is capable of adapting long-term and short-term interests by optimizing the model by the online regularizer.

Memory Cost Comparison. Consider that one key motivation of the work is to avoid recalling all previously seen data for model updates. To better show the results, we compare the number of interaction accesses in each update period to illustrate the memory cost of the best baselines (in I/O) of each category. Figure 2 shows the memory cost on Movielens and Yelp by periods.

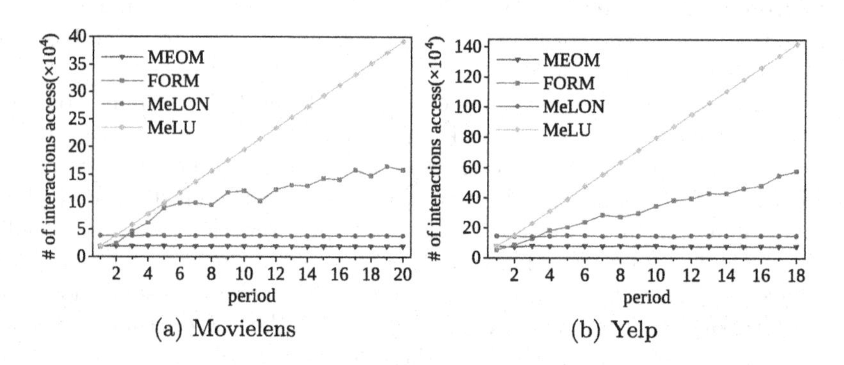

(a) Movielens (b) Yelp

Fig. 2. The number of interactions access over each time period on two datasets.

Related to the data partitioning we explained in Sect. 5.1, each period has an equal number of interactions. Our method outperforms other baselines in I/O cost because we only use new data for model updates. FORM requires more training interactions than our method, since it needs to recall all previously seen data and sample it to form training sets. The online recommendation method

MeLON access interactions more than MEOM about 18,800, as it requires to generate negative samples from items that have not interacted with to participate in model update and buffers fix-size historical interactions to measure the importance of new interactions. Offline meta-learning based method MeLU updates the model with the increasing number of interactions as it trains with all seen data to account for both long-term and short-term user preferences.

In summary, the regularizer can effectively preserve useful information without recalling previous data by time-varying model and historical task gradients.

Fast Model Update. In online settings, the model should update fast to serve for near feature recommendations. Thus, we compare the update time of users at each period on Yelp. The comparison results are shown in Table 3. We can see that our method obtains a lower cost of updating time than state-of-the-art FORM. The time cost of FORM is increasing which is caused by recalling previously seen data to update the model. The full-update method MeLU costs growing time because of the increased training data. Compared with meta-learning based methods, the online recommender MeLON is about 27 s lower than MEOM due to the need of training negative samples. Because Movielens-1M is small, there is no significant difference in update time.

Table 3. Updating time (seconds) at each training period on Yelp.

period	0	1	2	3	4	5	6	7
MeLON	75	76	75	74	75	76	77	75
MeLU	53	82	132	181	237	284	329	388
FORM	57	68	81	104	127	140	171	205
MEOM	48	49	48	49	48	47	47	46

Ablation Study. The effectiveness of MEOM comes from four key components: (1) the historical task gradient that is one term of online regularization to adjust the optimization direction of the model; (2) the time-varying model that ensures overall optimization direction of model update; (3) two adaptive learning rates that enrich the meaning of each task; (4) the preference modelling that enrich the preferences of cold-start users. To justify the designs in MEOM, we investigate the influences of each component by studying the performance of three variants:

- **MEOM-RG**, which removes the historical task gradients in regularization.
- **MEOM-RM**, which removes the time-varying model in regularization.
- **MEOM-TALR**, which removes two adaptive learning rates from local and global updates and uses fixed learning rates to replace them.
- **MEOM-CPM**, which removes the cluster-based preference modelling and predicts with only user and item profile.

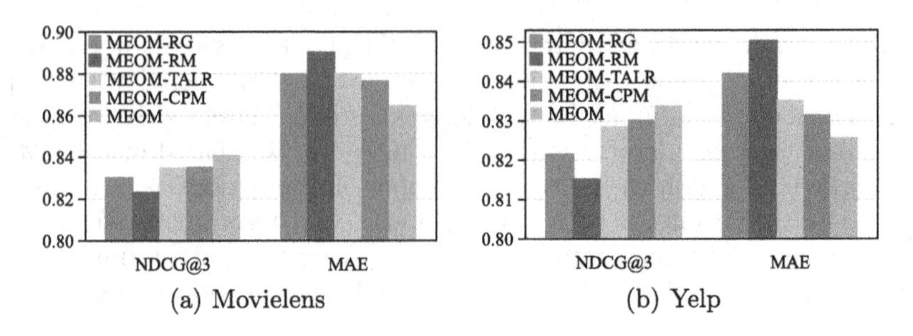

(a) Movielens (b) Yelp

Fig. 3. Recommendation performance of MEOM and its four variants.

We conduct experiments on MEOM and its four variants, and the performance is shown in Fig. 3. We have the following observation: (1) regarding the historical task gradients term, it can avoid overfitting since MEOM is better than MEOM-RG by 1.50% and 1.72% on average. (2) regarding the time-varying model term, MEOM performs better than MEOM-RM by 2.51% and 2.59% average on two datasets, which proves the effectiveness of the regularized update. (3) when adaptive learning rates are not considered, the average performance of MEOM-TALR is worse than MEOM by 1.22% and 0.89%. This signifies the benefit of adaptive learning rates. (4) when the preference modelling is not involved in the recommendation, the performance of MEOM-CPM drops by 1.01% and 0.57%. This might be caused by the lack of preferences from similar users.

Parameters Study. To illustrate the impact of embedding size and MLP layers of the base recommender, we test the performance of MEOM under the different embedding size and MLP layers. Figure 4(a) and 4(b) show the optimal MLP layers and embedding size of the recommender. We can see that the performance on Yelp is better than Movielens under the same embedding size and MLP layers because Movielens is more sparse than Yelp and the interaction of Yelp is richer.

We fix the structure of base recommender and change the number of clusters K on two datasets to test the performance of MEOM. As shown in Fig. 4(c).

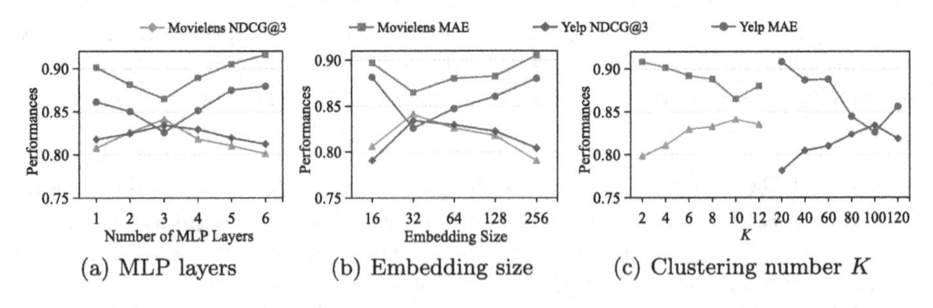

(a) MLP layers (b) Embedding size (c) Clustering number K

Fig. 4. Performance of MEOM w.r.t. different embedding size, MLP layers in the base recommender and the number of clusters K.

The optimal K on Movielens is smaller than Yelp because there are fewer users on Movielens. MEOM outperforms other baselines when K is greater than 8 on Movielens. K greater than 100 degrades performance due to overfitting on Yelp.

6 Conclusion

In this paper, we investigate the cold-start recommendation problem in an online setting and propose a novel MEOM model to tackle the encountered challenges. We propose an online regularization approach to implicitly leverage historical data by a time-varying model and historical task gradients. Based on this prior knowledge, we further utilize a regularizer as a penalty to guide model updates in a memory-efficient and accurate manner. Besides, to avoid overfitting, we apply dual-level adaptive learning rate based on learning shots and learning ability respectively. We conduct extensive experiments on two benchmark datasets that show the effectiveness of our approach in the online cold-start recommendation.

References

1. Acar, D.A.E., Zhu, R., Saligrama, V.: Memory efficient online meta learning. In: ICML 2021, pp. 32–42. PMLR (2021)
2. Al-Ghossein, M., Abdessalem, T., Barre, A.: A survey on stream-based recommender systems. ACM Comput. Surv. (CSUR) **54**(5), 1–36 (2021)
3. Chen, F., Luo, M., Dong, Z., et al.: Federated meta-learning with fast convergence and efficient communication. arXiv preprint arXiv:1802.07876 (2018)
4. Cheng, H.T., Koc, L., Harmsen, J., et al.: Wide & deep learning for recommender systems. In: DLRS 2016, pp. 7–10 (2016)
5. Dong, M., Yuan, F., Yao, L., et al.: Mamo: memory-augmented meta-optimization for cold-start recommendation. In: SIGKDD 2020, pp. 688–697 (2020)
6. Du, Z., Wang, X., Yang, H., et al.: Sequential scenario-specific meta learner for online recommendation. In: SIGKDD 2019, pp. 2895–2904 (2019)
7. Finn, C., Abbeel, P., Levine, S.: Model-agnostic meta-learning for fast adaptation of deep networks. In: ICML 2017, pp. 1126–1135. PMLR (2017)
8. Kim, M., Song, H., Shin, Y., et al.: Meta-learning for online update of recommender systems. arXiv preprint arXiv:2203.10354 (2022)
9. Lee, H., Im, J., Jang, S., et al.: MeLU: meta-learned user preference estimator for cold-start recommendation. In: SIGKDD 2019, pp. 1073–1082 (2019)
10. Li, J., Jing, M., Lu, K., et al.: From zero-shot learning to cold-start recommendation. In: AAAI 2019, vol. 33, pp. 4189–4196 (2019)
11. Li, Y., Xu, J., et al.: ATLRec: an attentional adversarial transfer learning network for cross-domain recommendation. J. Comput. Sci. Technol. **35**(4), 794–808 (2020)
12. Lv, Y., Xu, J., Zhou, R., Fang, J., Liu, C.: SSRGAN: a generative adversarial network for streaming sequential recommendation. In: Jensen, C.S., et al. (eds.) DASFAA 2021. LNCS, vol. 12683, pp. 36–52. Springer, Cham (2021). https://doi.org/10.1007/978-3-030-73200-4_3
13. Song, J., Xu, J., Zhou, R., et al.: CBML: a cluster-based meta-learning model for session-based recommendation. In: CIKM, pp. 1713–1722 (2021)

14. Sun, H., Xu, J., Zheng, K., et al.: MFNP: a meta-optimized model for few-shot next POI recommendation. In: Zhou, Z. (ed.) IJCAI 2021, pp. 3017–3023 (2021)
15. Sun, H., Xu, J., Zhou, R., et al.: HOPE: a hybrid deep neural model for out-of-town next POI recommendation. WWW **24**(5), 1749–1768 (2021)
16. Sun, X., Shi, T., Gao, X., et al.: FORM: follow the online regularized meta-leader for cold-start recommendation. In: SIGIR 2021, pp. 1177–1186 (2021)
17. Wang, L., Jin, B., Huang, Z., et al.: Preference-adaptive meta-learning for cold-start recommendation. In: IJCAI 2021, pp. 1607–1614 (2021)
18. Wang, Q., Yin, H., Hu, Z., et al.: Neural memory streaming recommender networks with adversarial training. In: SIGKDD 2018, pp. 2467–2475 (2018)
19. Wang, W., Yin, H., et al.: Streaming ranking based recommender systems. In: SIGIR 2018, pp. 525–534 (2018)
20. Xu, J., Song, J., Sang, Y., et al.: CDAML: a cluster-based domain adaptive meta-learning model for cross domain recommendation. WWW **26**(3), 989–1003 (2023)
21. Yu, B., Li, X., Fang, J., et al.: Memory-augmented meta-learning framework for session-based target behavior recommendation. WWW **26**(1), 233–251 (2023)
22. Yu, R., Gong, Y., He, X., et al.: Personalized adaptive meta learning for cold-start user preference prediction. In: AAAI 2021, vol. 35, pp. 10772–10780 (2021)
23. Zhang, Y., Feng, et al.: How to retrain recommender system? A sequential meta-learning method. In: SIGIR 2020, pp. 1479–1488 (2020)
24. Zhou, G., Zhu, X., Song, C., et al.: Deep interest network for click-through rate prediction. In: SIGKDD 2018, pp. 1059–1068 (2018)
25. Zhu, F., Wang, Y., Chen, C., et al.: Cross-domain recommendation: challenges, progress, and prospects. arXiv preprint arXiv:2103.01696 (2021)

A Social Bot Detection Method Using Multi-features Fusion and Model Optimization Strategy

Xiaohui Huang[1], Shudong Li[1,2](✉), Weihong Han[2], Shumei Li[3], Yanchen Xu[4], and Zikang Liu[5]

[1] Cyberspace Institute of Advanced Technology, Guangzhou University, Guangzhou 510006, China
hxh@e.gzhu.edu.cn

[2] Department of New Networks, Peng Cheng Laboratory, Shenzhen 518055, China
lishudong@gzhu.edu.cn

[3] Computer College, Jilin Normal University, Siping 136000, China

[4] School of Educational Science, LuDong University, Yantai 264025, China

[5] School of Cyber Science Engineering, Southeast University, Nanjing 211189, China

Abstract. Online Social Networks (OSNs) have become an indispensable part of our lives, providing a platform for users to access and share information. However, the emergence of malicious social bots has disrupted the normal functioning of OSNs, posing a threat to their healthy development. With the evolution of social bots making them increasingly difficult to distinguish from human users, social bot detection has become a significant challenge. The difficulty lies in constructing effective features that cater to detect multiple types of social bots, performance of a single model in detecting social bots, and handling the imbalanced distribution of social bots and human users in real environments. To address these challenges, this paper proposes a social bot detection method based on multi-features fusion and model optimization strategy. The proposed method analyzes differences in user profile, tweets content, temporal information, and activity behaviors to extract and fuse effective features. Weighted soft voting mechanism and a searching best detection threshold strategy are creatively introduced to improve the performance of model. The superiority of our method is confirmed on four real datasets. The effectiveness of different dimensional features on detecting social bots is also analyzed. Furthermore, our method achieves better performance on imbalanced datasets, indicating great robustness and its ability to detect social bots in real environments.

Keywords: Social bot detection · Feature engineering · Multi-features fusion · Model optimization

1 Introduction

Over the past few decades, Online Social Networks (OSNs) have experienced rapid growth. OSN platforms such as Twitter, Facebook, and Sina Weibo are

X. Song et al. (Eds.): APWeb-WAIM 2023, LNCS 14332, pp. 347–362, 2024.
https://doi.org/10.1007/978-981-97-2390-4_24

playing an increasingly important role in our daily lives and have a significant impact on public life [1]. With the widespread popularity of OSNs, the number of people interacting through these platforms has grown exponentially in recent years [2]. Despite their widespread use, the real-time, open, and anonymous nature of OSNs has given rise to a distinct class of program-controlled users, commonly known as social bots. Although some social bots are benign and used to assist users, such as those that can automatically respond to queries [3]. Unfortunately, the rise of malicious social bots has resulted in various detrimental effects on OSNs. The continued development of these malicious bots has led to issues such as opinion manipulation, dissemination of false information, and malicious comments, all of which pose a significant cybersecurity threat to the sustainable growth of OSNs [4,5]. Therefore, detecting these malicious social bots in OSNs has become a crucial task.

Current research on feature engineering-based approaches faces challenges to construct comprehensive features due to the complexity of different social bot attributes. Social bots often pretend to be normal users to avoid being detected; thus, it is necessary to construct features from multiple aspects to portray them more accurately. Many existing studies only focus on a single perspective [6–8], which leads to partial descriptions of social bots, low accuracy, or detection of only certain types of social bots. Furthermore, it is worthwhile to explore the differences between social bots and benign users from the data itself before extracting features. The aim is to uncover potential social bot patterns and utilize them to construct multi-dimensional features for effective detection of multiple types of social bots.

Existing machine learning methods for social bot detection have primarily focused on binary classification, distinguishing social bots from benign users without considering multiple types of social bots. Furthermore, current approaches tend to use a single model and an automated strategy that can achieve good detection results in ideal situations but fail to address the negative impact of uneven distribution of benign users and social bots in real environments, leading to serious model degradation. Therefore, adopting optimization strategies for base models is vital for improving the performance and robustness of the detection model.

To address the above background and motivation, this paper proposes a social bot detection method based on multi-features fusion and model optimization strategy. In summary, we highlight our contributions as follows: 1) we construct effective features set from multiple dimensions of user information, which can detect different types of social bots. 2) the mechanism of searching for best detection threshold and other model optimization strategies are used in the model, which effectively improve the model performance and robustness to detect multiple types of social bots. 3) the experimental results on four evaluation datasets demonstrate the superiority of our approach, and can better cope with the real environment with imbalanced distribution of benign users and social bots.

2 Related Work

2.1 Feature Engineering-Based Approaches

In early approaches to social bot detection [9], researchers used only simple features. In recent years, researchers have gradually used features from a couple of aspects. Fazil et al. [10] analyzed the detection abilities of different types of features and determined that user interaction-based and community-based features were most effective for spam detection. Yang et al. [6] proposed a social bot detection method combining temporal information features and personal information. Wu et al. [11] constructed behavior sequences from raw data and extracted key features from the behavior time series as a way to distinguish social bots from benign users. Alarfaj et al. [7] analyzed tweets posted by users, extracted content feature sets based on message, lexicality, and sentiment. But the aforementioned research works tend to focus on one or two aspects of features and just use a small number of features to build a detection model.

2.2 Machine Learning-Based Approaches

Currently, machine learning-based methods are the most widely used methods in the field of social robot detection. Machine learning methods can be divided into classical machine learning methods and deep learning methods.

Cai et al. [8] proposed an approach based on Extreme Learning Machine (ELM) to effectively detect social bots in Sina Weibo. Gannarapu et al. [12] performed a machine learning algorithm to detect bots in social networks. Li et al. [13] provided a framework combining machine learning and graph methods to detect malicious social bots in OSNs.

In the past few years, deep learning methods have been more widely used in social bot detection. Fazil et al. [14] presented an attention-aware deep neural network model, DeepSBD, for detecting social bots in OSNs. Long et al. [15] trained a Bidirectional Long Short-Term Memory model(Bi-LSTM) with an attention mechanism to perform sentiment calculation on the online text information of social accounts. In [16], a method is proposed by Pham et al. to apply Network Representation Learning (NRL) to social bots or spammer detection, called Bot2vec. Hayawi et al. [17] proposed DeeProBot, which is short for Deep Profile-based Bot detection Framework, utilized features including user's numerical or binary metadata and user's description, making the model more comprehensive. However, these methods mainly consider single model and only focus on detecting certain types of social bots.

3 Proposed Approach

In this study, we focus on multiple dimensions of users' information, including user profile, tweets content, temporal information, and activity behaviors in OSNs. We conduct effective feature engineering techniques that included feature extraction, fusion and selection to handle the complex features. Moreover, we employ machine learning approaches such as eXtreme Gradient Boosting(XGBoost) [18] and Light Gradient Boosting Machine(LightGBM) [19] to design detection model. To ensure robustness and enhance the performance of our models, we use weighted soft voting technique and a strategy for searching best detection threshold. Our proposed method offers an effective solution to detect different types of social bots and alleviate the problem of the imbalanced distribution of benign users and social bots in real-world settings. The proposed general framework is illustrated in Fig. 1.

3.1 Feature Engineering

The feature engineering part consists of three steps: feature extraction based on multi-dimensional information, feature fusion, and feature selection.

Profile-Based Features. The user profile is disparate between benign users and social bots. Benign users provide accurate personal information while social bots tend to use fabricated information that visibly displays some level of regularity [20]. As shown in Fig. 2(a), the disparity of *friends_count* on 0 and 1 (benign users by Label 0; social bots by Label 1) is significant, indicating that whether an account is a social bot is closely interrelated to this feature. Therefore, the extraction of profile features from the user profile can aid in detecting social bots. To accomplish this, we do different processing according to different types of features (including numeric, discrete, and text-based) to extract features adapted to detect social bots. 1) numeric features: numeric features such as *followers_count, friends_count, listed_count* etc. Based on the original feature values, we use feature intersection technique, such as calculating the ratio of *followers_count* to *friends_count*, to improve the non-linear expression of features. 2) discrete features: discrete features such as *verified, protected, default_profile* etc., reflecting the information of users' personal settings. Features with values that were unique are removed, as they have low variance, making them almost useless for the model. The remaining discrete features are encoded into numerical form using feature coding techniques to meet the requirements of the gradient boosting algorithms. 3) text-based features: text-based features such as *name, description, url* etc. After preprocessing, the remaining text-based features are not directly used for model training, here we use Term Frequency-Inverse Document Frequency(TF-IDF) [21] technique to obtain the TF-IDF feature vectors of the text-based features, which is applied on the principle that the more a word or a sentence appears in a user, while the less in others, the more representative it is as for the user, thereby exhibiting potential for distinguishing between benign users and social bots.

The TF-IDF value of each element (a word or a sentence) is equal to mulyiplying the values of TF and IDF, as shown in Eq. 1.

$$TF\text{-}IDF = TF * IDF \tag{1}$$

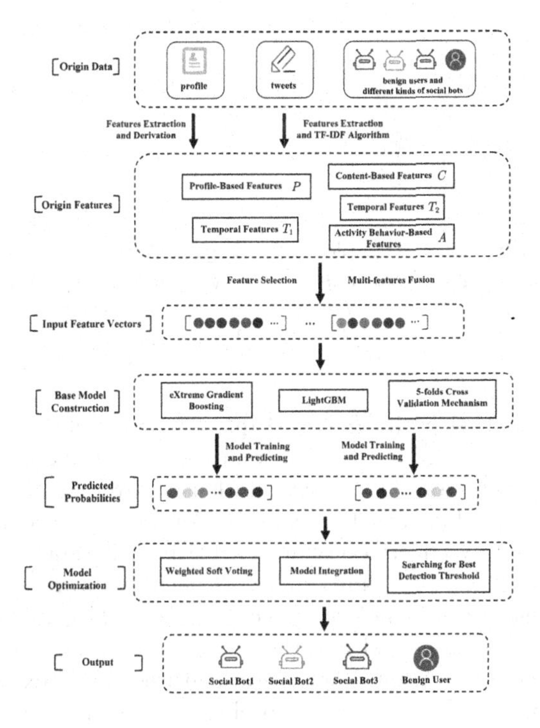

Fig. 1. The framework for a social bot detection method using multi-features fusion and model optimization strategy.

The TF and IDF are calculated as Eq. 2 and Eq. 3 respectively.

$$TF_{i,j} = \frac{n_{i,j}}{\sum_k n_{k,j}} \tag{2}$$

where $TF_{i,j}$ represents the frequency of element i in sample j ; $n_{i,j}$ is the times that element i emerges in the sample j, and $\sum_k n_{k,j}$ means the total count of emements in the sample j.

$$IDF_i = \log \frac{|D|}{1 + |\sum i \in j|} \tag{3}$$

where $|D|$ is the total amount of samples, and $|\sum i \in j|$ means the number of samples containing element i. In order to prevent the denominator from being 0, we add 1.

Once the features extracted and derived from each user on the three types of features are spliced and fused, we are able to obtain the profile-based features, or P, for each user.

Content-Based Features. In this work, we novelly obtain a numerical vector of statistical features and content based on tweet categories. Among all tweets, four categories of tweets are included, namely plain tweet, retweet (starting with

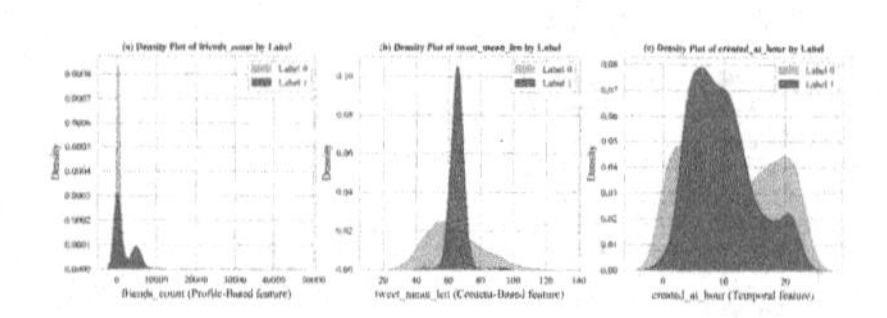

Fig. 2. Kernel Density Estimation plot with different dimensions of features. (a)*friends_count*. (b)*tweet_mean_len*. (c)*created_at_hour*.

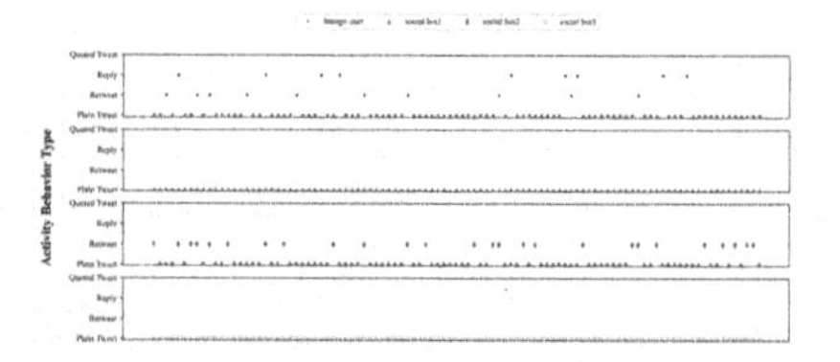

Fig. 3. Activity sequence behavior of a set of three social bots (one from each of the three social bot datasets) and a benign user

"RT@"), reply (starting with "@"), and quoted tweet (starting with "#"). We select the 100 most recently posted tweets for each user and compute statistical features, such as average length, maximum length, average and total number of words, for each of the four tweet categories. We find that the average length of tweets is a particularly effective feature to distinguish benign users from social bots, as shown in Fig. 2(b). Additionally, to better capture content differences between benign users and social bots, we extract TF-IDF feature vectors for each of the four tweet categories and fuse these results to obtain a text vectorization for each user. Finally, we fuse the statistical features of the text content with the digital vector splicing to obtain a comprehensive set of content-based features for each user, e.g. C.

Temporal Information-Based Features. Social bots exhibit some degree of consistency in their registration and activity time [14], which can be attributed to the way they are created by designers in batches. Conversely, benign users' registration and activity time are generally more random. Feature extraction begins with parsing the registration date and time of every user from the profile file. Here we extract temporal features such as the *hour*, textitdayofweek, *is_weekend*, *is_month_start*, *is_month_end*, etc.(T_1) Fig. 2(c) illustrates that the *created_at_hour* feature effectively reflects the difference between benign users and social bots. Next step involves scrutinizing the posting time of the last 100 tweets by each user and extracting the *hour* feature for each tweet across the four categories. Creative computations of the mean and standard deviation of the *hour* feature for all tweets within each category yields the T_2 feature. By fusing both T_1 and $T2$ features, the fusion produces a more effective temporal information-based feature set, is called T for social bot detection.

Activity Behavior-Based Features. Similarly, in OSNs, the activity behavior of social bots follows certain patterns. For example, different types of social bots are set to perform different activities. Some social bots perform retweeting tweets, while others perform replying to tweets [15]. However, the activity behavior of benign users tends to be random. Therefore, activity behavior is a good metric for social bot detection. In this paper, we visualize and analyze four

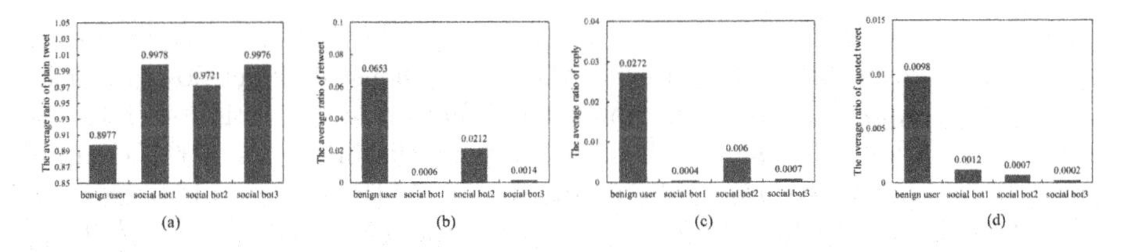

Fig. 4. Average ratio of different types of tweets on all users. (a)*plain tweet.* (b)*retweet.* (c)*reply.* (d)*quoted tweet.*

types of user activities in OSNs to identify the differences in activity behaviors between benign users and social bots. Specifically, we focus on the 100 most recent tweets posted by each user to encode their activity behavior sequence, which ensures that the length of the encoded behavior representation is equal for all users. We encode plain tweet by 0, retweet by 1, reply tweet by 2, and quoted tweet by 3. For instance, the 100 most recent tweets of a user can be used to construct the activity behavior-based encoding sequence, expressed as $\{0_1, 0_2, 0_3, 2_4, \cdots, 0_{99}, 1_{100}\}$, with 2_4 indicating that the user's fourth tweet is a retweet, representing the activity behavior of retweeting. Using this activity behavior coding sequence, we randomly select the activity behavior sequences of three social bots (of different types) and one benign user, as shown in Fig. 3. As can be seen from the figure, the sequence of activities of the social bots follows a certain pattern, where the 1st and 3rd social bots only post tweets and perform no other activities. In addition, observing the activity sequence of the 2nd social bot, an interesting phenomenon can be observed: the social bot periodically posts several tweets and then performs retweeting behavior. Moreover, we also calculated the average of the ratio of each type of tweets on all users in these 100 tweets, as shown in Fig. 4. Specifically: first, the average ratio of benign users' plain tweets is significantly lower than that of social bots, while the average ratio of other types of tweets is higher than that of social bots; second, the first and third types of social bots mainly perform the act of posting tweets, so the ratio of plain tweets is high; third, the average ratio of retweets is higher for the second type of social bots than for the remaining two types of social bots, which is consistent with the findings in Fig. 3. In conclusion, the ratios of different tweets show variability across benign users and different social bots. Therefore, we compute the ratio of plain tweet, retweet, reply, and quoted tweet posted by each user in these 100 tweets as the features A of each user based on activity behavior. This approach allows us to discern the differences in activity behaviors demonstrated by benign users and social bots.

Feature Fusion. After completing the feature extraction process for each dimension, this paper proposes the fusion of the extracted features to enhance their detection ability. In the previous section, we obtained four effective sets of features: profile-based (P), content-based (C), temporal information-based (T), and activity behavior-based (A). All the features of the above components are spliced and fused, i.e. $Features_{all} = [P\ C\ T\ A]$, so as to constitute the feature vector of the user in the comprehensive dimension. In order to study the importance of each dimensional feature for detecting social bots, in this paper, the features of each dimension are eliminated from the comprehensive dimensional feature combination to obtain the feature ablation analysis datasets, i.e. $Features_{C+T+A} = [C\ T\ A]$; $Features_{P+T+A} = [P\ T\ A]$; $Features_{P+C+A} = [P\ C\ A]$; $Features_{P+C+T} = [P\ C\ T]$. For instance, when investigating the significance of profile-based features, we exclude those features from the comprehensive feature set to obtain a dataset with only content-based, temporal information-based, and activity behavior-based features, i.e. $Features_{C+T+A} = [C\ T\ A]$. By comparing the performance of models trained on each of the remaining feature combinations to that of models using the full features set, we can ascertain the contribution of profile-based features to social bot detection.

Feature Selection. The feature vectors generated by TF-IDF processing may contain redundant information, which adversely impacts the model's performance. Consequently, this study explores feature selection of the vectors by using LightGBM, to assess the importance of each feature and eliminate any that are irrelevant or redundant. Notably, to achieve accurate calculation of feature importance, a five-fold cross-validation technique is applied to train the LightGBM model. The feature importance scores are averaged across the folds to provide a more reliable indication of the actual feature importance, which, in turn, ensures the effectiveness of the feature selection technique.

3.2 Model Construction

In the model building part, we use XGBoost and LightGBM as base models, respectively, and then further apply optimization strategies such as weighted fusion, weighted soft voting, and searching for best detection threshold to finally build social bot detection model with strong performance.

Base Model Construction. First, in this paper, based on the comprehensive set of social bot detection features on the dimensions obtained from the feature engineering part, these features are input into the XGBoost and LightGBM models for five-fold cross-validation training, and then the trained models are used to predict the training and test samples respectively, and the original prediction probabilities of the training and test sets are obtained. Second, for the dataset with only benign users and one kind of social bots, this paper creatively improves the prediction effect of the model by searching the best detection threshold based on the original prediction probability on the training set, and Algorithm 1 summarizes the main process of searching the best detection threshold.

Algorithm 1: Searching for best detection threshold

```
Input  : A training set predicted probability vector P_{n×1}, n is the number of training set
         samples; A matrix of training set true label y_{n×1}; Thresholds candidate set T;
Output: Training set predicted label y'_{n×1};Best threshold α; Best F1 score s;
Initialize: Best threshold α = 0; Best F1 score s = 0.
Train the base detection model first.
for t ∈ T do
    for i ← 0 to n do
        if P[i] < t then
            y'_i = 0
        else
            y'_i = 1
        end
        Calculate the F1 score between y and y';
    end
    if F1 > s then
        s = F1; α = t
    end
end
```

Model Integration. This paper employs a model weighting fusion mechanism to leverage the advantages of different models in learning various features, consequently enhancing the robustness of the final model. The final model encompasses weighted fusion of binary classification models and weighted soft voting algorithm of multi-classification models.

We use a weighted fusion mechanism to improve the performance of the base models for detecting only one type of social bots. Once the best detection threshold is obtained, the predicted probabilities on both the training and test sets are mapped to their respective labels, that is whether they are social bots. The performance of the two base models is evaluated, and based on that, different weights are set for each model. The weight is then multiplied by the prediction probabilities of the corresponding models on the training and test sets, and the results are added together to obtain the prediction probabilities after weighted fusion. The searching for best threshold mechanism is then reapplied to obtain the best detection threshold after model fusion. Finally, the prediction probabilities on the test set are mapped to labels to obtain the final prediction results.

In this paper, a method for constructing models that can detect multiple social bots is presented. The method involves using a weighted soft voting mechanism to combine the predictions of two models, XGBoost and LightGBM, that have previously been trained. The first step of the method involves calculating the log-loss values of the models based on their prediction probabilities on the training set. The negative log-loss values are then normalized to obtain the weights of the models. Once the weights of the models have been obtained, predictions are made on the test set. For each test set user, the prediction probabilities of each model are multiplied by the corresponding weights, and then add them to obtain the new prediction probabilities. The weighted average probability of each class is calculated, and the class with the highest weighted average probability is identified as the voting result for the user. The final prediction result for the user is selected from the predictions that agree with the voting

Table 1. Dataset statistics

Dataset	Number of Used User Accounts	
	Original	Used
SD1	991	991
SD2	3457	2104
SD3	464	461
GD	3474	1964

result and have the highest prediction probability among all models participating in the voting. By using this weighted soft voting mechanism, the performance of the models in detecting multiple types of social bots is improved.

4 Experiments

4.1 Experimental Setup

Dataset Preprocessing. We evaluate the method proposed in this study over four datasets constructed from a benchmark dataset provided by Cresci et al. [22]. The benchmark dataset contains 4 datasets, namely social spambots #1 (SD1), social spambots #2 (SD2), social spambots #3 (SD3), and genuine accounts (GD). The three social bot datasets (SD1, SD2, and SD3) represent different types of social bots. As the SD1, SD2, and SD3 consists of only social bot samples, we conducted a random selection of benign user samples from the real user dataset GD and merged them with the social bot samples from SD1, SD2, and SD3 to create the final evaluation datasets. Most of the existing social bot detection methods are competent enough in detecting specific classes of social bots, however, lack the capacity to detect other types of social bots. To assess the robustness of the proposed method in detecting various social bot classes, we created a fourth evaluation dataset (SD4) by choosing different types of social bot samples from SD1, SD2, and SD3, blended with a randomly chosen benign user samples from GD. For each dataset, only the latest 100 tweets were selected. Table 1 summarizes the statistics of the four datasets; *Original* records the number of users in the provided dataset, while *Used* indicates the number of selected users who posted more than 100 tweets. Previous studies demonstrated that social bots comprise about 9% to 15% of Twitter accounts, a popular online social network [23], indicating a lower percentage of social bots compared to benign users. Therefore, to test the effectiveness of our method in a genuine network atmosphere, we constructed evaluation datasets, including different ratios of benign users and social bots. Table 2 summarizes the figures that indicate the number of users included in each of the final four evaluation datasets with various percentages.

Evaluation Metrics. This paper employs four standard metrics to evaluate the model that detects only one type of social bots: detection rate (*DR*), precision (*Pr*), F1-Score (*F1*), and accuracy (*Acc*). For the model that detecing

Table 2. The number of users with different ratios of social bots and benign users in four datasets

Dataset		Number of User Account With Different Ratio of Social bots and Benign Users				
		1:1	1:2	1:3	1:4	1:5
SD1	Benign users	991	1964	1964	1964	1964
	Social Bot1	991	982	655	491	393
SD2	Benign users	1964	1964	1964	1964	1964
	Social Bot2	1964	982	655	491	393
SD3	Benign users	461	922	1383	1844	1964
	Social Bot3	461	461	461	461	461
SD4	Benign users	1964	1964	1964	1964	1964
	Social Bot1	547	274	183	137	110
	Social Bot2	1162	581	388	291	233
	Social Bot3	255	127	85	64	51

multiple types of social bots, this paper also uses four standard metrics, namely macro-detection rate(*Macro-DR*), macro F1-score (*Macro-F1*), multi-class log-loss (*logloss*) and accuracy (*Acc*). Multi-class log-loss *logloss* is the negative log-likelihood of the predicted sample given its true class, which is given as Eq. 4.

$$logloss = -\frac{1}{n}\sum_{i=1}^{n}\sum_{j=1}^{m} y_{ij}\log\left(p_{ij}\right) \tag{4}$$

where n is the number of samples, m is the number of classes. For y_{ij}, when the sample i belongs to the class j, it values 1, otherwise 0. p_{ij} is the probability that sample i belongs to the class j.

4.2 Performance Evaluation Results

For experimental evaluation, we used 80% data for training and remaining 20% data for the validation purpose. We performed several comparison experiments on four evaluation datasets to assess the performance of our approach and validate its ability to detect social bots from both the model and feature perspectives.

Effectiveness of Models. To verify the efficiency of the adopted model constructing strategies, we designed a set of comparison experiments using all the engineered features $Features_{All}$, as the input of the model. We employed various strategies to construct the detection models and evaluated their performance on four datasets (the ratio of social bots and benign users is 1:2). The results of the evaluation are tabulated in Table 3 and Table 4.

As shown in Table 3 and 4, by observing the different models and their results, we can draw two conclusions. On the one hand, the weighted fusion approach that introduces a mechanism to search for the best detection threshold contributes significantly to the improvement of the performance, with all evaluation metrics

Table 3. Detection performance of model constructed strategies(SD1, SD2, SD3)

Dataset	Different Models + All Features	DR	Pr	F1	Acc
SD1	XGBoost Base Model	0.967	1.0	0.983	0.990
	LightGBM Base Model	0.989	1.0	0.994	0.997
	Weighted Model Integration	0.973	1.0	0.984	0.992
	Weighted Model Integration + Best threshold	**0.995**	**1.0**	**0.997**	**0.998**
SD2	XGBoost Base Model	0.994	0.995	0.995	0.997
	LightGBM Base Model	0.995	1.0	0.997	0.997
	Weighted Model Integration	0.995	1.0	0.992	0.995
	Weighted Model Integration + Best threshold	**0.995**	**1.0**	**0.997**	**0.998**
SD3	XGBoost Base Model	0.980	**1.0**	0.990	0.993
	LightGBM Base Model	1.0	0.981	0.990	0.993
	Weighted Model Integration	1.0	0.981	0.990	0.993
	Weighted Model Integration + Best threshold	**1.0**	0.990	**0.995**	**0.996**

Table 4. Detection performance of model constructed strategies (SD4)

Dataset	Different Models + All Features	DR	Macro-F1	Acc	Multi-class log-loss
SD4	XGBoost Base Model	0.981	0.989	0.993	0.0392
	LightGBM Base Model	**0.985**	**0.991**	0.995	0.0410
	Weighted soft voting + Model Integration	0.981	0.990	**0.996**	**0.0379**

being best, compared to the two base models and simple weighted fusion. On the other hand, the model fusion method that introduces a weighted soft voting mechanism can reduce the multi-class log-loss value in the task of detecting multiple types of social bots. Moreover, other metrics also show that the method can perform well in the task of detecting multiple types of social bots instead of only detecting specific classes of social robots, indicating that the proposed method exhibits stronger robustness.

Effectiveness of Features. From the Feature Engineering part, it can be seen that our method constructs Profile-Based, Content-Based, Temporal Information-Based, and Activity Behavior-Based features. To demonstrate the effectiveness of feature engineering, we also established a set of comparison experiments and used feature ablation analysis to verify the effectiveness and importance of features on the model, and the results are presented in Table 5 and Table 6.

The full dimensional features achieves the best performance on all evaluated datasets, while the feature combination with one dimension removed fails to attain the desired performance, which implies the effectiveness of our multi-dimensional feature construction and fusion. Moreover, examining the effects of removing other feature combinations with one dimension implies a significant decrease in model performance after removing Profile-Based and Content-Based features. This finding suggests that these two aspects of features are the most useful in detecting social bots.

4.3 Results on Different Ratio of Social Bots and Benign Users

In this section, the evaluation results are given for different ratios of social bots to benign users, i.e., 1:1, 1:2, 1:3, 1:4, and 1:5, and the number of social bots to

Table 5. Detection performance of different feature sets (SD1, SD2, SD3)

Dataset	Best Model + Different Feature Sets	DR	Pr	F1	Acc
SD1	$Features_{C+T+A}$	0.986	1.0	0.993	0.995
	$Features_{P+T+A}$	0.978	1.0	0.989	0.993
	$Features_{P+C+A}$	0.989	1.0	0.994	0.997
	$Features_{P+C+T}$	0.990	1.0	0.995	0.997
	$Features_{All}$	**0.994**	**1.0**	**0.997**	**0.998**
SD2	$Features_{C+T+A}$	0.982	0.995	0.989	0.992
	$Features_{P+T+A}$	0.984	0.995	0.989	0.993
	$Features_{P+C+A}$	0.994	0.989	0.992	0.995
	$Features_{P+C+T}$	0.992	1.0	0.984	0.995
	$Features_{All}$	**0.995**	**1.0**	**0.997**	**0.998**
SD3	$Features_{C+T+A}$	0.979	**1.0**	0.990	0.993
	$Features_{P+T+A}$	0.950	0.979	0.965	0.975
	$Features_{P+C+A}$	0.989	0.989	0.989	0.993
	$Features_{P+C+T}$	0.990	**1.0**	0.994	0.996
	$Features_{All}$	**1.0**	0.990	**0.995**	**0.995**

Table 6. Detection performance of different feature sets (SD4)

Dataset	Best Model + Different Feature Sets	DR	Macro-F1	Acc	Multi-class log-loss
SD4	$Features_{C+T+A}$	0.976	0.986	0.990	0.0476
	$Features_{P+T+A}$	0.920	0.949	0.978	0.0807
	$Features_{P+C+A}$	0.981	0.989	0.993	0.0388
	$Features_{P+C+T}$	0.971	0.983	0.990	0.0515
	$Features_{All}$	**0.981**	**0.990**	**0.996**	**0.0379**

Table 7. Performance evaluation results of the method with different ratio of social bots and benign users

Ratio	Dataset															
	SD1				SD2				SD3				SD4			
	DR	Pr	F1	Acc	DR	Pr	F1	Acc	DR	Pr	F1	Acc	DR	Macro-F1	Acc	Multi-class log-loss
1:1	0.990	1.0	0.995	0.995	0.990	1.0	0.995	0.996	1.0	0.990	0.995	0.995	0.989	0.993	0.996	0.0185
1:2	0.995	1.0	0.997	0.998	1.0	0.990	0.997	0.998	1.0	0.990	0.995	0.996	0.981	0.990	0.996	0.0379
1:3	0.978	1.0	0.989	0.994	0.984	1.0	0.992	0.996	1.0	0.990	0.995	0.997	0.975	0.986	0.994	0.0428
1:4	0.977	1.0	0.989	0.996	0.960	1.0	0.980	0.990	0.990	1.0	0.995	0.998	0.947	0.971	0.990	0.0567
1:5	0.986	1.0	0.993	0.998	0.986	0.986	0.986	0.995	0.986	1.0	0.993	0.997	0.943	0.967	0.993	0.0266

benign users in each ratio is shown in Table 2. The result on the four datasets is shown in the Table 7.

On investigation, we find that in SD1, SD2, and SD3, the increased social bots to benign users ratio mildly affects the detection rate of our model. However, the impact on other three metrics is less significant, demonstrating the robustness of our method to some extent. Similarly, in SD4, we observes a more significant effect on the detection rate while the other metrics remains relatively stable. This finding suggests that imbalanced data could adversely affect the detection performance of our method in detecting multiple types of social bots. Nonetheless, our method manages to reduce the impact of data imbalance to a

Table 8. Comparison among our approach and others

Approach	SD1				SD2				SD3				SD4		
	DR	Pr	F1	Acc	DR	Pr	F1	Acc	DR	Pr	F1	Acc	DR	Macro-F1	Acc
DeeProBot [17]	0.960	0.983	0.971	0.982	0.989	0.967	0.978	0.986	0.921	0.976	0.947	0.963	0.902	0.907	0.955
Bot-Detective [24]	0.984	0.994	0.989	0.993	0.985	0.994	0.990	0.992	0.940	0.979	0.959	0.971	0.933	0.959	0.983
Rodriguez-Ruiz *et al.* [25]	0.973	1.0	0.986	0.992	0.984	1.0	0.992	0.995	0.940	0.989	0.964	0.975	0.947	0.954	0.981
ours	**0.995**	**1.0**	**0.997**	**0.998**	**0.995**	**1.0**	**0.997**	**0.998**	**1.0**	**0.990**	**0.995**	**0.996**	**0.981**	**0.990**	**0.996**

great extent and can better cope with the problem of imbalance between the ratio of benign users and social bots in the real environment.

4.4 Comparison with Previous Methods

To establish the superiority of the proposed method, we conducted a study on the current social bot detection techniques. We selected several representative methods for comparison and presented the outcome in Table 8. In their study [17], Hayawi et al. utilized user profile metadata from Twitter accounts. We used the parameters outlined in their paper to develop DeeProBot. Bot-Detective, proposed by Kouvela et al. [24], extracting 36 features from each account and using content features from the 20 most recent tweets to classify users. In a similar vein, Rodriguez-Ruiz et al. [25] created a one-class classification model that extracted 13 features from the social network and tweet information. The results demonstrates that our method outperforms all other evaluation metrics across the four evaluation datasets, indicating strong performance. Furthermore, while other methods perform well on SD1 and SD2, they exhibit low detection rates for social bots in SD3. Interestingly, our method distinctly outperforms all other methods in SD4, which contains multiple social bots. This suggests that current approaches are limited and can only detect specific classes of social bots, whereas our method is powerful in detecting multiple kinds of social bots.

5 Conclusions

In this paper, we present a novel social bot detection method based on multi-features fusion and model optimization strategy. Our method aims to improve model performance by constructing features that contain multiple dimensions. We introduce multiple optimization strategies in the model to achieve great performance. Our proposed method offers several notable advantages for both academic and industrial applications. 1) In the feature engineering phase, we analyze the differences between benign users and social bots directly from the data. We construct effective features that can detect multiple types of social bots. 2) We optimize the model construction phase by using weighted fusion, weighted soft voting, and searching for best detection threshold to enhance model performance. Our model has strong robustness, which make it suitable for detecting multiple types of social bots. 3) We conduct experiments on four datasets to confirm the effectiveness of our approach, from both model and feature perspectives. The results show that our method performs better than existing advanced deep learning and feature engineering-based methods. Our approach can be better applied to real environments where the ratio of social bots to benign users is imbalanced.

Acknowledgements. This research was supported by Key R&D Program of Guangzhou (No. 202206030001), NSFC (No. 62072131,61972106), 30602090501, Major Key Project of PCL (No. PCL2022A03-3, PCL2021A02), DongGuan Innovative Research Team Program (No. 2018607201008), Guangzhou Higher Education Innovation Group (No. 202032854).

References

1. Liang, H., Li, C.C., Jiang, G., Dong, Y.: Preference evolution model based on wechat-like interactions. Knowl.-Based Syst. **185**, 104998 (2019)
2. Li, S., Jiang, L., Wu, X., Han, W., Zhao, D., Wang, Z.: A weighted network community detection algorithm based on deep learning. Appl. Math. Comput. **401**, 126012 (2021)
3. Ferrara, E., Varol, O., Davis, C., Menczer, F., Flammini, A.: The rise of social bots. Commun. ACM **59**(7), 96–104 (2016)
4. Bessi, A., Ferrara, E.: Social bots distort the 2016 us presidential election online discussion. First monday **21**(11-7) (2016)
5. Li, S., Zhong, G., Jin, Y., Wu, X., Zhu, P., Wang, Z.: A deceptive reviews detection method based on multidimensional feature construction and ensemble feature selection. IEEE Trans. Comput. Soc. Syst. **10**(1), 153–165 (2022)
6. Yang, Z., Chen, X., Wang, H., Wang, W., Miao, Z., Jiang, T., et al.: A new joint approach with temporal and profile information for social bot detection. Secur. Commun. Networks **2022** (2022)
7. Alarfaj, F.K., Ahmad, H., Khan, H.U., Alomair, A.M., Almusallam, N., Ahmed, M.: Twitter bot detection using diverse content features and applying machine learning algorithms. Sustainability **15**(8), 6662 (2023)
8. Cai, C., Li, L., Zeng, D.: Detecting social bots by jointly modeling deep behavior and content information. In: Proceedings of the 2017 ACM on Conference on Information and Knowledge Management, pp. 1995–1998 (2017)
9. Lee, K., Caverlee, J., Webb, S.: Uncovering social spammers: social honeypots+ machine learning. In: Proceedings of the 33rd International ACM SIGIR Conference on Research and Development in Information Retrieval, pp. 435–442 (2010)
10. Fazil, M., Abulaish, M.: A hybrid approach for detecting automated spammers in twitter. IEEE Trans. Inf. Forensics Secur. **13**(11), 2707–2719 (2018)
11. Wu, J., Ye, X., Mou, C.: Botshape: a novel social bots detection approach via behavioral patterns. arXiv preprint arXiv:2303.10214 (2023)
12. Gannarapu, S., Dawoud, A., Ali, R.S., Alwan, A.: Bot detection using machine learning algorithms on social media platforms. In: 2020 5th International Conference on Innovative Technologies in Intelligent Systems and Industrial Applications (CITISIA), pp. 1–8. IEEE (2020)
13. Li, S., Zhao, C., Li, Q., Huang, J., Zhao, D., Zhu, P.: Botfinder: a novel framework for social bots detection in online social networks based on graph embedding and community detection. World Wide Web **26**(4), 1793–1809 (2023). https://doi.org/10.1007/s11280-022-01114-2

14. Fazil, M., Sah, A.K., Abulaish, M.: DeepSBD: a deep neural network model with attention mechanism for socialbot detection. IEEE Trans. Inf. Forensics Secur. **16**, 4211–4223 (2021)
15. Long, G., Lin, D., Lei, J., Guo, Z., Hu, Y., Xia, L.: A method of machine learning for social bot detection combined with sentiment analysis. In: Proceedings of the 2022 5th International Conference on Machine Learning and Natural Language Processing, pp. 239–244 (2022)
16. Pham, P., Nguyen, L.T., Vo, B., Yun, U.: Bot2vec: a general approach of intra-community oriented representation learning for bot detection in different types of social networks. Inf. Syst. **103**, 101771 (2022)
17. Hayawi, K., Mathew, S., Venugopal, N., Masud, M.M., Ho, P.H.: DeeProBot: a hybrid deep neural network model for social bot detection based on user profile data. Soc. Netw. Anal. Min. **12**(1), 43 (2022)
18. Chen, T., Guestrin, C.: XGBoost: a scalable tree boosting system. In: Proceedings of the 22nd ACM SIGKDD International Conference on Knowledge Discovery and Data Mining, pp. 785–794 (2016)
19. Ke, G., et al.: LightGBM: a highly efficient gradient boosting decision tree. In: Advances in Neural Information Processing Systems, vol. 30 (2017)
20. Kudugunta, S., Ferrara, E.: Deep neural networks for bot detection. Inf. Sci. **467**, 312–322 (2018)
21. Salton, G., Wong, A., Yang, C.S.: A vector space model for automatic indexing. Commun. ACM **18**(11), 613–620 (1975)
22. Cresci, S., Di Pietro, R., Petrocchi, M., Spognardi, A., Tesconi, M.: The paradigm-shift of social spambots: evidence, theories, and tools for the arms race. In: Proceedings of the 26th International Conference on World Wide Web Companion, pp. 963–972 (2017)
23. Varol, O., Ferrara, E., Davis, C., Menczer, F., Flammini, A.: Online human-bot interactions: detection, estimation, and characterization. In: Proceedings of the International AAAI Conference on Web and Social Media, vol. 11, pp. 280–289 (2017)
24. Kouvela, M., Dimitriadis, I., Vakali, A.: Bot-detective: An explainable twitter bot detection service with crowdsourcing functionalities. In: Proceedings of the 12th International Conference on Management of Digital EcoSystems. pp. 55–63 (2020)
25. Rodríguez-Ruiz, J., Mata-Sánchez, J.I., Monroy, R., Loyola-Gonzalez, O., López-Cuevas, A.: A one-class classification approach for bot detection on twitter. Comput. Secur. **91**, 101715 (2020)

Benefit from AMR: Image Captioning with Explicit Relations and Endogenous Knowledge

Feng Chen[1,2], Xinyi Li[2], Jintao Tang[2], Shasha Li[2], and Ting Wang[2](✉)

[1] China Academy of Engineering Physics, Mianyang 621000, China
[2] National University of Defense Technology, Changsha 410073, China
{chenfeng15a,tangjintao,tingwang}@nudt.edu.cn

Abstract. Recent advanced image captioning methods mostly explore implicit relationships among objects by object-based visual feature modeling, while failing to capture the explicit relations and achieve semantic association. To tackle these problems, we present a novel method based on Abstract Meaning Representation (AMR) in this paper. Specifically, in addition to implicit relationship modeling of visual features, we design an AMR generator to extract explicit relations of images and further model these relations during generation. Besides, we construct an AMR-based endogenous knowledge graph, which helps extract prior knowledge for semantic association, strengthening the semantic expression ability of the captioning model without any external resources. Extensive experiments are conducted on the public MS COCO dataset, and results show that the AMR-based explicit semantic features and the associated semantic features can further boost image captioning to generate higher-quality captions.

Keywords: image captioning · AMR · explicit relation modeling · endogenous knowledge association

1 Introduction

As a cross-domain research field, image captioning arouses the interest of many researchers in recent years, which generates natural language descriptions of images based on visual understanding. Inspired by the successful application of encoder-decoder framework in machine translation, early image captioning methods [19,27] mostly adopt this framework. To dynamically focus on visual features that relate to the current generate word, Xu et al. [31] and Gao et al. [10] resort to attention mechanisms. However, the above methods fail to capture semantic information of images. Therefore, Anderson et al. [1] further propose to extract object-aware visual features by pre-training with object detection task, which improves the quality of image representation vectors. Although the above methods successfully capture object-based semantic information, they are still lacking in explicit semantic representation (such as attributes, concepts, etc.). Hence,

X. Song et al. (Eds.): APWeb-WAIM 2023, LNCS 14332, pp. 363–376, 2024.
https://doi.org/10.1007/978-981-97-2390-4_25

some methods [16,30,35–37] propose to take advantage of the explicit semantic information. Among them, Zhang et al. [37] use Part of Speech (PoS) features as semantic guidance to enhance the POS perception ability of the captioning model. Moreover, Huang et al. [16] propose an attribute-based method to enhance semantic perception, which selects attribute sub-sequences by calculating the similarity between attributes and target objects, and utilizes attribute sub-sequences to generate the current word. However, these methods are limited to explore the explicit relationships among semantic features and further improve semantic awareness.

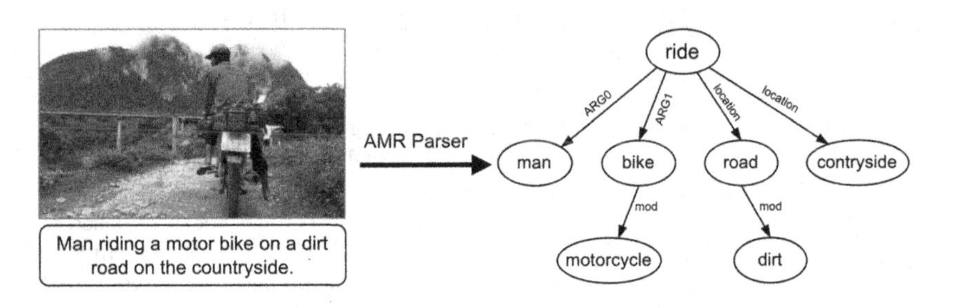

Fig. 1. The AMR parsing result of a given image caption.

In this work, we consider taking advantage of structured semantics to make up for the limitation of explicit relationship expression of unstructured semantics and further achieving semantic association based on endogenous knowledge during generation. To our knowledge, Abstract Meaning Representation (AMR) [2] is a rooted, directed, edge-labeled, and leaf-labeled graph that is used to represent the meaning of a sentence. The concepts and relations of a sentence can be built in AMR, which are crucial for understanding the meaning of a sentence. As Fig. 1 shows, a caption of the image is transformed to AMR, in which the notional words are abstracted as concept nodes and the structural relationships are abstracted as edges. It shows that AMR effectively captures the syntactic and the semantic information, the important concepts, and the relationships among concepts of the given image. Different from explicit semantic representations, the AMR-based semantic information not only covers important concepts, but also the explicit relationships among concepts, which has stronger semantic expression ability.

Based on the above analysis, we propose to generate image captions with the guidance of AMR-based features. Specifically, from the perspective of explicit semantic modeling, we utilize AMR features as the carrier of explicit relationships, which are then combined with implicit relationship modeling of visual features to explore the complementary between these two kinds of features for captioning. Furthermore, from the perspective of semantic association, we utilize a pre-constructed endogenous knowledge graph to extract associated semantic features based on concept pairs, which help the captioning model perceive richer relevant semantics when generating words, and explore the role of semantic association.

Contributions of this work are as follows:

- We are the first to use AMR features to provide explicit relations and associated semantic information for image captioning;
- We propose to construct an AMR-based endogenous knowledge graph to gather prior knowledge for semantic association;
- This study provides important insights into AMRs for image captioning. Extensive experiments demonstrate the effectiveness of the proposed method.

2 Related Work

Just as the name implies, image captioning aims to generate natural language descriptions of images. Different from traditional visual tasks, e.g., image retrieval [7] and object detection [25], image captioning is more challenging because it requires comprehensive visual understanding and natural language generation ability.

To produce descriptive sentences based on visual understanding, early works for image captioning apply the encoder-decoder framework that is widely used in machine translation [3], in which a CNN encodes the image into a single vector or a series of vectors and an RNN translates the image representation into natural language sentences. However, the encoders fail to capture fine-grained information of images and the decoders can't selectively focus on relevant features or regions during caption generation. To tackle the problems, attention-based methods [1,10,15,29,31] arise at the historic moment, which extracts the fine-grained image features (usually based on sub-regions), and dynamically pays attention to these features according to the current state with certain weighting rules, so as to optimize the alignment between the generated word and the image features.

Recent advanced research methods based on the visual attention mechanism [1,4,5,10,23,28] depend on the previously generated word and the visual attention result to predict the current word. For deeper image understanding, researchers have explored the semantic information of images for captioning and verified its effectiveness. Wu et al. [30] use attribute features of images to generate captions and verify that the attribute features are more expressive than the visual features; Yao et al. [35] enhance image representations with attribute-based semantic information, which make up for the lack of expression ability of visual-based vectors; You et al. [36] use a novel attention mechanism to encode both the visual features and the semantic information, which further improves the captioning model. Rather than utilize high-level semantic information to boost the model, Yao et al. [33] define the relations between different visual sub-regions and the relations between different targets by certain calculation rules, encode these relationships to deepen the visual understanding ability of the captioning model. Besides, Yang et al. [32] bridge the gap between symbolic reasoning and end-to-end feature mapping by scene graphs, in which a scene graph captioner captures structural semantics. Instead, Chen et al. [6] use Abstract Scene Graph (ASG) as the control signal to produce intention-aware captions. Specifically, ASGs are easier to be obtained automatically because they are constructed based on visual features.

Currently, it's a common rule to boost the captioning model by Reinforcement Learning (RL) strategy [1], which solves both the exposure bias and the non-differentiable metric problems. These methods [1,14,15,34,40] maximize the expected reward based on CIDEr [26] with policy gradient method.

3 Methodology

In this paper, we propose an AMR-based Captioning method (dubbed as AMRC), which takes advantage of structural semantics of AMRs to deepen visual understanding and achieve semantic association during generation. Figure 2 illustrates its overall framework.

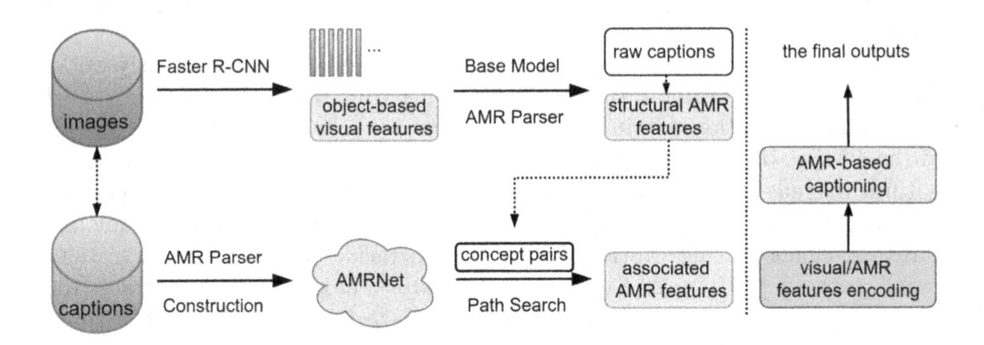

Fig. 2. Framework of AMRC.

3.1 Feature Extraction

To fully explore the role of AMR-based semantic information for image captioning, we use two kinds of features in AMRC: object-based visual features and AMR features (the structural AMR features and the associated AMR features).

Object-Based Visual Features. Object-based visual features we use are the same as in UpDown [1], which are extracted by a Faster R-CNN [24]. Specifically, the Faster R-CNN is pre-trained with object detection on Visual Genome [20], in which sub-regions with different bounding boxes are represented as V after an RoI pooling layer, whose probabilities of various object categories exceed a confidence threshold.

Structural AMR Features. To acquire features of attributes and descriptive words with actual meanings of images, and further explore explicit relations among these notional words. Firstly, this work uses a pre-trained base model [1] to generate raw captions with relatively complete semantics. After that, we apply JAMR [9] to initially parse the raw captions into AMRs (including word segmentation, part of speech, word stem, etc.). Then, we use STOG [38] to generate the amendatory AMR representations. Finally, to facilitate encoding in the captioning model, the final AMRs are represented as triplet sequences S.

Associated AMR Features. To further introduce endogenous knowledge perception into the captioning model, we propose to achieve the AMR-based semantic association. To the beginning, the endogenous knowledge graph needs to be constructed:

- For each caption in the training set, JAMR is used for preliminary parsing, and then STOG is used for further parsing to obtain the final AMR, which is then represented as a sequence of triplets;
- Gathering AMRs of all the annotated captions to construct the endogenous knowledge graph AMRNet. Specifically, there are about 3,560,000 triplets. After removing the duplicate items, remaining about 520,000 triplets, which are then used to construct AMRNet.

After that, construct associated sub-graphs based on path-searching algorithms:

- Gathering nodes in S as a concept sequence N, which contains $len(N)$ concepts;
- Construct the set of concept pairs P based on N. For each concept in N, the ith concept n_i is matched with the $(i+1)^{th}$ concept to the last one;
- For each concept pair p in P, searching paths [8] that less than 5-hops from AMRNet as candidates, gathering them to construct the associated endogenous knowledge sub-graph G;
- Extracting nodes from G to get the associated knowledge of notional concepts E.

3.2 AMR-Based Captioning

In the process of caption generation, according to the characteristics of different semantic features, we explore the roles of the structural AMR features and the associated AMR features respectively.

Since the attention mechanism is used several times in this work, here, we briefly introduce it. Based on the current attention query vector q and the input features $F = \{f_1, f_2, \cdots, f_n\}$, the attention layer calculates as follows:

$$a_i = W_1(\sigma((W_2F) \oplus (W_3q))) \tag{1}$$

$$\alpha_i = \frac{e^{a_i}}{\sum_{i=0}^{n} e^{a_i}} \tag{2}$$

$$\hat{f} = \sum_{i=0}^{n} \alpha_i f_i \tag{3}$$

where W_1,W_2, and W_3 are learned parameters of linear transformations, $\oplus$ means element-wise add, σ represents tanh activation layer, and n is the number of input features.

In the following sections, attention layers are respectively represented by f_{attV}, f_{attS}, and f_{attE}. The parameters of each attention layer are not shared.

Captioning with the Structural AMR Features. Intuitively, when compared with object-based visual features, the AMR-based structural semantic features capture explicit relations among objects. In order to explore the advantage of structural AMR features for captioning, we propose a novel LSTM-based method that utilizes both the visual features and the AMR features. Specifically, the visual features V and the structural AMR features S are modeled with attention layers, which helps the model fully use both features to generate higher-quality captions.

The structural AMR semantic features of a given image are stored as a sequence of triplets, in which the ith triple $tri_i = (h_i, r_i, t_i)$ contains a relationship r_i between the head concept h_i and the tail concept t_i. We encode and update tri_i as a vector based on the embedding matrix W_{amr} of AMR:

$$tri_i = w_t([W_{amr}h_i, W_{amr}r_i, W_{amr}t_i]) \tag{4}$$

$$S = \{tri_1, tri_2, \cdots, tri_n\} \tag{5}$$

where $[\cdot]$ means the concatenation operation of vectors, and w_t is the learned parameters of a linear transformation that maps tri_i to the dimension of the LSTM's hidden layer.

To effectively calculate visual query q_t for the visual attention block at step t, we take the previous hidden state of the language LSTM h^2_{t-1}, the global image representation $\bar{v}$ (the mean pooling of the object-based visual features), the previously generated word w'_{t-1}, and the previous state of the attention LSTM (h^1_{t-1}, c^1_{t-1}) into account:

$$w_{t-1} = W_{emb}w'_{t-1} \tag{6}$$

$$x_t = [h^2_{t-1}, \bar{v}, w_{t-1}] \tag{7}$$

$$h^1_t, c^1_t = f_{att-lstm}(x_t, (h^1_{t-1}, c^1_{t-1})) \tag{8}$$

$$q_t = h^1_t \tag{9}$$

where W_{em} is the word embedding matrix of the vocabulary of captions, and $f_{att-lstm}$ is the visual attention LSTM layer.

Intuitively, not all of the input features contribute equally to the current word, thus we resort to the attention mechanism to dynamically focus on the visual features based on q_t. At first, a multi-head attention layer [14] $MultiHead$ is applied to update V, which models implicit relations among the objected-based visual features. Then the attention result of visual features is calculated as:

$$V = MultiHead(V) \tag{10}$$

$$\hat{v} = f_{attV}(q_t, V) \tag{11}$$

To further take advantage of the structural AMR features, we calculate the explicit relation query q_t^s by $f_{amr-lstm}$, which is the attention LSTM layer. Then, calculate the attention result of the structural AMR features by f_{attS}:

$$h_t^s, c_t^s = f_{amr-lstm}(x_t, (h_{t-1}^s, c_{t-1}^s)) \tag{12}$$

$$q_t^s = h_t^s \tag{13}$$

$$\hat{s} = f_{attS}(q_t^s, S) \tag{14}$$

Based on the attention queries, the encoded visual feature $\hat{v}$, and the encoded structural AMR feature $\hat{s}$, a language LSTM is utilized to generate the current word:

$$\tilde{x_t} = [q_t, q_t^s, \hat{v}, \hat{s}] \tag{15}$$

$$h_t^2, c_t^2 = f_{lan-lstm}(\tilde{x_t}, h_{t-1}^2, c_{t-1}^2) \tag{16}$$

$$p_t = softmax(h_t^2) \tag{17}$$

where $f_{lan-lstm}$ is the language LSTM layer, and p_t is the word distribution on the pre-defined vocabulary of captions.

Captioning with the Associated AMR Features. In this section, we further explore the role of AMR-based associated semantic information for image captioning. We use an attention-based method to selectively pay attention to salient regions and concepts that relate to the current word, so as to enhance the semantic perception of the captioning model. Notably, the associated semantic information $E = \{e_1, e_2, \cdots, e_m\}$, which is represented as a series of related concepts, is different from S in encoding (Sect. 3.2). Instead, concepts in E are vectorized based on W_{amr}:

$$e_i = W_{amr} e_i \tag{18}$$

$$E = \{e_1, e_2, \cdots, e_m\} \tag{19}$$

We apply the same model structure as that of the structural AMR features, except that the input semantic features S are replaced by E.

Optimization. The proposed model is first trained to minimize negative log-likelihood estimation loss (XE stage) and then optimized by computing the gradient of the pre-defined CIDEr [26] reward (RL stage [1]).

$$loss = -\frac{1}{N}\sum_{i=1}^{N}\log p(T^i|V^i, \overline{v}^i, S^i/E^i; \theta) \tag{20}$$

$$\nabla_\theta \mathbb{E} \approx (R(c) - R(\hat{c}))\nabla_\theta log_{p_\theta}(c) \tag{21}$$

where i means the ith image of N, T is the target caption, θ is the learned parameter, R is the CIDEr reward, and $c/\hat{c}$ are random/max sampled captions.

Table 1. Results on the public MS COCO test split [19] at the XE stage.

Methods	Metrics							
	B-1	B-2	B-3	B-4	M	C	R	S
CNet [39]	73.1	54.7	40.5	29.9	25.6	107.2	53.9	–
AttNet [30]	74.0	56.0	42.0	31.0	26.0	94.0	–	–
LSTM-Att [35]	73.4	56.7	43.0	32.6	25.4	100.2	54.0	–
Adaptive-Att [22]	74.2	58.0	43.9	33.2	26.6	108.5	-	–
VSDA [12]	75.3	59.1	45.1	34.4	26.5	106.3	55.2	–
Stack-Cap [11]	76.2	60.4	46.4	35.2	26.5	109.1	–	–
RFNet [18]	76.4	60.4	46.6	35.8	27.4	112.5	56.5	20.5
Up-Down [1]	77.2	-	-	36.2	27.0	113.5	56.4	20.3
STMA [17]	**77.4**	**61.5**	**47.6**	36.5	27.4	114.4	56.8	20.5
Ours_S	76.9	61.0	47.2	36.2	28.1	116.5	57.0	**21.4**
Ours_E	77.1	61.3	**47.6**	**36.7**	**28.1**	**117.2**	**57.1**	21.3

4 Experiments

4.1 Dataset, Preprocess, and Metrics

MS COCO dataset[1] is the largest and most popular dataset for image captioning. It contains 123,287 images and each image is annotated with at least 5 captions.

A vocabulary of 10,369 words is constructed by converting the words in the annotated captions to lowercase and replacing the words that occur less than 5 times with a special character "UNK". Meanwhile, we resort to the JAMR parser and the STOG model to convert the captions into AMR form. Some suffixes (such as the numerical coding suffixes of actions) and low-frequency words in

[1] https://cocodataset.org/.

the concept node set are removed, and an AMR vocabulary containing 16,228 concepts and 109 relations is constructed.

We utilize publicly used metrics[2] to report the results, which contains BLEU (B-N), METEOR (M), CIDEr (C), ROUGE-L (R), and SPICE (S), and scores are represented as percentages (%).

4.2 Comparison with State-of-the-art Methods

Table 2. Results on the public MS COCO test split in the RL stage.

Methods	Metrics							
	B-1	B-2	B-3	B-4	M	C	R	S
IENet [13]	79.2	64.0	48.9	37.1	26.9	118.2	57.3	-
Stack-Cap [11]	78.6	62.5	47.9	36.1	27.4	120.4	56.9	20.9
RFNet [18]	79.1	63.1	48.4	36.5	27.7	121.9	57.3	21.2
Up-Down [1]	79.8	–	–	36.3	27.7	120.1	56.9	21.4
CAVP [21]	–	–	–	**38.6**	28.3	126.3	**58.5**	21.6
DeRF [10]	79.9	–	–	37.5	**28.5**	125.6	58.2	**22.3**
STMA [17]	**80.2**	64.4	49.7	37.7	28.2	125.9	58.1	21.9
Ours_S	79.9	64.6	50.2	38.5	28.3	125.7	58.4	21.9
Ours_E	**80.2**	**64.8**	**50.3**	38.4	28.4	**127.4**	58.4	**22.0**

For comprehensive comparisons, we show results of both the XE (Table 1) and the RL (Table 2) stages. At the XE stage, the proposed model outperforms existing knowledge-driven methods [30,39]. When compared with recent methods [1,11,12,17,18,22,35], Ours_S with the structural AMR features and Ours_E with the associated AMR features achieve consistency improvements. It shows slightly better scores on B-N (BLEU), but shows obvious improvement on METEOR, CIDEr, ROUGR-L, and SPICE, especially CIDEr (117.2) and SPICE(21.4). B-N is used to evaluate the matching of N-tuples without considering semantic information and distinguishing the semantic specificity of N-tuples fragments. However, other metrics address the deficiencies of BLEU from different perspectives. Therefore, higher scores of METEOR, CIDEr, ROUGR-L, and SPICE indicate better semantic consistency. Results preliminarily demonstrate the effectiveness of the introduced AMR semantics for caption generation.

At the RL stage, Ours_S/Ours_E outperform existing knowledge-driven method [13] and recent methods [1,10,11,17,18,21] on most metrics. To be detailed, except METEOR and SPICE, Ours_S outperforms DeRF [10] and STMA [17], especially on B-4 and CIDEr. Ours_E makes the absolute improvement over the previous leading methods, e.g., obtains a CIDEr score of 127.4,

[2] https://github.com/tylin/coco-caption.

which indicates higher-quality captions. These results further show the superiority of the proposed model in semantic consistency.

The above analysis shows that the proposed method is superior to the advanced methods in most evaluation metrics, especially in CIDEr. CIDEr is to evaluate the semantic consistency, a higher score indicates better semantic consistency of the generated caption and the image. Therefore, it demonstrates that the AMR-based features effectively provide supplementary semantic information for the captioning model to generate high-quality captions.

4.3 Ablation Studies

Table 3. Results of ablation studies. Base: the base model [1]; S: use the structural AMR features; V1:use a shared vocabulary for both captions and AMRs; E: use the associated AMR features; R: use the visual feature refining trick [14]; GT_AMR: use the ground truth AMR features.

Methods	Metrics							
	B-1	B-2	B-3	B-4	M	C	R	S
Base-XE	76.7	60.7	46.9	36.1	27.8	114.1	56.7	20.8
+S-V1	76.5	60.7	47.1	36.4	27.9	115.1	57.0	21.0
+S	76.6	60.7	47.1	36.2	27.8	115.1	56.6	20.9
+E	76.6	60.6	46.9	36.1	27.9	114.8	56.8	21.2
+S+R(Ours_S)	76.9	61.0	47.2	36.2	**28.1**	116.5	57.0	**21.4**
+E+R(Ours_E)	**77.1**	**61.3**	**47.6**	**36.7**	**28.1**	**117.2**	**57.1**	21.3
Base-RL	79.6	64.0	49.1	36.9	28.0	123.2	57.8	21.4
+S-V1	**80.9**	64.5	49.5	37.2	28.2	122.2	58.1	21.6
+S	79.6	63.8	49.2	37.3	28.0	124.7	57.9	21.6
+E	79.8	64.2	49.6	37.7	28.0	124.9	58.1	21.4
+S+R(Ours_S)	79.9	64.6	50.2	**38.5**	28.3	125.7	**58.4**	21.9
+E+R(Ours_E)	80.2	**64.8**	**50.3**	38.4	**28.4**	**127.4**	**58.4**	**22.0**
+GT_AMR	90.9	80.3	70.3	61.9	44.2	173.1	75.9	31.5

To figure out contributions of different features and factors, we conduct extensive experiments. Results are shown in Table 3.

- Semantic features. +S-V1 uses a shared vocabulary for both captions and AMRs during training and predicting. Compared with Base, +S-V1 gains higher scores except on CIDEr, e.g., at the RL stage, boosts B1 to 80.9. We consider that the shared word embeddings are updated based on two types of information, it is beneficial for the generation of related N-tuples but not the generation of specific contents. +S exceeds Base by 1.5 in CIDEr, which indicates the AMR structured semantic information effectively boosts

Base. Considering that CIDEr measures semantic relevance, we construct vocabularies of captions and AMRs respectively. +E exceeds Base by 1.7 in CIDEr, which shows that semantic association based on known AMR concepts further improves the captioning model.
- Visual feature refining. At both the XE and the RL stages, +S+R/+E+R boost +S/+E. Results show that visual feature refining (+R) further improves the performance because it models implicit relationships among object-based visual features.
- The structural AMR features and the associated AMR features. At both the XE and the RL stages, +E/+E+R boost +S/+S+R. +E that utilizes the associated AMR features performs better, we suspect that the reason is the structural AMR features capture the explicit relation information but have weak semantic awareness.

Besides, to further verify the role of the AMR features for captioning, we encode the ground truth AMRs of images as the semantic features (+GT_AMR, highlighted in grey). +GT_AMR achieves amazingly good results, which demonstrates that it is a feasible research idea to take advantage of AMR to further boost image captioning.

4.4 Case Study

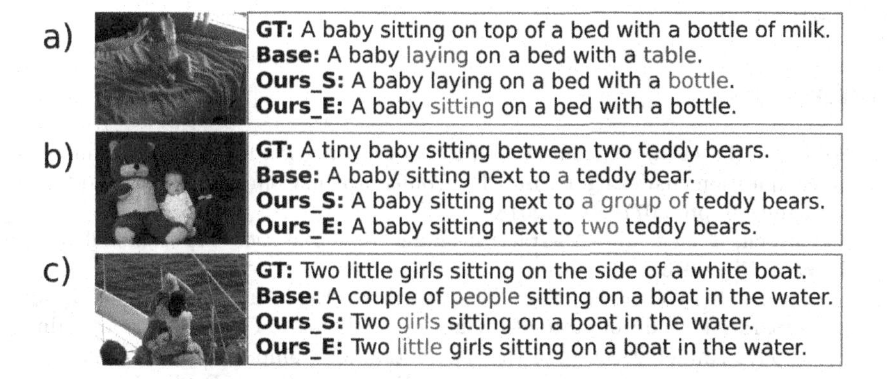

Fig. 3. Qualitative examples. We show ground truth captions (GT, annotated by humans), generated captions of the base model (Base), and the proposed AMR-based models (Ours_S/Ours_E).

Figure 3 shows some generated captions of the base model (Base) and the proposed AMR-based models (Ours_S/Ours_E). For example, in Fig. 3 a), Base correctly predicts "a baby" and "bed", while generating "table", which is a wrong object. By taking advantage of the AMR-based semantic features, Ours_S generates "bottle" and Ours_E further corrects the posture of the baby as "sitting".

In Fig. 3 b), Base predicts the toy in this image is "a teddy bear" but gives a wrong number "a". In this case, Ours_S gives "a group of" and Ours_E recognizes the right number "two". Similarly in Fig. 3 c), Base recognizes the scene depicted in the image "on a boat in the water", however, the description of the girls is too simple "a couple of people". In this case, Ours_S correctly predicts the subject "girl" and Ours_E further generates an adjective "little".

It shows that the AMR-based methods perform better image understanding and expression abilities. By inner-knowledge modeling, the proposed model achieves a broader semantic awareness, which helps the model generate more accurate captions.

5 Conclusion

From the perspective of explicit relationship perception and semantic association of images, we propose an AMR-based method for captioning. To our knowledge, we are the first to use AMR features to provide explicit relations and associated semantic information for image captioning. Specifically, in addition to implicit relationship modeling of visual features, we use structural AMR features to introduce explicit relation modeling into the captioning model. Furthermore, we strengthen the semantic awareness of the captioning model without any external resources by semantic association based on a pre-constructed endogenous knowledge graph. This study provides important insights into AMRs for image captioning. Results on the MS COCO dataset show the advantages of the proposed method.

References

1. Anderson, P., et al.: Bottom-up and top-down attention for image captioning and visual question answering. In: In IEEE Conference on Computer Vision and Pattern Recognition, pp. 6077–6086 (2018)
2. Banarescu, L., et al.: Abstract meaning representation for sembanking. In: LAW@ACL (2013)
3. Chatterjee, R., Weller, M., Negri, M., Turchi, M.: Exploring the planet of the apes: a comparative study of state-of-the-art methods for MT automatic post-editing. In: In 53rd Annual Meeting of the Association for Computational Linguistics and the 7th International Joint Conference on Natural Language Processing, vol. 2, pp. 156–161 (2015)
4. Chen, F., Xie, S., Li, X., Li, S., Tang, J., Wang, T.: What topics do images say: a neural image captioning model with topic representation. In: In IEEE International Conference on Multimedia & Expo Workshops (ICMEW), pp. 447–452. IEEE (2019)
5. Chen, L., et al.: SCA-CNN: spatial and channel-wise attention in convolutional networks for image captioning. In: In IEEE Conference on Computer Vision and Pattern Recognition, pp. 6298–6306. IEEE (2017)
6. Chen, S., Jin, Q., Wang, P., Wu, Q.: Say as you wish: fine-grained control of image caption generation with abstract scene graphs. In: In IEEE/CVF Conference on Computer Vision and Pattern Recognition, pp. 9962–9971 (2020)

7. Dong, G., Zhang, X., Lan, L. et al. Label guided correlation hashing for large-scale cross-modal retrieval. Multimed. Tools Appl. **78**, 30895–30922 (2019). https://doi.org/10.1007/s11042-019-7192-5
8. Feng, Y., Chen, X., Lin, B.Y., Wang, P., Yan, J., Ren, X.: Scalable multi-hop relational reasoning for knowledge-aware question answering. In: In Conference on Empirical Methods in Natural Language Processing (2020)
9. Flanigan, J., Thomson, S., Carbonell, J., Dyer, C., Smith, N.A.: A discriminative graph-based parser for the abstract meaning representation. In: In Annual Meeting of the Association for Computational Linguistics (2014)
10. Gao, L., Fan, K., Song, J., Liu, X., Xu, X., Shen, H.T.: Deliberate attention networks for image captioning. In: In AAAI Conference on Artificial Intelligence (2019)
11. Gu, J., Cai, J., Wang, G., Chen, T.: Stack-captioning: Coarse-to-fine learning for image captioning. In: AAAI Conference on Artificial Intelligence (2018)
12. He, C., Hu, H.: Image captioning with visual-semantic double attention. ACM Trans. Multimed. Comput. Commun. Appl. **15**(1), 26 (2019)
13. Huang, F., Li, Z., Chen, S., Zhang, C., Ma, H.: Image captioning with internal and external knowledge. In 29th ACM International Conference on Information and Knowledge Management (2020)
14. Huang, L., Wang, W., Chen, J., Wei, X.Y.: Attention on attention for image captioning. In: In IEEE International Conference on Computer Vision, pp. 4634–4643 (2019)
15. Huang, L., Wang, W., Xia, Y., Chen, J.: Adaptively aligned image captioning via adaptive attention time. In: In Advances in Neural Information Processing Systems, pp. 8942–8951 (2019)
16. Huang, Y., Chen, J., Ouyang, W., Wan, W., Xue, Y.: Image captioning with end-to-end attribute detection and subsequent attributes prediction. IEEE Trans. Image Process. **29**, 4013–4026 (2020)
17. Ji, J., Xu, C., Zhang, X., Wang, B., Song, X.: Spatio-temporal memory attention for image captioning. IEEE Trans. Image Process. **29**, 7615–7628 (2020)
18. Jiang, W., Ma, L., Jiang, Y.G., Liu, W., Zhang, T.: Recurrent fusion network for image captioning. In: In European Conference on Computer Vision, pp. 499–515 (2018)
19. Karpathy, A., Fei-Fei, L.: Deep visual-semantic alignments for generating image descriptions. In: IEEE Conference on Computer Vision and Pattern Recognition, pp. 3128–3137 (2015)
20. Krishna, R., et al.: Visual genome: connecting language and vision using crowdsourced dense image annotations. Int. J. Comput. Vision **123**(1), 32–73 (2017)
21. Liu, D., Zha, Z.J., Zhang, H., Zhang, Y., Wu, F.: Context-aware visual policy network for sequence-level image captioning. In: In 26th ACM international conference on Multimedia, pp. 1416–1424 (2018)
22. Lu, J., Xiong, C., Parikh, D., Socher, R.: Knowing when to look: adaptive attention via a visual sentinel for image captioning. In: IEEE Conference on Computer Vision and Pattern Recognition, vol. 6, p. 2 (2017)
23. Lyu, N.H.M.F.F.: TSFNET: triple-steam image captioning. IEEE Trans. Multimedia **25**, 1–14 (2022)
24. Ren, S., He, K., Girshick, R., Sun, J.: Faster R-CNN: towards real-time object detection with region proposal networks. In: Advances in Neural Information Processing Systems, pp. 91–99 (2015)
25. Tan, H., Zhang, X., Lan, L., Huang, X., Luo, Z.: Nonnegative constrained graph based canonical correlation analysis for multi-view feature learning. Neural Processing Letters, pp. 1–26 (2018)

26. Vedantam, R., Zitnick, C.L., Parikh, D.: Cider: consensus-based image description evaluation. In: IEEE Conference on Computer Vision and Pattern Recognition, pp. 4566–4575 (2015)
27. Vinyals, O., Toshev, A., Bengio, S., Erhan, D.: Show and tell: Lessons learned from the 2015 MSCOCO image captioning challenge. IEEE Trans. Pattern Anal. Mach. Intell. **39**(4), 652–663 (2016)
28. Vinyals, O., Toshev, A., Bengio, S., Erhan, D.: Show and tell: a neural image caption generator. In: IEEE Conference on Computer Vision and Pattern Recognition, pp. 3156–3164 (2015)
29. Wang, Y., Xu, J., Sun, Y.: A visual persistence model for image captioning. Neurocomputing **468**, 48–59 (2022)
30. Wu, Q., Shen, C., Wang, P., Dick, A., Hengel, A.V.: Image captioning and visual question answering based on attributes and external knowledge. IEEE Trans. Pattern Anal. Mach. Intell. **40**, 1367–1381 (2018)
31. Xu, K., et al.: Show, attend and tell: neural image caption generation with visual attention. In: Computer Science, pp. 2048–2057 (2015)
32. Yang, X., Tang, K., Zhang, H., Cai, J.: Auto-encoding scene graphs for image captioning. In: IEEE/CVF Conference on Computer Vision and Pattern Recognition, pp. 10685–10694 (2019)
33. Yao, T., Pan, Y., Li, Y., Mei, T.: Exploring visual relationship for image captioning. In: European Conference on Computer Vision, pp. 684–699 (2018)
34. Yao, T., Pan, Y., Li, Y., Mei, T.: Hierarchy parsing for image captioning. In: IEEE/CVF International Conference on Computer Vision, pp. 2621–2629 (2019)
35. Yao, T., Pan, Y., Li, Y., Qiu, Z., Mei, T.: Boosting image captioning with attributes. In: IEEE International Conference on Computer Vision, pp. 22–29 (2017)
36. You, Q., Jin, H., Wang, Z., Fang, C., Luo, J.: Image captioning with semantic attention. In: IEEE Conference on Computer Vision and Pattern Recognition, pp. 4651–4659 (2016)
37. Zhang, J., Mei, K., Zheng, Y., Fan, J.: Integrating part of speech guidance for image captioning. IEEE Trans. Multimedia **23**, 92–104 (2020)
38. Zhang, S., Ma, X., Duh, K., Durme, B.V.: Amr parsing as sequence-to-graph transduction. In: Annual Meeting of the Association for Computational Linguistics (2019)
39. Zhou, Y., Sun, Y., Honavar, V.G.: Improving image captioning by leveraging knowledge graphs. In: IEEE Winter Conference on Applications of Computer Vision (WACV), pp. 283–293 (2019)
40. Zhou, Y., Wang, M., Liu, D., Hu, Z., Zhang, H.: More grounded image captioning by distilling image-text matching model. In: IEEE/CVF Conference on Computer Vision and Pattern Recognition, pp. 4777–4786 (2020)

SSCAN:Structural Graph Clustering on Signed Networks

Zheng Zhao[1], Wei Li[1,2](✉), Xiangxu Meng[1], Xiao Wang[1], and Hongwu Lv[1]

[1] College of Computer Science and Technology, Harbin Engineering University, Harbin 150001, China
{zhaozheng,mxx,wangxiao1999,lvhongwu}@hrbeu.edu.cn
wei.li@hrbeu.edu.cn

[2] Modeling and Emulation in E-Government National Engineering Laboratory, Harbin Engineering University, Harbin 150001, China

Abstract. *Structural graph clustering* (SCAN) is a foundational problem about managing and profiling graph datasets, which is widely experienced across many realistic scenarios. Due to existing work on structural graph clustering focused on unsigned graphs, existing SCAN methods are not applicable to signed networks that can indicate friendly and antagonistic relationships. To tackle this problem, we investigate a novel structural graph clustering model, named SSCAN. On the basis of SSCAN, we propose an online approach that can efficiently compute the clusters for a given signed network. Furthermore, we also devise an efficient index structure, called SSCAN-Index$^+$, which stores information about core vertices and structural similarities. The size of our proposed index can be well bounded by $O(m)$, where m is the total amount of edges in an input signed network. Following the new index SSCAN-Index$^+$, we develop an index-based query method designed to avoid invalid scans of the entire network. Extensive experimental testings on eight real signed networks prove the effectiveness and efficiency of our proposed methods.

Keywords: Structural graph clustering · signed networks · hubs and outliers

1 Introduction

As a fundamental graph data analysis problem, graph clustering serves to discover hidden structural knowledge in graph applications and has received increasing attention in recent years (*e.g.*, [2,7,8,10]). Instinctively, a cluster in graph is a vertices set densely connected within the same set and sparsely connected to vertices between different sets. The main goal of graph clustering is to find all clusters, and this graph analysis is often found a wide range of scenarios, such as identifying communities in large networks [4], finding structural information in social networks [5] and recommendation systems [1]. For real graphs, except vertices as members of clusters, some vertices are hubs that bridge different clusters and some are outliers that lack strong ties to any clusters. However, most

X. Song et al. (Eds.): APWeb-WAIM 2023, LNCS 14332, pp. 377–392, 2024.
https://doi.org/10.1007/978-981-97-2390-4_26

graph clustering approaches ignore the role identity of each vertex. To overcome this shortcoming, SCAN is proposed in [10], which can distinguish the different roles of the vertices and uncover overlapping clusters.

Motivation. Although the effectiveness of SCAN and its improved work in identifying clusters as well as roles of vertices has been witnessed in many applications, SCAN only focuses on unsigned graphs. However, interactions over the networks encompass not only *positive* relationships (*i.e.*, friendships) but also *negative* relationships (*i.e.*, adversarial relationships). Compared with traditional graphs, signed graphs expand the representational capability by incorporating the relationships between entities or vertices through *positive* and *negative* edge signs [9], such as social networks, opinion networks, trust networks. If we use SCAN directly for signed graphs (*i.e.*, assuming that the edge signs are ignored and positive and negative edges are treated equally), the obtained clusters will lack stability (*i.e.*, a cluster involve numerous unbalanced triangles).

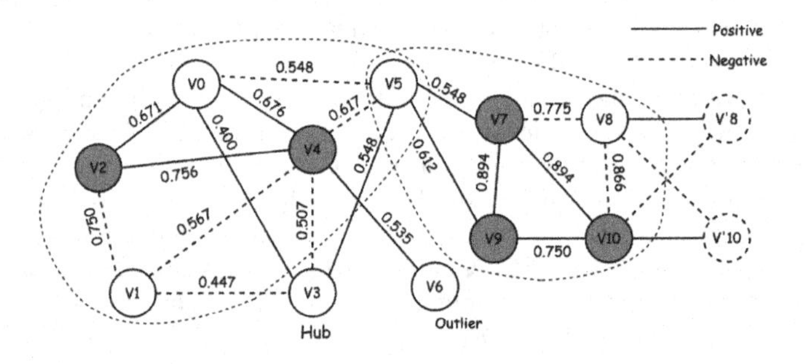

Fig. 1. An example signed graph G ($\epsilon_b = 0.6$ and $\mu = 4$).

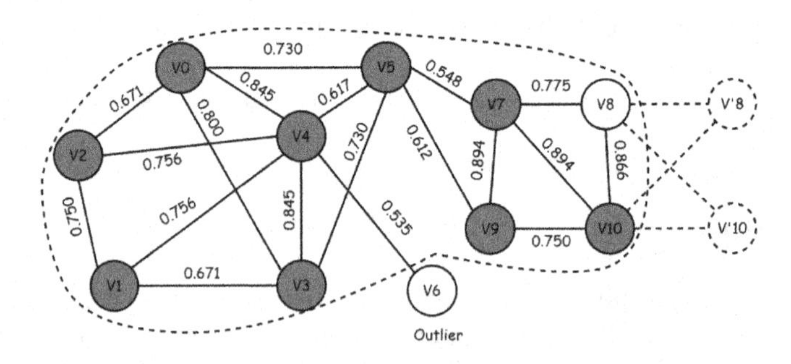

Fig. 2. An example graph G' ($\epsilon = 0.6$ and $\mu = 4$).

Then, we consider a simple case study about the signed example G shown in Fig. 1 with our proposed model, and the corresponding unsigned one G' is shown in Fig. 2. For signed graph G, it is evident that $\{v_0, v_1, v_2, v_4, v_5\}$ and $\{v_5, v_7, v_8, v_9, v_{10}\}$ are two more densely clusters. However, SCAN is unable to consider the signs of edges like its unsigned counterpart G' and makes the whole subgraph $\{v_0, v_1, v_2, v_3, v_4, v_5, v_7, v_8, v_9, v_{10}\}$ as a cluster. Note that, the dark nodes in above figures are *core* vertices.

Taking inspiration from this, we aim to develop a novel method for structural clustering in signed graphs that usually preserve the signs properties.

Our Approaches. There are two primary challenges about solving this problem.

1. How to design the structural clustering model in terms of structural balance of the signed graph?
2. How to efficiently compute clustering queries over massive signed graphs?

In our paper, we propose a new structural graph clustering model in signed graphs by considering the edge signs and maintain the same clustering style as SCAN. Based on *structural balance theory* [9], there are two types of triangles in a signed graph G, *i.e.*, balanced triangles and unbalanced triangles. As we all know, the key point of SCAN is the computation of *structural similarity*, which can be regard as the count of triangles that is constituted by vertices u, v, and their neighbors. Inspired by this, we propose a signed structural graph clustering model which is SSCAN (Signed SCAN) for signed networks. With the *balanced theory*, there are many balanced triangles and few unbalanced triangles in a cluster when finding clusters in signed networks. Therefore, we define *balanced structural similarity* between two vertices u and v with an edge by considering the count of balanced triangles form by u, v and their common neighbors (See Definition1). Furthermore, we define a vertex as a *core* vertex and the definition of *hub* and *outlier* is the same as that of SCAN.

We reconsider the three cluster results in Fig. 1 and Fig. 2. The SSCAN regards v_3 as a *hub* vertex which does not belong to any of two clusters while SCAN regards v_3 as a *core* vertex in a *cluster*. However, we can find that v_3 and some vertices in cluster form unstable structure (here, $\{v_0, v_3, v_5\}$ and $\{v_0, v_3, v_4\}$ are two unbalanced triangles). Compared with SCAN, it only considers the common neighbors ignoring positive/negative signs, which makes SCAN unable to find the stable structure in clusters. It can thus be demonstrated that our SSCAN model can get more balanced and meaningful clusters over signed networks.

Following the SSCAN model, we propose an online structural graph clustering approach, denoted SSCAN-Basic, based on efficient balanced triangles counting method. We also give a proof of the time complexity of SSCAN-Basic can be bounded by $O(\alpha \cdot m)$ which is same as SCAN. In addition, to avoid frequent optimization of parameter settings, we design a space-bounded index which is named SSCAN-Index$^+$. With the SSCAN-Index$^+$, we propose the corresponding index-based algorithm to answer the query for any possible ϵ_b and μ. All clusters can be found in $O((\alpha + \log n) \cdot m)$ with our query algorithm.

Contributions. As far as we are aware, this is the first study to investigate the structural graph clustering over signed networks. Our main contributions can be outlined as follows.

- We formally define a new structural clustering model, called SSCAN, for clustering the signed graphs.
- We propose an online approach to compute the clusters based on balanced triangles counting algorithm.
- We devise a space-bounded index SSCAN-Index$^+$ and propose an efficient query algorithm based on the index. We also propose an efficient algorithm to construct the index.

We conduct comprehensive empirical evaluations on large real and synthetic graphs. The empirical results validate the efficiency of our models and algorithms.

2 Problem Definition

In our paper, we give an *undirected signed graph* $G = (V, E)$, where V is the vertices set, and E denotes the signed edges set (*resp.* E^+ is positive edges set and E^- is negative edges set). We denote the count of vertices and edges by n and m, respectively, *i.e.*, $n = |V|$ and $m = |E| = |E^+| + |E^-|$. The $N(u)$ is defined as $\{v \in V | (u, v) \in E\}$ and $N[u] = N(u) \cup \{u\}$. To promote clarity, we simply refer to an unweighted signed graph as a graph when the context is clear.

Structural balance theory [9] is a basic theory for signed graph analysis (*i.e.*, the friend of my friend is my friend, and the enemy of my enemy is my friend). Based on this, we think of a triangle in a signed graph is *balanced* if it consists of odd number of positive edges, and an *unbalanced* otherwise. Intuitively, a cluster in signed networks ought to encompass more balanced triangles(denoted as $\triangle^+$) and less unbalanced ones(denoted as $\triangle^-$). Thus, we define the *Balanced Structural Similarity* between two vertices u and v.

Definition 1. Balanced Structural Similarity. *The balanced structural similarity between vertices u and v, denoted by $\sigma_b(u, v)$, is defined as the number of balanced triangles formed by u, v and common neighbors, normalized by the geometric mean of their cardinalities; that is,*

$$\sigma_b(u, v) = \frac{\sum_{w \in N[u] \cap N[v]} \triangle^+_{uvw}}{\sqrt{|N[u]| \cdot |N[v]|}} \tag{1}$$

And then, unlike SCAN, when faced with sets $\{u, v, u\}$ and $\{u, v, v\}$, we treat u or $v \in N[u] \cap N[v]$ as virtual vertices u' and v', and u, v have positive edge with itself. Furthermore, the sign between u' and v is identical to that between u and v.

Given a balanced structural similarity threshold $\epsilon_b (0 < \epsilon_b \leq 1)$, the ϵ_b-*neighborhood*, denoted by $N_{\epsilon_b}[u]$, is a vertex set in $N[u]$ whose similarities with u are at least ϵ_b (*i.e.*, $N_{\epsilon_b}[u] = \{v \in N[u] | \sigma_b(u, v) \geq \epsilon_b\}$). Here, due to $\sigma_b(u, u) = 1$, $N_{\epsilon_b}[u]$ will include u itself. We say a vertex u is a *core* vertex if $|N_{\epsilon_b}[u]| \geq \mu$, where integer $\mu \geq 2$; otherwise, it is a *non-core* vertex. After that, we neglect the positive and negative signs of the edges and define *structurally reachable* between two vertices u and v in the same way as SCAN. A *cluster* is a group of vertices that are *structurally reachable* from any *core* vertex.

Example 1. Consider the signed graph shown in Fig. 1, given $\epsilon_b = 0.6$ and $\mu = 4$, $N_{\epsilon_b}[v_2] = \{v_2, v_0, v_1, v_4\}$, and $N_{\epsilon_b}[v_4] = \{v_4, v_0, v_2, v_5\}$. Thus, v_2 and v_4 are two core vertices. Similarly, we can get v_7, v_9 and v_{10} are core vertices as well. There are two clusters: $C_1 = \{v_0, v_1, v_2, v_3, v_4, v_5\}$ and $C_2 = \{v_5, v_7, v_8, v_9, v_{10}\}$. v_3 is a hub vertex and v_6 is outlier.

Problem Statement. Given a signed graph G, a balanced structural similarity threshold $\epsilon_b (0 < \epsilon_b \leq 1)$ and an integer $\mu (\mu \geq 2)$, in this paper we investigate the efficient computation for the clusters set $\mathbb{C}$ in G.

3 Our SSCAN-Basic Approach

This section, we give our baseline method about structural graph clustering on signed networks. We firstly introduce an algorithm for counting balanced triangles. And then, we give a straightforward approach based on the above balanced triangles computation.

3.1 Balanced Triangles Counting

To avoid repetition in determining whether any three vertices form a balanced triangle, we have following lemmas.

Lemma 1. *Given a signed graph G, each balanced triangle has at least one positive edge.*

Lemma 2. *Two vertices u and v, if there is a common positive neighbor w for a positive edge(u, v). The id of three vertices should satisfy $u < v < w$.*

Following Lemmas 1 and 2, the pseudocode of our efficient algorithm for counting balanced triangles is shown in Algorithm 1. For each positive edge (u, v), we compute a common neighbor w from the neighbors of vertex v which has a smaller degree (Lines 3–9). If w is a common friend of u and v, we should guarantee that the id of vertex w is the largest among the three vertices according to Lemma 2 (Lines 10–11). Then, the number of balanced triangles corresponding all three edges are added by 1 (Lines 12–14). Conversely, if the vertex w is a common foe of u and v, the corresponding three edges are directly added by 1 (Lines 15–18).

Theorem 1. *The time complexity of Algorithm 1 can be bounded by $O(m^{1.5})$.*

Proof. In Algorithm 1, we only check all positive edges and iteratively check the vertices of smaller degree to calculate the balanced triangle. So the time complexity is $O(\sum_{(u,v)\in E^+} min\{|N(u)|, |N(v)|\} \leq \sum_{(u,v)\in E} min\{|N(u)|, |N(v)|\})$. According to [3], the $O(\sum_{(u,v)\in E} min\{|N(u)|, |N(v)|\}) \leq O(\alpha \cdot m)$ and α is the arboricity of G, which is the minimum number of spanning forests needed to cover all the edges of the graph G and $\alpha \leq \sqrt{m}$.

Algorithm 1: countBalancedTri$^+$

```
Input: A signed graph G(V, E)
Output: t_b(e) for each edge e ∈ E
1  for each edge(u, v) ∈ E do
2      t_b(u, v) ← 2;
3  for each u ∈ V do
4      for each v ∈ N(u) do
5          if u < v and (u, v) ∈ E+ then
6              if |N(u)| < |N(v)| then
7                  swap(u,v);
8              for each w ∈ N(v) do
9                  if w in N(u) then
10                     if (u, w) ∈ E+ and (v, w) ∈ E+ then
11                         if w > v then
12                             t_b(u, v) + +; t_b(v, u) + +;
13                             t_b(u, w) + +; t_b(w, u) + +;
14                             t_b(v, w) + +; t_b(w, v) + +;
15                     if (u, w) ∈ E- and (v, w) ∈ E- then
16                         t_b(u, v) + +; t_b(v, u) + +;
17                         t_b(u, w) + +; t_b(w, u) + +;
18                         t_b(v, w) + +; t_b(w, v) + +;
```

3.2 SSCAN-Basic

Based on the balanced triangle counting in Sect. 3.1, now, we present our baseline signed structural graph clustering approach, called SSCAN-Basic, as shown in Algorithm 2, which consists of several steps: firstly, counting all balanced triangles by procedure countBalancedTri$^+$ (Line 1). And then, for edge (u, v), it computes the balanced structural similarity between u and v following Definition 1 (Lines 2–3). After that, we calculate the number of positive ϵ_b-*neighborhood*, denoted as $N^+_{\epsilon_b}[u]$ for each vertex u (Lines 4–6), whose definition is as follows.

Definition 2. ***Positive*** *ϵ_b-neighborhood. The positive ϵ_b-neighborhood for a vertex u, denoted by $N^+_{\epsilon_b}[u]$, in which every vertex v satisfies $\sigma_b(u, v) \geq \epsilon_b$ and is positive neighbor for vertex u. i.e., $N^+_{\epsilon_b}[u] = \{v \in N[u] | \sigma_b(u, v) \geq \epsilon_b$ and $(u, v) \in E^+\}$.*

To cluster core vertices, for vertex u, we check the number of positive ϵ_b-*neighborhood* or ϵ_b-*neighborhood* is equal or larger than μ. All clusters are stored in the set $\mathbb{C}$ (Lines 17–18). The time complexity of Algorithm 2 is $O(\alpha \cdot m)$. Based on the previous analysis, it is easy to see that lines 2–6 cost $O(m)$ time and lines 7–18 take $O(m)$ time. Thus, counting balanced triangles is the dominating cost in Algorithm 2.

Algorithm 2: SSCAN-Basic

Input: A signed Graph G and two parameters $0 < \epsilon_b \leq 1$, $\mu \geq 2$
Output: The set $\mathbb{C}$ of clusters in G

```
1  countBalancedTri+(G);
2  for each edge(u, v) ∈ E do
3      Compute σ_b(u, v);
4      if σ_b(u, v) ≥ ε_b then
5          |N+_{ε_b}[u]| + +;
6          |N+_{ε_b}[v]| + +;
7  for each unexplored vertices u ∈ V do
8      C ← {u};
9      for each unexplored vertices v ∈ C do
10         Mark v as explored;
11         if |N+_{ε_b}[v]| ≥ μ then
12             C ← C ∪ N_{ε_b}[v];
13         else
14             Compute |N_{ε_b}[v]|;
15             if |N_{ε_b}[v]| ≥ μ then
16                 C ← C ∪ N_{ε_b}[v];
17     if |C| > 1 then
18         ℂ ← ℂ ∪ {C};
19 return ℂ;
```

4 An Index-Based Approach

To better understand clustering, it is essential to identify the *core* vertices within a graph. Unfortunately, the direct computation method through Algorithm 2 can be quite expensive, especially to compute the *balanced structural similarity*. To overcome this challenge, we can index all the values of the *balanced structural similarity* to speed up the computation of *core* vertices.

4.1 SSCAN Indexing

A straightforward idea is computing and storing all balanced structural similarities for edges (u, v) and (v, u), denoted as SSCAN-Index. Based on this naive index, we have to check all edges in G, and the time complexity will take up to $O(m)$. Therefore, we design a new index to reduce the computation costing and regard the *core* vertices based on the following lemma.

Lemma 3. *Given a vertex v in a signed graph G, if the number of structural neighborhood for v is less than integer μ (i.e., $|N[v]| < \mu$), v must not be a core vertex.*

Algorithm 3: SSCAN-Query$^+$

Input: SSCAN-Index$^+$ and two parameters $0 < \epsilon_b \leq 1$, $\mu \geq 2$
Output: The set $\mathbb{C}$ of clusters in G

```
1  for each |N[*]| ∈ [μ, deg_max + 1] do
2      for each vertex u ∈ MSO_|N[*]| do
3          if u is explored then Continue;
4          if |N[*]| = μ and ε_b > MS(u) then Break;
5          if |N[*]| > μ and ε_b > MS(u) and σ_b(u, NO_b[u][μ]) < ε_b then Continue;
6          C ← C ∪ {u}; Q ← ∅; mark u as explored; Q.push(u);
7          while Q ≠ ∅ do
8              v ← Q.pop();
9              for each w ∈ NO_b[v] do
10                 if σ_b(v, w) < ε_b then break;
11                 if w is explored then continue;
12                 Mark w as explored;
13                 C ← C ∪ {w};
14                 if σ_b(w, NO_b[w][μ]) ≥ ε_b then
15                     Q.push(w);
16                 else
17                     Mark w as the non-core vertex;
18         if |C| > 1 then ℂ ← ℂ ∪ {C};
19         for each non-core vertex v do
20             Make v as unexplored;
21 return ℂ;
```

Based on Lemma 3, we design to store the *minimal structural similarity* (see Definition 3) for each vertex in G.

Definition 3. ***Minimal Structural Similarity.*** *The minimal structural similarity for a vertex u, denoted by $MS(u)$, is the $N[u]$-th balanced similarity value among all structural neighborhoods of u. i.e., $MS(u) = \{v \in N[u] | min(\sigma_b(u, v))\}$.*

Through *minimal structural similarity*, we can further reduce unnecessary *core* vertices checking according to the following lemmas.

Lemma 4. *Given a clustering query ϵ_b and μ, for a vertex u, if $|N[u]| = \mu$ but $MS(u) < \epsilon_b$, u is not a core vertex.*

Lemma 5. *Given a signed graph G and two parameters ϵ_b and μ, a vertex u is a core vertex if (i) v is a core vertex; (ii) $|N[v]| \leq |N[u]|$, and (iii) $MS(v) \leq MS(u)$.*

And then, for each integer μ, we consider a vertex u is a *core* vertex if there exist not less than μ ϵ_b-*neighborhood* as the following lemma.

Lemma 6. *Given a signed graph G, an integer $\mu \geq 2$ and two balanced similarity thresholds $0 < \epsilon_b < \epsilon'_b < 1$, a vertex u is a core in a cluster obtained by μ and ϵ_b if it is a core in a cluster obtained by μ and ϵ'_b.*

According to Lemma 5, we can conclude that if v is *core* vertex, all vertices u with $|N[v]| \leq |N[u]|$ and $MS(v) \leq MS(u)$ must be *cores*. Therefore, to efficient obtain *core* vertices, we sort the vertices in non-increasing order with their *minimal structural similarity* for different number of structural neighborhood $|N[*]|(i.e., 1 \leq |N[*]| \leq deg_{max} + 1)$, where deg_{max} is the maximal degree in graph G. We define such order as follows.

Definition 4. ***Minimal Similarity Order.*** *For different values of $|N[*]|$, the minimal similarity order for all vertices, denoted by $MSO_{|N[*]|}$, is vertex order such that: (i) all vertices $v \in MSO_{|N[*]|}$ satisfy $|N[v]| = |N[*]|$; (ii) two*

Algorithm 4: SSCAN-Cons$^+$

```
1  countBalancedTri+ (G);
2  for each u ∈ V do
3      for each v ∈ N(u) do
4          if u > v then Continue;
5          σ_b(u, v) ← t_b(u, v)/√(|N[u]| · |N[v]|);
6          σ_b(v, u) ← t_b(v, u)/√(|N[v]| · |N[u]|);
7          NO_b[u] ∪ v; NO_b[v] ∪ u;
8      Sort NO_b[u] in non-increasing order of balanced structural similarity;
9  for each |N[*]| ∈ [2, deg_max + 1] do
10     MSO_|N[*]| ∪ {u ∈ V | |N[u]| = |N[*]|};
11     Sort vertices in MSO_|N[*]| according to Definition 4;
```

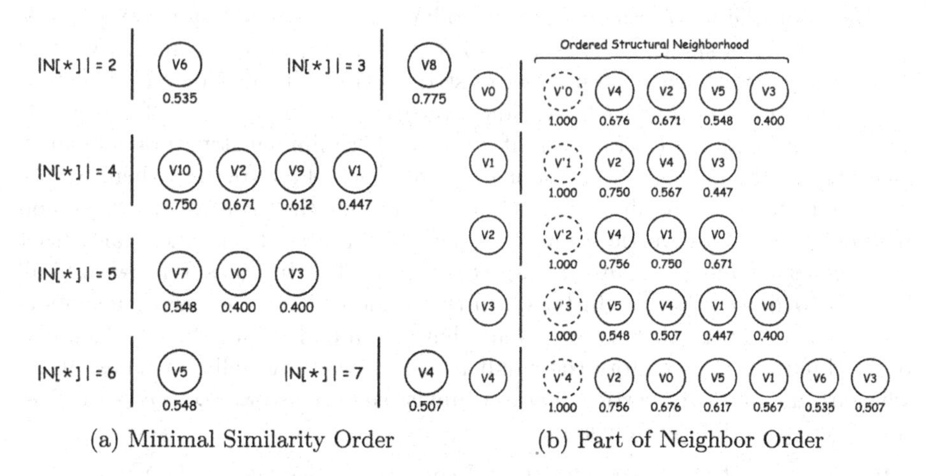

(a) Minimal Similarity Order (b) Part of Neighbor Order

Fig. 3. Two Parts of SSCAN-Index$^+$.

vertices u and v, u appears before v if the minimal structural similarity of u is not smaller than that of v.

Based on *minimal similarity order*, we can get all *core* vertices efficiently for different values of parameters ϵ_b and μ. We check *core* vertices from the first one and stop when the last vertex whose number of structural neighborhoods is equal to $deg_{max} + 1$. According to the Definition 3, if we want to find *minimal structural similarity* for vertex u, we should compute the similarity between u and its structural neighborhoods. After that, we sort these neighbors in a non-increasing order of their similarities and store the smallest one. In the meantime, if there is a vertex u with larger structural neighborhoods but smaller *minimal structural similarity* (*i.e.*, $|N[u]| \geq \mu$ and $MS(u) < \epsilon_b$), we should check its neighbors. So we also store this neighbor order in our new index, denoted as NO_b.

Example 2. Consider the signed graph shown in Fig. 1, the corresponding SSCAN-Index$^+$ shows part of neighbor order due to the limited space in Fig. 3. For example, given a clustering query $\epsilon_b = 0.6$ and $\mu = 3$, we directly know that v_2 is a *core* vertex with $|N[v_2]| > 3$ and $0.671 > 0.6$. However, for vertex v_0 with $|N[v_0]| > 3$ but $0.400 < 0.6$, we do not know if v_0 is *core* vertex from the *minimal structural similarity*. With neighbor order, we can directly get the *μ-th* neighbors of v_0 equal to $0.671 > 0.6$ and v_0 is a *core* vertex in above clustering query. After that, we can efficiently get the *ϵ_b-neighborhood* of *cores* from the neighbor order.

Clearly, neighbor order has the same space cost $O(m)$ with SSCAN-Index. For *minimal similarity order*, its space complexity can be bounded by $O(m)$ because we do not store those vertices whose number of structural neighborhood less than 2.

Theorem 2. *The space complexity of* SSCAN-Index$^+$ *is bounded by $O(m)$.*

Based on the SSCAN-Index$^+$, we introduce the corresponding query processing algorithm as shown in Algorithm 3. For a cluster query ϵ_b and μ, we find *cores* from vertices with the number of structural neighborhoods larger than or equal to μ (Line 1). For each candidate vertex u, we firstly check if u is a *core* vertex with Lemma 4 (Line 4) and if u has smaller minimal structural similarity than σ_b, we should use its neighbor order to check (Line 5). And then, we use a queue Q to compute all clusters (Lines 6–17). For each *core* vertex v, we add new discovered *core* vertices into the queue Q (Lines 14–15) and mark other *ϵ_b-neighborhood* as the *non-core* vertex (Lines 16–17). In Lines 9–13, we add all *ϵ_b-neighborhood* of v into cluster C. In the whole process, the visited neighbors are skipped because they are *cores* and have been added into the cluster previously (Lines 3,11). We store the result in $\mathbb{C}$ in Line 18. Finally, in Lines 19–20, we mark all *non-core* vertices back to unvisited because the *non-core* vertices may overlap in different clusters.

Theorem 3. *The time complexity of Algorithm 3 is bounded by $O(\sum_{u \in V_{|N[*]| \geq \mu}} |N[u]|)$.*

Proof. With SSCAN-Index$^+$, we can pass those vertices whose number of structural neighborhoods less than μ. For each satisfied vertex $u \in V_{|N[*]|>\mu}$, the time cost of checking its neighbors in Lines 9–17 is bounded by $O(|N(u)|)$ because we can pass some edges based on the neighbor order. So the total time cost for this query algorithm can be bounded by $O(\sum_{u \in V_{|N[*]| \geq \mu}} |N[u]|)$.

Example 3. Reconsider the SSCAN-Index$^+$ in the Fig. 3, if we give a cluster query $\epsilon_b = 0.6$ and $\mu = 4$, we firstly find that v_{10} is a core vertex which is marked dark with $|N[v_{10}]| = 4$ and $0.750 > 0.6$ (Lines 1–5). After marking v_{10} as explored, we put it into the queue Q and new cluster C (Line 6). After that, we continue to check the $u \in N_{\epsilon_b}[v_{10}]$ and put $\{v_7, v_8, v_9\}$ into the same cluster C (Lines 7–13). And then, v_7 and v_9 are also two *core* vertices, we put them into Q to find more cluster numbers (Lines 14–15). The remaining vertices v_8 and v_5 are marked as non-core vertex (Lines 16–17). Finally, we obtain the cluster $\{v_5, v_7, v_8, v_9, v_{10}\}$ and cluster $\{v_0, v_1, v_2, v_4, v_5\}$ can be obtained in a similar way.

After that, we also give an algorithm about constructing index, named SSCAN-Cons$^+$. The pseudocode of SSCAN-Cons$^+$ is shown in Algorithm 4. In Line 1, we use Algorithm 1 to get the number of balanced triangles for each edge. And then, we compute *balanced structural similarity* for each edge and store the structural neighborhoods(Lines 2–7). In neighbor order $NO_b[u]$ of vertex u, we sort the structural neighborhoods $v \in N[u]$ in a non-increasing order of $\sigma_b(u, v)$ (Line 8). We compute *minimal similarity order* for all possible $|N[*]|$ (Lines 9–11). Note that when the $|N[*]|$ equals to 1, each vertex u would be isolated and the result is meaningless. So we make the minimal value of $|N[*]|$ is equal to 2 and each vertex is stored in the corresponding number of structural neighborhoods in Line 10. Finally, the vertices are sorted by their *minimal structural similarity* under $|N[*]|$ (Line 11).

Theorem 4. *The time complexity of Algorithm 4 is* $O((\alpha + \log n) \cdot m)$.

Proof. Based on above analysis, we know Line 1 costs $O(\alpha \cdot m)$. And for each vertex u, sorting its neighbors costs $O(|N(u)| \cdot \log |N(u)|)$ time. So the time cost for all vertices is $O(\sum_{u \in V} |N(u)| \cdot \log |N(u)|) \leq O(2m \cdot \log n)$. Thus, the time complexity of Algorithm 4 is $O((\alpha + \log n) \cdot m)$.

5 Experiments

We have carried out thorough empirical studies to evaluate the effectiveness of our newly developed model, as well as the efficiency of our proposed algorithms. The following algorithms were specifically evaluated.

- SSCAN-Basic: Our baseline clustering algorithm in Sect. 3.
- SSCAN-Cons: Our basic index SSCAN-Index construction algorithm.
- SSCAN-Query: The query algorithm based on SSCAN-Index.
- SSCAN-Cons$^+$: Our proposed index SSCAN-Index$^+$ construction algorithm in Sect. 4.

* SSCAN-Cons$^+$-Naive: SSCAN-Index$^+$ construction algorithm based on naive balanced triangles counting (Compared with Algorithm 1, the naive balanced triangles counting method need to check all edges).
* SSCAN-Query$^+$: The query algorithm based on SSCAN-Index$^+$ in Sect. 4.

All algorithms are implemented in C++, and the experiments are conducted on a Windows server with i7-10700 and 128GB memory.

Datasets. We use 8 public real networks to evaluate our online and index-based algorithms. $\overline{deg}$ means average degree in graph G. Slashdot, Epinions and Wiki are three signed networks. DBLP is a co-authorship network, where vertices denote authors and each edge(u, v) means that the co-authored paper between authors u and v. We convert it to a signed graph same as [6]. The

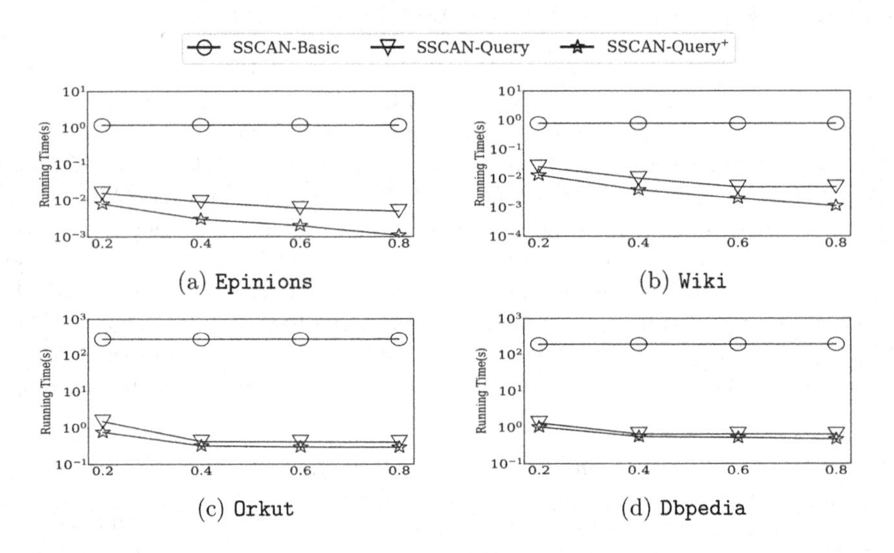

Fig. 4. (Eval-I) Query time for different $\epsilon_b(\mu = 7)$.

Table 1. GRAPH STATISTICS

Graph	$\|V\|$	$\|E\|$	$\|E^+\|$	$\|E^-\|$	$\frac{\|E^+\|}{\|E\|}$	$\overline{deg}$
Slashdot	82,140	500,481	381,648	118,833	0.763	12.2
Epinions	131,580	711,210	592,013	119197	0.832	10.8
Wiki	138,587	715,883	629,689	86,194	0.800	10.3
Pokec	1,632,803	22,301,964	15,611,375	6,690,589	0.700	27.3
DBLP	1,824,701	8,344,615	1,934,354	6,410,261	0.232	9.1
Orkut	3,072,441	117,185,083	82,029,559	35,155,524	0.700	76.3
Livejournal	4,846,609	42,851,237	29,995,866	12,855,371	0.700	17.7
Dbpedia	18,268,991	126,890,209	88,823,147	38,067,062	0.700	13.9

remaining four networks are all unsigned social networks and we randomly pick 30% of the edges as negative edge. `Wiki`, `DBLP` and `Dbpedia` are downloaded from Koblenz network collection[1]. Other datasets are downloaded from SNAP[2].

Parameters. For each graph, we have two parameters, $0 < \epsilon_b \leq 1$ and $\mu \geq 2$. For ϵ_b, we choose 0.2, 0.4, 0.6, and 0.8, with 0.6 as default. For μ, we choose 2, 5, 10, and 15, with 7 as default.

Eval-I: Varying ϵ_b. From Fig. 4, the query time of SSCAN-Basic is steady for different ϵ_b values due to exhaustively computing all structural similarities. Both SSCAN-Query and SSCAN-Query$^+$ take slightly more time for larger ϵ_b. SSCAN-Query$^+$ runs faster than SSCAN-Query for any ϵ_b as a result of the neighbor order in our proposed index SSCAN-Index$^+$.

Eval-II: Varying μ. As shown in Fig. 5, the time cost of SSCAN-Basic and SSCAN-Query does not decrease significantly when μ is increased by 2 to 15 since they both should check all core vertices. Moreover, SSCAN-Query$^+$ significantly outperforms both SSCAN and SSCAN-Query due to its *minimal similarity order* technique (see Sect. 4). Note, when $\mu = 2$, SSCAN-Query$^+$ skips only few vertices whose degree equal to 0, so SSCAN-Query$^+$ and SSCAN-Query have similar efficiency.

Eval-III: Query Efficiency on Different Datasets. Fig. 6 shows the time cost for three algorithms under default parameters $\epsilon_b = 0.6$, $\mu = 7$ on all datasets. It is easy to see SSCAN-Query$^+$ is more efficient than the other two algorithms. For example, the time cost of SSCAN-Query$^+$ in `Wiki` is about 2 milliseconds while SSCAN-Basic and SSCAN-Query will need 764 and 5 milliseconds respectively. In addition, the running time of all three algorithms increase with the number of input edges.

Eval-IV: Index Size on Different Datasets. The size of SSCAN-Index and SSCAN-Index$^+$ is shown in Fig. 7. Usually, as the number of edges increases, the size of both indexes gradually grows. It is important to note that the total size of SSCAN-Index$^+$ can be always bounded by 1.1 times the size of SSCAN-Index because SSCAN-Index only saves the balanced structural similarity for each edge while SSCAN-Index$^+$ need to store the *minimal similarity order*. Take `Wiki` as an example, its SSCAN-Index costs 10.92MB while SSCAN-Index$^+$ costs 11.53MB.

Eval-V: Indexing Construction Time on Different Datasets. The time cost of SSCAN-Cons, SSCAN-Cons$^+$ and SSCAN-Cons$^+$-Naive are reported in Fig. 8. Compared with SSCAN-Cons, we spend extra time sorting vertices and their neighbors according to similarity in SSCAN-Cons$^+$. Also, we can find that

[1] http://konect.cc/networks/
[2] https://snap.stanford.edu/

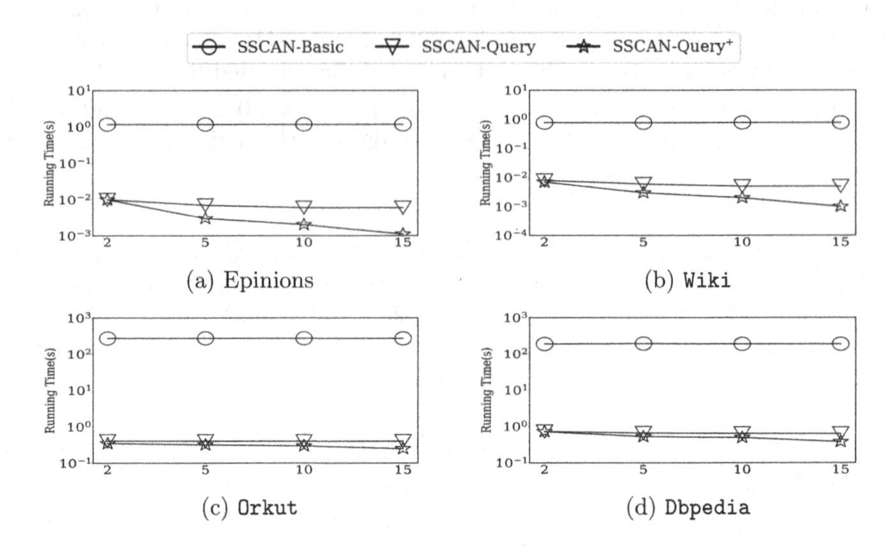

Fig. 5. (Eval-II) Query time for different $\mu(\epsilon_b = 0.6)$.

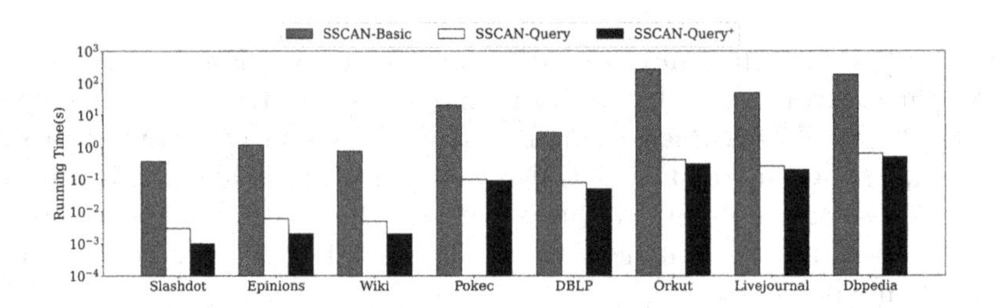

Fig. 6. (Eval-III) Query time on different datasets ($\epsilon_b = 0.6$, $\mu = 7$).

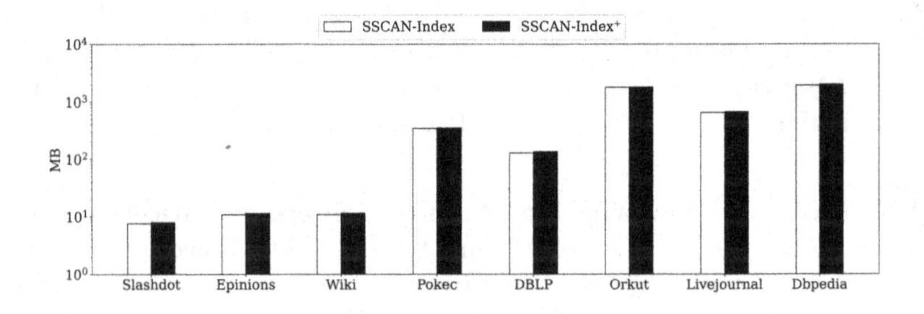

Fig. 7. (Eval-IV) Index size on different datasets.

SSCAN-Cons$^+$ is faster than SSCAN-Cons$^+$-Naive on all datasets, which proves that our pruning technique regarding counting balanced triangles is effective. Besides, the time of constructing index algorithms are almost linear to the number of input edges.

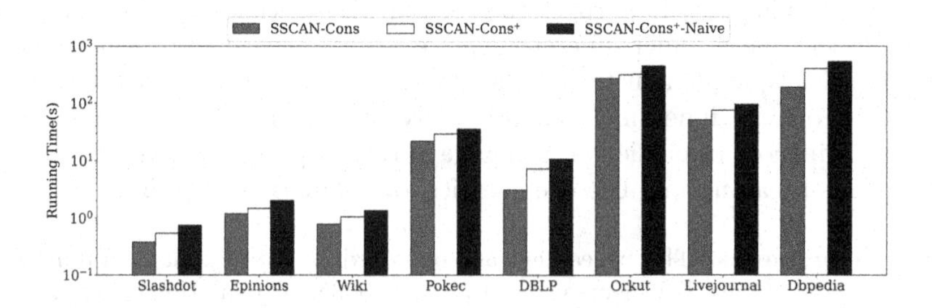

Fig. 8. (Eval-V) Index construction time on different datasets.

Eval-VI: Scalability Testing. Here, we test the scalability of our proposed algorithms. Due to space limitation, we choose two largest datasets (`Orkut` and `Dbpedia`) and sample nodes and edges respectively from 20% to 100%. The time cost of SSCAN-Cons and SSCAN-Cons$^+$ under different percentages are shown in Fig. 9. Generally, the running time of both SSCAN-Cons and SSCAN-Cons$^+$ tends to increase as the sampling ratio increases. However, the gap between them is relatively stable. Compared to the trend of the curve in Fig. 9a and Fig. 9b, the curve in Fig. 9c and Fig. 9d rise sharply with increasing sampling ratio.

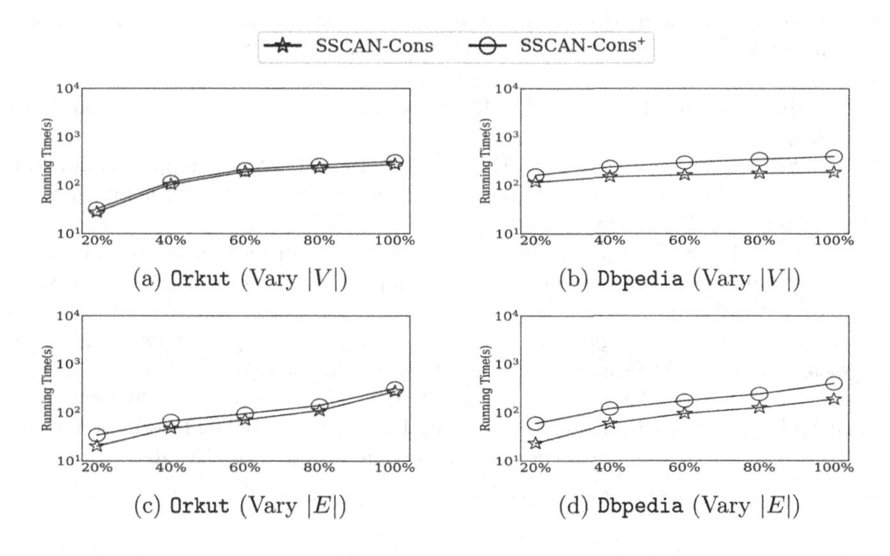

Fig. 9. (Eval-VI) Scalability testing.

6 Conclusion

In this paper, we formulated a problem of computing structural graph clustering over signed graphs. We designed a novel model in terms of structural balance and develop a basic online algorithm to compute clusters. Besides, we further proposed an index-based approach to speed up the checking of core vertices and answering the query under different given ϵ_b, μ. Extensive empirical studies on large signed graphs demonstrated the efficiency of our techniques. Our future work is to consider index maintenance in dynamic signed graphs based on our problem definition. In particular, a change in the positive/negative sign on edges (*i.e.*, friends break up and become enemies, or enemies become friends).

Acknowledgements. This research was sponsored by the Fundamental Research Funds for the Central Universities, 3072022TS0605, China University Industry University-Research Innovation Fund, 2021LDA10004, and National Natural Science Foundation of China, 62272126.

References

1. Bellogín, A., Parapar, J.: Using graph partitioning techniques for neighbour selection in user-based collaborative filtering. In: Proceedings of the Sixth ACM Conference on Recommender Systems, pp. 213–216 (2012)
2. Chang, L., Li, W., Lin, X., Qin, L., Zhang, W.: pSCAN: fast and exact structural graph clustering. In: 2016 IEEE 32nd International Conference on Data Engineering (ICDE), pp. 253–264. IEEE (2016)
3. Chiba, N., Nishizeki, T.: Arboricity and subgraph listing algorithms. SIAM J. Comput. **14**(1), 210–223 (1985)
4. Fortunato, S.: Community detection in graphs. Phys. Rep. **486**(3–5), 75–174 (2010)
5. Girvan, M., Newman, M.E.: Community structure in social and biological networks. Proc. Natl. Acad. Sci. **99**(12), 7821–7826 (2002)
6. Li, R.H., et al.: Efficient signed clique search in signed networks. In: 2018 IEEE 34th International Conference on Data Engineering (ICDE), pp. 245–256. IEEE (2018)
7. Lim, S., Ryu, S., Kwon, S., Jung, K., Lee, J.G.: LinkSCAN: overlapping community detection using the link-space transformation. In: 2014 IEEE 30th International Conference on Data Engineering, pp. 292–303. IEEE (2014)
8. Shiokawa, H., Fujiwara, Y., Onizuka, M.: Scan++ efficient algorithm for finding clusters, hubs and outliers on large-scale graphs. Proc. VLDB Endowment **8**(11), 1178–1189 (2015)
9. Tang, J., Chang, Y., Aggarwal, C., Liu, H.: A survey of signed network mining in social media. ACM Comput. Surv. (CSUR) **49**(3), 1–37 (2016)
10. Xu, X., Yuruk, N., Feng, Z., Schweiger, T.A.: SCAN: a structural clustering algorithm for networks. In: Proceedings of the 13th ACM SIGKDD International Conference on Knowledge Discovery and Data Mining, pp. 824–833 (2007)

ANSWER: Automatic Index Selector for Knowledge Graphs

Zhixin Qi, Haoran Zhang, Hongzhi Wang(✉), and Zemin Chao

Harbin Institute of Technology, Xidazhi Street 92, Harbin, China
{qizhx,wangzh,chaozm}@hit.edu.cn

Abstract. Efficient access to knowledge graphs is identified as the basic premise to make full use of knowledge graphs. Since the query processing efficiency is mainly affected by index configuration, it is necessary to create effective indexes for knowledge graphs. However, none of existing studies of index selection focuses on the characteristics of knowledge graphs. To fill this gap, we propose an automatic index selector for knowledge graphs based on reinforcement learning, named ANSWER, to select an appropriate index configuration according to the historical workloads. However, it is challenging a learn a well-trained index selection model due to the large action space of reinforcement learning model and the requirement of lightweight embedding strategies. To address this problem, we first develop a novel predicate filter, which not only determines which vertical partitioning tables are valuable to create indexes, but also reduces the action space of model. Based on the filtered predicates, we derive an effective and lightweight encoder to not only embed the main features of workloads into the model, but also guarantee the high-efficiency of ANSWER. Experimental results on real-world knowledge graphs demonstrate the effectiveness of ANSWER in terms of knowledge graph query processing.

Keywords: Index selection · Knowledge graph · Reinforcement learning · Vertical partitioning

1 Introduction

With the rapid development of knowledge extraction technologies, various domains construct plenty of domain knowledge graphs, such as geographical knowledge graph LinkedGeoData[1], proteinic knowledge graph UniProt[2], and the knowledge graph LUBM[3] consisting of university information. To take full advantage of these knowledge graphs, it is necessary to adopt effective access techniques for knowledge graphs.

[1] http://www.linkedgeodata.org/About.
[2] https://www.uniprot.org/help/about.
[3] http://swat.cse.lehigh.edu/projects/lubm/.

X. Song et al. (Eds.): APWeb-WAIM 2023, LNCS 14332, pp. 393–407, 2024.
https://doi.org/10.1007/978-981-97-2390-4_27

Relatively mature knowledge graph storage structure in these days is relation-based store, such as triple table, vertical partitioning, horizontal table, and etc [1,8,18,19,26,28]. Even though these stores are effective in knowledge graph management, the query processing efficiency of knowledge graphs is mainly affected by index configuration. Therefore, it is an urgent problem to determine the optimal index configuration based on the given storage structure. However, major researches focus on index recommendation on relational data [4,13,15,16,22], there is a research gap in index selection for knowledge graphs, which makes users hard to select an appropriate index configuration based on the current store. Motivated by this, we attempt to design an index selector for knowledge graphs, which not only helps users determine an appropriate index configuration, but also optimizes the query efficiency on knowledge graphs.

Note that even though index-free adjacent data model makes the query processing efficiency of graph databases higher than that of the relational databases [21], this paper focuses on the optimization of index configuration on relation-based knowledge graph stores, since the relational knowledge graph storage plans are more mature and applied more widely than native graph stores [31]. Therefore, in this paper, we discuss the indexes which are closely related to the knowledge graph stores and designed to optimize the access of knowledge graphs, rather than the indexes for generalized graph data.

Since vertical partitioning is a basic and representative relation-based knowledge graph store and handy to convert to other storage structures, we design the index selector for knowledge graphs based on vertical partitioning so that our selector is easy to be generalized to other knowledge graphs stores.

To test the query performance of different index configurations on vertical partitioning, we process the SPARQL query "SELECT ?a WHERE ?a lives In ?b, ?a has Academic Advisor ?d, ?d works At ?b" on all possible indexes, and the results of query costs are shown in Table 1. In Table 1, A B, and C denote the vertical partitioning table whose predicate is lives In, has Academic Advisor, and works At, respectively. $A.s$ represents creating index for the s (subject) attribute in table A, $B.o$ denotes creating index for o (object) attribute in table B, and $A.s{+}B.o$ represents creating indexes both for s attribute in A and o attribute in B.

From Table 1, we observe that the query costs of creating indexes on both subject attribute and object attribute of table A, B, or C are higher than those of creating index on subject or object attribute of table A, B, or C. This is because that in query processing, the optimizer traverses a vertical partitioning table only once and selects the index which costs less. Therefore, one index on a vertical partitioning table is enough. Moreover, the query costs of creating indexes on all vertical partitioning tables are lower than those of creating indexes on two tables or one table. The reason is that the indexes on each vertical partitioning table related to the query are useful to accelerate the query process.

According to the above observations, this paper aims to design an index selector based on vertical partitioning with the given knowledge graph and its

Table 1. Results of query costs under various index configurations (unit: s)

index	costs	index	costs	index	costs
No index	71.55	$A.s$+$C.o$	0.02	$B.s$+$B.o$	54.45
$A.s$	4.57	$A.o$+$B.s$	0.05	$C.s$+$C.o$	55.14
$A.o$	13.60	$A.o$+$B.o$	0.07	$A.s$+$B.s$+$C.s$	0.02
$B.s$	4.66	$A.o$+$C.s$	0.04	$A.s$+$B.s$+$C.o$	0.02
$B.o$	55.36	$A.o$+$C.o$	13.66	$A.s$+$B.o$+$C.s$	0.02
$C.s$	54.84	$B.s$+$C.s$	0.11	$A.s$+$B.o$+$C.o$	0.02
$C.o$	34.41	$B.s$+$C.o$	0.11	$A.o$+$B.s$+$C.s$	0.04
$A.s$+$B.s$	4.78	$B.o$+$C.s$	54.61	$A.o$+$B.s$+$C.o$	0.04
$A.s$+$B.o$	0.07	$B.o$+$C.o$	0.11	$A.o$+$B.o$+$C.s$	0.04
$A.s$+$C.s$	0.02	$A.s$+$A.o$	4.66	$A.o$+$B.o$+$C.o$	0.06

workloads. The index selector determines to create index on *which* attribute of *which* vertical partitioning table to minimize the knowledge graph query costs.

Even though index selection is a classical problem in the database community, major researches pay attention on recommending indexes for relational data [6,23–25], rare work concentrates on index selector on key-value data and document data [29,30], and none of existing researches focuses on the index recommendation problem for knowledge graphs. In addition, current index selection techniques are not able to be directly applied on recommending indexes for knowledge graphs. On the one hand, the frequent changes of knowledge graph and its workloads require the index recommendation to select indexes automatically based on the historical experience. However, existing heuristic methods are lack of memory ability so that they are not able to accumulate the index tuning experience for guidance. On the other hand, current index tuning approaches mainly consider the properties of relational data, key-value data or document data, while our index selector is designed according to the characteristics of knowledge graphs.

In this paper, we treat the knowledge graph index selection problem as a variant of knapsack problem [11]. Since the benefit of creating an index on different attributes of each vertical partitioning table is unknown and constantly changing, this problem is harder than the knapsack problem. Thus, the hardness of knowledge graph index selection problem is at least NP-hard. To solve this problem, we need to memorize the benefit of different attributes in each table from historical experience.

Inspired by the memory ability of machine learning [5,9,17,27], we adopt machine learning models to tune the index configuration for knowledge graphs. Since each tuning operation is affected by the results of last operation, we model the solution of this problem as a Markov decision process [20]. Given the state of current index configuration, the goal is to decide the next index tuning operation, i.e., creating index on which attribute of which vertical partitioning table, is the most valuable. Based on the observed performance and previous experience, we leverage reinforcement learning techniques to optimize the goal of Markov decision process and obtain the optimal index configuration [4,12,14].

The major contributions of this paper are summarized as follows.

- We propose an automatic index selector for knowledge graphs based on reinforcement learning, named ANSWER, to automatically determine the index configuration with the given knowledge graph and its workloads.
- To train the optimal index tuning strategy based on the historical workloads, we design a predicate filter which is able to not only determine to create indexes on the corresponding vertical partitioning tables of which predicates, but also reduce the action space of reinforcement learning model.
- To design an effective and lightweight embedding strategy, we design an encoder which embeds the main features of candidate index structures into the training model, but also guarantees the effectiveness of knowledge graph index selector.
- To demonstrate the effectiveness of ANSWER, we conduct extensive experiments on real-world knowledge graphs. Experimental results show that the index configuration selected by ANSWER is able to process the knowledge graph workloads effectively.

The rest of this paper is organized as follows. We review related work in Sect. 2. Section 3 introduces the framework of knowledge graph index selection. We discuss the proposed index selector ANSWER in Sect. 4. Experiments are conducted in Sect. 5. Section 6 concludes this paper.

2 Related Work

In this section, we review the research advances in index selection.

In the early years, index optimization concentrates on the index selection problem of self-adaptive database management system. Hammer and Chan obtain accurate models and predict data characteristics according to continuous detection and estimation techniques in order to choose indexes for a specific relation [6]. Chaudhuri and Narasayya propose a what-if index analysis tool which is applied on Microsoft SQL Server 7.0 database [2]. The tool is able to quantitatively analyze the impacts that existing indexes have on system performance, such as analyzing the database workloads, predicting the cost variations of a workload, so that many front-end tools and user interfaces are developed. These two researchers discuss the advances in automatic index selection of self-tuning database system based on their experience in AutoAdmin project, and present some opportunities and challenges in further researches [3].

In recent years, as the machine learning techniques develop, automatic index selection is continuously improved based on machine learning. Ding et al. observe that in the process of index tuning, it is an essential step to compare the execution costs of two plans which correspond to different index configurations [4]. However, traditional approaches usually leverage the prediction of optimizers to make this comparison. They formalize this comparison problem as a classification task in machine learning, and integrate the classifier into the index tuner. Evaluation results show that the classifier achieve a more accurate prediction result compared with optimizers. Kossmann et al. discuss 8 index selection algorithms based on different concepts and comprehensively compare them in terms of different dimensions, such as index quality, execution time, support for multi-dimension, and complexity [13]. Based on the evaluation results, they provide guidelines for applied scenarios of different index algorithms. Lan et al. propose five heuristic rules to generate candidate indexes [15]. They formalize the index selection problem as a reinforcement learning problem and leverage deep Q network to solve it. Experimental results show that heuristic rules reduce the number of dimensions in the action space and state space of reinforcement learning task dramatically. Also, the neural network in deep Q network models the relations between different indexes effectively. Licks and Meneguzzi present an automatic and dynamic database index architecture based on reinforcement learning [16]. The architecture determines to create or delete indexes according to evaluate the performance of query, insert, or update operations. Sadri et al. design an index recommender for read-only workloads in the replicated databases [22]. The recommender uses deep reinforcement learning methods to learn how to select an appropriate index configuration for nodes in a database cluster.

Existing researches of automatic index selection show that machine learning methods are able to be applied into the index optimization problem. Motivated by this, this paper adopts machine learning to select index configuration. Compared with the previous studies, our work focuses on selecting indexes according to the properties of knowledge graphs.

3 Framework Overview

In this section, we overview the framework of automatic index selector for knowledge graphs based on reinforcement learning. The index selector is designed for two goals. First, it needs to train the optimal index tuning strategy based on the historical workloads. Second, it is required to be effective and lightweight to embed candidate index configurations. For the first goal, we propose a predicate filter which is able to not only determine the vertical partitioning tables that correspond to which predicates are worthwhile creating indexes, but also reduce the action space of reinforcement learning models effectively. For the second goal, we design an encoding module which is able to embed the main features of index structures into the index selection model and guarantee the high effectiveness of the index selector.

The framework is sketched in Fig. 1. There are three components in the framework, that are predicate filter, encoder, and automatic index selector for knowledge graphs (ANSWER). First, the predicate filter chooses the effective predicates which are used to generate indexes according to the given workloads. The selected predicates are sent to the encoding module. The encoder extracts the main features of candidate indexes, encodes them into vectors, and inputs them into ANSWER. Then, ANSWER leverages the reinforcement learning model to obtain the index configuration after each index tuning operation and sends the configuration to the database execution engine. The engine computes the actual query costs according to the workloads and current index configuration, and then sends the results back to ANSWER. After several iterations, ANSWER outputs the selected index configuration.

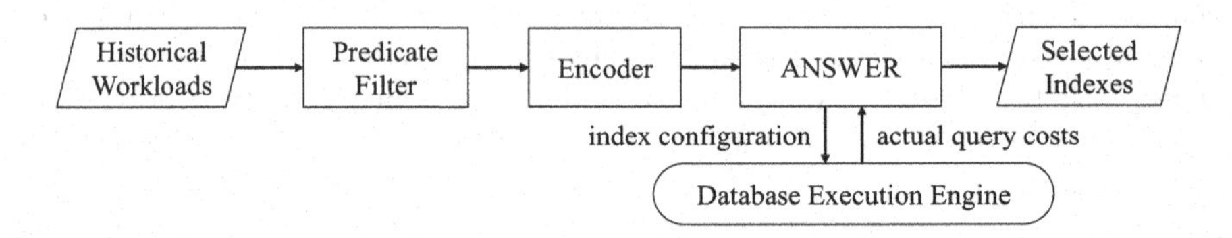

Fig. 1. Framework of automatic index selection for knowledge graphs

In the remaining part of this section, we introduce each component in the framework of automatic index selection for knowledge graphs.

3.1 Predicate Filter

The goal of this module is to determine the optimal index tuning strategy. In practice, a knowledge graph manager suspends the access to knowledge graphs periodically, reconfigures the knowledge graph indexes, and then restores the access service. Since the newly coming knowledge graph queries are most similar to the most recent batch of knowledge graph workloads, the predicate filter determine which predicates are valuable according to the different batches of historical workloads.

Assume there are N different predicates in a knowledge graph G, we compute a score for each predicate P_i ($1 \leq i \leq N$). If P_i occurs x_i times in the most recent batch, the score of P_i in this batch is $1 \times x_i$. If P_i occurs y_i times in the second most recent batch, the score of P_i in this batch is $0.5 \times y_i$. If P_i occurs z_i times in the third most recent batch, the score of P_i in this batch is $0.25 \times z_i$. The score of P_i which occurs before the most recent three batches is 0. Therefore, the score of each predicate P_i in the filter is $\sum_{1 \leq i \leq N} 1 \times x_i + 0.5 \times y_i + 0.25 \times z_i$.

According to the corresponding score of each predicate, we select the predicates whose scores are the first n highest to create indexes in the follow-up steps. In this way, we filter the predicates which are not important enough, so that the action space of the index selection model is reduced effectively. If the predicate filtering is skipped, that is, N predicates in the knowledge graph are all used to create index selection model, the action space is at least 3^N. If the number of predicates is too large, the corresponding action number of selecting indexes is too large, which leads to an unsatisfying accuracy of the index selection model. If the number of predicates is too small, the corresponding action number is too small, which makes it unable to create an effective index. According to the predicate filter, the action space is effectively reduced to 3^n from 3^N, where n is a parameter to be tuned.

As discussed above, the predicate filter is able to not only determine the vertical partitioning tables that correspond to which predicates, but also reduce the action space of reinforcement learning model to guarantee the model effectiveness.

3.2 Encoder

The goal of this component is to embed candidate index configurations into the index selection model effectively. To achieve this goal, we add an encoder in the automatic index selection framework. Since a complex embedder has impacts on the model performance, we attempt to design an effective and lightweight encoder to guarantee the effectiveness and high-efficiency of ANSWER. Motivated by this, we propose a lightweight encoding strategy to embed the main features of each candidate index structure into an index vector.

The encoding strategy encodes each candidate index structure into an n-dimensional index vector V_n, where n is the number of predicates which are selected by predicate filter. We use each bit vector of V_n to denote whether we create an index on the vertical partitioning table and the creation method. If there is not an index on the vertical partitioning table which the predicate corresponds to, the bit vector is 0. If there is an index on the subject attribute of the table which the predicate corresponds to, the bit vector is 1. If there is an index on the object attribute of the vertical partitioning table, the bit vector is 2. Even though we considers single index type in this paper, the index encoding strategy is easily extended to multi index types.

Therefore, the index encoding module not only retains the characteristics of each index configuration, but also guarantees the high-efficiency of ANSWER.

3.3 ANSWER

The goal of this module is to select an appropriate knowledge graph index configuration based on reinforcement learning. To achieve this goal, ANSWER obtains the index configuration after each index tuning operation of reinforcement learning model and sends the configuration to the database. The database execution engine computes the actual query costs according to the workloads and current index configuration, and then sends the results back to ANSWER. After several iterations, ANSWER produces the selected index configuration. We will discuss this component in detail in Sect. 4.

4 ANSWER

To accelerate the knowledge graph query process, we tune the index configurations according to dynamic workloads, In this section, we discuss the ANSWER module in Fig. 1.

We first define the knowledge graph index selection problem as follows.

Definition 1. *Given a knowledge graph historical workload W, a predicate set $P=\{p_1, p_2, ..., p_n\}$, the size budget of index structures B_G, the knowledge graph index selection problem is to compute an n-dimensional index vector $V_n=\{v_1, v_2, ..., v_n\}$, where v_i ($1 \le i \le n$) denotes the index creation method of the vertical partitioning table which p_i ($1 \le i \le n$) corresponds to, to minimize the query costs of W on this index configuration and the index size is within B_G.*

According to the above definition, the knowledge graph index selection problem is a variant of knapsack problem [11]. We treat the index configuration which v_i ($1 \le i \le n$) corresponds to as an item, the benefit of an index configuration as the weight of each item, and the size budget B_G as the maximum capacity of a knapsack. Since the benefit of each index configuration is unknown and changing, the knowledge graph index selection problem is harder than the knapsack problem. Therefore, the hardness of knowledge graph index selection is at least NP-hard.

To solve this problem, we model the knowledge graph index selection problem as a Markov decision process as follows.

(1) State: For each v_i ($1 \le i \le n$) in V_n, the state space is $\{0, 1, 2\}$.
(2) Action: Treat every change of each bit vector v_i ($1 \le i \le n$) as an action of index updating.
(3) Reward: The reward of each action is the cost difference between the query cost of W on index vector V_n after an action and those before this action.
(4) Policy: The final value of V_n denotes the index creating policy for knowledge graphs.

The pseudo codes of ANSWER are shown in Algorithm 1. The inputs are knowledge graph workloads W, index size budget B_G, the number of predicates after filtering n, random probability ϵ, learning rate α, discount factor γ, and each learning size extracted from experience pool *batch size*. The algorithm produces a learned index selection model M.

First, we initialize the parameters in evaluation network and target network of reinforcement learning model M (Line 1). In each training episode, we input current state s of index vector into evaluation network (Lines 2–3). According to different Q values outputted from evaluation network, we select the action whose Q value is maximum and label it as a (Line 4). By executing the selected action a under the current state s, we obtain the reward r on workloads W and the next state s_{next} (Line 5). When the stored samples in experience pool reach the upper limit, we extract *batch size* samples from the experience pool, input quaternary experience (s, a, r, s_{next}) into the target network, and update the parameters in model M (Lines 6–9).

The time complexity of Algorithm 1 is related to the training episode number and action space size. Since the action space is reduced by predicate filter effectively, the time complexity of model training is accordingly reduced. Therefore, the predicate filter is an important component in the framework of automatic index selection for knowledge graphs.

In the process of knowledge graph index selection, when we obtain the next state based on current state and action, we need to adopt index replacement

Algorithm 1. Model Training for ANSWER

Require: knowledge graph workloads W, index size budget B_G, the number of predicates after filtering n, random probability ϵ, learning rate α, discount factor γ, each learning size *batch size* extracted from experience pool
Ensure: learned index selection model M
1: initialize parameters in evaluation network and target network of model $M(\epsilon, \alpha, \gamma)$
2: **for** each *episode* **do**
3: input current state s of index vector into evaluation network
4: select action a whose Q value is maximum in evaluation network
5: s_{next}, $r \leftarrow W(s, a, B_G)$
6: **if** stored samples in experience pool reach the upper limit **then**
7: extract *batch size* samples from experience pool
8: $M \leftarrow (s, a, r, s_{next})$
9: **end if**
10: **end for**
11: return M

strategies due to the budget of index size. Frequently-used replacement strategies include First In First Out (FIFO) strategy, Last In First Out (LIFO) strategy, and Least Used First Out (LUFO) strategy. We will discuss the selection of index replacement strategies in our experimental evaluation.

5 Experimental Evaluation

In this section, we experimentally evaluate the proposed automatic index selector based on reinforcement learning ANSWER. First, we describe experimental settings in Subsect. 5.1. Then, in Subsect. 5.2, we compare ANSWER with other knowledge graph index selectors to demonstrate the advantages of our proposed method in terms of knowledge graph query processing. In Subsect. 5.3, we test the impacts of different replacement strategies on ANSWER with the constraint of index size budget. Finally, we tune parameters in ANSWER in Subsect. 5.4.

5.1 Settings

In this subsection, we introduce the knowledge graphs, workloads, evaluation metric, and experimental environment in the experiments.

Knowledge Graphs and Workloads. We select three public knowledge graphs, that are YAGO[4], WatDiv[5], and Bio2RDF[6]. The information of knowledge graphs are shown in Table 2. YAGO contains all the triples in YagoFacts and the triples whose predicate is "hasGivenName" or "hasFamilyName". We generate WatDiv by setting 14,634,621 as the number of triples. Bio2RDF consists of

[4] https://yago-knowledge.org/.
[5] https://dsg.uwaterloo.ca/watdiv/.
[6] https://download.bio2rdf.org/.

the knowledge from Interaction Reference Index, Online Mendelian Inheritance in Man, Pharmacogenomics Knowledge Base, and PubMed.

Table 2. Knowledge graph information

name	size	#-triples	#-(subject, object)	#-predicates	#-queries
YAGO	679.3M	16,418,085	5,593,541	39	30
WatDiv	2.08G	14,634,621	1,396,039	86	30
Bio2RDF	7.64G	60,241,165	8,914,390	161	30

We generate workloads from the query templates of YAGO and Bio2RDF provided in [7] and the public WatDiv query templates[7]. In this paper, we set each batch of queries as $\frac{1}{6}$ knowledge graph query workloads. To avoid the cold start of databases, the first batch of queries are used to pre-heat the databases and the query costs of this batch are not recorded in the total execution time.

Evaluation Metric. Our evaluation metric is time-to-insight, named *TTI* [10], which is the total elapsed time from a batch of workload submission to completion. *TTI* evaluates the costs of our online knowledge graph query process.

Experimental Environment. All experiments are conducted on a machine with Intel i7@2.60GHz, 16G memory, and a 1T hard disk, under Windows 10. All algorithms are implemented in Python and the used database is PostgreSQL 14.2.

5.2 Comparisons of Index Selectors

To test the effectiveness of the proposed ANSWER, we compare the performance of ANSWER with random index selector and fixed index selector on YAGO, WatDiv, and Bio2RDF. Random index selector creates indexes on random attributes of the vertical partitioning tables which the predicates selected by predict filter correspond to. Before each batch of queries comes, we reconfigure the indexes randomly. Fixed index selectors include creating fixed indexes on subject attribute of vertical partitioning tables and creating indexes on object attribute of tables. The indexes are created according to the ascending index size until the upper limit is reached. When the index size achieves the budget, ANSWER and random index selector adopt FIFO replacement strategy. The comparison results are shown in Table 3.

In Table 3, we observe that on YAGO, WatDiv, and Bio2RDF, the total *TTI* of ANSWER is lower than that of random index selector and two fixed index selectors. From the observations of *TTI* values in each batch, it can be seen that random index selector and fixed index selectors incur expensive large costs in

[7] https://dsg.uwaterloo.ca/watdiv/basic-testing.shtml.

Table 3. Comparison results of index selectors (unit: s)

data set	method	batch 1	batch 2	batch 3	batch 4	batch 5	total *TTI*
YAGO	RANDOM	52.33	29.56	0.31	5.33	9.72	97.25
	FIXED-S	98.89	65.40	0.42	16.88	25.68	207.27
	FIXED-O	78.73	58.30	0.43	17.95	26.68	182.09
	ANSWER	23.61	12.80	0.40	18.37	16.87	72.04
WatDiv	RANDOM	0.88	0.23	0.15	0.22	0.27	1.74
	FIXED-S	1.05	0.35	0.24	0.08	0.07	1.79
	FIXED-O	2.77	0.18	0.11	0.07	0.34	3.47
	ANSWER	0.95	0.20	0.07	0.04	0.30	1.56
Bio2RDF	RANDOM	0.22	0.42	2.70	0.28	0.03	3.65
	FIXED-S	0.01	0.01	4.40	0.43	0.06	4.91
	FIXED-O	0.10	0.64	4.69	0.41	0.02	5.86
	ANSWER	0.13	0.31	2.61	0.39	0.02	3.47

many batches, which causes big cost difference compared with ANSWER. By contrast, the *TTI* values of ANSWER are stable. Even though the query costs of ANSWER are larger than those of random selector and fixed selectors in some batches, the cost difference is little.

Experimental results demonstrate that the performance of ANSWER is better than that of random index selector and two fixed index selectors.

5.3 Comparisons of Replacement Strategies

To determine the appropriate replacement strategy for ANSWER, we compare the performance of First In First Out (FIFO), Last In First Out (LIFO), and Least Used First Out (LUFO) strategies on YAGO, WatDiv, and Bio2RDF. The comparison results are shown in Table 4.

From Table 4, we see that on YAGO, WatDiv, and Bio2RDF, the performance of FIFO is better than that of LIFO and LUFO. This is because that the features

Table 4. Comparison results of index replacement strategies (unit: s)

data set	strategy	batch 1	batch 2	batch 3	batch 4	batch 5	total *TTI*
YAGO	FIFO	23.61	12.80	0.40	18.37	16.87	72.04
	LIFO	26.98	18.80	0.41	17.51	26.22	89.91
	LUFO	24.38	24.37	0.42	17.54	11.09	77.80
WatDiv	FIFO	0.95	0.20	0.07	0.04	0.30	1.56
	LIFO	1.01	0.22	0.09	0.06	0.34	1.73
	LUFO	1.04	0.30	0.09	0.08	0.36	1.86
Bio2RDF	FIFO	0.13	0.31	2.61	0.39	0.02	3.47
	LIFO	0.11	0.32	3.41	0.41	0.02	4.27
	LUFO	0.15	0.33	3.43	0.42	0.02	4.35

of newly coming workloads are most similar to those of the most recent workloads and the new index configuration is created by the main features of most recent batch of queries. Therefore, the query performance achieves the best when we retain the latest indexes and replace the indexes which are created the earliest.

Thus, we recommend users to consider FIFO replacement strategy first when using ANSWER.

Table 5. Results of parameter tuning

parameter name	parameter value	TTI (s)
predicate number	4	148.8952
	5	161.6822
	6	130.5021
	7	175.0430
	8	267.3235
budget ratio	15%	128.9137
	20%	102.6831
	25%	145.9639
	30%	141.7626
	35%	113.5041
random probability	0.6	144.1609
	0.7	146.6438
	0.8	86.2498
	0.9	116.0678
	1.0	145.3699
learning rate	0.15%	136.8115
	0.2%	79.0571
	0.25%	153.4904
	0.3%	132.3348
	0.35%	136.6510
discount factor	0.75	153.1401
	0.8	117.0759
	0.85	99.2030
	0.9	74.6927
	0.95	121.9446
batch size	1	104.5400
	2	116.2073
	3	72.9175
	4	154.3170
	5	149.8909

5.4 Parameter Tuning

There are 6 parameters to be tuned in ANSWER, that are the number of predicates generated by predicate filter *predicate number*, the ratio of index size budget to total size *budget ratio*, random probability in deep Q network *random probability*, learning rate *learning rate*, discount factor *discount factor*, and the learning size extracted from experience pool once *batch size*.

To determine the optimal value of these parameters, we vary the value of each parameter and test the *TTI* values on YAGO. When a parameter value is changed, the other parameters are set to default values. The default value of *predicate number* is 5, *budget ratio* is 25%, *random probability* is 0.9, *learning rate* is 0.1%, *discount factor* is 0.9, and *batch size* is 3. The experimental results are shown in Table 5.

From Table 5, when the value of *predicate number* is 6, *budget ratio* is 20%, *random probability* is 0.8, *learning rate* is 0.2%, *discount factor* is 0.9, and *batch size* is 3, the value of *TTI* is the least. Therefore, to achieve the best training performance of ANSWER, we set *predicate number* to 6, *budget ratio* to 20%, *random probability* to 0.8, *learning rate* to 0.2%, *discount factor* to 0.9, and *batch size* to 3.

6 Conclusion

In this paper, we propose an automatic index selector for knowledge graphs based on reinforcement learning. To train the optimal index tuning strategy based on historical workloads, we design a predicate filter to select the vertical partitioning tables which are worthwhile creating indexes. To guarantee the high efficiency of index selector, we present a lightweight encoding strategy to embed the main features of index structures into the training model. Based on the vertical partitioning tables selected by predicate filter and the index vectors embedded by encoder, we obtain the knowledge graph index configuration. Evaluation results demonstrate the effectiveness of our proposed index selector.

Acknowledgements. This paper was partially supported by NSFC grant U1866602. Haoran Zhang and Zhixin Qi contributed to the work equally and should be regarded as co-first authors.

References

1. Abadi, D.J., Marcus, A., Madden, S.R., Hollenbach, K.: SW-store: a vertically partitioned DBMS for semantic web data management. VLDB J. **18**(2), 385–406 (2009)
2. Chaudhuri, S., Narasayya, V.: Autoadmin what-if index analysis utility. ACM SIGMOD Rec. **27**(2), 367–378 (1998)
3. Chaudhuri, S., Narasayya, V.: Self-tuning database systems: a decade of progress. In: Proceedings of the 33rd International Conference on Very Large Data Bases, pp. 3–14 (2007)

4. Ding, B., Das, S., Marcus, R., Wu, W., Chaudhuri, S., Narasayya, V.R.: AI meets AI: leveraging query executions to improve index recommendations. In: Proceedings of the 2019 International Conference on Management of Data, pp. 1241–1258 (2019)
5. Dutt, A., Wang, C., Nazi, A., Kandula, S., Narasayya, V., Chaudhuri, S.: Selectivity estimation for range predicates using lightweight models. Proc. VLDB Endowment **12**(9), 1044–1057 (2019)
6. Hammer, M., Chan, A.: Index selection in a self-adaptive data base management system. In: Proceedings of the 1976 ACM SIGMOD International Conference on Management of Data, pp. 1–8 (1976)
7. Harbi, R., Abdelaziz, I., Kalnis, P., Mamoulis, N., Ebrahim, Y., Sahli, M.: Accelerating SPARQL queries by exploiting hash-based locality and adaptive partitioning. VLDBJ (2016)
8. Harris, S., Gibbins, N.: 3store: Efficient bulk RDF storage. In: Proceedings of the 1st International Workshop on Practical and Scalable Semantic Systems, pp. 81–95 (2004)
9. Hasan, S., Thirumuruganathan, S., Augustine, J., Koudas, N., Das, G.: Deep learning models for selectivity estimation of multi-attribute queries. In: Proceedings of the 2020 ACM SIGMOD International Conference on Management of Data, pp. 1035–1050 (2020)
10. Hopkins, M.S.: Big data, analytics and the path from insights to value. Sloan Management Review (2011)
11. Horowitz, E., Sahni, S.: Computing partitions with applications to the knapsack problem. J. ACM (JACM) **21**(2), 277–292 (1974)
12. Kara, K., Eguro, K., Zhang, C., Alonso, G.: ColumnML: column-store machine learning with on-the-fly data transformation. Proc. VLDB Endowment **12**(4), 348–361 (2018)
13. Kossmann, J., Halfpap, S., Jankrift, M., Schlosser, R.: Magic mirror in my hand, which is the best in the land? An experimental evaluation of index selection algorithms. Proc. VLDB Endowment **13**(12), 2382–2395 (2020)
14. Kraska, T., Beutel, A., Chi, E.H., Dean, J., Polyzotis, N.: The case for learned index structures. In: Proceedings of the 2018 International Conference on Management of Data, pp. 489–504 (2018)
15. Lan, H., Bao, Z., Peng, Y.: An index advisor using deep reinforcement learning. In: Proceedings of the 29th ACM International Conference on Information & Knowledge Management, pp. 2105–2108 (2020)
16. Licks, G.P., Meneguzzi, F.: Automated database indexing using model-free reinforcement learning (2020). arXiv preprint arXiv:2007.14244
17. Müller, M., Moerkotte, G., Kolb, O.: Improved selectivity estimation by combining knowledge from sampling and synopses. Proc. VLDB Endowment **11**(9), 1016–1028 (2018)
18. Neumann, T., Weikum, G.: RDF-3X: a RISC-style engine for RDF. Proc. VLDB Endowment **1**(1), 647–659 (2008)
19. Pan, Z., Heflin, J.: DLDB: extending relational databases to support semantic web queries. In: Proceedings of the 1st International Workshop on Practical and Scalable Semantic Systems, pp. 109–113 (2004)
20. Puterman, M.L.: Markov decision processes. Handbooks Oper. Res. Manag. Sci. **2**, 331–434 (1990)
21. Robinson, I., Webber, J., Eifrem, E.: Graph databases: new opportunities for connected data. "O'Reilly Media, Inc." (2015)

22. Sadri, Z., Gruenwald, L., Lead, E.: DRLindex: deep reinforcement learning index advisor for a cluster database. In: Proceedings of the 24th Symposium on International Database Engineering & Applications, pp. 1–8 (2020)
23. Sattler, K.U., Geist, I., Schallehn, E.: QUIET: continuous query-driven index tuning. In: Proceedings 2003 VLDB Conference, pp. 1129–1132. Elsevier (2003)
24. Schkolnick, M.: The optimal selection of secondary indices for files. Inf. Syst. **1**(4), 141–146 (1975)
25. Stonebraker, M.: The choice of partial inversions and combined indices. Int. J. Comput. Inf. Sci. **3**(2), 167–188 (1974)
26. Sun, W., Fokoue, A., Srinivas, K., Kementsietsidis, A., Hu, G., Xie, G.: SQLGraph: an efficient relational-based property graph store. In: Proceedings of the 2015 ACM SIGMOD International Conference on Management of Data, pp. 1887–1901 (2015)
27. Wang, X., Qu, C., Wu, W., Wang, J., Zhou, Q.: Are we ready for learned cardinality estimation? Proc. VLDB Endowment **14**(9), 1640–1654 (2021)
28. Wilkinson, K., Wilkinson, K.: Jena property table implementation. In: Proceedings of the 2nd International Workshop on Scalable Semantic Web Knowledge Base Systems, pp. 35–46. Citeseer (2006)
29. Yan, Yu., Wang, H.: General model for index recommendation based on convolutional neural network. In: Zeng, J., Jing, W., Song, X., Lu, Z. (eds.) ICPCSEE 2020. CCIS, vol. 1257, pp. 3–15. Springer, Singapore (2020). https://doi.org/10.1007/978-981-15-7981-3_1
30. Yan, Y., Yao, S., Wang, H., Gao, M.: Index selection for NoSQL database with deep reinforcement learning. Inf. Sci. **561**, 20–30 (2021)
31. Zhang, R., Liu, P., Guo, X., Li, S., Wang, X.: A unified relational storage scheme for RDF and property graphs. In: Ni, W., Wang, X., Song, W., Li, Y. (eds.) WISA 2019. LNCS, vol. 11817, pp. 418–429. Springer, Cham (2019). https://doi.org/10.1007/978-3-030-30952-7_41

A Long-Tail Relation Extraction Model Based on Dependency Path and Relation Graph Embedding

Yifan Li[1], Yanxiang Zong[2], Wen Sun[2], Qingqiang Wu[1,2,3](✉), and Qingqi Hong[1](✉)

[1] School of Film, Xiamen University, Xiamen 361005, China
liyifan422@stu.xmu.edu.cn,
{wuqq,Hongqq}@xmu.edu.cn
[2] School of Informatics, Xiamen University, Xiamen 361005, China
[3] Key Laboratory of Digital Protection and Intelligent Processing of Intangible Cultural Heritage of Fujian and Taiwan Ministry of Culture and Tourism, Xiamen University, Xiamen, China

Abstract. Distant supervision, a method for relation extraction, leverages knowledge base triples to label entities and relations in text, but this leads to noisy labels and long-tail problems. Among long-tail dependency structures, the hierarchy tree of relations is the most classical and has demonstrated great efficacy in information extraction. However, the hierarchical tree of relations presents a challenge in obtaining sufficient information representation in cases where there is no sibling node or parent node without sibling node. To address this challenge, the use of constraint graphs has been proposed, but such approaches neglect the hierarchical information in the relations. To overcome this limitation, we propose a model based on dependency paths and relational graph embeddings. The model utilizes two relational graph structures, the constraint graph and the relation hierarchy tree, for relation learning, with the aim of transferring the knowledge learned in the data-rich relation to the long-tail relation. Additionally, the model leverages the shortest dependency path between entity pairs to increase the discriminative power of entity pairs in different bags for multi-instance learning. Experimental results show that the model achieves an AUC of 54.3% on the NYT-10 dataset and 86.3% on Hit@15 (<100).

Keywords: Distant Supervision · Relation Extraction · Relation Hierarchy Tree · Constraint Graph

1 Introduction

Relation extraction (RE) plays a critical role in information extraction, dedicated to distinguishing the relation between two entities based on their semantics in unstructured text. However, it requires sufficient labeled corpus, which are costly and laborious. To address this, Distantly Supervised Relation Extraction (DSRE) [1] has been proposed, which leverages knowledge graphs (KGs) to

X. Song et al. (Eds.): APWeb-WAIM 2023, LNCS 14332, pp. 408–423, 2024.
https://doi.org/10.1007/978-981-97-2390-4_28

annotate raw text corpus for relation extraction. It assumes given two entities, all sentences containing both entities will express the semantics of the relation between them in KGs. However, it is too strong and leads to certain problems.

The first challenge in DSRE is the label noise problem. Noise generally refers to information that is irrelevant to the relation, such as unrelated text between entities. To address this challenge, the most popular framework is a combination of multi-instance learning [4] and attention mechanisms. Multi-instance learning aggregates sentences containing two common entities into a bag and leverages the information of the sentences in the bag to classify entity pairs. Meanwhile, the attention mechanism is applied to filter out noisy information by assigning smaller weights to the noisy sentences [13–15]. Additionally, incorporating auxiliary information such as entity type and relation alias to construct features is also a common approach to reducing noise.

Another major challenge in DSRE is the long-tail problem, wherein a small fraction of relations constitutes the majority of training samples. For instance, in the training set of the NYT-10 dataset [12], only 40% of the 52 positive relations have more than 100 training examples. While long-tail relations are critical for constructing and augmenting knowledge graphs, the severe class imbalance in the dataset distribution poses a limitation on relation extraction performance.

To address the long-tail problem, recent work [2,20] in DSRE has proposed a relation hierarchical structure considering the semantic dependency between relations and the fact that most relations in KGs have a hierarchical structure. The approach learns from data-rich relations and transfers this knowledge to long-tail relations, leveraging coarse-grained overlaps [5] between them. Furthermore, some works [6,10] connect the long-tail relations and data-rich relations on the same node through a constraint graph, which facilitates more efficient information transfer. The constraint graph also restricts the type of entities, effectively eliminating noise sentences that do not conform to these constraints.

Although constraint graphs enrich the representation of long-tail relations by defining the type of head and tail entities, they ignore the hierarchical information in the relations. To provide a more effective representation of long-tail relations by fusing these two kinds of information, we propose a model that effectively integrates relation constraint information and hierarchical information, facilitating the flow of more effective information to long-tail relations. Our model utilizes a graph attention network (GAT) [28] as a graph encoder to extract interaction information from both the constraint graph and relation hierarchy. The GAT network can combine the characteristics of neighbor nodes to promote information propagation between nodes, and the attention mechanism calculates the importance of adjacent nodes, focusing on the representation of more relevant neighbor nodes, which effectively enhances the representation of long-tail relations. Our contributions can be summarized as follows:

- Our model combines dependency paths and relation graph embeddings to enhance the representation of relations. It achieves this by merging hierarchical trees and constraint graphs through relational graph fusion. Additionally, by using a combination of entity type and dependency path,

the model improves the discrimination of entity pairs across different sentences.
- We evaluate the effectiveness of our model on the NYT-10 dataset. The results demonstrate that our model can achieve excellent performance in extracting long-tail relations while also having better anti-noise interference ability than the other models.

2 Approach

In this section, we provide a detailed introduction to our model, as illustrated in Fig. 1. The model comprises five main components: a sentence encoder, a dependency path encoder, a relation encoder, an aggregator, and a classifier.

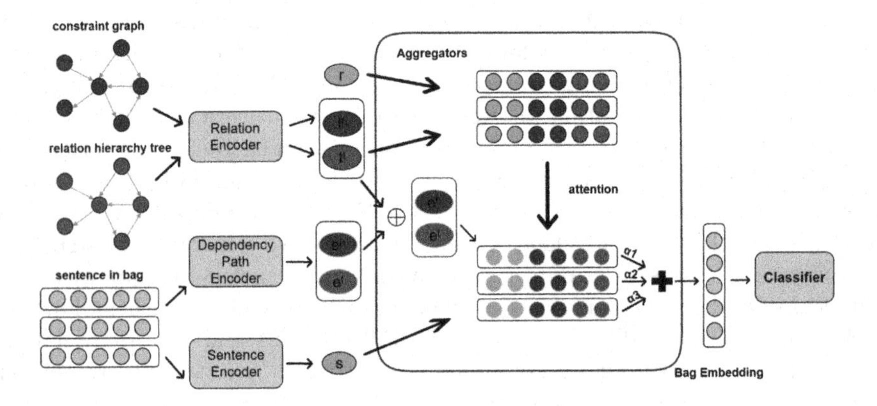

Fig. 1. Overall architecture of the model.

2.1 Task Definition

Given a bag of sentences $\{S_1, S_2, ..., S_m\}$ containing the same entity pair $\langle e_h, e_t \rangle$ and a set of pre-defined relation classes $\{r_1, r_2, ..., r_n\}$, the goal of DSRE is to identify the relations between the given entity pairs based on an entity-pair bag, and the long-tail problem refers to the difficulty of finding and extracting information about relationships that are not commonly mentioned or occur rarely in the available data.

2.2 Sentence Encoder

The sentence encoder consists of two parts: the embedding layer and the encoding layer. The embedding layer maps the word sequence in the sentence to a distributed semantic vector, while the encoding layer extracts a high-dimensional representation from the input sequence. In our model, the embedding layer uses

Entity-aware Word Embedding [7], while the encoding layer uses the Piecewise Convolutional Neural Network (PCNN) model [25].

Given a sentence $s = \{w_1, w_2, ..., w_m\}$, we first use the pre-trained word vector to map each word w_i into a d_v-dimensional vector v_i. The head entity e_h and tail entity e_t in the sentence are mapped to v^{e_h} and v^{e_t}, respectively.

$$x_i^p = [v_i; p_i^{e_h}; p_i^{e_t}] \in R^{d_v+2d_p} \tag{1}$$

$$x_i^e = [v_i; v^{e_h}; v^{e_t}] \in R^{3d_v} \tag{2}$$

After that, $X^e = \{x_1^e, x_2^e, ..., x_m^e\}$ and $X^p = \{x_1^p, x_2^p, ..., x_m^p\}$ are used to compute the entity-aware word vector representation of the sentence.

$$A^e = sigmoid(\lambda \cdot (W_1 X^e + b_1)) \tag{3}$$

$$\tilde{X}^p = tanh(W_2 X^p + b_2) \tag{4}$$

$$X = A^e \odot X^e + (1 - A^e) \odot \tilde{X}^p \tag{5}$$

where W_1 and W_2 are weight matrices, b_1 and b_2 are bias vectors. λ is a hyper-parameter that balances the weight of word information and entity information.

Following the embedding layer, the sentence $s = \{w_1, w_2, ..., w_m\}$ is transformed into a vector sequence $X = \{x_1, x_2, ..., x_m\}$. Then, the PCNN model divides the convolution result into three parts based on the positions of the head and tail entities in the sentence and and performs maximum pooling on these three parts of the vector through segmental maximum pooling.

$$p_{ij} = max(c_{ij})\{1 \le i \le n, 1 \le j \le 3\} \tag{6}$$

Among them, c_i is the output result of the i-th convolution kernel in the convolution layer, n is the number of convolution kernels, and j represents the j-th segment after segmentation according to the head and tail entities.

Finally, the pooled results are concatenated and passed through a non-linear transformation to obtain the final vector s of the sentence: $s = \rho(p_{1:n})$.

2.3 Dependency Path Encoder

The dependency syntax tree of a sentence can be obtained by natural language processing tools. By traversing the dependency tree, we can identify the dependencies between the words in the tree. An undirected edge is added between any two words that have a dependency relationship. The resulting undirected graph is then used to find the shortest path between the head entity and the tail entity. The words along this path are connected to form the shortest dependency path.

For a given sentence, we can obtain the shortest dependency path, denoted as *sdp*, which has a length of l and consists of the words $\{w_{e_h}, w_{a2}, ..., w_{e_t}\}$ that are associated with the head and tail entities. To obtain a semantic representation of the *sdp*, we use a dependency path encoder, which passes the word sequence through an embedding layer and an encoding layer. The input to the embedding layer is the *sdp*, and the output is a vector sequence $V_{sdp} = \{v_{e_h}, v_{a2}, ..., v_{e_t}\}$, where each word is mapped to a vector representation. Next, the encoding layer uses a one-dimensional CNN to obtain the final dependency path encoding vector *sdp* by applying operations such as convolution and pooling to V_{sdp}.

For entity pairs without the shortest dependency path, we assume that there is a direct dependency path between the head and tail entities. In this case, the shortest dependency path can be represented as $sdp = \{w_{c_h}, w_{c_t}\}$.

2.4 Relational Encoder

The relation encoder comprises two relation structures, namely the constraint graph and the relation hierarchy, which can both be modeled as graph structures. Next, we will elaborate on the process of converting these structures into vector representations and how to encode and fuse relations within the graph structures.

Constraint Graph Embedding Representation: The definition of a constraint graph is [6]: a constraint graph can be represented by a triple $\mathcal{G}_1 = \{\mathcal{T}, \mathcal{R}_c, \mathcal{C}\}$, where $\mathcal{T}, \mathcal{R}_c, \mathcal{C}$ represent the entity type set, relation set and constraint rule respectively. Each element in $\mathcal{C}$ represents a constraint $<t_{r_c}^{e^h}, r_c, t_{r_c}^{e^t}>$, which means that for relation r_c, the type of the head entity can be $t_{r_c}^{e^h} \in T$, the type of tail entity can be $t_{r_c}^{e^t} \in T$.

According to the given constraint graph $\mathcal{G}_1 = \{\mathcal{T}, \mathcal{R}_c, \mathcal{C}\}$, we can define the node set $\mathcal{V}_c = \mathcal{T} \cup \mathcal{R}_c$ and the edge set $\mathcal{E}_c$ is an empty set. For each constraint $<t_{r_c}^{e^h}, r_c, t_{r_c}^{e^t}>$ in $\mathcal{C}$, $<t_{r_c}^{e^h}, r_c>$ and $<r_c, t_{r_c}^{e^t}>$ is added to $\mathcal{E}_c$. After traversing the constraint rule set $\mathcal{C}$, the constraint graph $\mathcal{G}_1 = \{\mathcal{V}_c, \mathcal{E}_c\}$ can be obtained.

After obtaining the constraint graph $\mathcal{G}_1 = \{\mathcal{V}_c, \mathcal{E}_c\}$, we first construct the adjacency matrix $\hat{A}_c \in \mathbb{R}^{m \times m}(m = |\mathcal{V}_c|)$ through the edge set $\mathcal{E}_c$.

$$\hat{A}_{ij}^c = \begin{cases} 1 & if(v_i, v_j) \in \mathcal{E}_c \\ 0 & otherwise \end{cases} \tag{7}$$

Then, an embedding vector $v_i^{c(0)}$ of dimension d_v is randomly initialized for each node $v_i \in \mathcal{V}_c$. At this point, the constraint graph $\mathcal{G}_1$ has been converted into a readable vector form.

Relation Hierarchical Tree Embedding Representation: The relation hierarchy is expressed as $\mathcal{G}_2 = \{\mathcal{R}_h, \mathcal{E}_h\}$, where $\mathcal{R}_h$ and $\mathcal{E}_h$ represent the relation set and the edge set, respectively. The meaning of each element $<r_h^1, r_h^2>$ in $\mathcal{E}_h$ is that there is a hierarchical relation between the relation $r_h^1 \in \mathcal{E}_h$ and the relation $r_h^2 \in \mathcal{E}_h$. According to the given hierarchical relation graph $\mathcal{G}_2$, we first construct the adjacency matrix $\hat{A}^h \in \mathbb{R}^{n \times n}(n = |\mathcal{R}_h|$ through the edge set $\mathcal{E}_h$.

$$\hat{A}_{ij}^h = \begin{cases} 1 & if(r_i, r_j) \in \mathcal{R}_h \\ 0 & otherwise \end{cases} \tag{8}$$

Then, an embedding vector $r_i^{h(0)}$ of dimension d_v is randomly initialized for each node $r_i \in \mathcal{R}_h$. So far, the hierarchical relation tree $\mathcal{G}_2$ has been converted into a vector form.

Relation Encoding and Fusion: The initial embeddings of the constraint graph and the relation hierarchy tree are encoded using two-layer GAT models [28]. The GAT network is well-suited for processing graph structures and overcomes the limitations of traditional Graph Convolutional Network (GCN) models [27] when dealing with directed graphs. It achieves this by assigning varying weights to neighboring nodes, allowing for more effective information extraction from adjacent nodes. One GAT network is applied to the hierarchical relation graph, while the other is applied to the constraint graph.

After two layers of GAT, the node vector of the obtained relation hierarchy graph is $R^{h(2)} \in \mathbb{R}^{m \times d_h}$, and d_h is the vector dimension. Only leaf nodes are selected from them, that is, the n_r relations that actually exist, which are expressed as $R^h \in \mathbb{R}^{n_r \times d_h}$. The node vector of the obtained constraint graph is $V^{c(2)} \in \mathbb{R}^{n \times d_h}$, and GAT divides $V^{c(2)}$ into entity types according to whether the node is an entity type or a relationship to represent $T \in \mathbb{R}^{n_t \times d_h}$ and the relation denoted $R^c \in \mathbb{R}^{n_r d_h}$.

Finally, the relation vectors obtained from the two relation structures are combined through the dot product operation to obtain the final relation vector representation $R = R^h \odot R^c$.

2.5 Aggregators

The aggregator is primarily composed of two components, namely the vector aggregated representation and the attention mechanism. The former encompasses the ultimate vector representation of every sentence in the bag and the concluding vector representation of the attention mechanism. The attention mechanism pertains to acquiring the final bag vector representation by means of an attention mechanism. The following expound on these two components in detail.

Vector Aggregation Representation: For a bag $\mathcal{B} = \{s_1, s_2, ..., s_{|\mathcal{B}|}\}$, constraint graph $\mathcal{G}_1 = \{\mathcal{V}_c, \mathcal{E}_c\}$ and relation hierarchy tree $\mathcal{G}_2 = \{\mathcal{R}_h, \mathcal{E}_h\}$ for a given dataset $\mathcal{D}$, by the sentence encoder, we can obtain the set of sentence vector representations in the package $\mathbf{S} = \{\mathbf{s}_1, \mathbf{s}_2, ..., \mathbf{s}_{|\mathcal{B}|}\}$; through the dependency path encoder, we can get the shortest dependency path encoding vector $\mathbf{SDP} = \{\mathbf{sdp}_1, \mathbf{sdp}_2, ..., \mathbf{sdp}_{|\mathcal{B}|}\}$; through the relation encoder, we can obtain the relation vector representation set $\mathbf{R} = \{\mathbf{r}_1, \mathbf{r}_2, ..., \mathbf{r}_{n_r}\}$ and the entity type vector representation set $\mathbf{T} = \{\mathbf{t}_1, \mathbf{t}_2, ..., \mathbf{t}_{n_t}\}$. Among them, the dimensions of $\mathbf{s}_i$, $\mathbf{r}_i$, and $\mathbf{t}_i$ are all d_h, and the dimension of $\mathbf{sdp}_i$ is $2d_h$.

For the sentences in bag $\mathcal{B}$, the entity types of the head and tail entities in the sentence can be identified by named entity recognition tools, and then the corresponding entity type vector representations can be obtained from the entity type vector representation set $\mathbf{T}$. Then, the entity type vector representations $\mathbf{t}_{s_i}^{e^h}$ and $\mathbf{t}_{s_i}^{e^t}$ of the head entity e^h and tail entity e^t of the sentence s_i are obtained. Then, concatenate the two as the entity type representation of sentence s_i.

$$\mathbf{t}_{s_i} = [\mathbf{t}_{s_i}^{e^h}; \mathbf{t}_{s_i}^{e^t}] \in \mathbb{R}^{2d_h} \tag{9}$$

As a coarse-grained information representation of head and tail entities, entity type has certain limitations to distinguish entity pairs between different sentences. The model uses the shortest dependency path between head and tail entities as the fine-grained representation of head and tail entities, and combines the two to increase the discrimination between different sentence entities. The entity information representation of the obtained sentence s_i is shown as follows:

$$\mathbf{f}_{s_i} = \gamma \cdot \mathbf{t}_{s_i} + (1-\gamma) \cdot \mathbf{sdp}_{s_i} \tag{10}$$

where γ is a hyperparameter that balances the weight between t_{s_i} and sdp_{s_i}.

Finally, the sentence representation and entity information representation are concatenated. The final vector representation of each sentence in the bag is as follows.

$$\mathbf{t}_{s_i} = [\mathbf{s}_i; \mathbf{f}_{s_i}] \in \mathbb{R}^{3d_h} \qquad 1 \leq i \leq |\mathcal{B}| \tag{11}$$

After obtaining the sentence vector representation with entity information added, a similar way is used to represent the relation vector. Through the constraint graph, for each relation $\mathbf{r}_i$, its head entity type $\mathbf{t}_{r_i}^{e^h}$ and tail entity type $\mathbf{t}_{r_i}^{e^t}$ can be obtained. If the head entity or the tail entity contains multiple entity types, the vector mean of the head entity or the tail entity is taken as the corresponding vector representation. As for the relation "NA", its head entity type and tail entity type are both represented by vectors using the mean of all entity type nodes. Finally, through the splicing method, the relation vector representation of adding entity constraint information is obtained.

$$\mathbf{c}_i = [\mathbf{r}_i; \mathbf{t}_{r_i}^{e^h}; \mathbf{t}_{r_i}^{e^t}] \in \mathbb{R}^{3d_h} \qquad 1 \leq i \leq n_r \tag{12}$$

Attention Mechanism: The model uses the semantic similarity between sentences and relations to calculate the attention of each sentence in the bag. For a given sentence representation $\mathbf{G} = \{\mathbf{g}_1, \mathbf{g}_2, ..., \mathbf{g}_{|\mathcal{B}|}\}$, the attention mechanism is used to obtain the weight of each sentence. The calculation process is as follows. $\mathbf{g}_i$ and $\mathbf{c}_r$ are sentence vectors and relation vectors combined with constraint information, respectively. Multiply $\mathbf{g}_i$ and $\mathbf{c}_r$ to get $\mathbf{e}_i$, the unnormalized attention score: $\mathbf{e}_i = \mathbf{g}_i \mathbf{c}_r$. After that, according to the weight of each sentence, the sentences in the bag are weighted to obtain the vector representation of the bag as follows:

$$\mathbf{z} = \sum_{i=1}^{|\mathcal{B}|} \alpha_i \mathbf{g}_i \tag{13}$$

$$\alpha_i = \frac{exp(e_i)}{\sum_{j=1}^{|\mathcal{B}|} exp(e_j)} \tag{14}$$

2.6 Classifier

Multi-instance learning models the task of distantly supervised relation extraction as a bag-level classification task. Through the attention mechanism, the vector representation $\mathbf{z}$ of the whole bag is obtained from all the sentences in the bag. Finally, through a softmax activation function, the relation probability distribution of bag $\mathcal{B}$ in the relation set $\mathcal{R}$ is calculated as follows:

$$p(R|B;\theta) = sigmoid(\mathbf{Wr} + \mathbf{b}) \tag{15}$$

where θ represents all parameters in the model, $\mathbf{W}$ is the weight matrix of the classifier, and $\mathbf{b}$ is the bias vector of the classifier.

The model uses the cross-entropy function as the loss function of the model, which is calculated as follows:

$$J(\theta) = -\frac{1}{n}\sum_{i=1}^{n}(logp(r_i|B_i;\theta)) \tag{16}$$

where n is the total number of bags in dataset $\mathcal{D}$, and r_i is the relation label corresponding to bag $\mathcal{B}_i$.

3 Experiments

3.1 Experimental Setup

Datasets and Evaluation Metrics: To evaluate the proposed model, we have used the widely adopted dataset in the domain of DSRE, namely the NYT-10 dataset. This dataset was constructed by [12] by aligning the relations in the Freebase knowledge base with the New York Times Corpus (NYT). The training set includes sentences from 2005–2006, while the test set includes sentences from 2007. The dataset comprises 53 relations, and while the data-rich relations have abundant instances, the long-tail relations are severely underrepresented. We used the dataset provided by OpenNRE [8]. Following prior research, we use P-R curve, AUC and P@N to evaluate the classification performance of the model on the label noise problem, and use Hits@K on the long-tail problem.

Implementation Detail: To ensure the experimental results are fair and comparable to prior research, we kept most of the hyperparameters consistent with previous studies. Specifically, we initialized the weight matrices in the network using the Xavier initialization, while setting the initial values of the bias vectors to zero. For word vectors, we employed the Word2vec pre-trained in the OpenNRE framework. The constraint graph used in the model was sourced from [6]. To prevent overfitting, we applied dropout. A detailed breakdown of the model's parameter settings is presented in Table 1.

Table 1. Hyperparameter settings for the model.

Hyperparameter	Value	Hyperparameter	Value
word embedding dimension	50	CNN window size	3
position embedding dimension	5	maximum sentence length	120
PCNN convolution kernel	230	maximum relative position length	100
PCNN window size	3	binding coefficient	0.85
smoothing factor	20	batch size	160
GAT embedding layer dimensions	700	Iterations	20
GAT first layer dimension	950	learning rate	0.5
CNN convolution kernel	230	dropout probability	0.5

3.2 Analysis of Noise Experiment

To evaluate the anti-noise effect of the model proposed, we select 6 baseline models for comparison:

MINTZ [1] is a multi-class logistic regression model that uses textual features such as lexical and syntactic features, and connection features. PCNN+ATT [9] is a piecewise CNN model with sentence-level attention, which is the first to apply the attention mechanism to DSRE. RESIDE [10] is a neural network model using GCN and Bi-GRU, which uses auxiliary information such as entity types to improve the model effect. DISTRE [11] uses pre-trained GPT to capture semantic and syntactic features. REDSandT [3], utilizes the pre-trained model of BERT [26] to capture the semantic and syntactic features of sentences by transferring commonsense knowledge in the pre-trained model. At the same time, the model merges entity types into 18 types and embeds them into the Bert input layer. CGRE [6], uses PCNN as a sentence classifier, which is the first paper to propose a constrained graph structure. The model uses GCN to extract semantic information from the constraint graph. Based on this model, we replace GCN with GAT to construct a new baseline CGRE(GAT).

The comparison results of our model and other models are shown in Fig. 2, where the Precision-Recall (P-R) curves are presented. To facilitate the comparison of the models, Table 2 summarizes the Area Under the Curve (AUC) values, as well as the precision at 100, 200, and 300, and their respective mean values. The following points can be seen:

- When the recall rate is below 0.05, all models exhibit accuracy above 0.5. However, the Mintz model [1] performs the worst. As recall increases to 0.2, Mintz's accuracy drops to 0.3, while other models maintain accuracy above 0.6. As recall increases to 0.3, the accuracy of other models still remains above 0.5, whereas Mintz's accuracy drops below 0.2.
- In contrast to CGRE (GAT), our model demonstrates improved P@100, P@200, and P@300 values, leading to an average improvement of 1.1%, despite a decrease of 0.4% in AUC. These results suggest that the use of GAT in place of GCN leads to increased accuracy for high-confidence entity pairs.

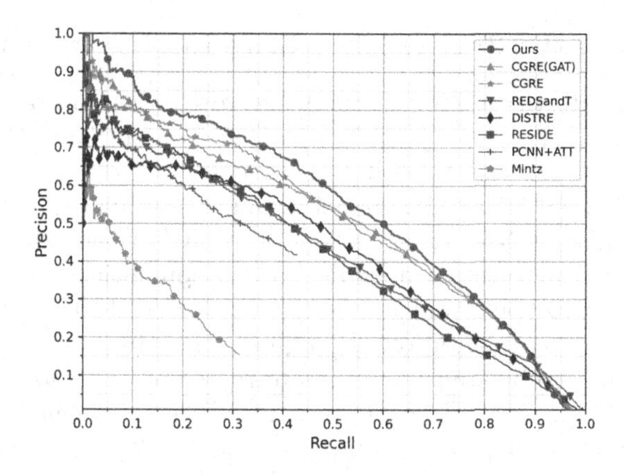

Fig. 2. The P-R curves on NYT-10

- The results demonstrate that our model achieves high performance. Specifically, when the recall is 0.35, the accuracy of the proposed method exceeds 0.7, while other methods perform lower than 0.65. Additionally, when the recall is below 0.9, the accuracy of the proposed method remains higher than 0.7. Notably, the proposed method consistently outperforms the baseline model across the entire recall range. The AUC value of the proposed method is 54.3%, which is 3.5% points higher than the second-ranked CGRE model. Furthermore, the proposed method achieves higher values of P@100, P@200, and P@300 than the baseline model.

Table 2. AUC values and P@N results of NYT-10(%).

	AUC	P@100	P@200	P@300	Mean
MINTZ	10.7	52.3	50.2	45	49.2
PCNN+ATT	34.1	73	68	67.3	69.4
RESIDE	41.5	81.8	75.4	74.3	77.2
DISTRE	42.2	68	67	65.3	66.8
REDSandT	42.4	78	75	73	75.5
CGRE	50.8	91	85.9	83.3	86.7
CGRE(GAT)	50.4	91.8	87.8	83.8	87.8
Ours	**54.3**	**92.8**	**91**	**87.4**	**90.4**

3.3 Analysis of the Results of the Long-Tail Experiment

In order to evaluate the ability of the model to handle long-tail relations, this section selects 7 baseline models for comparison. In addition to the CGRE mentioned above, there are also the following models:

PCNN+HATT [2] is the first model to apply relation hierarchy to long-tail relation extraction. PCNN+KATT [20] utilizes GCN to encode relation hierarchy and uses knowledge graph embedding for relation initialization. PA-TRP [23] learns relation prototypes from unlabeled text and embeds in relation hierarchies to enrich the representation of long-tail relations. CoRA [29] is a model that uses sentence embeddings to generate representations for each node in a relation hierarchy. PCNN+HRS [30] uses PCNN as a sentence encoder, treats relation extraction as a path planning task on a relation hierarchy tree, and shares information from related instances between head relations and long-tail relations semantics. HiRAM [31], this model uses multi-granularity entity types to enrich distantly supervised sentence representations and aligns sentences with hierarchies to enrich long-tail relation representations.

Table 3 lists the long-tail effect evaluation results of this model and baselines. The following conclusions can be drawn from the table:

Table 3. The top-K hit rate for long-tail relationships of NYT-10(%)

Training Instances	<100			<200		
Hits@K	10	15	20	10	15	20
PCNN+HATT	29.6	51.9	61.6	41.4	60.6	68.2
PCNN+KATT	35.3	62.4	65.1	43.2	61.3	69.2
PCNN+HRS	36.8	64.0	68.8	44.8	62.0	71.5
PA-TRP	63.9	70.3	72.2	66.7	72.3	73.8
CoRA	59.7	63.9	73.6	65.4	69.0	77.4
HiRAM	72.2	96.3	96.3	77.3	96.9	96.9
CGRE	72.2	77.8	87.0	77.3	81.8	89.4
CGRE(GAT)	78.5	88.9	91.1	82.4	90.9	92.7
Ours	**86.3**	**97.8**	**98.9**	**88.8**	**98.2**	**99.1**

- Compared with CGRE, CGRE (GAT) has obvious improvement in 6 indicators of long-tail relations. Notably, Hits@15 when the number of instances is less than 100 shows the most significant improvement, with an 11.1% increase. Even the smallest improvement, as shown in Hits@20 (<200), is still a 3.32% increase. These results suggest that the use of GAT in place of GCN, by allowing different weights to be assigned to neighbor nodes, more efficiently transfers information from the data-rich relation of the constraint graph to the long-tail relation.

- Apart from HiRAM, the models CGRE and CGRE(GAT) that utilize constraint graphs exhibit superior results compared to PCNN+HRS, PA-TRP, and CoRA, which employ relation hierarchical trees. This indicates that the constraint graph has better performance in addressing long-tail issues.
- The results demonstrate that our model outperforms all the baseline models, regardless of the number of instances being less than 100 or less than 200. This indicates that the proposed model effectively integrates the constraint graph and the relation hierarchy tree model, and obtains more informative features, thereby enhancing the representation of long-tail relations.

3.4 Ablation Experiment

To further investigate the impact of each component on the model, this section presents ablation experiments on our model, in which the relation hierarchy tree and dependency path are separately removed from the original model. The experimental results are shown in Table 4.

Table 4. Results of ablation experiments on NYT-10(%)

	AUC	P@100	P@200	P@300	Mean
w/o hierarchical	52.4	92.0	89.8	87.1	89.6
w/o path	53.8	92.6	90.5	**87.4**	90.1
Ours	**54.3**	**92.8**	**91.0**	**87.4**	**90.4**

The ablation experiments conducted on our model revealed that the removal of either the relation hierarchy tree or dependency path resulted in a significant impact on model performance. Specifically, the removal of the relation hierarchy tree led to a 1.9% decrease in AUC and a 0.8% decrease in the average value of P@100, P@200, and P@300, whereas the removal of the dependency path caused a 0.5% decrease in AUC and a 0.3% decrease in the average value of P@100, P@200, and P@300. These results indicate that both the relation hierarchy tree and dependency path components have a significant impact on model performance, with the former exhibiting a more pronounced effect.

Effect of the Number of GAT Layers in the Relation Hierarchy Tree: Based on the original model, we modified the number of GAT layers in the relation hierarchy tree to one, two, and three, and conducted experiments to explore their impact on the model performance. The results are presented in Table 5. We observed that the AUC and the values of P@100, P@200, and P@300 were higher when using a two-layer GAT compared to the other two cases. This suggests that the two-layer GAT can extract more relevant information and avoid the overfitting problem that can arise from using too many layers.

Table 5. Results for different CLS layers on NYT-10 and GDS

	AUC	P@100	P@200	P@300	Mean
One	53.5	91.6	90.1	87.1	89.6
Two	**54.3**	**92.8**	**91.0**	**87.4**	**90.4**
Three	53.9	91.8	90.0	85.3	89.0

3.5 Case Study

In order to better illustrate the effect of our model, we visually display two detailed cases and compares it with the results of CGRE (GAT).

The first case pertains to the relation "*/business/company/founders*" with head and tail entities "*StevenSpielberg*" and "*DreamWorks*", respectively. Table 6 displays the attention assignments for the four instances in the bag. The table indicates that $s1$, $s2$, and $s4$ express this relation correctly, whereas $s3$ does not effectively demonstrate a "*founders*" relation between "*StevenSpielberg*" and "*DreamWorks*" despite implying that "*StevenSpielberg*" is working in "*DreamWorks*." The attention scores suggest that our model assigns more attention to the correct instances and ignores the wrong ones better than the CGRE (GAT) model. This outcome results from our model's ability to learn more information from the extensive "*/business/company/**" relations in the relation hierarchy tree and the dependency paths of the head and tail entities. Our model also allocates relatively more attention to the correct instances, even when it assigns more attention to the incorrect ones. For instance, Table 7 shows an example of a long-tail relation "*/people/person/ethnicity*." In comparison to the CGRE (GAT) model, our model assigns s6 an additional 10 percent attention to the correct instances, and this 10 percent difference enables the bag to be classified correctly.

4 Related Work

Label Noise Problem: To solve this problem, some work has developed a multi-instance learning (MIL) framework [4,12]. In addition, [9] improved the sentence encoder by using the PCNN [9] model. Then designed a selective attention between sentences/instances at the sentence level. Inspired by this, different aspects of attention model are proposed [13–18].

Long-tail Relations: [19] proposed an Explanation-Based Learning (EBL) approach to effectively learn relation extraction rules using unlabeled data. However, it still has the disadvantages of poor robustness. [2] obtained advanced performance by introducing a hierarchical attention scheme to obtain additional bag-level features. [20] used TransE [21] and GCN [22] to obtain relation embedding, and designed a new attention network along the relation hierarchy. [5]

attempt to mine semantic dependency between relations from relation hierarchical tree and carried out knowledge transfer through relation hierarchical. [24] Use entity types to implicitly transfer information between different relations. However, there is still a lot of room for improvement in these tasks.

Table 6. Case description of "/business/company/founders".

ID	Sentence	Correct	Score	
			Ours	CGRE (GAT)
s1	A billionaire with manifest interests, he has a lot of time on his hands now that **DreamWorks**, the mini-major studio he formed with **Steven Spielberg** and Jeffrey Katzenberg, has been bought and folded into Paramount	Yes	0.3544	0.0604
s2	Identity crisis for **Dreamworks** – On paper, 2007 is shaping up to be one of DreamWorks' most promising years. The 12-year-old studio, co-founded by the director **Steven Spielberg**	Yes	0.2052	0.0002
s3	David Geffen, Jeffrey Katzenberg and **Steven Spielberg** of **DreamWorks** are the hosts - with a private dinner afterward at Mr. Geffen's for those who sold 20 tickets or more	No	0.0188	0.8526
s4	It is hard to say whether the unusual heat behind the evening owes more to interest in Mr. Obama or to the three men who spearheaded the fund-raiser: the **DreamWorks** co-founders David Geffen, Jeffrey Katzenberg and **Steven Spielberg**	Yes	0.084	0.0009

Table 7. Case description of "/people/person/ethnicity"

ID	Sentence	Correct	Score	
			Ours	CGRE (GAT)
s5	How is it that **John McCain** now believes **American** lives are being wasted, yet he so stubbornly supports the president's plan to escalate the war in Iraq and put more American lives in harms way	No	0.6371	0.7331
s6	**John McCain** is a remarkable **American** hero who has experienced war, and in the post-9/11 period, what better kind of leader could we have than that?	Yes	0.3603	0.2604

5 Conclusion

This paper proposes a novel approach for addressing the long-tail problem in DSRE. The model leverages dependency paths and relation graph embeddings to extract information from the relation hierarchy tree and constraint graph through GAT and combines the embedded representations of the two to learn rich information representation for long-tail relations. To mitigate the deterioration of the distinction between head and tail entities in different sentences in the same bag caused by entity type constraints in the constraint graph, the model employs a combination of dependency paths and entity types to effectively increase the discrimination between different sentences. Experimental results demonstrate that the proposed model outperforms 7 baselines that solve the long-tail problem and 6 baselines that solve the label noise problem, showcasing its superiority in solving long-tail problems and improving anti-noise performance.

Acknowledgements. This research was supported by the Dongbo Future Artificial Intelligence Research Institute Co., Ltd. Joint Laboratory (School Agreement No. 20223160C0026), Xiaozhi Deep Art Artificial Intelligence Research Institute Co., Ltd. Computational Art Joint Laboratory (School Agreement No. 20213160C0032), and Xiamen Yinjiang Smart City Joint Research Center (School Agreement No. 20213160C0029).

References

1. Mintz, M., Bills, S., Snow, R., Jurafsky, D.: Distant supervision for relation extraction without labeled data. In: ACL-AFNLP, pp. 1003–1011 (2009)
2. Han, X., Yu, P., Liu, Z., Sun, M., Li, P.: Hierarchical relation extraction with coarse-to-fine grained attention. In: Proceedings of EMNLP, pp. 2236–2245 (2018)
3. Christou, D., Tsoumakas, G.: Improving distantly-supervised relation extraction through BERT-based label and instance embeddings. IEEE Access **9**, 62574–62582 (2021)
4. Hoffmann, R., Zhang, C., Ling, X., Zettlemoyer, L., Weld, D.S.: Knowledge-based weak supervision for information extraction of overlapping relations. In: Proceedings of ACL HLT, pp. 541–550 (2011)
5. Yu, E., Han, W., Tian, Y., Chang, Y.: ToHRE: a top-down classification strategy with hierarchical bag representation for distantly supervised relation extraction. In: Proceedings of COLING, pp. 1665–1676 (2020)
6. Liang, T., Liu, Y., Liu, X., Zhang, H., Sharma, G., Guo, M.: Distantly-supervised long-tailed relation extraction using constraint graphs. IEEE Trans. Knowl. Data Eng. **35**(7), 6852–6865 (2022)
7. Li, Y., et al.: Self-attention enhanced selective gate with entity-aware embedding for distantly supervised relation extraction. Proceedings of the AAAI **34**(05), 8269–8276 (2020)
8. Han, X., Gao, T., Yao, Y., Ye, D., Liu, Z., Sun, M.: OpenNRE: An open and extensible toolkit for neural relation extraction (2019). arXiv preprint arXiv:1909.13078
9. Lin, Y., Shen, S., Liu, Z., Luan, H., Sun, M.: Neural relation extraction with selective attention over instances. In: Proceedings of ACL, pp. 2124–2133 (2016)
10. Vashishth, S., Joshi, R., Prayaga, S.S., Bhattacharyya, C., Talukdar, P.: RESIDE: Improving distantly-supervised neural relation extraction using side information (2018). arXiv preprint arXiv:1812.04361
11. Alt, C., Hübner, M., Hennig, L.: Fine-tuning pre-trained transformer language models to distantly supervised relation extraction (2019). arXiv preprint arXiv:1906.08646
12. Riedel, S., Yao, L., McCallum, A.: Modeling relations and their mentions without labeled text. In: Balcázar, J.L., Bonchi, F., Gionis, A., Sebag, M. (eds.) ECML PKDD 2010. LNCS (LNAI), vol. 6323, pp. 148–163. Springer, Heidelberg (2010). https://doi.org/10.1007/978-3-642-15939-8_10
13. Qu, J., Ouyang, D., Hua, W., Ye, Y., Li, X.: Distant supervision for neural relation extraction integrated with word attention and property features. Neural Netw. **100**, 59–69 (2018)

14. Du, J., Han, J., Way, A., Wan, D.: Multi-level structured self-attentions for distantly supervised relation extraction (2018). arXiv preprint arXiv:1809.00699
15. Yuan, Y., et al.: Cross-relation cross-bag attention for distantly-supervised relation extraction. In: Proceedings of the AAAI, pp. 419–426 (2019)
16. Ye, Z.-X., Ling, Z.-H.: Distant supervision relation extraction with intra-bag and inter-bag attentions (2019). arXiv preprint arXiv:1904.00143
17. Dai, L., Xu, B., Song, H.: Feature-level attention based sentence encoding for neural relation extraction. In: Tang, J., Kan, M.-Y., Zhao, D., Li, S., Zan, H. (eds.) NLPCC 2019. LNCS (LNAI), vol. 11838, pp. 184–196. Springer, Cham (2019). https://doi.org/10.1007/978-3-030-32233-5_15
18. Yu, B., Zhang, Z., Liu, T., Wang, B., Li, S., Li, Q.: Beyond word attention: using segment attention in neural relation extraction. In: IJCAI, pp. 5401–5407 (2019)
19. Gui, Y., Liu, Q., Zhu, M., Gao, Z.: Exploring long tail data in distantly supervised relation extraction. In: Lin, C.-Y., Xue, N., Zhao, D., Huang, X., Feng, Y. (eds.) ICCPOL/NLPCC -2016. LNCS (LNAI), vol. 10102, pp. 514–522. Springer, Cham (2016). https://doi.org/10.1007/978-3-319-50496-4_44
20. Zhang, N., et al.: Long-tail relation extraction via knowledge graph embeddings and graph convolution networks (2019). arXiv preprint arXiv:1903.01306
21. Bordes, A., Usunier, N., Garcia-Duran, A., Weston, J., Yakhnenko, O.: Translating embeddings for modeling multi-relational data. In: Advances in Neural Information Processing Systems, vol. 26 (2013)
22. Defferrard, M., Bresson, X., Vandergheynst, P.: Convolutional neural networks on graphs with fast localized spectral filtering. In: Advances in Neural Information Processing Systems, vol. 29 (2016)
23. Cao, Y., Kuang, J., Gao, M., Zhou, A., Wen, Y., Chua, T.-S.: Learning relation prototype from unlabeled texts for long-tail relation extraction. IEEE Trans. Knowl. Data Eng. **35**(2), 1761–1774 (2021)
24. Gou, Y., Lei, Y., Liu, L., Zhang, P., Peng, X.: A dynamic parameter enhanced network for distant supervised relation extraction. Knowl.-Based Syst. **197**, 105912 (2020)
25. Zeng, D., Liu, K., Chen, Y., Zhao, J.: Distant supervision for relation extraction via piecewise convolutional neural networks. In: Proceedings of EMNLP, pp. 1753–1762 (2015)
26. Devlin, J., Chang, M.-W., Lee, K., Toutanova, K.: BERT: Pre-training of deep bidirectional transformers for language understanding (2018). arXiv preprint arXiv:1810.04805
27. Kipf, T.N., Welling, M.: Semi-supervised classification with graph convolutional networks (2016). arXiv preprint arXiv:1609.02907
28. Veličković, P., Cucurull, G., Casanova, A., Romero, A., Lio, P., Bengio, Y.: Graph attention networks (2017). arXiv preprint arXiv:1710.10903
29. Li, Y., Shen, T., Long, G., Jiang, J., Zhou, T., Zhang, C.: Improving long-tail relation extraction with collaborating relation-augmented attention (2020). arXiv preprint arXiv:2010.03773
30. Wang, J.: RH-Net: improving neural relation extraction via reinforcement learning and hierarchical relational searching (2020). arXiv preprint arXiv:2010.14255
31. Li, Y., Long, G., Shen, T., Jiang, J.: Hierarchical relation-guided type-sentence alignment for long-tail relation extraction with distant supervision (2021). arXiv preprint arXiv:2109.09036

Multi-token Fusion Framework for Multimodal Sentiment Analysis

Zhihui Long, Huan Deng, Zhenguo Yang(✉), and Wenyin Liu

Guangdong University of Technology, Guangzhou, China
{yzg,liuwy}@gdut.edu.cn

Abstract. In this paper, we design a multi-token fusion (MTF) framework to process inter-modality and intra-modality information in parallel for multimodal sentiment analysis. Specifically, a tri-token transformer (TT) module is proposed to extract three tokens from each modality where one of them retains the unimodal feature and the other two tokens learn multi-modal features from the other two modalities respectively. Furthermore, a module based on the hierarchical element-wise self-attention (HESA) is used to process the three tokens of each modality extracted by TT. As a result, the important elements of tokens will be given more attention. Finally, we conduct extensive experiments on two public datasets, which prove the effectiveness and scalability of our network.

Keywords: multimodal fusion · emotion · multimodal representation learning

1 Introduction

In recent years, the multimodal sentiment analysis has attracted a lot of attention [22]. On one hand, human beings are naturally able to engage with world through multimodal information. On the other hand, the recent works [6,12] have proved that neural networks based on multi-modality perform better than neural networks based on single modality. If the machine is given the ability of emotion recognition through a well designed network, then it can be applied in many fields to save labor cost, e.g., the detection of depression, video understanding and so on.

Finding a proper way to fuse features from different modalities is one of the primary tasks in multimodal data fusion. Quiet a few methods have been devised to achieve this goal. For example, Zhao et al. [29] use several techniques to extract features from images, while the features of image and audio are concatenated directly, ignoring the interactions between two modalities. Therefore, many works have been focused on modeling both the intra-modality and inter-modality dynamics. Mai et al. [12] propose an adversarial encoder-decoder-classifier network to prevent the unimodal information from being lost, then a fusion method based on graph is used to merge the trimodal information. However, the features are trained to hold two kinds of information at the same time:

X. Song et al. (Eds.): APWeb-WAIM 2023, LNCS 14332, pp. 424–438, 2024.
https://doi.org/10.1007/978-981-97-2390-4_29

intra-modality information and inter-modality information, which means the two kinds of information are both compromised. Another challenge is to pay attention to important characteristics and less attention to insignificant ones when training. For example, the facial expression is a more important feature than the background image in the visual features. To accomplish this, many methods based on attention mechanisms have been explored. Guo et al. [5] divide existing attention methods into four basic categories: channel attention, spatial attention, temporal attention and branch attention. These attention methods have offered great benefits in both natural language processing and many visual tasks. However, recent works [8,28] have shown that it's helpful to apply attention mechanism at a fine-grained level to select the more important elements of a channel or frame.

In this paper, we design a multi-token fusion (MTF) framework including a tri-token transformer (TT) module and a hierarchical element-wise self-attention (HESA) module to acquire reliable joint multimodal representations. To be specific, the framework can be divided into two stages. First, three tokens instead of one are extracted by the TT module from each modality. The three tokens are divided into two categories: one token of each modality is dedicated to preserve unimodal information where a regression loss is used as an additional constraint accordingly, while the other two tokens are employed to collect the interaction between modalities. In this manner, these two kinds of information are captured and processed in parallel which means these two kind of information will not influence each other. Second, the HESA module is responsible for paying more attention to important elements of these tokens extracted by the TT module. It consists of several element-wise self-attention (ESA) components used in unimodal, bimodal and trimodal level. The self-attention mechanism used here is different from the previous work where the weights of feature elements are computed by the other feature. We calculate the weights of the token elements by themselves where half of the token elements are used to compute the weight of the other half of elements. In conclusion, the main contributions of our paper are listed below:

- We propose a module based on tri-token transformer to give consideration to both intra-modality information and inter-modality information, where the first token is responsible for intra-modality information and the rest two tokens are in charge of inter-modality information.
- We offer a module that uses a hierarchical element-wise self-attention mechanism, which is designed to merge the tokens that diverts greater attention to significant token elements.
- We carry out thorough experiments on two public datasets MOSEI and MOSI, and the results validate the stability and reliability of the proposed MTF.

The rest of this paper is organized in the following manner. Section 2 presents a brief review of the related work. In Sect. 3, we discuss the proposed methods in detail. The experiment results and comparison with other methods are shown in Sect. 4. Finally, we conclude our work in Sect. 5.

2 Related Work

In this section, we will have a brief presentation of previous works on multimodal sentiment analysis and other multimodal fusion methods based on attention mechanism.

2.1 Multimodal Sentiment Analysis

In recent years, sentiment analysis using multimodal information has received much attention. Zhao et al. [29] explore several techniques for extracting features from images and temporal attentions for collecting features from audio, then these features are concatenated directly for classification. Mai et al. [12] believe that it's important to match the distribution of different modalities before fusion operation, thus an adversarial encoder-decoder network combined with reconstruction loss and classification loss is introduced. Some recent works have paid attention to refine the input feature to filter out noise. Pham et al. [15] offer a method to fusion different modalities by translating between modalities, together with a cycle consistency loss to ensure maximal information being retained, the learned joint representation is less sensitive to noisy or missing modalities at test time. The M3ER network proposed by Mittal et al. [13] filters out the noise in a different way. The extracted features of different modalities are checked by Canonical Correlational Analysis to distinguish an ineffectual and effectual input modality. Besides, many related works have been focused on finding a proper network to learn joint representations of different modalities, which contain both modality specific information and relations between modalities. Hazarika et al. [6] devise a novel framework where the feature of each modality is projected into two subspaces: modality-invariant and modality-specific and several loss functions are used to make sure the partition. In spite of all the efforts, the problem of capturing both modality specific information and joint representation between modalities remains challenging.

2.2 Attention Mechanisms in Multimodal Fusion

Attention mechanisms have been very successful in many deep learning tasks and transformer is the most famous architecture which has impressive performance on the result and speed for its non-recurrent structure. Lin et al. [9] use a hierarchical attention where the multimodal inputs are first processed by a shared transformer and then two separate transformers are used for each modality. Some works [1,14] explore the idea of bottleneck fusion through transformer where the cross-modal attention is restricted via a small set of fusion bottlenecks, which improve the performance together with a lower computational cost. The aforementioned methods based on attention mechanism known as spatial attention mechanism have gained great performance in many tasks, however, recent works have revealed the importance of the attention mechanism at attribute

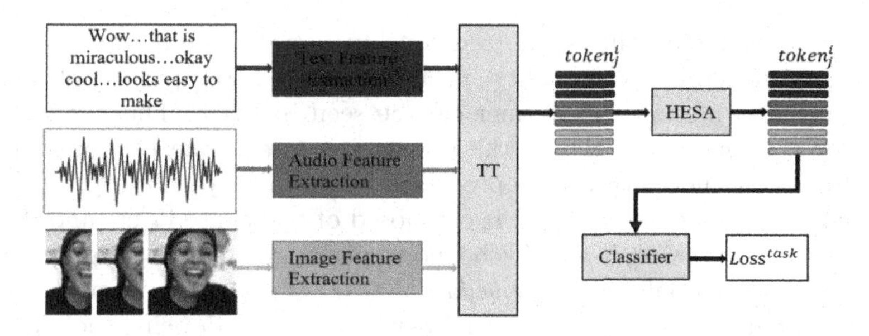

Fig. 1. The framework of MTF. $token_j^i$ represents the three extra learnable tokens, $i \in \{l, a, v\}, j \in \{1, 2, 3\}$.

level. Zhang et al. [28] devise a question-guided image attribute-wise attention and an image-guided question feature-wise attention to put more attention to important attributes of images and questions. Lai et al. [8] calculate the weight of modality feature in a different way. The weight of each modality is calculated by different neural networks with the same input data which is the concatenation of all modality features. In contrast, we calculate the weight of each modality by the feature themselves.

3 Proposed Method

An overview of the whole MTF framework is given in Fig. 1. It can be divided mainly into two parts: the tri-token transformer (TT) module and the hierarchical element-wise self-attention (HESA) module. In TT module, we conduct multimodality fusion through the tri-token transformer where both modality-specific information and modality interactions are perceived. Finally, a module based on element-wise attention mechanism is used to process the tokens to focus more on the important attributes of a feature.

3.1 Tri-token Transformer (TT) Module

Given the preprocessed aligned features, we aim to capture both intra-modality information and inter-modality information in this module as revealed in Fig. 2. It can be divided into two parts, the first of which involves gathering intra-modality data and the second of which focuses on inter-modal data.

As the two public datasets CMU-MOSI [27] and CMU-MOSEI [25] are both collection of video clips which are typical sequence data, so we choose the transformer instead of LSTM for its inherent superiority in parallelism and processing long data sequence in the first step. However, inspired by the work of [14], we believe that one token only is not enough to contain the entire sequence features. So three learnable tokens here are used to represent each modality where two tokens are used for capturing the relationship between modalities and one of them is used to retain the modality-specific features. A comparison of tri-token transformer and normal transformer can be seen in Fig. 3. Thus we get three tokens for each modality. Before the tokens are sent to the next module, the tokens from each modality will be processed by a Multilayer Perceptron (MLP) respectively. In practice, the MLP is composed of two dense layers and dropout layers activated by $LeakyReLU$. To be more specific, we denote the preprocessed features of three modalities as $Embedding^i \in R^{d_i}, i \in \{l, a, v\}$ where l, a, v represent the language, acoustic and visual respectively and d_l being the dimension size of $Embedding^l$ and so on. Tri-token transformer and fully connected layer for each modality are denoted as $TTrans^i, FC^i$. The tokens and embeddings extracted from each modality are $token^i_j, Embedding^i, j \in \{1, 2, 3\}$. Put it all together, the first step can be defined as:

$$TTrans^i(Q, K, V) = Softmax\left(\frac{W^q Q W^k K}{\sqrt{d_q}}\right) W^v V \tag{1}$$

$$TTrans^i(X) = TTrans^i(X, X, X) \tag{2}$$

$$Input = Concat(token^i_j, FC^i(Embedding^i)) \tag{3}$$

$$(token^i_j, Embedding^i) = TTrans^i(Input) \tag{4}$$

where the W^q, W^k and W^v are trainable query-projection matrices, key-projection matrices and value-projection matrices used in the attention mechanism [19].

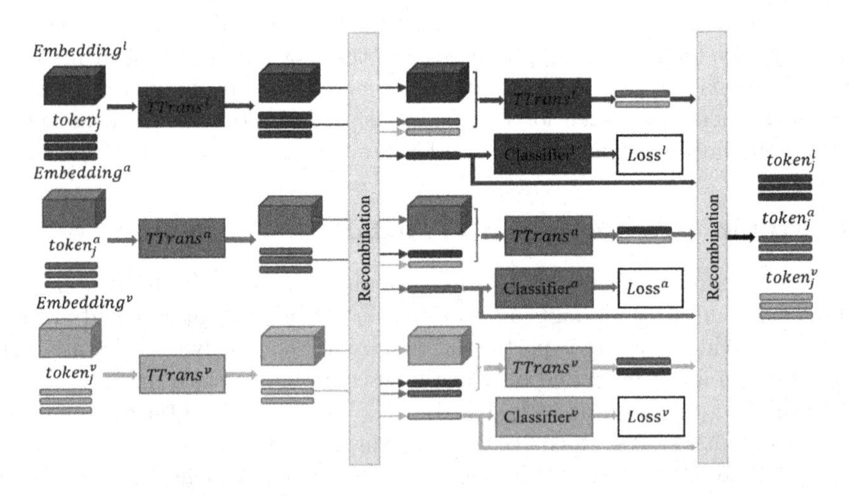

Fig. 2. The architecture of TT. $TTrans^i$ represents the tri-token transformer, $i \in \{l, a, v\}$.

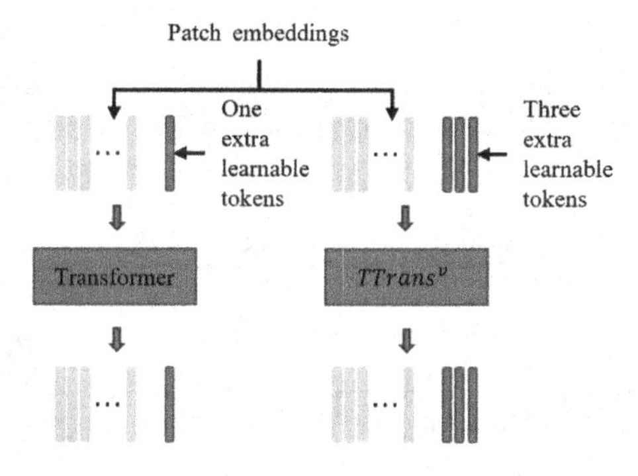

Fig. 3. Comparison of normal transformer (left) and tri-token transformer (right).

Given the three tokens extracted from each modality in the previous step, we aim to capture inter-modality features in the second step. In this regard, many works [12,13] focus on mining interactive information between modalities. For example, Mai et al. [12] propose a hierarchical graph fusion network to fuse features from different modalities, before the fusion process, one decoder for each modality is used to prevent the unimodal information from being lost. However, we believe it's needed to use both unimodal information and cross-modal information to make the final prediction, cross-modal information only will reduce the unimodal information one feature contained. Therefore, in this step, three tokens extracted from each modality are divided into two parts. On one hand, to preserve the intra-modality features, $token_1^i$ are sent to a MLP respectively to make the prediction where Mean Squared Error (MSE) is used as the loss function. In practice, the MLP contains one hidden dense layer and dropout layer activated by the GELU function. Thus we get three losses $loss_i$ for each modality.

$$Pred^i = MLP^i(token_1^i) \tag{5}$$

$$Loss^i = MSE(Pred^i, Label) \tag{6}$$

On the other hand, to get the inter-modality features, $token_j^i, j \in \{2,3\}$ are used to extract information from the other two modalities respectively by the transformer. For example, to extract information from text modality with the audio and video token, $Embedding^l$, $token_2^a$ and $token_2^v$ are sent to the transformer $TTrans^l$ where $token_2^a$ and $token_2^v$ are concatenated as the query feature and the $Embedding^l$ is used as the key and value feature for the transformer.

$$Q = Concat(token_2^a, token_2^v) \tag{7}$$

$$Q = TTrans^l(Q, Embedding^l, Embedding^l) \tag{8}$$

In this module, we get 6 tokens (two for each modality) which contain cross-modal information, together with 3 tokens (one for each modality which does not participate the fusion process) which are used to retain the modality-specific feature. As a result, we are able to maintain the modality-specific characteristic as much as possible while concurrently examining cross-modal information.

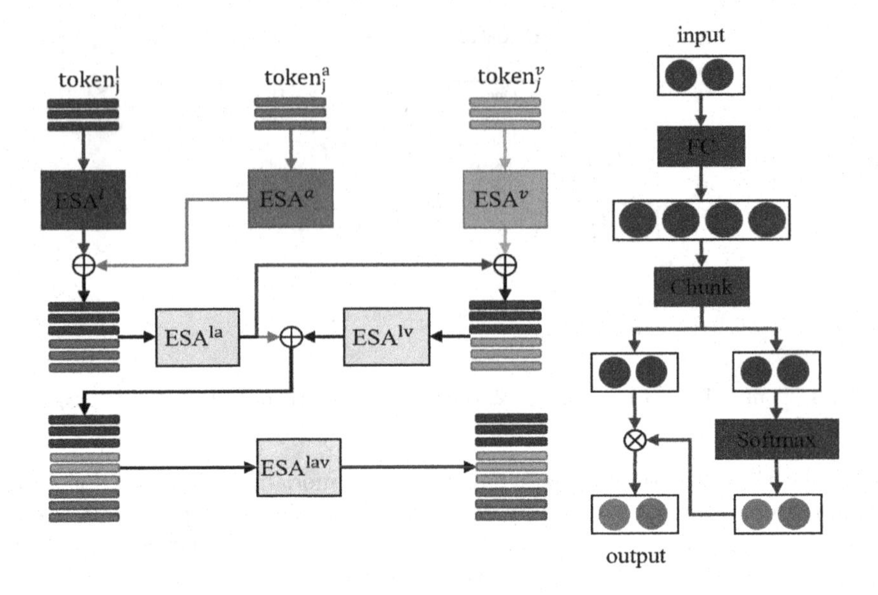

Fig. 4. The architecture of HESA (left) and ESA block (right).

3.2 Hierarchical Element-Wise Self-Attention (HESA) Module

In this section, we propose a hierarchical element-wise self-attention module to pay attention to important elements of a feature while suppressing the unimportant elements. The whole framework of HESA and the structure of its basic unit are presented in Fig. 4. We will introduce the ESA component first and then the HESA module. We apply a novel but simple way to achieve self-attention mechanism. Inspired by the work of Liu et al. [10], we split the feature into two parts on average and the weight of half of the feature is calculated by the other half of the feature attributes as defined in Eq.(10-11). More specifically, given the input $token^i \in R^d, i \in \{l, a, v\}$, a FC layer for each modality is used to expand the feature dimension first, so we get $\overline{token^i} \in R^{2d}$ as defined in Eq.(9). Then the $\overline{token^i}$ is chunked into two equal parts at the feature dimension where one part is used to calculate the weight of the other part as defined in Eq.(10). To get the new weighted tokens, we use the element-wise multiplication. Finally, a LayerNorm operation is applied to the tokens as show in Eq.(12). The unimodal level ESA^i can be defined as:

$$\overline{token^i} \in R^{2d} = FC^i(token^i) \tag{9}$$

$$token^i_1, token^i_2 \in R^d = \overline{token^i} \tag{10}$$

$$weight^i = Softmax(token^i_2) \tag{11}$$

$$token^i = LN(token^i_1 \odot weight^i) \tag{12}$$

where $i \in \{l, a, v\}$ and LN indicates a layer normalization operation. By repeating this process in unimodal, bimodal and trimodal level, we get the HESA module. In the bimodal level, the same process is employed to the tokens concatenated from two modalities which include text-audio modalities and text-vision modalities which are denoted as ESA^{la} and ESA^{lv} in Fig. 4. For example, the bimodal level ESA^{la} can be defined as:

$$token^{la} = Concat(token^{l}, token^{a}) \tag{13}$$

$$token^{la} = ESA^{la}(token^{la}) \tag{14}$$

It should be noticed that the bimodal process is done sequentially, which means the token^{l} outputed by ESA^{la} are sent to ESA^{lv}. In the last, all the tokens are concatenated and sent to the ESA^{lav} block which can be defined as follows:

$$token^{lav} = Concat(token^{lv}, token^{a}) \tag{15}$$

$$token^{lav} = ESA^{lav}(token^{lav}) \tag{16}$$

3.3 Objective Function

For the sentiment regression tasks, we use the mean squared error loss as the follows.

$$Loss^{task} = -\frac{1}{N}\sum_{i=0}^{N} \left\| y_i - \hat{y_i} \right\|^2 \tag{17}$$

where $\hat{y_i}$ is the predicted sentiment label for the i-th sample, y_i is the corresponding ground truth, and N is the total number of samples in a batch. Combined with single modality loss mentioned in Sect. 3.2, we get the final loss as:

$$Loss^{final} = Loss^{task} + \alpha Loss^{l} + \beta Loss^{a} + \gamma Loss^{v} \tag{18}$$

where α, β and γ is the hyperparameter to balance the effect of different losses.

4 Experiments

4.1 Datasets

CMU-MOSI. It [27] contains a collection of 2,199 opinion video clips. Each video clip is labeled with sentiment in the range [−3,3]. To make sure the effect of the model, we split the dataset into train, validation and test sets as the previous works [6,22] did where 1,284 videos are treated as training set, 229 videos are treated as valid set and 686 videos are test set.

CMU-MOSEI. It [25] is the largest dataset in multimodal sentiment analysis and emotion recognition area so far, include 23,453 sentence utterance videos from more than 1,000 online YouTube speaker whose gender is also well balanced and contains 250 different topics. Samely, 16,326 utterances are treated as training set, 1,871 utterances are treated as valid set, and 4,659 utterances are treated as test set like the previous work [6,22] did.

4.2 Evaluation Metrics

For both CMU-MOSI and CMU-MOSEI, the task is a standard regression task. We adopt the metrics as the previous works [6,22]. We will report accuracy on two and seven classes (Acc-2 and Acc-7), F_1 score, mean absolute error (MAE) and correlation (Corr) on both CMU-MOSI and CMU-MOSEI. Specifically, there are two ways to calculate Acc-2 and F_1 score. The first one is a classification between negative and non-negative samples where the labels of the non-negative samples are greater than or equal to 0. The second one is a classification between negative and positive samples where the labels of the positive samples are greater than 0. Both of them are reported in our results.

4.3 Experimental Details

Our model is developed on Pytorch. The classic loss function Mean Square Error is used for the network together with Adam [7] being the optimizer. The initial learning rate of Adam is set to $1e-2$ and the weight decay is set to $1e-2$ too. The batch size is set to 30 for CMU-MOSI and 100 for CMU-MOSEI. In feature preprocessing stage, for both CMU-MOSI and CMU-MOSEI dataset, we use the same preprocessed aligned features extracted by Self-MM [22] where the sequence length of the extracted features is 50 for all modalities. More specifically, the language features are extracted by a pretrained BERT [4]. For the visual features, the information about face is extracted by FACET[1] for it's the most important part in the visual information. As for the acoustic features, COVAREP [2], a acoustic analysis network is used where key aspects about emotions in human voices can been extracted.

4.4 Baselines

A fair comparison with the following baselines is conducted to verify the performance of the our model:

(i) Tensor fusion models: To explicitly combine unimodal, bimodal, and trimodal interactions, Tensor Fusion Network (TFN) [23] performs outer products between modalities representations. In order to improve the TFN framework, Liu et al. [11] introduce a low-rank multimodal fusion technique dubbed LMF by concurrently decomposing the tensor and weights. MTFN [21] is another multi-tensor network which focuses on the fusion of the intra-modal and cross-modal representations instead of on the concatenation of unimodal representations.

[1] iMotions 2017. https://imotions.com/

Table 1. Results of methods on CMU-MOSEI. [1] is from [22] and [2] is from [6]. In F_1 and Acc-2, the left of the "/" is calculated as "negative/non-negative" and the right is calculated as "negative/positive"

Methods	F_1	Acc-2	Acc-7	MAE	Corr
TFN[1]	-/82.1	-/82.5	50.2	0.593	0.700
LMF[1]	-/82.1	-/82.0	48.0	0.623	0.677
MTFN	-/81.3	-/80.8	50.1	0.589	0.674
MFN[1]	76.0/-	76.0/-	–	–	–
MFM[1]	-/84.3	-/84.4	–	0.568	0.717
MISA[1]	82.7/84.0	**82.6**/84.2	–	0.568	0.724
RAVEN[1]	79.5/-	79.1/-	50.0	0.614	0.662
DFMR	-/83.5	-/83.8	48.9	0.594	0.713
MulT[2]	-/82.3	-/82.5	51.8	0.580	0.703
LMF-MulT	-/80.3	-/80.3	50.2	0.616	0.662
MTF	**82.9/86.2**	82.4/**86.2**	**53.0**	**0.546**	**0.758**

(ii) Temporal models: Memory Fusion Network (MFN) [24] applies a special attention mechanism to find cross-modal interaction and a dynamic memory module to store the crucial cross-modal interaction. A variant of LSTM is used by MARN [26] to model modality-specific dynamics while the hybrid memory unit is employed to record important cross-modality dynamics related to that modality.

(iii) Representation learning models: MFM [18] factorizes multimodal representations into two groups of independent elements: modality-specific and multimodal discriminative factors. MISA [6] uses the similar idea where each modality representation is two distinct subspaces: modality-invariant and modality-specific and three additional loss functions are used to make sure the projection well functioning. RAVEN [20] focuses on the intricate organization of nonverbal subword sequences and dynamically changes word representations in response to nonverbal behaviors identified from the other modalities. ARGF [12] provides an adversarial encoder-decoder to reduce the gap between the distributions of various modalities. DFMR [3] exploits stacked dense fusion blocks to model cross-modal interactions and a multimodal residual module to retain the low-level information.

(iv) Transformer-based models: In order to pay attention to interactions between multimodal sequences, MulT [17] employs paired cross-modal attention. LMF-MulT [16] conducts cross attention between low rank and modality-specific representations via the multimodal transformer.

Table 2. Results of methods on CMU-MOSI. [1] is from [22] and [2] is from [6]. In F_1 and Acc-2, the left of the "/" is calculated as "negative/non-negative" and the right is calculated as "negative/positive"

Methods	F_1	Acc-2	Acc-7	MAE	Corr
TFN[1]	-/80.7	-/80.8	34.9	0.901	0.698
LMF[1]	-/82.4	-/82.5	33.2	0.917	0.695
MTFN	-/81.0	-/80.9	38.9	0.891	0.691
MFN[1]	77.3/-	77.4	34.1	0.965	0.632
MARN	-/77.0	-/77.1	34.7	0.968	0.625
MFM[1]	-/81.6	-/81.7	35.4	0.877	0.706
MISA[1]	80.8/82.0	80.8/82.1	–	**0.804**	**0.764**
RAVEN[1]	-/76.6	-/78.0	33.2	0.915	0.691
ARGF	-/81.5	-/81.4	–	–	–
DFMR	-/79.2	-/79.3	30.3	0.961	0.663
MulT[2]	-/82.8	-/83.0	**40.0**	0.871	0.698
LMF-MulT	-/78.5	-/78.5	34.0	0.957	0.681
MTF	**81.9/83.5**	**81.9/83.5**	37.6	0.872	0.719

4.5 Results and Discussions

The comparison results with other methods on datasets CMU-MOSEI and CMU-MOSI are shown in Table 1 and Table 2. As we can see that our method MTF performs better than other methods in almost all the metrics on dataset CMU-MOSEI which reflects the effectiveness of our method. More specifically, MTF improves 2.2 points on F_1 score, 2 points on Acc-2, 1.2 points on Acc-7 and achieves obvious improvement on the regression evaluation metrics. As for the dataset CMU-MOSI, MTF achieves the best results on F_1 score and Acc-2. More specifically, MTF improves 0.7 points on F_1 score, 0.5 points on Acc-2 and comparable results on the regression evaluation metrics. The smaller sample amounts compare to CMU-MOSEI dataset may cause the not so good results.

4.6 Ablation Study

In this section, we conduct ablation studies on the CMU-MOSEI and CMU-MOSI datasets to study the influence of each proposed module and explore the significance of each modality.

The Influence of Each Proposed Module. In this section, several variants of our proposed method are used on the same dataset while the hyperparameters stay the same:

(i) MTF (base) is a baseline version where only one token is used to represent each modality, then the three tokens are sent to a new transformer to explore the inter-modality information.

Table 3. Ablation study results on CMU-MOSEI dataset

Methods	F_1	Acc-2	Acc-7	MAE	Corr
base	80.7/84.0	80.1/84.0	51.8	0.562	0.733
2 tokens	81.8/85.3	81.2/85.3	52.7	0.551	0.749
3 tokens+concat	82.7/83.5	82.7/83.8	53.4	0.554	0.752
3 tokens+ch	83.0/85.0	82.7/85.2	51.3	0.557	0.747
3 tokens+attn	82.7/85.4	82.3/85.5	**53.8**	**0.539**	**0.759**
MTF	**82.9/86.2**	**82.4/86.2**	53.0	0.546	0.758

(ii) MTF (2 tokens) is a variant where only two tokens for each modality are used to capture the inter-modality information, then the six tokens are also sent to a new transformer.
(iii) MTF (3 tokens+concat) is a variant where three tokens for each modality are used where two tokens are applied to capture the inter-modality information and one token is applied to get the intra-modality information, then the nine tokens are concatenated directly.
(iv) MTF (3 tokens+ch) is the same with MTF (3 tokens+concat) except the concatenation operation of nine tokens is replaced with an ESA block for each modality tokens only which means only unimodal level ESA is used.
(v) MTF (3 tokens+attn) is the same with MTF (3 tokens+concat) except the concatenation operation of nine tokens is replaced with sending the nine tokens to a new transformer.

The results of ablation study on dataset CMU-MOSEI are revealed in Table 3. It can be seen from the comparison results between MTF (base) and MTF (2 tokens) that two tokens added only for inter-modality information capture can make a great improvement. From the comparison of MTF (3 tokens+attn) and MTF (2 tokens) we can see the effectiveness of the additional one token used for intra-modality capture. The results verify the strength of two kinds of tokens. In the end, we offer the results of three variants of 3 tokens to check the effectiveness of HESA module. As we can see that one layer of ESA blocks only is not enough for the task from the comparison of MTF (3 tokens+ch) and MTF. And MTF performs better than MTF (3 tokens+attn) on F_1 score and Acc-2 while achieving comparable results in the other metrics which demonstrates the HESA module performs better than the transformer. Similar results can also be found on dataset CMU-MOSI.

Effectiveness of Multimodalities. To demonstrate the importance of each modality, we conduct several experiments with single modality, two modalities and three modalities. The results are summarized in Table 4 and they are quite similar to previous works [3,6]. First of all, with respect to single modality, language modality works better than audio and vision modalities. Two reasons have been pointed out by previous work [6], one of them is that language has

Table 4. Performance of MTF using one modality, two modalities and three modalities on CMU-MOSEI dataset

Modalities	F_1	Acc-2	Acc-7	MAE	Corr
l	82.0/84.4	76.5/77.3	50.3	0.561	0.751
a	59.0/48.5	71.0/62.9	32.4	0.941	0.114
v	66.2/61.9	68.0/65.0	41.6	0.817	0.245
la	82.3/84.3	82.0/84.4	51.9	0.553	0.751
lv	82.5/84.4	82.3/84.7	**53.1**	0.552	0.745
av	64.1/63.2	63.2/63.6	42.1	0.818	0.248
MTF	**82.9/86.2**	**82.4/86.2**	53.0	**0.546**	**0.758**

better data quality than audio and vision. In contrast, audio and vision modality might be influenced by trivial noise data. The other one is that the BERT model which is used to extract the language feature is a more mature and powerful one than the model used to extract audio and vision modality. Secondly, with regard to two modalities, we can see the performance is improved if the modality is combined with language modality which indicates that text information helps the understanding of non-text information. Finally, the experiment conducted with three modalities score highest on almost all the metrics demonstrates the usefulness of multimodal sentiment analysis.

4.7 Case Study

In this section, we present some failure cases of MTF on CMU-MOSI dataset in Fig. 5. For simplicity, we omit the presentation of audio here. As we can see that the framework misclassifies these samples into opposite semantics.

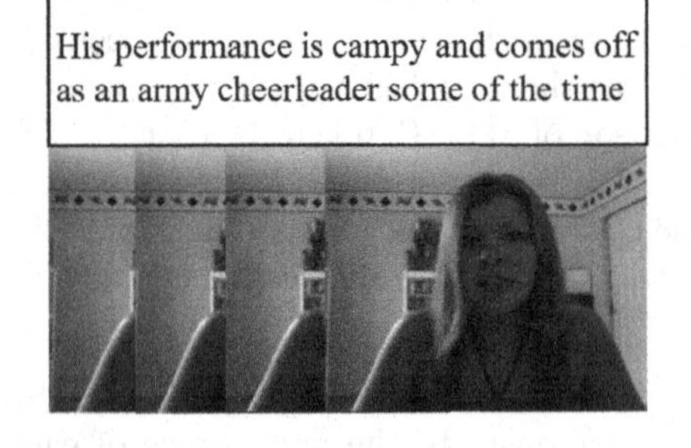

(a) Truth: -2.00, Prediction: 1.86

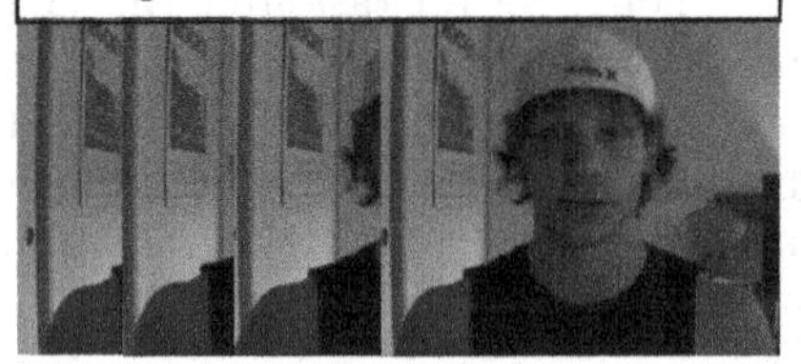

(b) Truth: -1.60, Prediction: 0.57

Fig. 5. Failure examples on CMU-MOSI dataset.

The first sample shows an extreme case where the semantics expressed by different modalities are contradictory, focus only on visual modality will lead to wrong prediction. In terms of the second sample, we can observe the text and visual modality have not shown discriminative information about negative semantics which may confuse the model.

5 Conclusions

In this paper, we design a new multimodal framework for multimodal sentiment analysis which mainly include TT module and HESA module. With the tri-token transformer, modality-specific information and cross-modal information are extracted and processed parallely in the TT module. Furthermore, the HESA module weights the feature in element-wise way to pay more attention to important elements. Numerous experiments show that the MTF framework performs competitively on two benchmark datasets.

Acknowledgements. This work is supported by the Youth Talent Support Programme of Guangdong Provincial Association for Science and Technology (No. SKXRC202305) and the Huangpu International Sci&Tech Cooperation Foundation under Grant 2021GH12.

References

1. Cheng, M., et al.: Vista: vision and scene text aggregation for cross-modal retrieval. In: CoRR abs/2203.16778 (2022)
2. Degottex, G., Kane, J., Drugman, T., Raitio, T., Scherer, S.: COVAREP - a collaborative voice analysis repository for speech technologies. In: ICASSP, pp. 960–964 (2014)
3. Deng, H., Kang, P., Yang, Z., Hao, T., Li, Q., Liu, W.: Dense fusion network with multimodal residual for sentiment classification. In: ICME, pp. 1–6 (2021)
4. Devlin, J., Chang, M., Lee, K., Toutanova, K.: BERT: pre-training of deep bidirectional transformers for language understanding. In: NAACL-HLT, pp. 4171–4186 (2019)
5. Guo, M., et al.: Attention mechanisms in computer vision: a survey. Comput. Vis. Media **8**(3), 331–368 (2022)
6. Hazarika, D., Zimmermann, R., Poria, S.: MISA: modality-invariant and -specific representations for multimodal sentiment analysis. In: ACM Multimedia, pp. 1122–1131 (2020)
7. Kingma, D.P., Ba, J.: Adam: a method for stochastic optimization. In: ICLR (2015)
8. Lai, H., Yan, X.: Multimodal sentiment analysis with asymmetric window multi-attentions. Multimedia Tools Appl. **81**(14), 19415–19428 (2022)
9. Lin, J., Yang, A., Zhang, Y., Liu, J., Zhou, J., Yang, H.: InterBERT: vision-and-language interaction for multi-modal pretraining. CoRR abs/2003.13198 (2020)
10. Liu, H., Dai, Z., So, D.R., Le, Q.V.: Pay attention to MLPs. In: NeurIPS, pp. 9204–9215 (2021)
11. Liu, Z., Shen, Y., Lakshminarasimhan, V.B., Liang, P.P., Zadeh, A., Morency, L.: Efficient low-rank multimodal fusion with modality-specific factors. In: ACL, pp. 2247–2256 (2018)

12. Mai, S., Hu, H., Xing, S.: Modality to modality translation: an adversarial representation learning and graph fusion network for multimodal fusion. In: AAAI, pp. 164–172 (2020)
13. Mittal, T., Bhattacharya, U., Chandra, R., Bera, A., Manocha, D.: M3ER: multiplicative multimodal emotion recognition using facial, textual, and speech cues. In: AAAI, pp. 1359–1367 (2020)
14. Nagrani, A., Yang, S., Arnab, A., Jansen, A., Schmid, C., Sun, C.: Attention bottlenecks for multimodal fusion. In: NeurIPS, pp. 14200–14213 (2021)
15. Pham, H., Liang, P.P., Manzini, T., Morency, L., Póczos, B.: Found in translation: learning robust joint representations by cyclic translations between modalities. In: AAAI, pp. 6892–6899 (2019)
16. Sahay, S., Okur, E., Kumar, S.H., Nachman, L.: Low rank fusion based transformers for multimodal sequences. CoRR abs/2007.02038 (2020)
17. Tsai, Y.H., Bai, S., Liang, P.P., Kolter, J.Z., Morency, L., Salakhutdinov, R.: Multimodal transformer for unaligned multimodal language sequences. In: ACL, pp. 6558–6569 (2019)
18. Tsai, Y.H., Liang, P.P., Zadeh, A., Morency, L., Salakhutdinov, R.: Learning factorized multimodal representations. In: ICLR (2019)
19. Vaswani, A., et al.: Attention is all you need. In: NeurIPS, pp. 5998–6008 (2017)
20. Wang, Y., Shen, Y., Liu, Z., Liang, P.P., Zadeh, A., Morency, L.: Words can shift: dynamically adjusting word representations using nonverbal behaviors. In: AAAI, pp. 7216–7223 (2019)
21. Yan, X., Xue, H., Jiang, S., Liu, Z.: Multimodal sentiment analysis using multi-tensor fusion network with cross-modal modeling. Appl. Artif. Intell. **36**(1), 2000688 (2022)
22. Yu, W., Xu, H., Yuan, Z., Wu, J.: Learning modality-specific representations with self-supervised multi-task learning for multimodal sentiment analysis. In: AAAI, pp. 10790–10797 (2021)
23. Zadeh, A., Chen, M., Poria, S., Cambria, E., Morency, L.: Tensor fusion network for multimodal sentiment analysis. In: EMNLP, pp. 1103–1114 (2017)
24. Zadeh, A., Liang, P.P., Mazumder, N., Poria, S., Cambria, E., Morency, L.: Memory fusion network for multi-view sequential learning. In: AAAI, pp. 5634–5641 (2018)
25. Zadeh, A., Liang, P.P., Poria, S., Cambria, E., Morency, L.: Multimodal language analysis in the wild: CMU-MOSEI dataset and interpretable dynamic fusion graph. In: ACL, pp. 2236–2246 (2018)
26. Zadeh, A., Liang, P.P., Poria, S., Vij, P., Cambria, E., Morency, L.: Multi-attention recurrent network for human communication comprehension. In: AAAI, pp. 5642–5649 (2018)
27. Zadeh, A., Zellers, R., Pincus, E., Morency, L.: Multimodal sentiment intensity analysis in videos: facial gestures and verbal messages. IEEE Intell. Syst. **31**(6), 82–88 (2016)
28. Zhang, W., Yu, J., Wang, Y., Wang, W.: Multimodal deep fusion for image question answering. Knowl. Based Syst. **212**, 106639 (2021)
29. Zhao, S., et al.: An end-to-end visual-audio attention network for emotion recognition in user-generated videos. In: AAAI, pp. 303–311 (2020)

Generative Adversarial Networks Based on Contrastive Learning for Sequential Recommendation

Li Jianhong(✉), Wang Yue, Yan Taotao, Sun Chengyuan, and Li Dequan

School of Artificial Intelligence, Anhui University of Science and Technology, Huainan 232001, China
1659117121@qq.com

Abstract. Generative Adversarial Networks(GAN) has made key breakthroughs in computer vision and other fields, so some scholars have tried to apply it to sequential recommendation. However, the recommendation performance of GAN-based algorithms is unsatisfactory. The reason for this is that the discriminator cannot distinguish the original data from the generated data well if it only relies on the target function. Based on this, we propose Generative Adversarial Networks based on Contrastive Learning for sequential recommendation (shortened to CtrGAN). Firstly, the generator generates item sequences that the user may be interested in. Additionally, the true item sequences of the user are subjected to a mask operation, which means that the sequences with mask operation are fake. Therefore, both generative sequences and fake sequences can be used in Contrastive Learning to train the generator. The true sequences and their mask operations are then combined with the generative sequences to employ the discriminator for distinguishing them. Finally, the contrastive loss and discriminative loss are combined to guide the generator to generate item sequences that the user may be interested in. Compared with existing sequential recommendation algorithms, experimental results illustrate that CtrGAN has better recommendation accuracy.

Keywords: Generative Adversarial Networks · Contrastive Learning · Triplet Loss · Sequential Recommendation

1 Introduction

Collaborative Filtering (CF) [22] has been used in many recommendation tasks like food recommendation, movie recommendation, and so on [33]. Some companies like Amazon have deployed it and acquired great economic benefits. To receive better recommendation accuracy, many methods based on CF have been proposed, such as Matrix Factorization [14]. However, the improvement of these recommendation methods is poor. With the development of Artificial Intelligence, deep learning [15] has been made breakthrough in computer vision and natural language processing [17]. Then some researchers use deep learning methods to improve recommendation accuracy. Typical algorithms include

X. Song et al. (Eds.): APWeb-WAIM 2023, LNCS 14332, pp. 439–453, 2024.
https://doi.org/10.1007/978-981-97-2390-4_30

Neural Collaborative Filtering (NCF) [9] and Collaborative Noise Auto-Encoder (CDAE) [34], and so on [19,25]. As a variant of deep neural networks, Generative Adversarial Networks (GAN) [7] have made major breakthroughs in computer vision. By learning from the data distribution, the GAN obtains correlations between the data. Therefore, the GAN is now widely used in image generation, image inpainting [20] and other fields. Then researchers employ GAN for sequential recommendation. Wang et al. [29] proposed the Generative Adversarial Information Retrieval Model (IRGAN) to apply GAN to recommend item sequence for the specific-user. GraphGAN [28] is designed to construct graph representations of users and items. Chen et al. [4] created the CascadingDQN algorithm to learn the user's dynamic behavior and reward mechanism. Li et al. [5] designed Wide & deep generative adversarial networks for recommendation system, the core idea of which is to employ wide & deep model in generator. However, most of them need to rely on reinforcement learning mechanisms and require the results of the discriminator as the reward mechanism of the generator [3]. This means that the recommendation accuracy will be affected if the reward mechanism doesn't design well. In addition, some researchers have designed adversarial networks methods for recommendation system like Adversarial Collaborative Neural Network for the robust recommendation (FG-ACAE) [35], Prioritize Long And Short-Term Informationusing Adversarial Training (PLASTIC) [36] and Recurrent Generative Adversarial Networks for Recommendation Systems (RecGAN) [1]. Chae et al. [2] proposed a Collaborative Filtering based on Generative Adversarial Networks algorithm(CFGAN) and applied it to sequence recommendation. It is pity that these methods requires a large amount of data for learning and the recommendation accuracy is unsatisfied. In summary, GANs-based recommendation methods exists some problems: 1) the discriminator can't distinguish the original data from the generated data well if it only relies on the target function; 2) the item sequence by the generator generated is unsatisfactory.

To solve those problems we have mentioned, Generative Adversarial Networks based on Contrastive Learning for sequential recommendation(CtrGAN) is proposed, and the framework is as shown as in Fig. 2. First, the generator can be designed to generate sequences of items that the user may be interested in. Then, the truth item sequence of the user will be added mask operation, which means the item sequence will become fake sequence. Therefore, based on Contrastive Learning theory [18], the generator can be trained by using generative sequence and fake sequence. After that, the triplet loss in discriminator is designed to distinguish the truth sequence, fake sequence and generative sequence. Finally, the contrastive loss and the triplet loss are combined to optimize and guide generator to generate item sequence that the user might be interested in. The contributions of this paper can be summarized as follows:

1. We propose a new sequential recommendation framework called CtrGAN. It is based on Contrastive Learning and GAN. As far as we know, this is the first time to combine Contrastive Learning with GAN for sequential recommendation.

2. The generator of CtrGAN is to generate the item sequence that the user may be interested in. In addition, the truth item sequence of the user will be added mask operation as fake sequence, so fake sequence and generative sequence adopt contrastive loss to train generator.
3. The triplet loss in discriminator is to capture difference of three sequences(fake sequence, generative sequence and truth sequence). Then, the triplet loss will get training information feedback and guide generator to generate better item sequence.

The remainder of this paper is organized as follows: In Sect. 2, it includes sequential recommendation and GAN. CtrGAN will be described in Sect. 3. Section 4 contains the description of the datasets, measurement metrics, experimental results and analysis. We provide a brief conclusion and future work in Sect. 5.

2 Preliminaries

2.1 Sequential Recommendation

According to Wang et al. [33] description, an sequential recommendation takes a sequence of user-item historic interactions as the input and predict the subsequent user-item interactions that may happen in the near future. Namely, given a sequence of user-item interactions, a recommendation list consisting of top ranked candidate items are generated by maximizing a utility function value, the experssion can be described by:

$$R = argmax\, F(UI) \tag{1}$$

Here, R is recommendation item sequence, UI is interactions records of user-item. F is function of output a rank score for items. Here $U{=}(u_1{,}u_2{,}u_3{,}...{,}u_n)$ is set of users, and $I{=}(i_1{,}i_2{,}i_3{,}...{,}i_n)$ is set of items.

2.2 GAN

The GAN, which is an unsupervised model, has attracted widespread attention from both academia and industry. It includes two models: a generator(shortly G) and a discriminator(shortly D). The training processing can be described as: G learns to generate data that conform to the distribution of real data as much as possible; D is to distinguish between the real data and those generated by G. The two models compete against each other and optimize themselves via feedback loops until D cannot discriminate both of them. The following equation shows the process of calculating:

$$\underset{G}{min}\underset{D}{max}V(D,G) = ((E_{x\sim p_{data}(x)}[logD(x)] + E_{z\sim p_z(z)}[log(1 - D(G(z)))] \tag{2}$$

where D represents the discriminator, and G is the generator. x,z are true data and fake data.

3 CtrGAN

For GAN-based sequential recommendation(*i.e.*, IRGAN,CFGAN and so on [16]), they train Generator(G) as follows as in Fig. 1. It shows that G generates items for the specific-user, and Discriminator(D) discriminates the difference between generative seqeuence and the truth sequence of user. Then, D guides G to generate a better sequence. However, there are two disadvantages: **(1)** G's results only rely on D; **(2)** the item sequence needs to use reinforcement learning to constrain G generated sequence, which means recommendation performance will be affected by the reward mechanism. To solve these problems, the CtrGAN are proposed and given a detailed introduction in the following description.

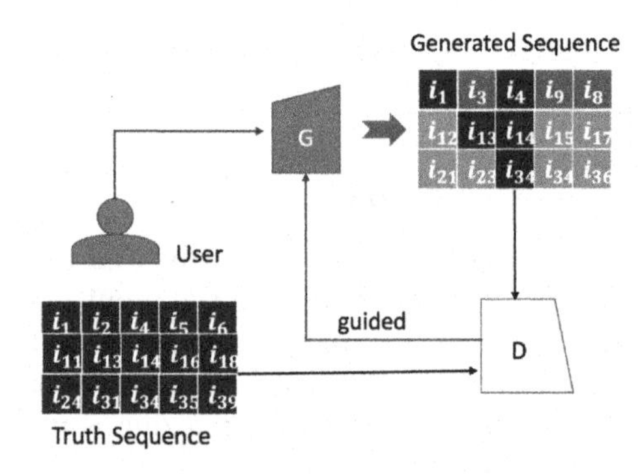

Fig. 1. The GAN-based algorithms for item sequence generation.

Input Processing. Based on Eq.(1), the dataset information needs to transform into an interactive matrix M from user and item interactive information. The user information is represented as a vector representation using embedding, and the operation can be defined as follows.

$$S = Embedding(M) \tag{3}$$

Here S is the item sequence vector representation of the specific-user u_k, and all values are 1. *Embedding* is function of embedding.

3.1 Generative Model(G)

Ensure that the item sequence is for the specific-user that she/he may be interested in. Here, we employ Deep Neural Network(DNN) to generate item sequence. The whole process can be shown as follows.

$$x^{(l+1)} = f(x^{(l)}w^{(l)} + b^{(l)}) \tag{4}$$

where $x^{(l+1)}$ means output results of $l+1$-layer of DNN.$w^{(l)}$ and $b^{(l)}$ are DNN weight, and DNN bias. f is activation. It is noted that we employ DNN as generator because of generalization. The output can be written by:

$$S_g = \sigma(x^{(l+1)}) \tag{5}$$

where S_g is output result of G and σ is *sigmoid* function.

According to the theory of Contrastive Learning, the sequence of items is fake and has some differences from G. Therefore, we employ the mask operation and the expression can be represented by the following formula:

$$S_f = S \otimes mask \tag{6}$$

Here, *mask* means the random noise sequence, $\otimes$ is the multiply operation.

As the target function of G for specific-user u_k, $V(G)$ can be expressed by:

$$V(G)_{\mu} = \sum_{\mu} ||S_f|u_k - S_g|u_k|| \tag{7}$$

where μ is parameters, $V(G)$ represents sum of G loss, S_g is generative item sequence of user u_k, and S_f represents fake item sequence of user u_k. The essential thing to note here is that the loss is different from other GAN-based methods. The reason is that S_g and S_f should keep quite different, so the S_g is similar to S as much as possible. Obviously, the larger the loss, the better the generative capacity. || || is euclidean metric.

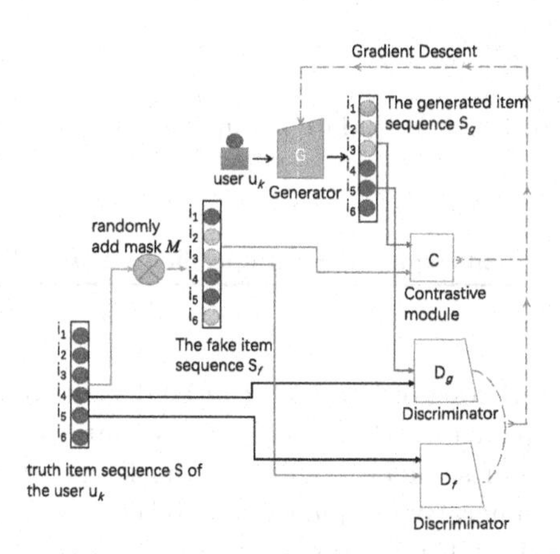

Fig. 2. CtrGAN sequential recommendation algorithm framework.

3.2 Adversarial Nets(D)

In this section, we will describe Adversarial Nets, Discriminator(D) namely. It includes two parts: D_f and D_g. D_g is to discriminate the item sequence by G generated S_g and truth item sequence S; D_f distinguish the difference between fake item sequence S_f and truth item sequence S of specific-user u_k. Obviously, it is triplet loss, and the loss function of D can be expressed by:

$$\underset{\theta}{V(D)} = D_g + D_f = \sum_{\theta} [||D(S_g|u_k) - D(S|u_k)|| + ||D(S_f|u_k) - D(S|u_k)||] \tag{8}$$

Here, θ is parameters.

Algorithm 1: CtrGAN

Input: Item sequence(S), Users(U),Batch_size(B), parameters μ and θ.
1 **for** $u_k \in U$ **do**
2   According to Eq.(6), Mask the truth sequence of u_k and some values changed from 1 to 0 as S_f;
3   **while** *G in step* **do**
4   Sample B for u;
5   According to Eq.(5),Generate items of user may like $S_g|u_k = \left\{\hat{i_1}, \hat{i_2}, ..., \hat{i_n}\right\} \leftarrow$ G;
6   According to Eq.(7),Calculate the difference of $S_f|u_k$ and $S_g|u_k$;
7   According to Eq.(11),Update parameters of G: $\mu \leftarrow \mu$ -$\nabla \underset{\mu}{V(G)}$;
8   Real items from ground truth: $\{i_1, i_2, ..., i_k, ..., i_n\} \leftarrow S$;
9   **end**
10   **while** *D in step* **do**
11   Sample B for u;
12   According to Eq.(8), D($S_g|u_k \approx S$) ? and get $\underset{\theta}{V(D)}$;
13   According to Eq.(11), Update parameters of D: $\theta \leftarrow \theta$ -$\nabla \underset{\theta}{V(D)}$;
14   **end**
15 **end**
Output: Generate item sequence $S_g$for specific-user u_k.

In summary, the core idea of CtrGAN can be described as: the Generator(G) will generate item sequence of the specific-users that they might be interested in. Then, the truth item sequences will add mask operation, so the output is fake. After that, the generative sequence and fake sequence will be contrasted. In addtion, the Discrimnator(D) not only distingushs the difference between generative sequences and truth, but also discriminates the fake sequence whether is true or not based on the ground truth value. All the information will be feedback to G and guide it to generate item sequence that is similar to truth as far as possible. The whole algorithm is as shown as in Algorithm 1.

3.3 Training

CtrGAN combines (G,D) to train and achieve better results, the whole training process can be expressed by the formula:

$$\begin{aligned} \underset{\mu}{min}\,\underset{\theta}{max}V(G,D) = &\sum_{\theta}||D(S_g|u_k) - D(S|u_k)|| + ||D(S_f|u_k) - D(S|u_k)|| \\ &+ \sum_{\mu}||S_f|u_k - S_g|u_k|| \end{aligned} \tag{9}$$

To keep the generated vector not to be loose or dense, we adopt l_2 loss [2] to construct constraint loss, the vector does not contain many negative or positive samples. The objective function of the constraint loss can be expressed as follows.

$$loss = \sum[\alpha(S_g|u_k - a)^2 + \beta(S_g|u_k - b)^2] \tag{10}$$

Here a is negative value 0, b represents positive value 1. α and β are penalty coefficient.

Therefore, the target function of CtrGAN can be expressed by:

$$\begin{aligned} \underset{\mu}{min}\,\underset{\theta}{max}V(D,G) = &\sum_{\mu}\underbrace{||S_f|u_k - S_g|u_k|| + [\alpha(S_g|u_k - a)^2 + \beta(S_g|u_k - b)^2]}_{G} \\ &+ \sum_{\theta}\underbrace{||D(S_g|u_k) - D(S|u_k)|| + ||D(S_f|u_k) - D(S|u_k)||}_{D} \end{aligned} \tag{11}$$

Here, μ and θ are learning parameters. For user u_k, G of CtrGAN will generate the item sequence that the u_k might be interested in. D_g and D_f will discriminate against ground truth with item sequence of the G generated and fake sequence. All the loss information will be feedback to G. Here, we use *Adam* function to optimize G and D. CtrGAN will update parameters based on target function and receive best recommendation performance.

Table 1. Statistics of the experimental datasets

	Ciao	MovieLens 100K	MovieLens 1M
Users	996	943	6,040
Items	1,927	1,682	3,883
Records	18,648	100,000	1,000,209
Sparsity	98.72%	93.7%	95.8%

4 Experiments and Analysis

In this section, the experiments and analysis will be introduced. To evaluate correctly the performance of CtrGAN, the experiments are setting in three different datasets.

4.1 Datasets

The datasets include Ciao, MovieLens 100K and MovieLens 1M[1], all of them are very popular in sequential recommendation task. Three datasets information can be shown in Table 1. Each of them will be split into two parts: 20% for test and the rest will be setting for training. It needs to pay attention that CtrGAN never use any auxiliary context information of user and item.
Ciao: it includes 996 users and 1,927 items. The interactive records are 18,648.
MovieLens 100K: it contains 100,000 interactive records from 943 users and 1,682 items.
MovieLens 1M: the 1,000,209 interactive records from 6,040 users on 3,883 items.

4.2 Implementation Details and Metrics

The implement of CtrGAN is based on TensorFlow-1.15[2], and it deploys on a NVIDIA RTX 2080 Ti GPU with 11 GB of memory. The OS is Ubuntu 18.04 LTS server, and memory is 64 GB. Recall, Normalized Discounted Cumulative Gain (NDCG), Mean Reciprocal Rank (MRR), and Precision are popular metrics in sequential recommendation. Thus, we employ these metrics to evaluate the performance of CtrGAN.

Comparison Algorithms. To better evaluate recommendation performance of our proposed model, the experimental parameters are based on CFGAN [2] algorithm. The experiments will be run 5 times and the results will then be taken from the mean values. Comparison algorithms with the following:

1. **ItemPop**. It inherits from CF, and the core idea is based on popularity of user.
2. **BPR**. Bayesian personalized ranking algorithm with implicit feedback.
3. **FISM**. Factored item similarity models, it is based on CF method.
4. **CDAE**. Collaborative Denoise AutoEncoder.
5. **IRGAN**. Information Retrieval Generative Adversarial Nets for sequential recommendation.

[1] https://grouplens.org/datasets/movielens/.
[2] https://www.tensorflow.org.

6. **GraphGAN**. Generative Adversarial Nets based on graph representation for recommendation .
7. **CAAE**. Collaborative Adversarial AutoEncoder.
8. **CFGAN**. Collaborative Filtering algorithm based on conditional Generative Adversarial Nets.
9. **PLASTIC**. Sequential recommendation using adversarial training and long-short information.
10. **FG-ACAE**. Adversarial neural network robust recommendation recommendation algorithm.
11. **CollaGAN**. Generative adversarial network with collaborative method for recommendation systems.
12. **NGCF** [30]. Neural Graph Collaborative Filtering.
13. **GAT** [27]. Graph Attention Networks for sequentail recommendation.
14. **MCREC** [10]. Leveraging Meta-path based Context with A Neural Co-Attention Model.
15. **DGRBR** [25]. Generative Recommendation sequence method Based on Deep List-Wise Ranking.
16. **SEMI-FL-MV-DSSM** [11]. A Federated Deep Learning Recommendations Framework.
17. **FED-MVMF** [6]. Matrix factorization combined with Federated multi-view learning for personalized recommendations.

4.3 Performance Comparison with Baselines

The experimental results of the Ciao dataset are shown in Table 2. This clearly shows that CtrGAN receives better results, but not so well in Recall@5. Based on the information of Table 1, we think the reason is that Ciao dataset is small, which means we proposed method can't fully mine user information if item sequence is short. Even so, CtrGAN is able to receive better recommendation in Recall@20. This proves that the method we proposed can mine efficiently user preference when the dataset of user-item interactive information is small.

We can see that CtrGAN has an advantage over the other methods on MovieLens 100K dataset. Compared to other algorithms, our method achieves the best results in terms of top-5 and top-20 sequential recommendation. It illustrates that the CtrGAN utilizes fully the interactive information and receives great recommendation accuracy.

We can find that the CtrGAN algorithm performance is higher than the comparison methods on MovieLens 1M dataset. The reason is similar to the MovieLens 100K dataset, the larger the dataset, the more information will be mined. Thus, all these results reveal that CtrGAN efficiently uses interaction information to accurately recommend a sequence of items that the user might like.

Table 2. Top-k experimental performance of the CtrGAN network and the baselines on the different datasets when $k \in \{5,20\}$.

Ciao	Prec@5	Prec@20	Recall@5	Recall@20	NDCG@5	NDCG@20	MRR@5	MRR@20
Pop	0.031	0.024	0.040	0.127	0.047	0.065	0.056	0.067
BPR	0.036	0.025	0.040	0.141	0.052	0.066	0.066	0.078
FISM	0.062	0.040	0.072	0.178	0.079	0.109	0.127	0.147
CDAE	0.061	0.042	0.075	0.185	0.081	0.108	0.127	0.151
GraphGAN	0.026	0.017	0.041	0.100	0.041	0.058	0.057	0.068
IRGAN	0.035	0.023	0.042	0.111	0.046	0.066	0.082	0.088
CFGAN	0.053	0.032	0.055	0.140	0.068	0.093	0.124	0.144
CAAE	0.067	0.042	0.079	0.187	0.086	0.120	0.144	0.164
CFWGAN	0.068	0.044	0.074	0.190	0.085	0.122	0.142	0.168
CtrGAN	**0.068**	**0.045**	0.076	**0.197**	**0.088**	**0.125**	**0.147**	**0.173**

MovieLens-100K	Prec@5	Prec@20	Recall@5	Recall@20	NDCG@5	NDCG@20	MRR@5	MRR@20
Pop	0.181	0.138	0.102	0.251	0.163	0.195	0.254	0.292
BPR	0.348	0.236	0.116	0.287	0.370	0.380	0.556	0.574
FISM	0.426	0.285	0.140	0.353	0.462	0.429	0.674	0.685
CDAE	0.433	0.287	0.144	0.353	0.465	0.425	0.664	0.674
GraphGAN	0.212	0.151	0.102	0.260	0.183	0.249	0.282	0.312
IRGAN	0.312	0.221	0.107	0.275	0.342	0.368	0.536	0.523
CFGAN	0.430	0.286	0.146	0.345	0.465	0.426	0.671	0.683
CAAE	0.435	0.289	0.151	0.348	0.475	0.432	0.686	0.697
CFWGAN	0.446	0.297	0.154	0.360	0.480	0.440	0.684	0.697
PLASTIC	0.312	-	-	-	0.331	-	-	-
CtrGAN	**0.452**	**0.298**	**0.158**	**0.361**	**0.488**	**0.442**	**0.693**	**0.704**

MovieLens-1M	Prec@5	Prec@20	Recall@5	Recall@20	NDCG@5	NDCG@20	MRR@5	MRR@20
Pop	0.157	0.121	0.076	0.197	0.154	0.181	0.252	0.297
BPR	0.341	0.252	0.077	0.208	0.349	0.362	0.537	0.556
FISM	0.420	0.302	0.107	0.270	0.443	0.399	0.637	0.651
CDAE	0.419	0.307	0.108	0.272	0.439	0.401	0.629	0.644
GraphGAN	0.178	0.194	0.070	0.179	0.205	0.184	0.281	0.316
IRGAN	0.263	0.214	0.072	0.166	0.264	0.246	0.301	0.338
CFGAN	0.430	0.308	0.107	0.268	0.455	0.405	0.647	0.661
CAAE	0.436	0.310	0.109	0.275	0.461	0.414	0.653	0.666
FG-ACAE	-	-	-	-	0.458	-	-	-
CollaGAN	0.428	-	-	-	0.417	-	-	-
CFWGAN	0.435	0.311	0.108	0.270	0.459	0.409	0.649	0.663
CtrGAN	**0.438**	**0.315**	**0.111**	**0.275**	**0.463**	**0.414**	**0.657**	**0.670**

Recently, Graph Neural Networks(GNNs) [32] has an advantage in network representation of the social network, therefore some researchers adopt it into recommendation [30] [10]. Due to the sparsity of the data, the experimental performance is measured with NDCG@10 of MovieLens 100K. The results of NDCG@10 are shown in Table 3. We can find that some GNN-based algorithms such as NGCF can achieve better recommendation effects. Nevertheless, Ctr-

Table 3. Experimental performance of CtrGAN in NDCG@10 on the MovieLens 100K dataset.

	MovieLens 100K NDCG@10
BMF	0.408
NMF	0.234
GAT	0.256
MCRec	0.262
NGCF	0.418
DGRBR	0.327
SEMI-FL-MV-DSSM	0.317
FED-MVMF	0280
CtrGAN	**0.452**

GAN also achieves the best recommendation accuracy. Other MF-based methods (BMF, NMF, etc.) can not get satisfactory recommendation results. It also proves that CtrGAN is efficient in mining social network data.

4.4 The Different Loss Function of CtrGAN

To express the influence of different loss function based on euclidean metric(G and D) in CtrGAN, we choose three typical loss function: Wasserstein loss [8],Cross Entropy loss and Smooth L1 loss (Smooth l_1 loss, namely). The results of different loss functions in CtrGAN on different datasets are as shown in Fig. 3, respectively. Next, we will introduce it in detail.

The different results of the Ciao dataset has been shown in Fig. 3(a). Compared to other different loss functions, CtrGAN with Smooth L1 loss can possess better recommendation accuracy. It is obviously shown that different loss functions will affect the recommendation accuracy of the Ciao dataset. Similarly to Ciao dataset, MovieLens 100K results have an average improvement of 0.5% in Fig. 3(b). Based on Table 1, we think that MovieLens 100K is bigger than Ciao, then the effective information can be received as much as possible. Figure 3(c) also demonstrates the experimental results of CtrGAN with different loss function in MovieLens 1M, the average improvement of CtrGAN is about 0.1%. According to Table 1, MovieLens 1M is larger than the other two datasets. Therefore, the effective information can be easily mined by different recommendation algorithms. It should be noted that the Smooth L1 loss has been received best results in three loss function, and it creates a criterion that uses a squared term if the absolute element-wise error falls below beta and an l_1 term otherwise. Obviously, the Smooth L1 loss is appropriate for CtrGAN.

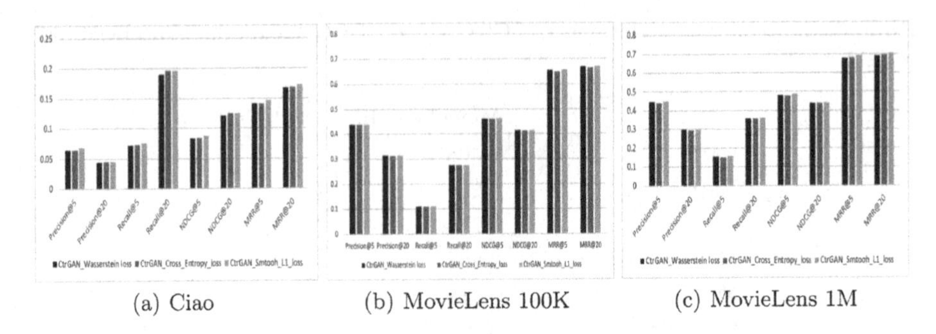

(a) Ciao (b) MovieLens 100K (c) MovieLens 1M

Fig. 3. The Experimental results of CtrGAN with different loss function.

Table 4. Ablation study of CtrGAN on the several dataset.

		NCDG@5	MRR@5
Ciao	*w/o C*	0.083	0.140
	w/o D_f	0.083	0.140
	w/o both	0.080	0.137
	w/ both	**0.088**	**0.147**
MovieLens 100K	*w/o C*	0.484	0.686
	w/o D_f	0.483	0.687
	w/o both	0.480	0.682
	w/ both	**0.488**	**0.693**
MovieLens 1M	*w/o C*	0.461	0.651
	w/o D_f	0.460	0.653
	w/o both	0.457	0.648
	w/ both	**0.463**	**0.657**

4.5 Ablation Study

To verify the effectiveness of the different components of CtrGAN, *i.e.*, we remove contrastive module(*w/o C*), drop discriminator of the fake sequence module(*w/o* D_f) or both of them(*w/o both*). The results have been shown in Table 4. It can see that the different components have an influence on recommendation accuracy, for example, on the Ciao dataset, if *w/o C* is employed, its performance decreases by 0.5% on NDCG@5 and 0.7% on MRR@5. When *w/o C* is used on the MovieLens datasets (100K,1M), it decreases almost 1% in both NDCG and MRR. *w/o* D_f and *w/o both* also have a similar situation on three datasets. Apparently, several components of CtrGAN have been designed are rational.

4.6 Stability of CtrGAN

The stability of CtrGAN at three datasets has been revealed in Fig. 4 in different training epochs. It is clearly see that CtrGAN keeps stable situation after a few hundreds epochs. For example, Ciao dataset have achieved the best results about 300 epochs; The best recommendation accuracy of Recall metrics of MovieLens 100K has obtained in about 150 epochs; the best result of MovieLens 1M of Recall meet in 150 epochs. It should be noted that the Ciao dataset is different from Moivelens 100K and Moivelens 1M, we think the reason is that Ciao dataset is small and need to much times to train. In one word, The CtrGAN method not only obtains a good recommendation effect, but also possesses better stability.

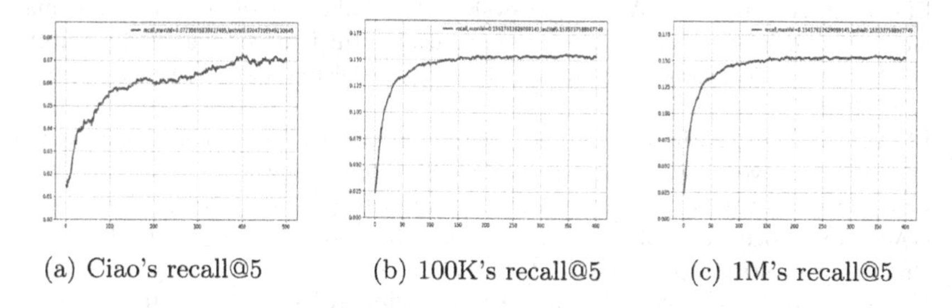

(a) Ciao's recall@5 (b) 100K's recall@5 (c) 1M's recall@5

Fig. 4. The Recall@5 results of Epochs on three datasets

5 Conclusion

In this paper, we present Generative Adversarial Networks based on Contrastive Learning (CtrGAN). The generator of CtrGAN is to generate the item sequence that the user may be interested in. The user's sequence is added to the mask operation, and the model can be trained using Contrastive Learning for the generator. We design triplet loss in discriminative model. Therefore, the contrastive loss and triplet loss are combined to optimize CtrGAN. Experimental results demonstrate the effectiveness of CtrGAN on several datasets. In addition, the model-based Transformer can mine semantic information from data. In the next work, we will utilize these methods to improve the performance of CtrGAN.

Acknowledgements. This work was partially supported by University-level key projects of Anhui University of Science and Technology(Grants #xjzd2020-15), Scientific Research Foundation for introduced talents of Anhui University of Science and Technology(Grants #13200426), Directive Science and technology plan projects in 2021 of Huainan City(Grants #2021003, #2021136), and the Supported projects (natural science) of Anhui University of Science and Technology(Grants #xjyb2020-13).

References

1. Bharadhwaj, H., Park, H., Lim, B.: RecGAN: recurrent generative adversarial networks for recommendation systems. In: Proceedings of the 12th ACM Conference on Recommender Systems, pp. 372–376 (2018)
2. Chae, D., Kang, J., Kim, S., et al.: CFGAN: a generic collaborative filtering framework based on generative adversarial networks. In: Proceedings of the 27th ACM International Conference on Information and Knowledge Management, pp. 137–146 (2018)

3. Chae, D., Shin, J., Kim, S.: Collaborative adversarial autoencoders: an effective collaborative filtering model under the GAN framework. IEEE Access **7**, 37650–37663 (2019)
4. Chen, X., Li, S., Li, H., et al.: Generative adversarial user model for reinforcement learning based recommendation system. In: Proceedings of the International Conference on Machine Learning, pp. 1052–1061 (2019)
5. Li, J., Li, J., Wang, C., et al.: Wide and deep generative adversarial networks for recommendation system. Intell. Data Anal. **27**(1), 121–136 (2023)
6. Flanagan, A., Oyomno, W., Grigorievskiy, A., et al.: Federated multi-view matrix factorization for personalized recommendations. In: Proceedings of the European Conference on Machine Learning and Knowledge Discovery in Databases, pp. 324–347 (2020)
7. Li, Y., Wang, Q., Zhang, J.: The theoretical research of generative adversarial networks: an overview. Neurocomputing **435**, 26–41 (2021)
8. Gulrajani, I., Ahmed, F., Arjovsky, M., et al.: Improved training of Wasserstein GANs. In: Proceedings of the 31st International Conference on Neural Information Processing Systems, pp. 5769–5779 (2017)
9. He, X., Liao, L., Zhang, H., et al.: Neural collaborative filtering. In: Proceedings of the 26th International Conference on World Wide Web, pp. 173–182 (2017)
10. Hu, B., Shi, C., Zhao, W., et al.: Leveraging meta-path based context for top-n recommendation with a neural co-attention model. In: Proceedings of the 24th ACM SIGKDD International Conference on Knowledge Discovery and Data Mining, pp. 1531–1540 (2018)
11. Huang, M., Li, H., Bai, B., et al.: A federated multi-view deep learning framework for privacy-preserving recommendations. arXiv preprint arXiv:2008.10808(2020)
12. Kabbur, S., Ning, X., Karypis, G.: FISM: factored item similarity models for top-n recommender systems. In: Proceedings of the 19th ACM SIGKDD International Conference on Knowledge Discovery and Data Mining, pp. 659–667 (2013)
13. Plaat, A., Kosters, W., Preuss, M.: High-accuracy model-based reinforcement learning, a survey. Artif. Intell. Rev. **56**(9), 9541–9573 (2023). https://doi.org/10.1007/s10462-022-10335-w
14. Koren, Y., Bell, R., Volinsky, C.: Matrix factorization techniques for recommender systems. Computer **42**(8), 30–37 (2009)
15. LeCun, Y., Bengio, Y., Hinton, G.: Deep learning. Nature **521**(7553), 436–444 (2015)
16. Lu, G., Zhao, Z., Gao, X., et al.: SRecGAN: pairwise adversarial training for sequential recommendation. In: Proceedings of the International Conference on Database Systems for Advanced Applications, pp. 20–35 (2021)
17. Peng, S., Zeng, R., Liu, H., et al.: Emotion classification of text based on BERT and broad learning system. In: Proceedings of the APWeb-WAIM International Joint Conference on Web and Big Data, pp. 382–396 (2021)
18. Karimi, H., Barthe, G., Scholkopf, B.: A survey of algorithmic recourse: contrastive explanations and consequential recommendations. ACM Comput. Surv. **55**(5), 1–29 (2022)
19. Qian, F., Huang, Y., Li, J., et al.: DLSA: dual-learning based on self-attention for rating prediction. Int. J. Mach. Learn. Cybern. **12**(7), 1993–2005 (2021)
20. Qian, F., Li, J., Du, X., et al.: Generative image inpainting for link prediction. Appl. Intell. **50**(12), 4482–4494 (2020)
21. Rendle, S., Freudenthaler, C., Gantner, Z., et al.: BPR: Bayesian personalized ranking from implicit feedback. arXiv preprint arXiv:1205.2618(2012)

22. Sarwar, B., Karypis, G., Konstan, J., et al.: Item-based collaborative filtering recommendation algorithms. In: Proceedings of the 10th International Conference on World Wide Web, pp. 285–295 (2001)
23. Weerakody, P., Wong, K., Wang, G.: A review of irregular time series data handling with gated recurrent neural networks. Neurocomputing **441**, 161–178 (2021)
24. Shi, J., Ji, H., Shi, C., Wang, X., Zhang, Z., Zhou, J.: Heterogeneous graph neural network for recommendation. arXiv preprint arXiv:2009.00799(2020)
25. Sun, X., Liu, H., Jing, L., et al.: Deep generative recommendation based on list-wise ranking. J. Comput. Res. Dev. **57**(8), 1697–1706 (2020)
26. Tong, Y., Luo, Y., Zhang, Z., et al.: Collaborative generative adversarial network for recommendation systems. In: Proceedings of the IEEE 35th International Conference on Data Engineering Workshops, pp. 161–168 (2019)
27. Wang, X., He, X., Cao, Y.: KGAT: knowledge graph attention network for recommendation. In: Proceedings of the 25th ACM SIGKDD International Conference on Knowledge Discovery and Data Mining, pp. 950–958 (2019)
28. Wang, H., Wang, J., Wang, J., et al.: GraphGAN: graph representation learning with generative adversarial nets. In: Proceedings of the AAAI Conference on Artificial Intelligence (2018)
29. Wang, J., Yu, L., Zhang, W., et al.: IRGAN: a minimax game for unifying generative and discriminative information retrieval models. In: Proceedings of the 40th International ACM SIGIR Conference on Research and Development in Information Retrieval, pp. 515–524 (2017)
30. Wang, X., He, X., Wang, M., et al.: Neural graph collaborative filtering. In: Proceedings of the 42nd International ACM SIGIR Conference on Research and Development in Information Retrieval, pp. 165–174 (2019)
31. Wu, Y., DuBois, C., Zheng, A., et al.: Collaborative denoising auto-encoders for top-n recommender systems. In: Proceedings of the Ninth ACM International Conference on Web Search and Data Mining, pp. 153–162 (2016)
32. Wu, Z., Pan, S., Chen, F., et al.: A comprehensive survey on graph neural networks. IEEE Trans. Neural Netw. learn. syst. **32**(1), 4–24 (2020)
33. Wang, S., Hu, L., Wang, Y., et al.: Sequential recommender systems: Challenges, progress and prospects. In: Proceedings of the 28th International Joint Conference on Artificial Intelligence, pp. 6332–6338 (2019)
34. Yuan, F., Yao, L., Benatallah, B.: Adversarial collaborative auto-encoder for top-n recommendation. In: Proceedings of the International Joint Conference on Neural Networks (IJCNN), pp. 1–8 (2019)
35. Yao, W., DuBois, C., Alice, Zheng., et al.: Collaborative denoising auto-encoders for top-n recommender systems. In: Proceedings of the Ninth ACM International Conference on Web Search and Data Mining, pp. 153–162 (2016)
36. Zhao, W., Wang, B., Ye, J., et al.: PLASTIC: prioritize long and short-term information in top-n recommendation using adversarial training. In: Proceedings of the Proceedings of International Joint Conference on Artificial Intelligence, pp. 3676–3682 (2018)

Multimodal Stock Price Forecasting Using Attention Mechanism Based on Multi-Task Learning

Haoyan Yang(✉)

Beijing Normal University-Hong Kong Baptist University United International College, Zhuhai 519087, China
p930026145@mail.uic.edu.cn

Abstract. This paper proposes a Multi-Task Attention-based Stock Prediction Model (MTASPM) to tackle the challenges of stock price forecasting in the Chinese market, characterized by solid volatility and numerous influencing factors. Employing multimodal information from stock correlation, historical trading data, company news, and government policies, MTASPM leverages multi-task learning and attention mechanism to enhance predictive accuracy and capture data patterns for stock price forecasting. Experimental results on the Shanghai Exchange Stock Price Dataset (SHESPD) and the Shenzhen Exchange Stock Price Dataset (SZESPD) demonstrate that MTASPM outperforms eight baseline models. Specifically, MTASPM achieves improvements of 42.16% in MSE, 25.18% in RMSE, and 6.88% in MAE on SHESPD, and improvements of 16.95% in MSE, 8.64% in RMSE, and 6.12% in MAE on SZESPD. Overall, this study presents an effective approach for accurate stock price prediction that considers multiple influencing factors and utilizes multimodal information.

Keywords: Stock Prediction · Attention Mechanism · Multi-Task · Multimodal · Chinese Market

1 Introduction

The stock market presents an enticing yet risky investment opportunity, with investors seeking methods to enhance returns and mitigate risks. Due to its rapid growth, China's stock market has witnessed a substantial increase in the number of shareholders, surpassing 200 million in 2022 [6]. Consequently, trading stocks has become increasingly significant in the investment landscape and there is a growing interest in accurately predicting stock prices and trends. Traditional approaches rely on subjective analysis and mathematical models, but these methods have limitations in terms of information utilization and the analysis of diverse stock characteristics. With the introduction and widespread use of machine learning and deep learning, these techniques have gained popularity in stock price prediction, including linear regression (LR), Support Vector

X. Song et al. (Eds.): APWeb-WAIM 2023, LNCS 14332, pp. 454–468, 2024.
https://doi.org/10.1007/978-981-97-2390-4_31

Machine (SVM), Recurrent Neural Network (RNN), and Long Short-Term Memory (LSTM) [8,19]. Nonetheless, these methods face challenges that hinder their effectiveness. Firstly, the inter-connectivity among stocks and their mutual influence pose challenges for prediction models. Secondly, the critical issue of stock price volatility requires models that can identify underlying patterns. Lastly, numerous economic, political, and trading factors influence stock prices, necessitating the incorporation of multimodal information into prediction models.

To address these challenges, we propose the Multi-Task Attention-based Stock Prediction Model (MTASPM). This novel model leverages the influence of four factors: stock correlation, historical trading data, company news, and government policies. By employing multi-task learning and attention mechanism techniques, MTASPM effectively integrates these diverse information sources for accurate stock price prediction. Our extensive experiments, conducted using the Shanghai Exchange Stock Price Dataset (SHESPD) and Shenzhen Exchange Stock Price Dataset (SZESPD), demonstrate the superior performance of MTASPM compared to baseline models.

The remainder of this paper is structured as follows: Sect. 2 provides a review of related work in mathematical models, machine learning, and deep learning for stock price prediction. Section 3 presents the architecture of MTASPM, followed by a detailed explanation of its design. In Sect. 4, we describe the datasets used for evaluation. The evaluation results are then presented and discussed. Finally, in Sect. 5, we conclude our work and highlight future directions in stock price prediction research.

2 Related Work

Researchers have made significant efforts to predict stock prices using a variety of models, including mathematical, machine learning, and deep learning approaches. Table 1 presents a specific overview of research studies conducted on the stock price prediction using three distinct types of models.

Despite their usefulness, different types of models used for stock price prediction have their limitations. Mathematical models like ARIMA and machine learning models such as KNN and SVR may struggle to predict long-term trends and overlook external factors. Deep learning models like perceptron, LSTMs, and CNN-BiLSTM-AM can suffer from overfitting issues when relying solely on historical transaction data. To overcome these limitations, we emphasize the importance of considering multiple factors such as historical stock data, stock correlations, company news, and government policies. Our research aims to incorporate multi-modal factors comprehensively, establishing MTASPM to predict the stock price in the Chinese stock market.

Table 1. Overview of Related Work on the Stock Price Prediction.

Model Type	Researcher	Model	Data Used	Result
Mathematic Model	Al-Shiab et al. [1]	Univariate autoregressive integrated moving average (ARIMA)	Amman Stock Exchange (ASE)	ASE followed efficient markets hypothesis in weak form
	Ariyo et al. [3]	ARIMA	New York Stock Exchange and Nigerian Stock Exchange	ARIMA has great potential for short-term forecasting
	Hushanni et al. [7]	ARIMA, vector autoregression, LSTM and nonlinear autoregressive Exogenous	National Association of Securities Dealers Automated Quotations	Math technical analysis combined with the deep learning model is better
Machine Learning Model	Alkhatib et al. [2]	K-Nearest Neighbor (KNN)	Jordan Stock Exchange	KNN algorithm was robust and had a small error rate
	Reddy et al. [13]	SVM	IBM Stock	SVM does not give a problem of over fitting and machine learning methods are practical
	Nair et al. [11]	Decision Tree	Bombay Stock Exchange	Better performance compared to Artificial Neural Network and Naive Bayes
Deep Learning Model	Schöneburg et al. [16]	Perceptron	German stocks	Achieved good results with 90% accuracy in 10-day prediction time
	Yu et al. [24]	LSTMs	Multiple stock indexes	Can predict different stock indexes with high accuracy
	Lu et al. [10]	CNN-BiLSTM-AM	Shanghai Composite Index	Performed the best among 7 methods

3 Methodology

3.1 MTASPM Architecture

MTASPM incorporates historical trading data X_1, company news X_2, and government policy X_3 within a 30-day period T. It consists of three components: stock correlation graph construction, the main task, and the auxiliary task. Figure 1 illustrates the architecture of it.

The stock correlation graph is constructed as an undirected graph without weights, representing the correlations between stocks. Each stock is assigned a unique id, which is used to embed the stock correlation graph into the multi-modal attention-based stock prediction block (MMASPB) in the main task. To embed the stock correlation graph, the unsupervised graph embedding approach called struc2vec [14] is employed, transforming the graph into a D_{model} dimensional embedding matrix. This matrix is then utilized as the initial weight for an embedding layer in the MMASPB, effectively incorporating the stock correlation graph.

In addition, the main task of MTASPM is to predict the closing price of stock P_{price}^{T+1} at time $T+1$. Historical trading data is used as input X_1 to the MMASPB, which leverages an attention mechanism. This attention mechanism assigns varying degrees of attention to features based on their significance, enabling the model to focus on relevant information for stock price prediction.

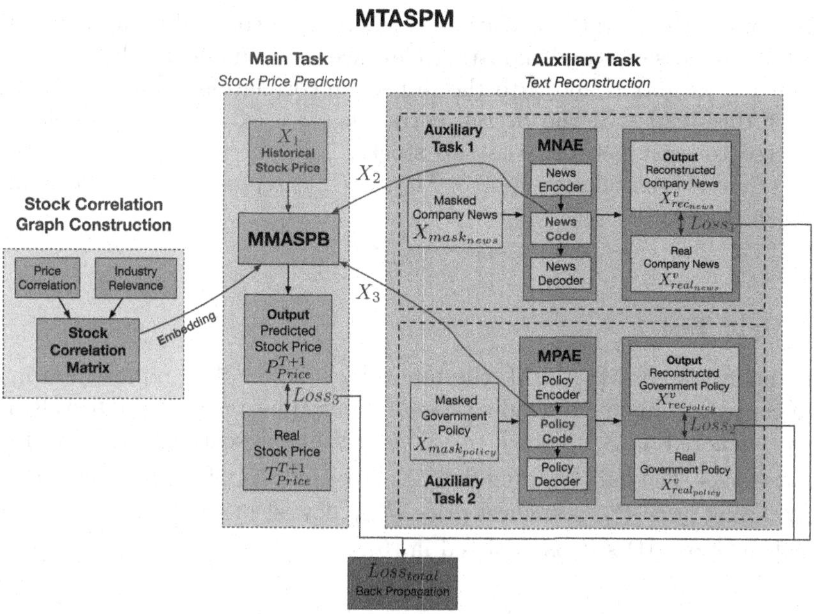

Fig. 1. Architecture of MTASPM

Moreover, the auxiliary task of MTASPM aims to reconstruct the text of the corresponding company news and government policies. The decoded text information from the masked news autoencoder (MNAE) and masked policy autoencoder (MPAE) serves as input X_2 and X_3 for the MMASPB in the main task.

3.2 Stock Correlation Graph Construction

The stock correlation graph construction of MTASPM considers the correlation between stocks from two perspectives. One is the correlation coefficient, which is calculated based on the closing price series of each stock. The correlation coefficient, denoted as $\rho^1_{i,j}$, measures the correlation between the closing price series of stocks S_i and S_j.

$$\rho^1_{\mathrm{i,j}} = \left| \frac{\mathrm{Cov}\left(\mathrm{S_i}, \mathrm{S_j}\right)}{\sqrt{\mathrm{Var}\left[\mathrm{S_i}\right] \mathrm{Var}\left[\mathrm{S_j}\right]}} \right| \tag{1}$$

Another aspect is the industry relevance between companies. Textual similarity of company profiles is used to evaluate the degree of business overlap between companies. The text-similarity, denoted as $\rho^2_{i,j}$, between any two word embeddings of company profiles, I_1 and I_2, is expressed as:

$$\rho^2_{\mathrm{i,j}} = \frac{\mathrm{I_1} \cdot \mathrm{I_2}}{\|\mathrm{I_1}\| \, \|\mathrm{I_2}\|} \tag{2}$$

To obtain the overall correlation $\rho_{i,j}$ between two stocks, the correlations from both price series and industry relevance are combined. This is achieved by applying weights α and β to the respective correlations $\rho^1_{i,j}$ and $\rho^2_{i,j}$. Using a preset threshold ξ, if the overall correlation $\rho_{i,j}$ exceeds ξ, a correlation is considered to exist between the two stocks. This threshold approach enables the creation of a correlation matrix $M_0 \in \mathbb{R}^{m\times m}$ representing the correlations among m stocks.

$$\rho_{\mathrm{i,j}} = \alpha\rho^1_{\mathrm{i,j}} + \beta\rho^2_{\mathrm{i,j}} \tag{3}$$

3.3 Main Task

In the main task of MTASPM, the input $X_1 \in \mathbb{R}^{n\times 30\times 6}$ is represented in (4), which consists of historical trading data organized into n 30-day intervals. Each interval, denoted as $X_1^{T_i}$ shown in (5), is a list representing the daily trading data of stocks. The daily trading data $X_1^{T_{i,j}}$ includes six features: opening price P^j_{open}, closing price P^j_{close}, highest price P^j_{high}, lowest price P^j_{low}, trading volume V^j and the stock ID sid, as depicted in (6).

$$\mathrm{X_1} = \left[\mathrm{X_1^{T_1}, X_1^{T_2}, \ldots, X_1^{T_{n-1}}, X_1^{T_n}}\right] \tag{4}$$

$$\mathrm{X_1^{T_i}} = \left[\mathrm{X_1^{T_{i,1}}, X_1^{T_{i,2}}, \ldots, X_1^{T_{i,29}}, X_1^{T_{i,30}}}\right] \tag{5}$$

$$\mathrm{X_1^{T_{i,j}}} = \left[\mathrm{P^j_{open}}, \mathrm{P^j_{close}}, \mathrm{P^j_{high}}, \mathrm{P^j_{low}}, \mathrm{V^j},\ \mathrm{sid}\right]^{\mathrm{T}} \tag{6}$$

As shown in the MMASPB structure of Fig. 2, the stock ID feature is separated from the other features to incorporate the stock correlation graph and represented as $X_1^{sid} \in \mathbb{R}^{n\times 30\times 1}$. It is one-hot encoded and embedded using the stock correlation matrix M_0 through stock construction graph (SCG) embedding layer, resulting in an embedded representation $X_1^{embed} \in \mathbb{R}^{n\times 30\times D_{model}}$. The remaining features, referred to as $X_1^{others} \in \mathbb{R}^{n\times 30\times 5}$, are projected into a D_{model} dimensional space using a fully connected layer (FC), resulting in $X_1^{projected} \in \mathbb{R}^{n\times 30\times D_{model}}$. In addition to X_1, the input also includes X_2 and X_3, which represent encoded company news and government policies. These inputs are concatenated with $X_1^{projected}$ to create $X_{c1} \in \mathbb{R}^{n\times 30\times (D_{model}\times 3)}$.

X_{c1} is defined as a composite of 30 patch embeddings $PE_i \in \mathbb{R}^{n\times 1\times (D_{model}\times 3)}$ along with an additional learnable embedding $PE_0 \in \mathbb{R}^{n\times 1\times (D_{model}\times 3)}$ added at the head of X_{c1}, as shown in (7) and (8). This configuration enables the prediction of the closing stock price P^{T+1}_{price} at time $T+1$, using the learnable embedding.

$$\mathrm{X_{c1}} = [\mathrm{PE_0}, \mathrm{PE_1}, \ldots, \mathrm{PE_{30}}] \tag{7}$$

$$\mathrm{PE_i} = X_1^{\mathrm{projected}\ i}\mathrm{E} + \mathrm{X_2^i} + \mathrm{X_3^i}(\mathrm{i} = 1, 2, \ldots, 30) \tag{8}$$

where $E \in \mathbb{R}^{5\times (D_{model}\times 3)}$.

The augmented X_{c1} is further processed by introducing patch embeddings and positional embeddings. The resulting matrix undergoes training through an attention layer (AL) consisting of a multi-head self-attention layer (MHSAL) and convolutional layers (Conv1D). The MHSAL performs multiple rounds of self-attention on X_{c1} and concatenates the results, which then undergo a linear transformation. The calculation process is detailed in (9) to (12).

$$\mathrm{Q_i = QW_i^Q, W_i = KW_i^K, V_i = VW_i^Q, i = 1, 2, \ldots, h} \tag{9}$$

$$\text{head }_{\mathrm{i}} = \mathrm{Attention}\left(\mathrm{Q_i, W_i, V_i}\right) \tag{10}$$

$$\mathrm{Multihead(Q, K, V)} = \mathrm{Concat}\left(\text{ head }_1, \text{ head }_2, \ldots, \text{ head }_{\mathrm{i}}\right) \mathrm{W}^0 \tag{11}$$

$$\text{Attention } = \mathrm{softmax}\left(\frac{\mathrm{Q_i K_i^T}}{\sqrt{\mathrm{d_k}}}\right) \mathrm{V_i} \tag{12}$$

where Q, K, V are query parameter matrix, key parameter matrix, and value parameter matrix respectively; W_i^Q, W_i^K, W_i^V are weight parameter maxtrix for Q, K, V respectively; W^O is the output weight parameter matrix; h is the number of attention heads.

Finally, the output from AL is concatenated with X_1^{embed} to derive $X_{c2} \in \mathbb{R}^{\mathrm{n} \times 31 \times (D_{model} \times 4)}$. This combined matrix is passed through a multi-layer perceptron head (MLPH) consisting of two FC layers to reduce the dimensionality to $\mathbb{R}^{n \times 31 \times 1}$. The head of X_{c2}, represented by the trained PE_0, is extracted to obtain the predicted closing stock price P_{price}^{T+1}.

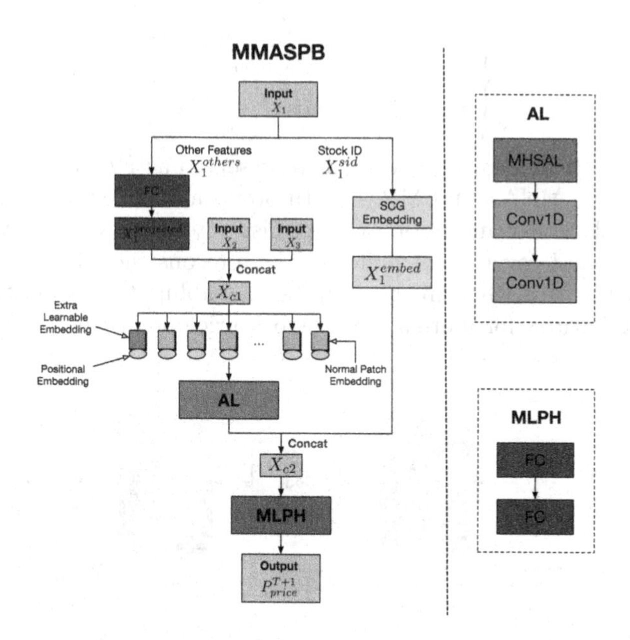

Fig. 2. Structure of MMASPB

3.4 Auxiliary Task

In the auxiliary task of text reconstruction for company news and government policies, we utilize MNAE and MPAE. These autoencoders consist of an encoder and a decoder, as illustrated in Fig. 3 and Fig. 4. To input the masked text for the autoencoders, we randomly replace 15% of the tokens with the symbol "[mask]". This masking technique helps prevent overfitting in the autoencoders. The masked text, denoted as X_{mask}, is tokenized and transformed into the vector representation X^{v}_{mask}.

The encoder structure comprises multiple identical layers, each consisting of a MHSAL and a sub-layer with a Dropout layer and a FC layer. Residual connections are employed between these sub-layers to enhance information flow. A FC layer is added between the encoder and decoder for connection purposes, reducing the data dimension from $\mathbb{R}^{D_{\text{model}}}$ to $\mathbb{R}^{D_{\text{decoder}}}$.

The decoder, with the same layer structure as the encoder, reconstructs the text vector from D_{decoder} dimensions to match the dimension of the original text vector X^{v}_{real}. This is achieved through stacked identical layers, with a flatten layer and a FC layer added at the end of the decoder. The reconstructed text vector X^{v}_{rec} is obtained.

To incorporate positional information, as shown in (13), we use a positional embedding method to map the position of each token to a one-dimensional $\mathbb{R}^{D_{\text{pos}}}$ position vector. This approach ensures that the autoencoder structure accounts for the meaningful order of tokens in the text.

$$\begin{cases} \mathrm{PE}_{2\mathrm{i}}(\mathrm{p}) = \sin\left(\frac{\mathrm{p}}{10000^{\frac{2\mathrm{i}}{\mathrm{D_{pos}}}}}\right) \\ \mathrm{PE}_{2\mathrm{i}+1}(\mathrm{p}) = \cos\left(\frac{\mathrm{p}}{10000^{\frac{2\mathrm{i}}{\mathrm{D_{pos}}}}}\right) \end{cases} \tag{13}$$

where the i-th element of a vector is represented as $PE_i(p)$

By employing MNAE and MPAE with positional embedding, we can reconstruct the masked text and obtain the reconstructed text vectors X^{v}_{rec} for both company news and government policies. These reconstructed text vectors serve as inputs for the MMASPB in the main task, enabling the model to utilize the encoded text features for more accurate stock price prediction.

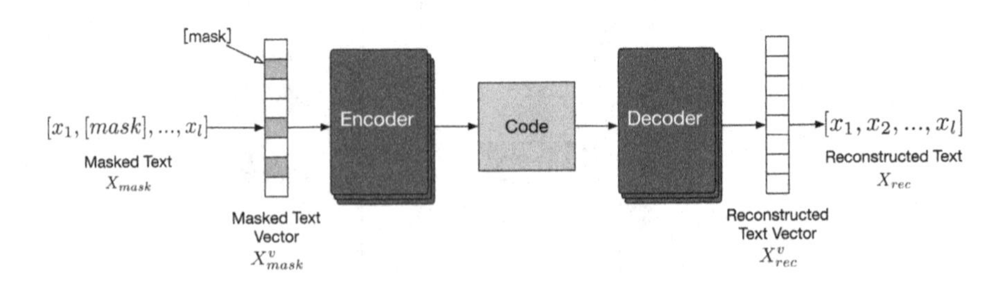

Fig. 3. Architecture of MNAE and MPAE

3.5 Loss Function

The loss functions for the auxiliary tasks and the main task of MTASPM are as follows:

For the two auxiliary tasks of text reconstruction, the loss functions $Loss_1$ and $Loss_2$ can be calculated using the CategoricalCrossentropy loss:

$$\mathrm{Loss}_1 = \mathrm{CategoricalCrossentropy}\left(X^{\mathrm{v}}_{\mathrm{rec}_{\mathrm{news}}}, X^{\mathrm{v}}_{\mathrm{real}_{\mathrm{news}}}\right) \tag{14}$$

$$\mathrm{Loss}_2 = \mathrm{CategoricalCrossentropy}\left(X^{\mathrm{v}}_{\mathrm{rec}_{\mathrm{policy}}}, X^{\mathrm{v}}_{\mathrm{real}_{\mathrm{policy}}}\right) \tag{15}$$

$$\mathrm{CategoricalCrossentropy} = \frac{1}{N}\sum_{i}^{N} -\log\left(\frac{e^{\mathrm{s_i}}}{\sum_{\mathrm{j}}^{\mathrm{C}} \mathrm{e}^{\mathrm{s_j}}}\right) \tag{16}$$

where X^v_{rec} is the reconstructed text vector and X^v_{real} is the real text vector

For the main task of stock price prediction, the loss function $Loss_3$ is determined using the Mean Square Error (MSE) loss:

$$\mathrm{Loss}_3 = \frac{1}{\mathrm{N}}\sum_{\mathrm{i=1}}^{\mathrm{N}}\left(\mathrm{P}^{\mathrm{T+1}}_{\mathrm{price}} - \mathrm{T}^{\mathrm{T+1}}_{\mathrm{price}}\right)^2 \tag{17}$$

where P^{T+1}_{price} is the predicted stock price and T^{T+1}_{price} is the real stock price at time $T + 1$

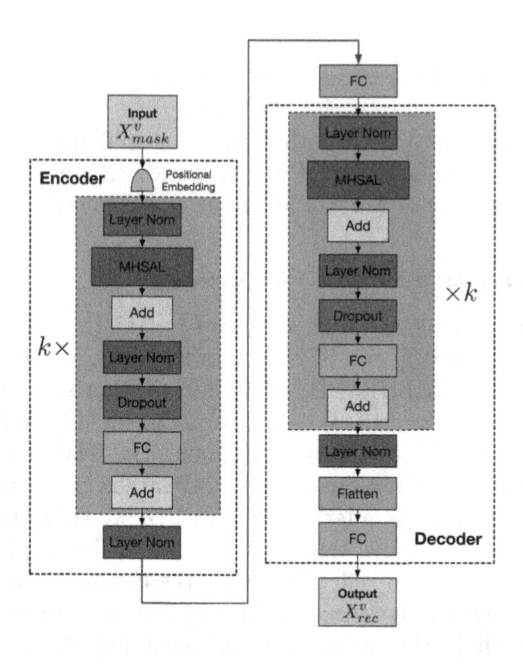

Fig. 4. Specific Structures of Encoder and Decoder.

The total loss function $Loss_{\text{total}}$ of MTASPM is a weighted combination of the auxiliary tasks and the main task, with weights of 0.05, 0.05, 0.9, respectively:

$$\text{Loss}_{\text{total}} = 0.05\,\text{Loss}_1 + 0.05\,\text{Loss}_2 + 0.9\,\text{Loss}_3 \tag{18}$$

4 Experiments

4.1 Datasets and Data Processing

We create two datasets, namely the Shanghai Exchange Stock Price Dataset (SHESPD) and the Shenzhen Exchange Stock Price Dataset (SZESPD), for our experiments. SHESPD consists of daily historical trading data, company news, and government policies of 52 sample stocks [17] listed on the Shanghai Stock Exchange from January 1, 2020, to June 30, 2022. The trading data is obtained from Netease Stock [20], while the company news text is collected from Meijing Website [23], and the government policy text is crawled from the Flush Website [22]. Each stock in SHESPD has 603 trading records, resulting in a dataset size of 29,796. Similarly, SZESPD contains data from 26 sample stocks listed on the Shenzhen Stock Exchange during the same time period, with a dataset size of 14,898.

To ensure consistency and eliminate the impact of different stock characteristics on the model, we apply the Z-score standardization method to the historical trading data. The method is expressed as:

$$\boldsymbol{x}' = \frac{\boldsymbol{x} - \mu}{\sigma} \tag{19}$$

where $\boldsymbol{x}$ is the original data, $\boldsymbol{x}'$ is the standardized data, μ is the average, and σ is the standard deviation of each data feature.

The pre-processed data from both datasets is split into a training set (80% of the data) and a testing set (20% of the data). These datasets are used to evaluate the performance of MTASPM in our experiments.

4.2 Stock Correlation Graph Visualization

We apply the threshold method with a value of $\xi = 0.5$ to construct the stock correlation graphs for SHESPD and SZESPD. In Fig. 5(a), we present the complete stock correlation graph for SHESPD, while Fig. 5(b) shows a sub-graph with Ping An of China (Stock Code: 601318) as the core, displaying all the related stocks. It is worth noting that PICC (601319) is among the related stocks, confirming the meaningfulness of the obtained stock correlation graph and its alignment with subjective cognition. Similarly, for SZESPD, we construct the stock correlation graph using the same method, as shown in Fig. 6. The visualizations of the results demonstrate the effectiveness of the threshold method-based stock correlation graph, accurately reflecting the relationship between different stocks. These findings highlight the significance and reliability of the constructed stock correlation graphs in capturing the interconnections and dependencies among stocks in the datasets.

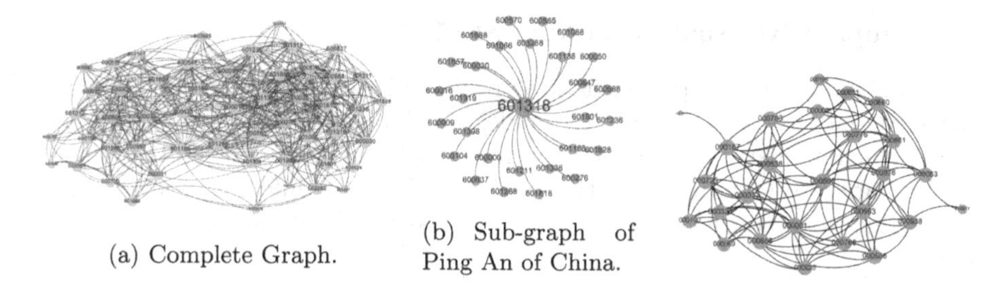

(a) Complete Graph.

(b) Sub-graph of Ping An of China.

Fig. 5. Stock Correlation Graph on SHESPD.

Fig. 6. Stock Correlation Graph on SZESPD.

4.3 Baseline Models and Metrics

To evaluate the performance of MTASPM, we compare it with eight baseline models for benchmarking purposes. The baseline models are as follows:

- HA: Historical Average model. We use the average value of the 30-day stock closing prices to predict the closing price of the next day.
- MLR [4]: Multiple Linear Regression. The average, median, standard deviation, maximum and minimum values of the 30-day closing prices will be used to predict the next day's closing price through multiple linear regression analysis.
- SVR [5]: Support Vector Regression. Support vector regression analysis uses the same five statistical characteristics as MLR to predict the closing price.
- RNN [12]: Recurrent Neural Network.
- TCN [9]: Temporal Convolutional Network. A convolution network applied to time series composed of multilayer extended and causal 1D convolutions.
- GRU [18]: Gate-Recurrent-Unit Network, a special RNN model.
- LSTM [15]: Long-Short-Term-Memory Network. It's also a special RNN model.
- ConvLSTM [21]: Convolutional LSTM, a neural network combining the convolution layer and LSTM layer.

To assess the performance of these models, we employ three evaluation metrics: mean square error (MSE), root mean square error (RMSE), and mean absolute error (MAE). MSE, RMSE, and MAE are calculated as shown in (17), (20), and (21) respectively.

$$RMSE = \sqrt{\frac{1}{N}\sum_{i=1}^{N}\left((\mathrm{P}_{\mathrm{price}}^{\mathrm{T}+1} - \mathrm{T}_{\mathrm{price}}^{\mathrm{T}+1}\right)^2} \tag{20}$$

$$\mathrm{MAE} = \frac{1}{N}\sum_{i=1}^{N}\left|\mathrm{P}_{\mathrm{price}}^{\mathrm{T}+1} - \mathrm{T}_{\mathrm{price}}^{\mathrm{T}+1}\right| \tag{21}$$

4.4 Comparative Analysis of MTASPM

Results of comparative experiments are presented in Table 2, which reveals that MTASPM outperforms all baseline models, including traditional mathematical and machine learning models, as well as deep learning models. The multimodal nature of MTASPM allows it to effectively leverage different types of input data, leading to better learning of stock price patterns and improved prediction accuracy.

4.5 Parameter Sensitivity Analysis

The performance of MTASPM is evaluated under various parameter settings, including D_{model}, D_{decoder}, the number of attention heads (h), and the number of layers (k) for the encoder and decoder in MNAE and MPAE. The results of the parameter sensitivity analysis are presented in Fig. 7 and Fig. 8. These figures demonstrate the impact of different parameter values on the performance of MTASPM using the SHESPD and SZESPD datasets. In general, the analysis

Table 2. Results of Comparative Experiments.

Dataset	Model	MSE	RMSE	MAE
SHESPD	HA	255.84	16.00	5.52
	MLR	132.99	11.53	4.07
	SVR	219.20	14.81	8.45
	RNN	120.42	10.97	2.63
	TCN	<u>67.53</u>	<u>8.22</u>	2.19
	GRU	78.24	8.85	1.98
	LSTM	74.84	8.65	1.92
	ConvLSTM	68.45	8.27	<u>1.89</u>
	MTASPM	**39.06**	**6.25**	**1.76**
	Improvement	*42.16%*	*25.18%*	*6.88%*
SZESPD	HA	182.80	13.52	5.09
	MLR	52.16	7.22	3.09
	SVR	76.87	8.77	5.19
	RNN	17.34	4.16	1.64
	TCN	15.31	3.91	1.54
	GRU	15.28	3.91	1.50
	LSTM	14.98	3.87	1.48
	ConvLSTM	<u>14.63</u>	<u>3.82</u>	<u>1.47</u>
	MTASPM	**12.15**	**3.49**	**1.38**
	Improvement	*16.95%*	*8.64%*	*6.12%*

*<u>Underline</u> represents the best result of baseline models.
***Bold** represents the best result of all the models.

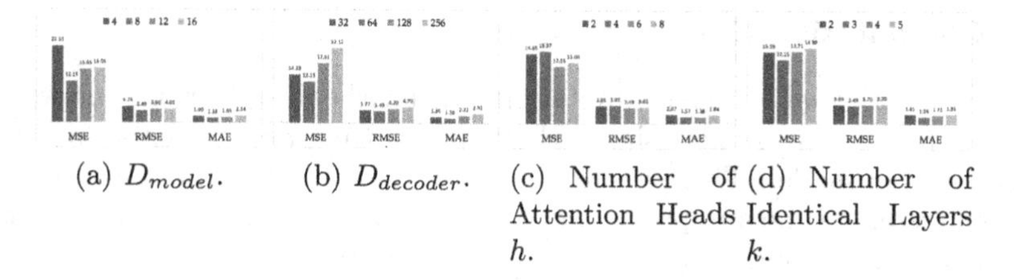

(a) D_{model}. (b) $D_{decoder}$. (c) Number of Attention Heads h. (d) Number of Identical Layers k.

Fig. 7. MTASPM Performance under Various Parameters on SHESPD.

emphasizes the importance of selecting appropriate parameter values to optimize the performance of MTASPM, which provides insights into the model's sensitivity to different parameter settings and can guide future researchers in selecting the most effective configurations for their specific datasets and prediction tasks.

4.6 Ablation Studies

The results of ablation studies are shown in Table 3, indicating that both the auxiliary task and stock correlation graph contribute to the overall improvement in the MTASPM performance. The findings from these ablation studies support the notion that constructing stock correlation graphs and implementing multi-task learning strategies are valuable approaches for enhancing stock price prediction models.

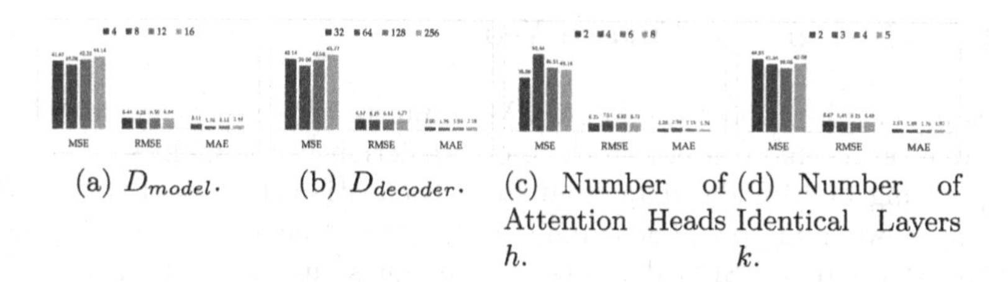

(a) D_{model}. (b) $D_{decoder}$. (c) Number of Attention Heads h. (d) Number of Identical Layers k.

Fig. 8. MTASPM Performance under Various Parameters on SZESPD.

Table 3. Results of Ablation Studies.

Dataset	Model	MSE	RMSE	MAE
SHESPD	Neither	48.64	6.97	1.88
	Without Auxiliary Task	45.77	6.77	1.84
	Without Correlation Graph	41.12	6.41	1.81
	MTASPM	**39.06**	**6.25**	**1.76**
SZESPD	Neither	14.34	3.79	1.47
	Without Auxiliary Task	13.85	3.72	1.45
	Without Correlation Graph	13.46	3.67	1.41
	MTASPM	**12.15**	**3.49**	**1.38**

*__Bold__ represents the best result of all the models.

4.7 Computation Cost

Lastly, We also analyze the computation cost for the deep learning models used in the experiments. Table 4 shows the training time and trainable parameter count for each model. These results indicate that MTASPM strikes a trade-off between training time and predictive accuracy. Despite the longer training times and increased computational requirements, it achieves remarkable accuracy in stock price prediction. This highlights the strength of MTASPM in effectively forecasting stock prices, demonstrating its capability to provide valuable insights for investors.

Table 4. Training Time and Trainable Parameter of Deep Learning Models.

Model	Training Time(s)	Trainable Parameter
RNN	347.68	17281
TCN	102.69	75457
GRU	63.32	51969
LSTM	67.15	68737
ConvLSTM	80.47	103553
MTASPM	604.25	199705

5 Conclusion and Future Work

In this paper, we propose the MTASPM, a multi-task learning model based on the attention mechanism, for stock price prediction in the Chinese market. By incorporating multimodal data from stock correlation, historical trading data, company news, and government policies, MTASPM achieves promising results on SHESPD and SZESPD datasets. The findings suggest that MTASPM has the potential to assist investors in mitigating risks and improving profits. The research validates the importance of considering multiple factors and leveraging multimodal data in stock price prediction, providing a valuable perspective and approach for future research. In future work, our focus will be on exploring efficient and streamlined deployment methods for the model, aiming to ensure both its effectiveness and ease of application. We will dedicate efforts to simplify the model and optimize the data sources, enabling more convenient and seamless integration of the proposed approach. By doing so, we aim to provide a practical and user-friendly solution that can be readily deployed in real-world scenarios.

References

1. Al-Shiab, M.: The predictability of the Amman stock exchange using the univariate autoregressive integrated moving average (ARIMA) model. J. Econ. Adm. Sci. **22**(2), 17–35 (2006)
2. Alkhatib, K., Najadat, H., Hmeidi, I., Shatnawi, M.K.A.: Stock price prediction using k-nearest neighbor (kNN) algorithm. Int. J. Bus. Humanit. Technol. **3**(3), 32–44 (2013)
3. Ariyo, A.A., Adewumi, A.O., Ayo, C.K.: Stock price prediction using the ARIMA model. In: 2014 UKSim-AMSS 16th International Conference on Computer Modelling and Simulation, pp. 106–112. IEEE (2014)
4. Asghar, M.Z., Rahman, F., Kundi, F.M., Ahmad, S.: Development of stock market trend prediction system using multiple regression. Comput. Math. Organ. Theory **25**, 271–301 (2019)
5. Bathla, G.: Stock price prediction using LSTM and SVR. In: 2020 Sixth International Conference on Parallel, Distributed and Grid Computing (PDGC), pp. 211–214. IEEE (2020)
6. Clear, C.: The number of investors in the securities market exceeded 200 million (Feb 25 2022). http://www.chinaclear.cn/zdjs/xgsdt/202202/e466e3e5bcaf4d3da6ddfbbff98bbe73.shtml
7. Hushani, P.: Using autoregressive modelling and machine learning for stock market prediction and trading. In: Yang, X.-S., Sherratt, S., Dey, N., Joshi, A. (eds.) Third International Congress on Information and Communication Technology. AISC, vol. 797, pp. 767–774. Springer, Singapore (2019). https://doi.org/10.1007/978-981-13-1165-9_70
8. Jiang, W.: Applications of deep learning in stock market prediction: recent progress. Expert Syst. Appl. **184**, 115537 (2021)
9. Liu, Y., Dong, H., Wang, X., Han, S.: Time series prediction based on temporal convolutional network. In: 2019 IEEE/ACIS 18th International Conference on Computer and Information Science (ICIS), pp. 300–305. IEEE (2019)
10. Lu, W., Li, J., Wang, J., Qin, L.: A CNN-biLSTM-AM method for stock price prediction. Neural Comput. Appl. **33**, 4741–4753 (2021)
11. Nair, B.B., Mohandas, V., Sakthivel, N.: A decision tree-rough set hybrid system for stock market trend prediction. Int. J. Comput. Appl. **6**(9), 1–6 (2010)
12. Pawar, K., Jalem, R.S., Tiwari, V.: Stock market price prediction using LSTM RNN. In: Rathore, V.S., Worring, M., Mishra, D.K., Joshi, A., Maheshwari, S. (eds.) Emerging Trends in Expert Applications and Security. AISC, vol. 841, pp. 493–503. Springer, Singapore (2019). https://doi.org/10.1007/978-981-13-2285-3_58
13. Reddy, V.K.S.: Stock market prediction using machine learning. Int. Res. J. Eng. Technol. **5**(10), 1033–1035 (2018)
14. Ribeiro, L.F., Saverese, P.H., Figueiredo, D.R.: struc2vec: learning node representations from structural identity. In: Proceedings of the 23rd ACM SIGKDD International Conference on Knowledge Discovery and Data Mining, pp. 385–394 (2017)
15. Roondiwala, M., Patel, H., Varma, S.: Predicting stock prices using LSTM. Int. J. Sci. Res. **6**(4), 1754–1756 (2017)
16. Schöneburg, E.: Stock price prediction using neural networks: a project report. Neurocomputing **2**(1), 17–27 (1990)

17. Securities, H.: Shanghai 50 concept listed company stock (July 1 2022). https://m.hx168.com.cn/stock/concept/BK2440.html
18. Sethia, A., Raut, P.: Application of LSTM, GRU and ICA for stock price prediction. In: Satapathy, S.C., Joshi, A. (eds.) Information and Communication Technology for Intelligent Systems. SIST, vol. 107, pp. 479–487. Springer, Singapore (2019). https://doi.org/10.1007/978-981-13-1747-7_46
19. Sharma, A., Bhuriya, D., Singh, U.: Survey of stock market prediction using machine learning approach. In: 2017 International Conference of Electronics, Communication and Aerospace Technology (ICECA), vol. 2, pp. 506–509. IEEE (2017)
20. Stock, N.: Netease stock quotes (July 1 2022). http://quotes.money.163.com/stock/
21. Wang, J., Sun, T., Liu, B., Cao, Y., Zhu, H.: Clvsa: A Convolutional LSTM based variational Sequence-to-sequence Model with Attention for Predicting Trends of Financial Markets. arXiv preprint arXiv:2104.04041 (2021)
22. Website, F.: Flush financial information (July 1 2022). http://stock.10jqka.com.cn
23. Website, M.: National business daily (July 1 2022). http://www.nbd.com.cn
24. Yu, P., Yan, X.: Stock price prediction based on deep neural networks. Neural Comput. Appl. **32**, 1609–1628 (2020)

Federated Trajectory Search via a Lightweight Similarity Computation Framework

Chen Wu[1] and Zhiyong Peng[1,2,3](✉)

[1] School of Computer Science, Wuhan University, Wuhan, China
[2] The Big Data Institute, Wuhan University, Wuhan, China
[3] Intellectual Computing Laboratory for Culture Heritage, Wuhan University, Wuhan, China
{chenwu,peng}@whu.edu.cn

Abstract. Contact tracing is one of the most effective ways of disease control during a pandemic. A typical method for contact tracing is to examine the spatio-temporal companion between the trajectories of patients and others. However, human trajectory data collected by mobile devices cannot be directly shared due to privacy. To utilize personal trajectory data in contact tracing, this paper presents a federated trajectory search engine called Fetra, which can efficiently process top-k search over a data federation composed of numerous mobile devices without uploading raw trajectories. To achieve this, we first propose a lightweight similarity measure LCTS based on spatio-temporal companion time to evaluate the similarity between trajectories. We then build a federated grid index named FGI via location anonymization. Given a query, a pruning strategy over FGI is applied to prune the candidate mobile devices dynamically. In addition, we propose a local optimization strategy to accelerate similarity computations in mobile devices. Extensive experiments on real-world dataset verify the effectiveness of LCTS and the efficiency of Fetra.

Keywords: Federated search · Trajectory similarity · Contact tracing

1 Introduction

With the rapid development of mobile internet and locating technologies, human mobility data are collected into trajectory data by locating modules. One direct application of the trajectory data is contact tracing [13] which aims to detect whether a person has come into direct contact with patients during a pandemic.

Existing trajectory data management systems, running on a single machine [29] or distributed clusters [16], suffer from data isolation [24] as these data are separately owned by multiple parties and it is illegal to share them directly. A promising solution to deal with data isolation is to perform federated search over a data federation [22]. Federated search has been adopted in spatial data management, however it has not been applied to trajectory data.

X. Song et al. (Eds.): APWeb-WAIM 2023, LNCS 14332, pp. 469–485, 2024.
https://doi.org/10.1007/978-981-97-2390-4_32

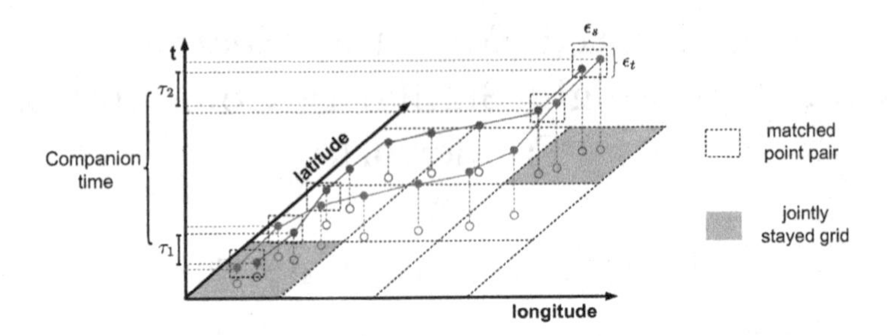

Fig. 1. An example of spatial temporal longest common subsequence (STLCSS) and spatio-temporal companion (STC).

As hardware continues to upgrade, the storage capacity and computing power of mobile devices have fulfilled daily usage. Apart from being used to collect trajectory data, mobile devices have been adopted in on-device works such as *stay-point detection* [18] and *online map-matching* [5]. These works inspire us to adopt mobile devices into personal trajectory management without uploading raw data and processing it in a central server. In other words, we aim to build a data federation over mobile devices.

The main challenge for federated trajectory search is how to measure the similarity between trajectories and migrate the processing into mobile devices. Existing similarity measures are mainly based on point-wise computation. A typical similarity measure is *spatial temporal longest common subsequence* (STLCSS) [25] shown in Fig. 1. STLCSS treats two points to be matched pair only if their spatial and temporal distances are less than two thresholds ϵ_s and ϵ_t, respectively. It requires point-wise distance computation between trajectories and thus cannot fit into mobile devices with limited resources. Another similarity measure is *spatio-temporal companion* (STC) [2] that splits the spatial range into grids and accumulates the staying time in jointly visited ones. If the accumulated staying time exceeds a threshold, the owner of the trajectory will be treated as a potential contact. However, this method considers the irrelevant outdoor trajectories while infections mainly happen in indoor buildings.

In this paper, we design and implement an efficient federated trajectory search engine, Fetra, for data federation over mobile devices. We design a lightweight federated grid index (FGI) which supports effective pruning in federated search. In addition, we propose the longest companion time similarity measure (LCTS) which is based on a lightweight trajectory representation and is well-suited for mobile devices. In our experiments, Fetra can answer a federated trajectory query in about 1 s over a data federation with 100,000 mobile devices.

In summary, we make the following contributions in this paper:

- We propose a lightweight similarity measure based on the companion time, which is efficient to compute for mobile devices (see Sect. 3).

- We propose a federated trajectory search engine for contact tracing. We design a federated index framework that supports global pruning and boosts local similarity computation (see Sects. 4 and 5).
- We conduct experiments over real-world datasets. The results show the effectiveness of LCTS and the efficiency of Fetra (see Sect. 6).

2 Related Work

Trajectory Similarity Measures. According to trajectory representations, existing trajectory similarity measures can be broadly divided into point-based and road network-based measures. Point-based similarity measures treat a trajectory as a set of points, such as Longest Common Subsequence (LCSS) [26], Edit Distance on Real sequence (EDR) [6], Fréchet distance [19] and Hausdorff distance [17]. These similarity measures focus on spatial dimension and can be extended with temporal information, i.e., STLCSS [25] and STLC [20]. Road network-based similarity measures can be used in traffic management. The vehicle trajectories will be mapped onto road networks and represented by a set of connected road segments. Typical examples of road network-based similarity measures include LORS [29] and EBD [28] which are employed for efficient trajectory query and trajectory clustering, respectively. However, they highly depend on map matching for pre-processing, which is infeasible in mobile devices.

Top-k Trajectory Search. Given a query trajectory, the top-k trajectory search aims to find the k most similar trajectories from a dataset, which is widely used in trajectory analysis tasks such as traffic control or route planning. Existing studies on top-k trajectory search focus on either centralized solutions, e.g., Torch [29], or distributed solutions including DITA [10] and UITraMan [21]. Centralized solutions put all trajectories in a single machine, which is not suitable for personal mobility data management due to privacy. On the other hand, existing distributed solutions partition trajectories over clusters horizontally. Further, they utilize distributed indexing mechanism to support large-scale trajectory search. However, these systems rely on high-performance clusters with large memory, which is beyond the scope of this paper.

Federated Search. Federated search works over a data federation [22] where a service provider interacts with multiple data owners and resolves external query requests. Federated search has been adopted in spatial data management called *federated spatial search* [24] which aims to tackle data isolation and support efficient queries over spatial data federation. Existing federated spatial search mainly adopts secure multi-party computation (SMC) to enable secure query processing. However, these works often consider distance-related operations on location points such as range queries [22], distance join, and kNN join and cannot capture the relevance of point sequences in trajectories. Hence, our work aims to enable trajectory similarity search in a federated manner.

3 Problem Definition

3.1 Problem Model

Definition 1 (Raw Trajectory). *A raw trajectory* T *consists of a sequence of chronologically ordered GPS points* $\{p_1, p_2, \cdots, p_n\}$. *Each point* $p = (lat, lng, t)$ *contains latitude* lat, *longitude* lng *and timestamp* t.

Definition 2 (Stay Point). *A stay point* $SP = (lat, lng, \tau)$ *stands for a geographic location where the user stayed during the time interval* $\tau = [t_s, t_e]$.

Definition 3 (Point of Interests (POI)). *A POI is a tuple* $\rho = (id, lat, lng)$, *where id is a unique identifier,* lat *and* lng *represent the latitude and longitude.*

Definition 4 (Visit). *A visit is a tuple* $\pi = (\rho, \tau)$ *which means the user stayed in POI* ρ *during the time interval* $\tau = [t_s, t_e]$.

Definition 5 (POI-Matched Trajectory). *A POI-matched trajectory* $\bar{T} = \{\pi_1, \pi_2, \cdots, \pi_n\}$ *is a set of visits in chronological order.*

The process that extracts stay points from a raw trajectory is called *stay point detection* [11]. In this paper, we adopt a cluster-based algorithm [15] in this process. Each visit indicates a user stayed on a POI for a while, so we can detect the relevant visit from the stay point by searching nearby POIs within a certain distance where the visit contains the same time interval as the stay point. This process is called *POI visit extraction.* Figure 2 shows three example trajectories and their time intervals of visits, where Q_0 is a patient's trajectory, and T_1 and T_2 are the trajectories of two pedestrians.

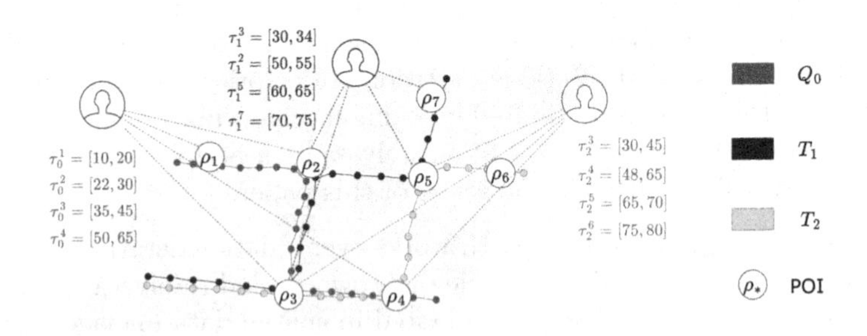

Fig. 2. Example POI-matched trajectories and the timetable of visits on POIs.

Definition 6 (Trajectory Data Federation). *The trajectory data federation* $S = \{s_1, s_2, \cdots, s_n\}$ *consists of* n *mobile devices. Each device* s_i *holds a local trajectory dataset* D_i.

Similar to previous studies on data federation [22], Fetra also utilizes a server, denoted s_0, as the coordinator that accepts external query requests and communicates with these mobile devices. In this paper, we assume the server s_0 can access the similarity values and anonymized location data from mobile devices.

Definition 7 (Federated Top-k Trajectory Search (FTS)). *Given a query trajectory Q, federation S, trajectory similarity measure M, and parameter k, the search returns the result set $S' \subseteq S$ with k mobile devices such that: $\forall s_i \in S', \forall s_j \in S - S', M(Q, D_i) > M(Q, D_j)$, where $M(Q, D_i)$ measures the similarity between query trajectory Q and the local trajectory dataset D_i.*

3.2 Lightweight Trajectory Similarity Measure

Companion Time as Similarity. Most existing studies define spatio-temporal similarity as the weighted sum of spatial and temporal distances, which is not suitable for contact tracing limited by critical spatial and temporal conditions. For example, as shown in Fig. 2, Q_0 and T_1 have an overlapped segment between ρ_2 and ρ_3, so they are similar in terms of spatial dimension. However, the common segments are not overlapped in the temporal range. So Q_0 and T_1 cannot fulfill the contact condition. On the other hand, we observe that Q_0 and T_2 have a shorter overlapped segment between ρ_3 and ρ_4, but more contact time in ρ_3 and ρ_4, which means T_2 has more infection risk. Such temporal contact relationship is not captured in existing spatio-temporal similarity measures. Inspired by LCSS [26] and LORS [29], we define the Longest Companion Time Similarity (LCTS) to measure the similarity between two POI-matched trajectories.

Definition 8 (LCTS). *Let $\bar{T_1}$ and $\bar{T_2}$ be two POI-matched trajectories of T_1 and T_2, the Longest Companion Time Similarity is defined as:*

$$\bar{M}(\bar{T_1}, \bar{T_2}) = \sum_{\rho \in P_{\bar{T_1}} \cap P_{\bar{T_2}}} |l^{\rho}_{\bar{T_1}} \cap l^{\rho}_{\bar{T_2}}| \tag{1}$$

where $P_{\bar{T_1}}$ is a set of POIs visited in $\bar{T_1}$, $l^{\rho}_{\bar{T_1}}$ is an interval list of visits at ρ in $\bar{T_1}$ and $|l^{\rho}_{\bar{T_1}} \cap l^{\rho}_{\bar{T_2}}|$ represents the overlapped time of two interval lists in POI ρ. As the visits in a single POI are temporally disjoint, we can compute $|l^{\rho}_{\bar{T_1}} \cap l^{\rho}_{\bar{T_2}}|$ in a sequential scan with linear complexity.

Example 1. Considering three trajectories in Fig. 2, we can observe that $P_{\bar{Q_0}} = \{\rho_1, \rho_2, \rho_3, \rho_4\}$, $P_{\bar{T_1}} = \{\rho_2, \rho_3, \rho_5, \rho_7\}$ and $P_{\bar{T_2}} = \{\rho_3, \rho_4, \rho_5, \rho_6\}$. Hence, $P_{\bar{T_1}} \cap P_{\bar{T_2}} = \{\rho_3, \rho_5\}$. Note that $l^{\rho_3}_{\bar{T_1}} = \{[30, 34]\}$, $l^{\rho_5}_{\bar{T_1}} = \{[60, 65]\}$, $l^{\rho_3}_{\bar{T_2}} = \{[30, 45]\}$ and $l^{\rho_5}_{\bar{T_2}} = \{[65, 70]\}$, we can derive $|l^{\rho_3}_{\bar{T_1}} \cap l^{\rho_3}_{\bar{T_2}}| = 4$ and $|l^{\rho_5}_{\bar{T_1}} \cap l^{\rho_5}_{\bar{T_2}}| = 0$, therefore, LCTS between $\bar{T_1}$ and $\bar{T_2}$ is $\bar{M}(\bar{T_1}, \bar{T_2}) = 4$. In the same way, we have $\bar{M}(\bar{Q_0}, \bar{T_1}) = 0$ and $\bar{M}(\bar{Q_0}, \bar{T_2}) = 25$.

Property 1. LCTS is a non-metric similarity measure.

This property can be verified by observing the example in Fig. 2. Note that $\bar{M}(Q_0, T_1) = 0$, $\bar{M}(Q_0, T_2) = 25$, $\bar{M}(T_1, T_2) = 4$, then $|\bar{M}(Q_0, T_1) + \bar{M}(T_1, T_2)| < \bar{M}(Q_0, T_2)$. Hence, LCTS does not obey triangular inequality and is non-metric, for which a metric index, e.g., M-tree [9], is infeasible for pruning. In the following sections, we will present our solution to federated search with LCTS.

4 Fetra Framework

Fetra is a federated trajectory search engine over a data federation composed of multiple mobile devices. Figure 3 presents the framework which consists of three main modules: *Local Processing*, *Federated Indexing*, and *Query Processing*. To support efficient search using LCTS, we propose a lightweight *federated grid index* framework (short as FGI) that consists of local indexes in mobile devices, and a global index in the server s_0.

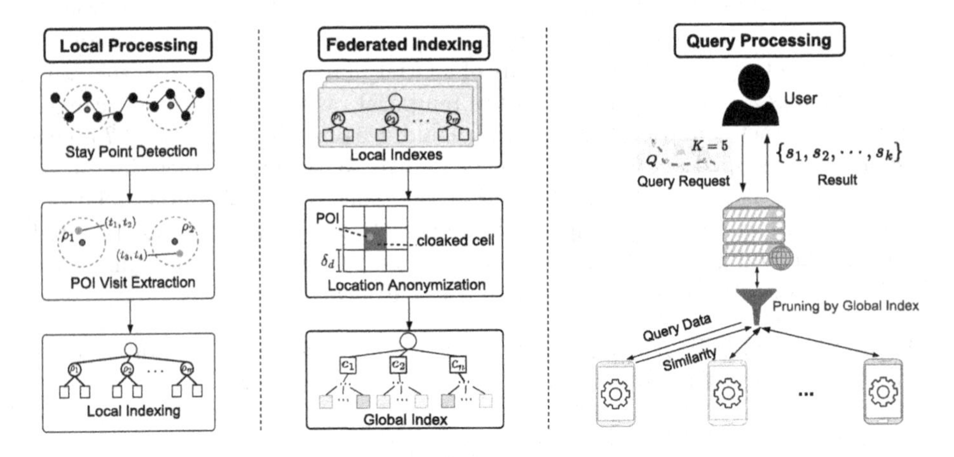

Fig. 3. Our proposed Fetra framework.

4.1 Local Processing on Devices

This module works in mobile devices, takes raw trajectories as input, transforms them into POI-matched trajectories and builds local indexes.

Pre-processing. In this paper, we propose a two-phase trajectory mapping method shown in Fig. 3. Firstly, we utilize a cluster-based algorithm [15] to detect stay points in raw trajectories. To detect stay points more fine-grained, we set smaller time and distance thresholds of 5 min and 50 m than the default values of 30 min and 200 m in [15] respectively. Then we extract the POIs around stay points within a distance threshold (default 50 m same as the distance threshold in stay point detection) as visited.

Local Indexing. After pre-processing, we group these visits by their POIs. Then we build a temporal index for each group. As all visits are temporally disjoint, the temporal index can be simplified as a sorted integer list and further compressed with *delta encoding* [14]. In this way, a local index in device s_i can be represented as a set of sub-indexes: $I_{s_i} = \{I_{s_i}^{\rho} | \rho \in P_{D_i}\}$, where $I_{s_i}^{\rho}$ is the temporal index for the group in ρ and P_{D_i} is the POI set visited in s_i. For federated indexing on the server, each device will anonymize its location by replacing POIs with cells in the spatial grid in Sect. 4.2.

4.2 Federated Indexing on Server

As a prerequisite for federated search, the server constructs the global index to find candidate mobile devices corresponding to the query dataset Q.

Definition 9 (Global Grid). *A global grid $G = \{c_1, c_2, \cdots, c_{m*n}\}$ partitions the spatial range R into $m * n$ cells, where $R = \{lat_{min}, lat_{max}, lng_{min}, lng_{max}\}$ is a bounding rectangle and each cell is a square with width δ_d.*

Definition 10 (Footprint). *A footprint $\sigma = (c, i, \tau)$ records the time interval τ of device s_i staying on cell c.*

Location Anonymization. It is infeasible to upload local index to the server directly in federated search due to personal privacy. We utilize location anonymization [7,27] to blur the exact location into a cloaked cell. In the federation, the server partitions the spatial range into multiple cells, each covering several POIs. In this way, each device will transform its visits in local trajectory dataset into footprints and upload them to the server. As the range R is fixed, the partition is only determined by the width δ_d.

Global Indexing. The global index maintains a two-layer tree structure as shown in Fig. 3. In the first layer, it partitions the spatial range into unified cells, called *spatial partitioning*. In the second layer, it utilizes a temporal index to index the time intervals of footprints in each cell, called *temporal indexing*. With the independence between spatial partitioning and temporal indexing, different temporal indexes, like interval tree, can be employed. Thus, a global index can be represented as a set of temporal indexes over the spatial grid: $I_S = \{I_S^c | c \in G\}$, where G is the spatial grid and I_S^c is the temporal index over cell c. Algorithm 1 illustrates the construction of global index. Firstly, the server broadcasts parameters δ_d and R to all devices (Line 1). Then each mobile device converts its visits to a footprint set and uploads it to the server (Line 2–3). Meanwhile, the server creates a temporal index for each cell (Line 6). Finally, the server traverses the footprint set and updates temporal indexes in each cell (Line 8–9).

Algorithm 1: GlobalIndexConstruction(δ_d, R, S)

Input: δ_d: cell width, R: range, S: federation
Output: I_S: the global index
1 Broadcast δ_d and R to devices in S;
2 **foreach** *device* $s_i \in S$ **do**
3 | $F_i \leftarrow LocalConvert(\delta_d, R, I_i)$; s_i uploads footprint set F_i to server s_0 ;
4 $m \leftarrow \lceil \frac{R.lat_{max} - R.lat_{min}}{\delta_d} \rceil$, $n \leftarrow \lceil \frac{R.lng_{max} - R.lng_{min}}{\delta_d} \rceil$;
5 Partition R into $m * n$ cells with width δ_d;
6 Create a temporal index I_S^c for each cell c;
7 $F \leftarrow \bigcup_{i=1}^{|S|} F_i$;
8 **foreach** *footprint* $\sigma \in F$ **do**
9 | Add σ to $I_S^{\sigma.c}$;
10 **return** $I_S = \{I_S^{c_1}, I_S^{c_2}, \cdots, I_S^{c_{m*n}}\}$;
11 ---
12 **Function** `LocalConvert`(δ_d, R, I_{s_i}):
  Input: δ_d: cell width, R: range, I_{s_i}: local index
  Output: F_i: footprint set
13 $F_i \leftarrow \emptyset$, $rownum \leftarrow \lceil \frac{R.lng_{max} - R.lng_{min}}{\delta_d} \rceil$;
14 **foreach** $I_{s_i}^{\rho_j} \in I_{s_i}$ **do**
15 $y \leftarrow \lceil \frac{\rho_j.lat - R.lat_{min}}{\delta_d} \rceil$, $x \leftarrow \lceil \frac{\rho_j.lng - R.lng_{min}}{\delta_d} \rceil$, $c \leftarrow y * rownum + x$;
16 **foreach** $\tau \in I_{s_i}^{\rho_j}$ **do**
17 | $\sigma \leftarrow (c, i, \tau)$, $F_i \leftarrow \{\sigma\} \bigcup F_i$;
18 **return** F_i;

4.3 Period Temporal Index

We further design the period temporal index for efficient retrieval in Fig. 4.

Construction. As the duration of human activity is rarely more than one day, we split the temporal range into several buckets with the length of 24 h. Each bucket is further divided into L levels, where the i-th level contains 2^{i-1} partitions. Each raw interval is aligned by linear scaling:

$$[t_s, t_e] \rightarrow [\lfloor \frac{t_s}{|B|} * (2^{L-1} - 1) \rfloor, \lceil \frac{t_e}{|B|} * (2^{L-1} - 1) \rceil], \tag{2}$$

where $|B|$ is the bucket length. To index time intervals and protect privacy, we align each interval using Eq. (2) and insert the aligned interval in the same way as [8]. The resolution is the partition length in the lowest level, $\delta_t = \frac{|B|}{2^{L-1}}$. The quantization error ϵ_e is defined as the length difference between the aligned interval and the raw interval, which can be bounded by the following equation

$$\epsilon_e \leq 2 * \delta_t = \frac{|B|}{2^{L-2}}. \tag{3}$$

When the interval length is greater than the bucket length, we split it into multiple segments and add each segment to the corresponding buckets.

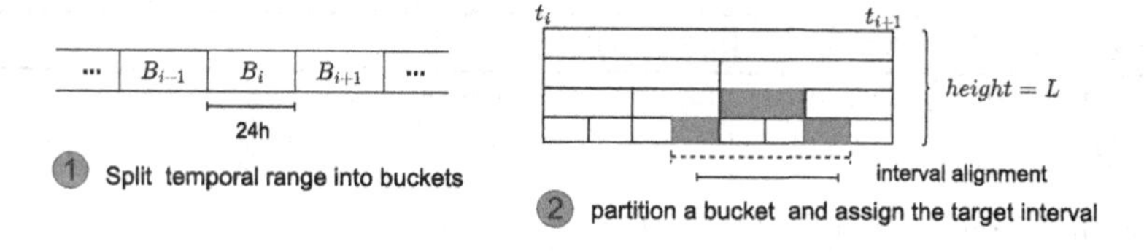

Fig. 4. Period Temporal Index (PTI).

Comparison-Free Search. Given a query interval, we check the overlapped temporal partitions for each level and add all the candidate devices to the result. As we do not store the raw intervals, there is no comparison during the search. In real usage, we set $L = 10$ and the quantization error is about 5.5 min (337 s) (using Eq. (3)).

5 Federated Top-k Search

We propose a baseline method called direct for federated top-k search using LCTS that the server s_0 directly broadcasts the query Q to all devices and accepts similarities from them. When computing LCTS between query trajectory Q and the local trajectory dataset, we only need to sum up the overlapped time intervals of visits. The baseline method requires high communication cost and local computation cost. We present our method for federated top-k search as follows.

5.1 Algorithm Overview

Similar to the dynamic pruning strategy in [29], our algorithm is composed of *filtering* (Line 4–11) and *refinement* (Line 16–21) shown in Algorithm 2. The principle of filtering is that candidate devices must have jointly visited POIs with query dataset Q. So the filtering phase will traverse visits in Q, identify the cell of a visit, and search for the footprints overlapped with the target visit in the same cell (Line 5–10). The server maintains a similarity upper bound list UB for candidates (Line 9). Termination occurs when newly accepted similarity is greater than the upper bound UB_{max} for the remaining devices (Line 18), and the server returns the top-k result set.

5.2 Pruning for LCTS

Recall from Property 1 that LCTS is non-metric, for which a metric index cannot be used for pruning. The main idea of our pruning is to terminate the query process in advance without waiting for the responses of all candidates.

Upper Bounding Similarity. For LCTS, after accessing temporal indexes in I_S for all visits in Q, we get potential visit lists $Q^- = \{Q_1^-, Q_2^-, \cdots, Q_{|S|}^-\}$ for all devices. A device s_i becomes a candidate only when Q_i^- is not empty, indicating it could have relevant visits with Q. The upper bound is the sum of overlapped time interval length (Line 9), where $|\pi.\tau \cap \sigma.\tau|$ is the overlapped time length.

Algorithm 2: FederatedTopKSearch(Q, k, S)

Input: Q: query dataset, k: parameter, S: federation
Output: W: Top-k result set

1 $can \leftarrow \emptyset$, $W \leftarrow \emptyset$;
2 Initialize a query data list $Q^- = \{Q_1^-, Q_2^-, \cdots, Q_{|S|}^-\}$;
3 Initialize upper bound list $UB = \{UB_1, UB_2, \cdots, UB_{|S|}\}$;
4 **foreach** $\pi \in Q$ **do**
5   Identify the cell c of $\pi.\rho$;
6     $F \leftarrow$ temporally overlapped footprints in I_S^c; ;
7   **foreach** $\sigma \in F$ **do**
8         $i \leftarrow$ device id of σ;
9         $UB_i \leftarrow UB_i + |\pi.\tau \cap \sigma.\tau|$;
10         $Q_i^- \leftarrow Q_i^- \bigcup \{\pi\}$;
11 $can \leftarrow \bigcup_{|Q_i^-| \neq 0} \{s_i\}$;
12 Send query data Q_i^- to $s_i \in can$;
13 **foreach** *device* $s_i \in can$ **do**
14   compute $\bar{M}(Q_i^-, D_i)$;
15   Send $\bar{M}(Q_i^-, D_i)$ to the server s_0;
16 **while** $can \neq \emptyset$ **do**
17   Accept $\bar{M}(Q_i^-, D_i)$ from s_i;
18   **if** $\bar{M}(Q_i^-, D_i) \geq UB_{max}$ **then**
19   break;
20     $UB_i \leftarrow \bar{M}(Q_i^-, D_i)$, $W \leftarrow W \bigcup \{s_i\}$;
21   Remove s_i from can;
22 Sort W by similarity;
23 **return** W_{1-k};

Lemma 1. *$\forall s_i \in$ candidate devices, $\bar{M}(Q, D_i) = \bar{M}(Q_i^-, D_i)$.*

Proof. This lemma can be derived from Eq. (1). Any visit in Q not located in the same cells with D_i, denoted as $Q - Q_i^-$, cannot affect the similarity as two visits have companion time only when they stay in the same POI.

Query Data Pruning. Besides filtering out unqualified mobile devices, we also accelerate the query process by dropping unrelated visits in Q. The server sends pruned query data $Q_i^- \subseteq Q$ to device s_i. Based on the above Lemma, device s_i can compute the similarity using pruned query dataset Q_i^-.

5.3 Local Boost for LCTS

A direct way to compute LCTS between Q and D_i is to traverse these two lists order by time. Note that there exist spatial-temporal companions only when two visits happened in the same POI. Inspired by this, we boost local similarity computation on LCTS. We first divide Q^- into multiple groups based on POIs.

Then, we calculate overlapped time length for each group by its corresponding sub-index. In the end, the results in all groups are accumulated as the final result. As all groups are independent, the process can be implemented in parallel.

6 Experiments

Table 1. Statistics of trajectory datasets.

Characteristics	Geolife	NYC	Tokyo
#trajectories	58,725	93,971	199,245
#POIs	625,910	400	385
#visits	385,639	228,511	575,996
sampling rate (s)	5	-	-
Space (MB) of $\mathcal{D}$	2,190	29.25	73.07

Table 2. Experimental parameters.

Parameter	Value
$\lvert S\rvert$	$\{10^1, 10^2, 10^3, 10^4, 10^5\}$
$\lvert D\rvert$	{100, 200, **300**, 400, 500}
$\lvert Q\rvert$	$\{1, 2, \mathbf{3}, 4, 5\} \times 0.001 \times \lvert\mathcal{Q}\rvert$
k	{10, 20, **30**, 40, 50}
δ_d	$\{1, 3, \mathbf{5}, 7, 9\} \times 0.001°$

6.1 Experimental Setup

Datasets. As it is hard to access personal trajectory data in daily life, we adopt three trajectory datasets close to real scenarios. We use the Beijing trajectory dataset of the Geolife project [30] and the New York City and Tokyo Check-ins data of FourSquare [1]. We utilize the method in Sect. 4.1 to transform raw trajectories of Geolife into POI-matched ones. Since Tokyo and NYC datasets just contain the check-in time, we add the check-out time by assigning the stay time which ranges from 10 to 60 min randomly and these check-in data denoted by the same user in one day are aggregated into a single POI-matched trajectory. Table 1 describes the statistics of three datasets. NYC and Tokyo are check-in data so they do not contain sampling rate and have smaller space sizes.

Implementation. Since it is difficult to conduct such an experiment in real life, we implement a simulation program that contains a server and multiple clients in Java. All raw trajectories are transformed into POI-matched trajectories before indexing. All experiments ran on a server with an Intel Xeon 8269CY CPU (8 cores) and 64 GB RAM running CentOS 7.6.

Experiment Goals. The effectiveness and efficiency of Fetra are evaluated in the next three subsections from the following aspects:

1. *Efficiency of Search* - whether FGI and our search algorithm answer federated top-k search using LCTS efficiently.
2. *Effectiveness of LCTS* - whether LCTS similarity is robust to the sampling rate, GPS error, and point shifting.

Baseline Methods. There are five methods in our experiments:

1. **direct** - The baseline method described in Sect. 5.
2. **copt-PTI** - It optimizes federated search with local boost in Sect. 5.3 using FGI based on the period temporal index.
3. **popt-PTI** - It optimizes federated search with global pruning in Sect. 5.2 using FGI based on the period temporal index.
4. **copt+popt-PTI** - It optimizes federated search with local boost and global pruning using FGI based on the period temporal index.
5. **copt+popt-IT** - It optimizes federated search with local boost and global pruning using FGI based on the interval tree index.

We also compare the performance of federated trajectory search with existing two spatio-temporal similarity measures STLCSS [26] and STLC [23]. For these two similarities, we build the hierarchical index [26] and the grid index [20] as the local index for mobile devices respectively. We also utilize filtering and refinement optimization in Algorithm 2 during the search.

Parameters. Table 2 shows the parameters used in experiments. When a parameter varies, other parameters are set to default values (highlighted in bold). We select 0.001° as the unit of grid width, where 0.001° is roughly 111 m and $\delta_d = 9$ means 0.009°, i.e., about 1000 m. The query dataset size $|Q|$ is set as the ratio of the trajectory dataset $\mathcal{Q}$. For each parameter condition, we execute the experiments 10 times and report the average values as the results.

6.2 Efficiency Evaluation of Search Algorithms

In a federation with $|S|$ mobile devices, we randomly assign $|D|$ trajectories to each device. We also randomly sample $|Q|$ trajectories from these datasets as the query dataset and report running time in Figs. 5(a), 5(b), and 5(c). We compare five methods in Sect. 6.1. We have the following observations:

1. With the increase of $|S|$, D and Q, it degrades the performance of all methods for heavier communication workload in the server and computation workload in mobile device. Smaller δ_d accelerates the query process because it partitions the spatial range R more precisely and makes the spatial pruning more accurate. However, it does not work notably on NYC and Tokyo because there are fewer POIs in these datasets thus the activity ranges of mobile devices are closer, which impairs the spatial pruning.
2. All methods are not sensitive to the increase of k, which is slightly different from the traditional top-k search. The reason is that the computation cost highly depends on the slowest device.
3. Our PTI improves the performance of FGI about 2 times compared with interval tree. For example, when $|S| = 100,000$ on Geolife, copt+popt-PTI takes 0.566 s while copt+popt-IT takes 1.203 s. This is because PTI employs a comparison-free search style and reduces the filtering time.

4. copt+popt-PTI significantly outperforms the other methods. For example, when $|S| = 100,000$ on Geolife, direct takes 17.109 s, copt-PTI takes 1.73 s, popt-PTI takes 4.509 s while copt+popt-PTI takes 0.566 s. The reasons are two-fold: *i)* copt+popt-PTI employs an efficient pruning strategy in Sect. 5.2 to reduce unnecessary computations while direct and copt-PTI do not; *ii)* Compared with direct and popt-PTI, copt+popt-PTI utilizes the local boost method in Sect. 5.3 to accelerate the local similarity computations in mobile devices.

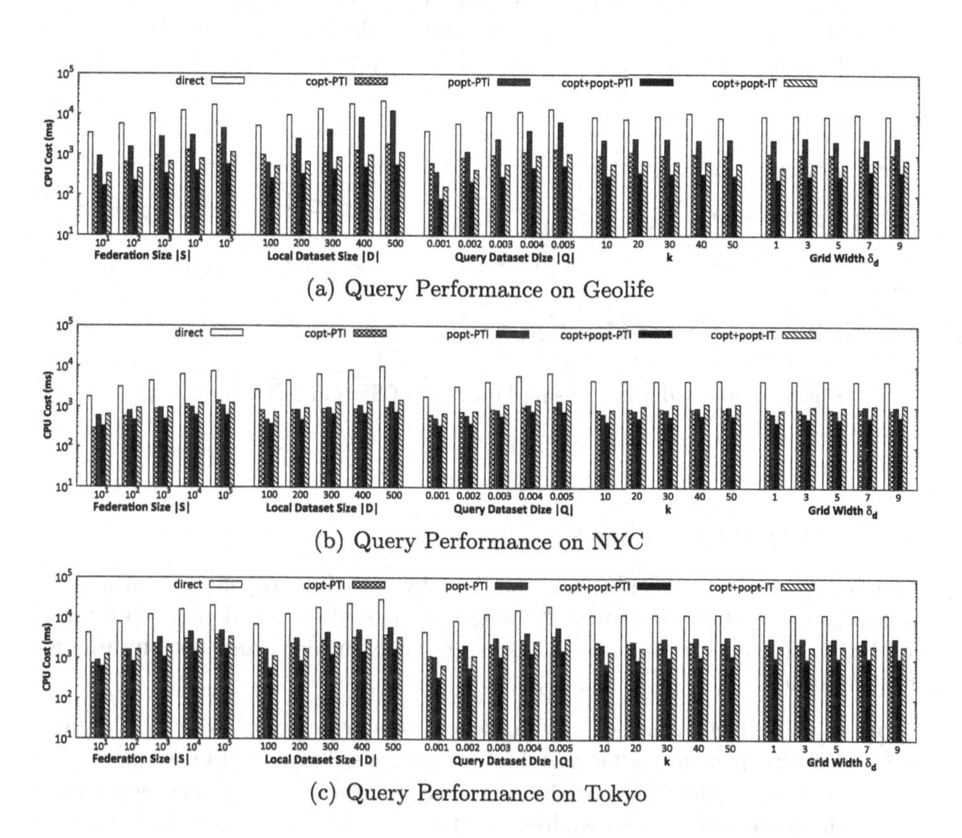

(a) Query Performance on Geolife

(b) Query Performance on NYC

(c) Query Performance on Tokyo

Fig. 5. Running Time Comparison of Federated Search on Different Datasets

6.3 Efficiency Evaluation of LCTS

We evaluate the efficiency of LCTS by conducting federated search experiments using LCTS, STLCSS and STLC. As STLCSS and STLC are defined on raw trajectories, we only compare these three similarity measures on Geolife. We utilize our filtering strategy for global pruning. As the indexes for STLCSS and STLC are not based on the spatial grid index, we mainly consider the other four parameters. The result is shown in Fig. 6. We have the following observations.

1. LCTS outperforms the other two similarity measures since it is computed over the transformed trajectories without complex spatial distance computation. STLC performs worse than STLCSS since it is a linear combination of spatial and temporal similarities, which has large computation cost.
2. The time cost of all similarity measures increases as $|S|$, $|D|$ and $|Q|$ increase. The query size $|Q|$ mainly affects the filtering cost while the other two parameters affects the local computation cost in the refinement phase.
3. The parameter k has little impact on the search performance. The reason is that the server waits for candidate mobile devices computing similarities locally, thus the search time highly depends on the slowest device.

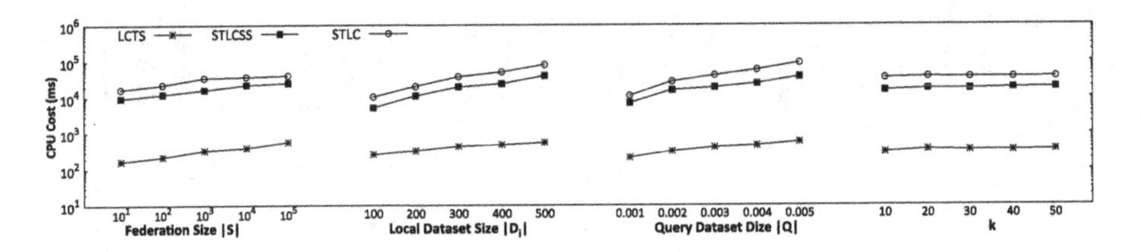

Fig. 6. Federated Search using LCTS, STLCSS and STLC on Geolife.

6.4 Effectiveness Evaluation

We employ a relevance judgement method by creating trajectories on a given POI sequence and checking whether they can be still retrieved as a top-k result on Geolife since the other two datasets are check-in data and not suitable for this evaluation.

Ground Truth Simulation. We choose 1,600 trajectories from the dataset and transform them into POI-matched trajectories. For each POI sequence, we generate k trajectories from the shortest paths planned by the navigation service [3] in walk mode and insert them into the dataset. To evaluate the effectiveness of spatio-temporal similarity measures, we mainly consider three features: sampling rate, GPS error and point shifting. For the sampling rate, we take each navigation path as base, sample a point every 1, 2, ..., k seconds along the path between two adjacent POIs, and add $\frac{\tau}{k}$ points around each POI to generate k trajectories, where τ is the staying time in the POI. For the GPS error, we re-sample each point and move it by a distance of 2–7.8 m as the GPS error of $\leq$7.8 m [4]. For point shifting, we shift all points of the query trajectory along the segments every two POIs by 1 m, 2 m, ..., $\frac{k}{2}$ in two directions. For all the search results of a query, we evaluate them with three grades: **2** for those trajectories in the generated ground truth of the query, **1** for those trajectories not in the generated ground truth of but temporally or spatially overlap with the query, and **0** for all the remaining trajectories in the dataset.

Comparison. The comparison is conducted by changing $k = 5, 10, 15$ using three simulated ground-truth sets. We measure the precision with top-k result precision (**P@k**) and *normalized discounted cumulative gain* (**NDCG@k**) [12]. We have the following observations.

1. LCTS has the highest search precision because LCTS contains no extra parameters than the other two similarity measures and irrelevant trajectories are not matched to the original POIs.
2. As k increases, the precision of all measures degrades since the results contain more trajectories which are not in the ground truth set.
3. LCTS is more robust to GPS error and point shifting since it is based on stay points which are mean points of clusters and less affected by spatial disturbance (Table 3).

Table 3. The evaluation result of sampling rate, GPS error, and point shifting.

		Sampling Rate			GPS Error			Point Shifting		
		LCTS	STLCSS	STLC	LCTS	STLCSS	STLC	LCTS	STLCSS	STLC
Geolife	P@5	0.971	0.905	0.903	0.975	0.925	0.918	0.955	0.915	0.909
	P@10	0.957	0.897	0.890	0.959	0.917	0.912	0.947	0.907	0.905
	P@15	0.948	0.889	0.876	0.953	0914	0.906	0.943	0.898	0.886
	NDCG@5	0.982	0.923	0.919	0.982	0.936	0.931	0.967	0.923	0.917
	NDCG@10	0.968	0.902	0.894	0.968	0.928	0.926	0.964	0.917	0.913
	NDCG@15	0.956	0.897	0.884	0.964	0.926	0.919	0.959	0.908	0.897

7 Conclusion and Future Work

This paper proposes a federated trajectory search engine called Fetra to efficiently answer similarity search over a federation composed of mobile devices. We assume trajectory data cannot be directly accessed and users' exact locations are blurred into cloaked cells. In future, we plan to support more similarity measures and larger data federation with millions of mobile devices.

References

1. Foursquare dataset (2014). https://www.kaggle.com/datasets/chetanism/foursquare-nyc-and-tokyo-checkin-dataset
2. Spatio-temporal companion (2021). https://www.bloomberg.com/news/articles/2021-11-08/people-you-don-t-know-and-can-t-see-are-close-contacts-in-china
3. Amap (2022). https://www.amap.com/
4. GPS accuracy (2022). https://www.gps.gov/systems/gps/performance/accuracy/

5. Chen, C., Ding, Y., Wang, Z., Zhao, J., Guo, B., Zhang, D.: VTracer: when online vehicle trajectory compression meets mobile edge computing. IEEE Syst. J. **14**(2), 1635–1646 (2020)
6. Chen, L., Özsu, M.T., Oria, V.: Robust and fast similarity search for moving object trajectories. In: SIGMOD, pp. 491–502 (2005)
7. Chow, C.Y., Mokbel, M.F., Aref, W.G.: Casper*: query processing for location services without compromising privacy. TODS **20**, 1–45 (2010)
8. Christodoulou, G., Bouros, P., Mamoulis, N.: Hint: a hierarchical index for intervals in main memory, pp. 1257–1270, June 2022
9. Ciaccia, P., Patella, M., Zezula, P.: M-tree: an efficient access method for similarity search in metric spaces. In: VLDB, pp. 426–435 (1997)
10. Ding, X., Chen, L., Gao, Y., Jensen, C.S., Bao, H.: UlTraMan: a unified platform for big trajectory data management and analytics. PVLDB **11**(7), 787–799 (2018)
11. Hu, Y., et al.: Salon: a universal stay point-based location analysis platform. In: SIGSPATIAL, pp. 407–410, November 2021
12. Järvelin, K., Kekäläinen, J.: Cumulated gain-based evaluation of IR techniques. ACM Trans. Inf. Syst. **20**(4), 422–446 (2002)
13. Kato, F., Cao, Y., Yoshikawa, M.: PCT-TEE: trajectory-based private contact tracing system with trusted execution environment. ACM Trans. Spatial Algorithms Syst. **8**(2), 1–35 (2022)
14. Lemire, D., Boytsov, L.: Decoding billions of integers per second through vectorization. Softw. Practice Exp. **45**(1), 1–29 (2015)
15. Li, Q., Zheng, Y., Xie, X., Chen, Y., Liu, W., Ma, W.Y.: Mining user similarity based on location history. In: SIGSPATIAL, pp. 1–10 (2008)
16. Li, R., et al.: TrajMesa: a distributed NoSQL-based trajectory data management system. IEEE Trans. Knowl. Data Eng. **14**(8), 1013–1027 (2021)
17. Nutanong, S., Jacox, E.H., Samet, H.: An incremental Hausdorff distance calculation algorithm. PVLDB **4**(8), 506–517 (2011)
18. Pérez-Torres, R., Torres-Huitzil, C., Galeana-Zapién, H.: Full on-device stay points detection in smartphones for location-based mobile applications. Sensors **16**(10), 1693 (2016)
19. Schmuck, M., Bazant, M.Z.: Computing the discrete Fréchet distance in subquadratic time. SIAM J. Comput. **75**(3), 1369–1401 (2015)
20. Shang, S., Chen, L., Wei, Z., Jensen, C.S., Zheng, K., Kalnis, P.: Trajectory similarity join in spatial networks. PVLDB **10**(11), 1178–1189 (2017)
21. Shang, Z., Li, G., Bao, Z.: DITA: distributed in-memory trajectory analytics. In: SIGMOD, pp. 725–740 (2018)
22. Shi, Y., Tong, Y., Zeng, Y., Zhou, Z., Ding, B., Chen, L.: Efficient approximate range aggregation over large-scale spatial data federation. IEEE Trans. Knowl. Data Eng. **35**(1), 418–430 (2021)
23. Su, H., Liu, S., Zheng, B., Zhou, X., Zheng, K.: A survey of trajectory distance measures and performance evaluation. VLDB J. **29**, 3–32 (2020)
24. Tong, Y., et al.: Hu-Fu: efficient and secure spatial queries over data federation. PVLDB **15**(6), 1159–1172 (2022)
25. Vlachos, M., Gunopulos, D., Kollios, G.: Robust similarity measures for mobile object trajectories. In: ICDE, pp. 721–726 (2002)
26. Vlachos, M., Kollios, G., Gunopulos, D.: Discovering similar multidimensional trajectories. In: ICDE, pp. 673–684 (2002)
27. Wang, H., Li, Y., Gao, C., Wang, G., Tao, X., Jin, D.: Anonymization and de-anonymization of mobility trajectories: dissecting the gaps between theory and practice. IEEE Trans. Mob. Comput. **20**(3), 796–815 (2021)

28. Wang, S., Bao, Z., Culpepper, J.S., Sellis, T., Qin, X.: Fast large-scale trajectory clustering. PVLDB **13**(1), 29–42 (2019)
29. Wang, S., Bao, Z., Culpepper, J.S., Xie, Z., Liu, Q., Qin, X.: Torch: a search engine for trajectory data. In: SIGIR, pp. 535–544 (2018)
30. Zheng, Y., Zhang, L., Xie, X., Ma, W.Y.: Mining interesting locations and travel sequences from GPS trajectories. In: WWW, pp. 791–800 (2009)

Central Similarity Multi-view Hashing for Multimedia Retrieval

Jian Zhu[1,2], Wen Cheng[2], Yu Cui[2], Chang Tang[3], Yuyang Dai[2], Yong Li[2], and Lingfang Zeng[2(✉)]

[1] University of Science and Technology of China, Hefei, China
ustczhuj@mail.ustc.edu.cn
[2] Zhejiang Lab, Hangzhou, China
{chengwen,yu.cui,daiyy,yonglii,zenglf}@zhejianglab.com
[3] China University of Geosciences, Wuhan, China
tangchang@cug.edu.cn

Abstract. Hash representation learning of multi-view heterogeneous data is the key to improving the accuracy of multimedia retrieval. However, existing methods utilize local similarity and fall short of deeply fusing the multi-view features, resulting in poor retrieval accuracy. Current methods only use local similarity to train their model. These methods ignore global similarity. Furthermore, most recent works fuse the multi-view features via a weighted sum or concatenation. We contend that these fusion methods are insufficient for capturing the interaction between various views. We present a novel Central Similarity Multi-View Hashing (CSMVH) method to address the mentioned problems. Central similarity learning is used for solving the local similarity problem, which can utilize the global similarity between the hash center and samples. We present copious empirical data demonstrating the superiority of gate-based fusion over conventional approaches. On the MS COCO and NUS-WIDE, the proposed CSMVH performs better than the state-of-the-art methods by a large margin (up to 11.41% mean Average Precision (mAP) improvement).

Keywords: Multi-view Hash · Central Similarity Learning · Multi-modal Hash · Multimedia Retrieval

1 Introduction

Multi-view hashing solves multimedia retrieval problems. The accuracy can be significantly increased with a well-crafted multi-view hashing method. Multi-view hashing, as opposed to single-view hashing, which only searches in a single view, can make use of data from other sources (e.g., image, text, audio, and video). Extraction of heterogeneous features from several views is accomplished via multi-view hashing representation learning. It fuses multi-view features to capture the complementarity of different views.

J. Zhu, W. Cheng and Y. Cui—These authors contributed equally to this work.

X. Song et al. (Eds.): APWeb-WAIM 2023, LNCS 14332, pp. 486–500, 2024.
https://doi.org/10.1007/978-981-97-2390-4_33

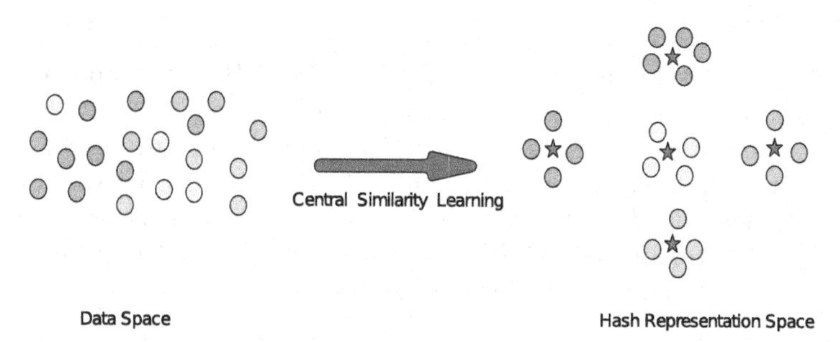

Fig. 1. The inputs are randomly distributed in the data space. Central similarity learning utilizes the hash center to separate samples from different classes while minimizing the class intra-distance.

Retrieval accuracy is currently poor for multi-view hashing methods. The reasons are as follows. First, current methods are fascinated by the information provided by the local similarity. The global similarity is not given enough credit. For example, Flexible Graph Convolutional Multi-modal Hashing (FGCMH) [16] is a GCN-based [22] multi-view hashing method. It constructs the edges of the graph based on similarity. The GCN then aggregates features of adjacent nodes. Hence, the global similarity is not taken into account during this approach. Second, it is not enough to fuse multiple views. To get a global embedding, common multi-view hashing methods (e.g., Deep Collaborative Multi-View Hashing (DCMVH) [29], Flexible Multi-modal Hashing (FMH) [30]) use weighted sum or concatenation to fuse the features. During the fusing process, the relationship between the text and image is disregarded, which results in a lack of expressiveness in the fused features. The aforementioned facts result in low overall retrieval accuracy.

This paper proposes *Central Similarity Multi-View Hashing* method termed CSMVH, which introduces central similarity learning to multi-view hashing. As shown in Fig. 1, samples are distributed randomly in the raw data space. Using central similarity learning, semantically similar samples are close to one another, while semantically different samples are pushed away from each other. This can improve the ability to distinguish sample semantic space. At the same time, because central similarity learning also has the advantage of linear complexity, it is a very effective method.

CSMVH takes advantage of the Gated Multimodal Unit (GMU) [1] to learn the interaction and dependency between the image and text. Different from typical methods, our method fuses multi-view features into a global representation without losing dependency. The semantics-preservation principle of hash representation learning governs the selection of the best embedding space.

We evaluate our method CSMVH on the multi-view hash representation learning tasks of MS COCO and NUS-WIDE datasets. The proposed method provides up to 11.41% mAP improvement in benchmarks. Our main contributions are as follows:

- We propose a novel multi-view hash method CSMVH. The proposed method achieves state-of-the-art results in multimedia retrieval.
- Central similarity learning is introduced to multi-view hashing for the first time. Our method has lower time complexity than typical pairwise or triplet similarity methods. And it converges faster than previous methods.
- We take advantage of the GMU to learn a better global representation of different views to address the insufficient fusion problem.

2 Related Work

Multi-view hashing [2,9,10,12,14,18,20,23,24,28] fuses multi-view features for hash representation learning. These methods use a graph to model the relationships among different views for hash representation learning. Multiple Feature Hashing (MFH) [20] not only preserves the local structure information of each view but also considers global information during the optimization. Multi-view Alignment Hashing (MAH) [12] focuses on the hidden semantic information and captures the joint distribution of the data. Multi-view Discrete Hashing (MvDH) [18] performs spectral clustering to learn cluster labels. Multi-view Latent Hashing (MVLH) [19] learns shared multi-view hash codes from a unified kernel feature space. Compact Kernel Hashing with Multiple Features (MFKH) [14] treats supervised multi-view hash representation learning as a similarity preserving problem. Different from MFKH, Discrete Multi-view Hashing (DMVH) [24] constructs a similarity graph based on Locally Linear Embedding (LLE) [8,17].

Recently, some deep learning-based multi-view hashing methods are proposed. Flexible Discrete Multi-view Hashing (FDMH) [13], Flexible Online Multi-modal Hashing (FOMH) [15], and Supervised Adaptive Partial Multi-view Hashing (SAPMH) [26] seek for a projection from input space to embedding space using nonlinear methods. The learned embeddings are fused into multi-modal embedding for multi-view hashing. Instead of seeking an embedding space, Deep Collaborative Multi-view Hashing (DCMVH) [29] directly learns hash codes using a deep architecture. A discriminative dual-level semantic method is used for their supervised training. FGCMH [16] is based on a graph convolutional network (GCN). It preserves both the modality-individual and modality-fused structural similarity for hash representation learning. To facilitate multi-view hash at the concept aspect, Bit-aware Semantic Transformer Hashing (BSTH) [21] explores bit-wise semantic concepts while aligning disparate modalities.

Compared with previous multi-view hashing methods, we use central similarity learning and gate-based fusion for the first time. First, current multi-view methods underrate the importance of global similarity and only obtain information provided by local similarity. Therefore, we introduce central similarity learning, which can take advantage of global similarity to make our method learn more helpful information. Second, current hashing methods fuse multi-view features insufficiently, leading to a weak expressiveness of the obtained global representation. To address this issue, we adopt gate-based fusion to learn the interaction and dependency between the image and text features.

3 The Proposed Methodology

The goal of CSMVH is to use central similarity learning to train a deep multi-view hashing network. We first propose the deep multi-view hashing network, which uses gate-based fusion for the multi-view features. Then the loss of central similarity learning is turned to illustrate. Eventually, CSMVH reduces the computational complexity.

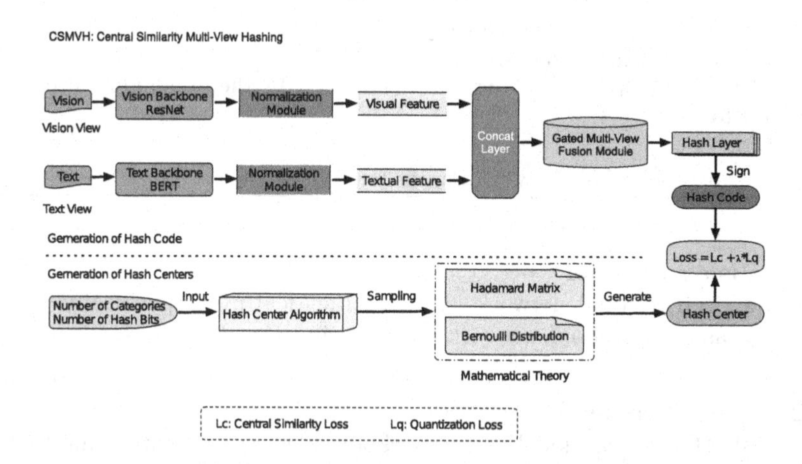

Fig. 2. Architecture Overview of CSMVH. CSMVH has two pipelines, one for the multi-view hash code generation and the other for the hash center generation. In the first pipeline, the image and text features are extracted by the backbones, and then the features are fused by the GMU and hashed by the hash layer. In the second pipeline, the hash center for each class is generated using the Hadamard matrix or Bernoulli distribution. Eventually, CSMVH is optimized by minimizing the distance between the hash code and the corresponding hash center while considering the quantization loss.

3.1 Deep Multi-view Hashing Network

The deep multi-view hashing network is made to transform multi-view data into hash code. As shown in Fig. 2, CSMVH comprises a vision backbone, text backbone, normalization modules, multi-view fusion module, and a hash layer. We utilize deep residual net (ResNet) [6] for image feature extraction and BERT-base [5] for text feature extraction. The extracted features are then passed to the fusion module. These modules are described in detail below. In order to better explain our method, we detailly write the steps of the CSMVH in Algorithm 1.

Algorithm 1. Central Similarity Multi-View Hashing

Input: Training Set $\left\{\mathbf{x_i}^{(m)}, \mathbf{Y_i}\right\}_{i=1}^{N}$, Number of views m, Number of training sets N, Hash Code Length K, Hash Centers $\mathcal{HC} = \{hc_i\}_{i=1}^{V} \subset \{-1, 1\}^K$, V is the number of categories of Y. Hyper-parameter λ, Training epochs T.

1: Initialize BERT parameters w_{BERT} and ResNet parameters w_{ResNet} by loading pre-trained model parameters.
2: Randomly initialize Normalization Module parameters w_{vnorm} and w_{tnorm}, Multi-View Fusion Module parameters w_{fusion} and Hash Layer parameters w_{hash}.
3: Initialize learning rate, mini-batch size, dropout, and λ.
4: **for** *epoch* = 1 to T **do**
5:     **for** $i = 1$ to N **do**
6:         Backbones extract embedding from view data.
7:         Normalize embedding to the same dimensions.
8:         Fuse multi-view embedding to the global representation.
9:         Hash Layer generates hash code.
10:         Update L_{total} by Eq.(13).
11:         Update w_{vnorm}, w_{tnorm}, w_{fusion}, and w_{hash} by the Adam Optimizer.
12:     **end for**
13: **end for**

Output: Multi-View Hash Network Parameters: $\mathbf{W}_K^{hash}$.

3.1.1 Vision Encoder

We use a deep residual network [6] to extract features. Specifically, we take ResNet-50 as the backbone network for acquiring visual features.

3.1.2 Text Encoder

We use BERT-base [5] model as the backbone network for extracting text features. We transform the semantic tag of images into descriptive sentences and then input it to the BERT-base model for feature extraction. Finally, the CLS Token embedding of the BERT model output layer is taken as the semantic vector of the image tag.

3.1.3 Normalization Module

The normalization module projects multi-view features into the same dimension, and the output range of features is also within a certain threshold. It is implemented through a fully connected layer network.

3.1.4 Gated Multi-view Fusion Module

In multimodal hashing, we need to fuse different information sources so that the fused one can provide more information than a single source. Fusion can be performed at the feature level or decision level. Gate mechanism can combine both feature and decision fusion. In this work, we utilize a gate-based fusion module called GMU. GMU is inspired by GRU [4] and LSTM [7]. It can work as an internal unit finding intermediate representation for multi-view data. The data from different sources is modulated. The gate will learn a prediction from input data to fusion weight. The modulated data is combined using weighted summation. Primarily, for image and text input, the equation governing can be represented as follows:

$$h_i = \tanh\left(W_i \cdot x_i\right) \tag{1}$$

$$h_t = \tanh\left(W_t \cdot x_t\right) \tag{2}$$

$$z = \sigma\left(W_z \cdot [x_i, x_t]\right) \tag{3}$$

$$h_f = z * h_i + (1 - z) * h_t \tag{4}$$

where x_i and x_t are the image and text input, h_i and h_t are the modulated feature. σ represents the sigmoid activation function. z is the predicted weight for summation. We let $\Theta = \{W_i, W_t, W_z\}$, where Θ is the learnable parameters. h_f is the global vector after multi-view fusion.

3.1.5 Hash Layer

The hash layer is a linear layer with a *tanh* activation, which is denoted by:

$$h_{\text{k-bit}} = \text{sgn}[\tanh(w_{\text{hash}} X_{\text{fusion}} + b_{\text{hash}})], \tag{5}$$

where *sgn* is the signum function. $w_{\text{hash}} \in \mathbb{R}^{n \times n}$ and $b_{\text{hash}} \in \mathbb{R}^n$ are network parameters. The output has exactly as many dimensions as the hash code has. Each dimension corresponds to a hash bit.

3.2 Central Similarity Learning

Let $\mathcal{X} = \left\{\left\{x_i{}^{(m)}\right\}, Y_i\right\}_{i=1}^{N}$ denotes the train dataset, where each $x_i{}^{(m)} \in \mathbb{R}^D$ is a multi-view instance, m is the number of views, and N is the total number of dataset samples. And $Y = \{y_1, y_2, \ldots, y_N\}$ is set, where y_i represents the label of x_i. We use the symbol $F : x \mapsto he \in \{-1, 1\}^K$ to signify the deep hash function from the input space $\mathbb{R}^D$ to K-bit Hamming space $\{-1, 1\}^K$. Like other deep hashing methods, we investigate a deep multi-view hashing method, which can generate hash code approximation for data points with the same semantic label in Hamming space.

The use of central similarity learning can cause data points from the same class to cluster together around a single hash center and those from other semantic categories to cluster together around several hash centers, respectively. It makes sense that hash centers maintain a suitable mutual separation from one another, which effectively keeps samples from various classes apart in the Hamming space. High-quality hash codes can be produced through central similarity learning in the deep hash function F by learning the overall similarity information between data pairs.

3.2.1 Hash Center Definition

We use a similar definition of hash centers as Central Similarity Quantization (CSQ) [25] with a useful modification. We replace the hash code 0 with -1, thus $\mathcal{HC} = \{hc_i\}_{i=1}^{V} \subset \{-1, 1\}^K$. V is the number of categories of Y. An elegant linear relationship exists between Hamming distance $\text{dist}_H(\cdot, \cdot)$ and inner product $\langle \cdot, \cdot \rangle$:

$$dist_H(hc_i, hc_j) = \frac{1}{2}(K - \langle hc_i, hc_j \rangle), \tag{6}$$

where K is the length of hash bits. We can easily have the following:

$$\frac{1}{\mathrm{T}} \sum_{i \neq j}^{V} \langle hc_i, hc_j \rangle \leq 0 \tag{7}$$

V is the number of hash centers, and T is the number of combinations of different hc_i and $hc_j \in \mathcal{HC}$.

3.2.2 Generation of Hash Centers

With the definition of hash center, we can naturally notice a boundary scenario:

$$\frac{1}{\mathrm{T}} \sum_{i \neq j}^{V} \langle hc_i, hc_j \rangle = 0 \tag{8}$$

If all hash centers are orthogonal to each other, this equation holds. The most intuitive way to generate orthogonal vectors is by leveraging the Hadamard matrix. The rows of the Hadamard matrix are orthogonal to each other. Thus, we can easily get the hash center with the Hadamard matrix and concatenated transpose. However, we need Bernoulli Distributions for the extra when we need more than $2K$ hash centers.

3.2.3 Loss Function

Given the generated centers $\mathcal{HC} = \{hc_i\}_{i=1}^{V} \subset \{-1, 1\}^K$ in the K-dimensional Hamming space for training data X with V categories, we obtain the semantic hash centers $\mathcal{HC}' = \{hc'_1, hc'_2, \ldots, hc'_N\}$ for single-label or multi-label data, where hc'_i denotes the hash center of the data sample x_i. Since hash centers are binary vectors, we use Binary Cross Entropy (BCE) to measure the Hamming distance between the hash code and its center, $dist_H(hc'_i, he_i) = BCE(hc'_i, he_i)$. We obtain the optimization objective of the central similarity loss L_c:

$$hc'_{i,k} = \frac{1}{2}(1 + hc'_{i,k}) \tag{9}$$

$$he_{i,k} = \frac{1}{2}(1 + he_{i,k}) \tag{10}$$

$$L_c = \frac{1}{N} \sum_{i=1}^{N} \sum_{k=1}^{K} \left[hc'_{i,k} \log he_{i,k} + (1 - hc'_{i,k}) \log (1 - he_{i,k}) \right] \tag{11}$$

The output of the neural network is continuous. However, in a Multi-view hashing problem, the output is binary code. Hence, an intuitive method uses a threshold to quantify the network output. Unfortunately, the foundation of the elegant linear relation between Hamming distance and inner product requires the outputs to be exact −1 and 1. Otherwise, it does not hold. Furthermore, quantifying the output using a threshold will introduce an uncontrollable quantization error, which can not pass through the backpropagation. Inspired by DHN [27], we utilize a quantization before constraining the network output. Different from DHN, we use the log cosh function rather than L1-norm for its smoothness, which can be represented as:

$$L_q = \frac{1}{N}\sum_{i=1}^{N}\sum_{k=1}^{K}\left(\log\cosh\left(|he_{i,k}| - 1\right)\right) \tag{12}$$

Eventually, we have a central similarity optimization problem:

$$L_{total} = L_c + \lambda L_q \tag{13}$$

where L_{total} is the total loss function of our algorithm. λ is the hyper-parameter obtained through grid search in our work.

3.2.4 Computational Complexity

Concerning time complexity, CSMVH only has a $O(n)$ value, where n is the number of samples. To achieve similar results, most approaches use pairwise or triplet data similarity, which has a time complexity of $O(n^2)$ or $O(n^3)$. As a result, CSMVH is more effective than the earlier method. For example, Flexible Graph Convolutional Multi-modal Hashing (FGCMH) [16] is a GCN-based multi-view hashing method. It requires an adjacency matrix to represent the relationship between the nodes. Therefore, the complexity of FGCMH is $O(n^2)$. In addition, Eq. (13) can also be concluded that the complexity of our method is $O(n)$.

Through the above analysis, we use central similarity learning to improve the computational complexity, which is that our CSMVH converges faster than the previous methods.

4 Experiments

We evaluate the proposed CSMVH on large-scale multimedia retrieval tasks in experiments. Two genetic datasets are adopted: NUS-WIDE [3] and MS COCO [11]. These datasets have been widely used for evaluating multimedia retrieval performance. We use the mean Average Precision (mAP) as the evaluation metric. The statistics of two datasets used in experiments are summarized in Table 1.

Table 1. General statistics of two datasets. The dataset size, number of categories, and feature dimensions are included.

Datasets	Training Size	Retrieval Size	Query Size	Categories	Visual Feature	Text Feature
MS COCO	18000	82783	5981	80	ResNet(768-D)	BERT(768-D)
NUS-WIDE	21000	193749	2085	21	ResNet(768-D)	BERT(768-D)

4.1 Evaluation Datasets

4.1.1 MS COCO

In our experiments, the MS COCO 2014 dataset is adopted. It contains 82,783 training samples and 40,504 validation samples. We randomly select 80 validation samples from each category as the query set to retrieve samples from the training set and use 18000 of them for training.

4.1.2 NUS-WIDE

NUS-WIDE contains 269,648 Flickr images with 81 ground-truth semantic concepts. In experiments, we select the 21 most common concepts. We randomly select 100 samples for each concept as the query set. The rest of the images are treated as the retrieval set. We utilize 21,000 samples from the retrieval set for training.

4.2 Evaluation Metrics

To evaluate the metric of multi-view hashing methods, mean Average Precision (mAP) is utilized. The Hamming ranking measure can be assessed with good steady using the mAP. It is described as follows:

$$AP(q) = \frac{1}{M}\sum_{r=1}^{R} P_q(r)\mathrm{Ind(r)} \tag{14}$$

$$mAP = \frac{1}{Q}\sum_{i=1}^{Q} AP\left(q_i\right) \tag{15}$$

where Q is the number of queries and P_q is the precision for query q when the top r^{th} similar searching results returned, $Ind(r)$ is an indicator function which is 1 when the rth result has the same label with q and otherwise 0, M is the number of same label samples of query q, and R is the size of the retrieval dataset.

4.3 Implementation Details

Our code is implemented on the PyTorch platform. A ResNet-50 pre-trained on ImageNet is employed as the backbone. We use the BERT-base model as the text backbone. The image and text feature outputs are set to 768-dimensional by normalization modules. The dropout probability is set to 0.1 to improve the generalization capability. The combination coefficient λ of the total loss function is 0.25.

4.4 Baseline

To evaluate the retrieval metric, the proposed CSMVH is compared with twelve comparable multi-view hashing methods, including four unsupervised methods (e.g., Multiple Feature Hashing (MFH) [20], Multi-view Alignment Hashing (MAH) [12], Multi-view Latent Hashing (MVLH) [19], and Multi-view Discrete Hashing (MvDH) [18]) and eight supervised methods (e.g., Multiple Feature Kernel Hashing (MFKH) [14], Discrete Multi-view Hashing (DMVH) [24], Flexible Discrete Multi-view Hashing (FDMH) [13], Flexible Online Multi-modal Hashing (FOMH) [15], Deep Collaborative Multi-View Hashing (DCMVH) [29], Supervised Adaptive Partial Multi-view Hashing (SAPMH) [26], Flexible Graph Convolutional Multi-modal Hashing (FGCMH) [16], and Bit-aware Semantic Transformer Hashing (BSTH) [21]).

4.5 Analysis of Experimental Results

Table 2. The comparable mAP results on NUS-WIDE and MS COCO. The experimental conditions of all methods are the same. The best results are bolded, and the second-best results are underlined. The * indicates that the results of our method on this dataset are statistical significance.

Methods	Ref.	NUS-WIDE*				MS COCO*			
		16 bits	32 bits	64 bits	128 bits	16 bits	32 bits	64 bits	128 bits
MFH	TMM13	0.3603	0.3611	0.3625	0.3629	0.3948	0.3699	0.3960	0.3980
MAH	TIP15	0.4633	0.4945	0.5381	0.5476	0.3967	0.3943	0.3966	0.3988
MVLH	MM15	0.4182	0.4092	0.3789	0.3897	0.3993	0.4012	0.4065	0.4099
MvDH	TIST18	0.4947	0.5661	0.5789	0.6122	0.3978	0.3966	0.3977	0.3998
MFKH	MM12	0.4768	0.4359	0.4342	0.3956	0.4216	0.4211	0.4230	0.4229
DMVH	ICMR17	0.5676	0.5883	0.6902	0.6279	0.4123	0.4288	0.4355	0.4563
FOMH	MM19	0.6329	0.6456	0.6678	0.6791	0.5008	0.5148	0.5172	0.5294
FDMH	NPL20	0.6575	0.6665	0.6712	0.6823	0.5404	0.5485	0.5600	0.5674
DCMVH	TIP20	0.6509	0.6625	0.6905	0.7023	0.5387	0.5427	0.5490	0.5576
SAPMH	TMM21	0.6503	0.6703	0.6898	0.6901	0.5467	0.5502	0.5563	0.5672
FGCMH	MM21	0.6677	0.6874	0.6936	0.7011	0.5641	0.5273	0.5797	0.5862
BSTH	SIGIR22	0.6990	0.7340	0.7505	0.7704	0.5831	0.6245	0.6459	0.6654
CSMVH	Proposed	**0.7360**	**0.7633**	**0.7740**	**0.7819**	**0.6028**	**0.7002**	**0.7485**	**0.7795**

The mAP results are presented in Table 2. The experimental comparisons of all methods are conducted according to the unified conditions of the train set, the retrieval set, and the query set in Table 1. The proposed CSMVH is

Table 3. Ablation experiments on two datasets. The mAP is shown in the table, indicating the performance impact of different modules.

Methods	NUS-WIDE				MS COCO			
	16 bits	32 bits	64 bits	128 bits	16 bits	32 bits	64 bits	128 bits
CSMVH-central	0.7352	0.7594	0.7683	0.7798	0.5892	0.6893	0.7361	0.7731
CSMVH-quant	0.3085	0.3085	0.3085	0.3085	0.3502	0.3502	0.3502	0.3502
CSMVH-text	0.4873	0.5176	0.5196	0.5242	0.5548	0.6134	0.6407	0.6784
CSMVH-image	0.7208	0.7512	0.7623	0.7679	0.5459	0.6454	0.6833	0.7197
CSMVH-concat	0.7283	0.7566	0.7647	0.7718	0.5844	0.6864	0.7282	0.7580
CSMVH	**0.7360**	**0.7633**	**0.7740**	**0.7819**	**0.6028**	**0.7002**	**0.7485**	**0.7795**

a top performer in multi-view hashing tasks. On the NUS-WIDE dataset, our method outperforms the prior state-of-the-art method by a large margin (up to 3.70%). However, the more complex dataset is where our method shines. MS COCO contains 80 categories, and the samples are more complex than other datasets. Our method beats the previous state-of-the-art 128-bit hashing result with only a 16-bit hash code. Our method shows a magnificent retrieval accuracy improvement with the number of hash bits increasing. The mAP of BSTH [21] only increased 8.23% from 16-bit to 128-bit, while CSMVH has a 17.67% absolute mAP increase, which indicates a great capability of solving very complex hashing problems. The absolute 128-bit mAP increase over BSTH is 11.47% on the MS COCO dataset.

The main reasons for these superior results come from two aspects:

- Central similarity learning can take advantage of global similarity to make CSMVH learn more useful information. It enhances the discriminative and semantic capability of hash codes. This can improve the accuracy of multimedia retrieval.
- The multi-view fusion module could deeply fuse the multi-view features into a global representation. It can fully explore the complementarity of multi-view features and improve the ability of feature expression. This can generate high-quality hash codes.

4.6 Ablation Studies

To evaluate the proposed CSMVH component by component, we perform ablation with different experiment settings of our method and report the performance. The experiment settings are as follows:

- *CSMVH-central*: The BCE loss of central similarity learning is used. The quantization loss is removed.
- *CSMVH-quant*: The quantization loss is used. The BCE loss of central similarity learning is removed.
- *CSMVH-image*: Only the visual feature is used. The text embedding is removed.

- *CSMVH-text*: Only the text embedding is used. The visual feature is removed.
- *CSMVH-concat*: The image and text features are fused with concatenation. The multi-view fusion module is removed.
- *CSMVH*: Our full framework.

The comparison results are presented in Table 3. Beginning with the loss function, when we solely use the quantization loss to train the proposed CSMVH, it can not learn anything just as we expected. The quantization loss can not perform any optimization on the embeddings. Because the CSMVH obtains data at random, all of the tasks have relatively poor mAP. On the contrary, the BCE loss of central similarity learning can help the CSMVH learn the embedding well. Due to the absence of a quantization constraint, CSMVH-central performs marginally worse than the entire framework.

Moving on to the multi-view features, CSMVH-text is barely superior to random selection (CSMVH-quant). In every task, CSMVH-image performs significantly better than CSMVH-text. It can be seen that the image features contain more useful information than text. Searching with images is more likely to find a related result. Our method already performs better with concatenated multi-view features than the state-of-the-art methods. But the multi-view fusion module takes it one step further, reaching even higher mAP.

To conclude, the multi-view fusion module can improve the complementarity and representation capability of the fused feature. And the central similarity learning enhances the discriminative capability of hash codes. Both image and text features can provide rich information for the multi-view hashing task.

4.7 Convergence Analysis

We do tests to see how well the CSMVH can generalize and converge. Using the MS COCO dataset, we perform hash benchmarks using various code lengths. The results are shown in Fig. 3. The graphic displays training loss and tests mAP. The loss gradually lessens as the training progresses. The loss stabilized at 40 epochs, indicating that the local minimum is attained. When the experiment starts, the mAP for the test metric quickly increases. The test mAP remains steady after 20 epochs. No worsening on the test MAP is seen after additional training, which suggests high generalization capacity and no overfitting. With other datasets, we notice a similar convergence outcome. The convergence of our method is promised on the popular datasets.

4.8 mAP@K and Recall@K

The matching mAP@K and Recall@K curves for the MS COCO dataset with increasing numbers of retrieved samples are shown in Fig. 4 for various code lengths. The recall curve exhibits quick linear growth, whereas the mAP of the four graphs somewhat declines as the $TOPK$ grows. The tendency demonstrates how well our method does in the retrieval tasks. The majority of people focus their attention on the first few results of the received data. In this

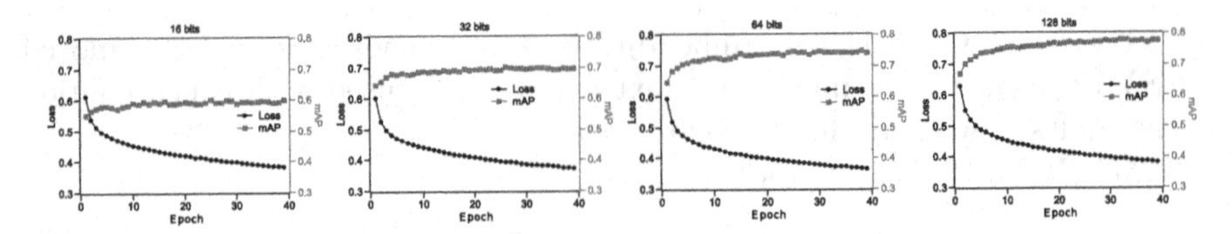

Fig. 3. The training loss curve and test mAP curve on MS COCO dataset. In this figure, the upper curve is the test mAP, and the bottom is the training loss. We notice that after around 40 epochs, our method converges, and the training loss no longer decreases. The test mAP increases rapidly during the first 10 epochs and slowly converges afterward. No overfitting is observed during the experiment.

case, the precision of our method is considerably higher. Compared to normal users, experts typically browse more results. With respect to the expansion of the retrieved data, our method can deliver a recall that grows linearly. When conducting their search, professionals may anticipate dependable, high-quality outcomes. The CSMVH can purposely give satisfying retrieval results for various user groups.

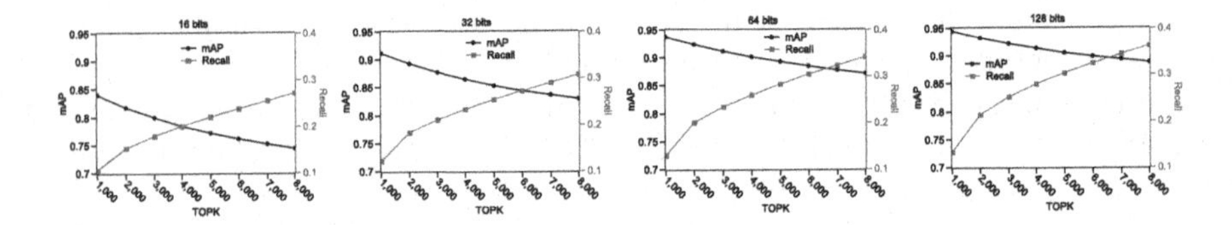

Fig. 4. The mAP@K and Recall@K curves on the MS COCO dataset. The recall curves show a nice linear increase as K increases, while the mAP only slightly decreases. We notice some sweet spots during the experiment. For example, in the 128-bit group, when $TOPK = 5000$, the mAP keeps the same as $TOPK = 4000$, but the recall keeps linearly increasing.

5 Conclusion and Future Work

We propose a new deep multi-view hashing method termed CSMVH. To address multi-view hashing issues, it makes use of central similarity learning. The CSMVH is proposed to conquer two main challenges of the multi-view hashing problem. In our experiment, CSMVH tends to yield consistent improvement in mAP with the growing length of hash code without any signs of performance degradation or overfitting. CSMVH is less computationally intensive than current methods. Under multiple experiment settings, it delivers an impressive performance increase over the state-of-the-art methods. Impressively, our method performs even better in more complex hashing tasks. We also find some problems during the process of our experiments. For instance, the performance improvement becomes less and less noticeable as the hash code length increases for some datasets. To further enhance the proposed method, we will work on these problems.

Acknowledgment. This work is supported in part by the Zhejiang provincial "Ten Thousand Talents Program" (2021R52007), the National Key R&D Program of China (2022YFB4500405), and the Science and Technology Innovation 2030-Major Project (2021ZD0114300).

References

1. Arevalo, J., Solorio, T., Montes-y Gómez, M., González, F.A.: Gated multimodal units for information fusion. arXiv preprint arXiv:1702.01992 (2017)
2. Chen, Y., Zhang, N., Yan, J., Zhu, G., Min, G.: Optimization of maintenance personnel dispatching strategy in smart grid. World Wide Web **26**(1), 139–162 (2023)
3. Chua, T.S., Tang, J., Hong, R., Li, H., Luo, Z., Zheng, Y.: NUS-WIDE: a real-world web image database from national university of Singapore. In: Proceedings of the ACM International Conference on Image and Video Retrieval, pp. 1–9 (2009)
4. Chung, J., Gulcehre, C., Cho, K., Bengio, Y.: Empirical evaluation of gated recurrent neural networks on sequence modeling. arXiv preprint arXiv:1412.3555 (2014)
5. Devlin, J., Chang, M.W., Lee, K., Toutanova, K.: BERT: pre-training of deep bidirectional transformers for language understanding. arXiv preprint arXiv:1810.04805 (2018)
6. He, K., Zhang, X., Ren, S., Sun, J.: Deep residual learning for image recognition. In: Proceedings of the IEEE Conference on Computer Vision and Pattern Recognition, pp. 770–778 (2016)
7. Hochreiter, S., Schmidhuber, J.: Long short-term memory. Neural Comput. **9**(8), 1735–1780 (1997)
8. Hou, Y., Zhang, P., Xu, X., Zhang, X., Li, W.: Nonlinear dimensionality reduction by locally linear inlaying. IEEE Trans. Neural Netw. **20**(2), 300–315 (2009)
9. Kang, Y., Kim, S., Choi, S.: Deep learning to hash with multiple representations. In: 2012 IEEE 12th International Conference on Data Mining, pp. 930–935. IEEE (2012)
10. Kim, S., Choi, S.: Multi-view anchor graph hashing. In: 2013 IEEE International Conference on Acoustics, Speech and Signal Processing, pp. 3123–3127. IEEE (2013)
11. Lin, T.-Y., et al.: Microsoft COCO: common objects in context. In: Fleet, D., Pajdla, T., Schiele, B., Tuytelaars, T. (eds.) ECCV 2014. LNCS, vol. 8693, pp. 740–755. Springer, Cham (2014). https://doi.org/10.1007/978-3-319-10602-1_48
12. Liu, L., Yu, M., Shao, L.: Multiview alignment hashing for efficient image search. IEEE Trans. Image Process. **24**(3), 956–966 (2015)
13. Liu, L., Zhang, Z., Huang, Z.: Flexible discrete multi-view hashing with collective latent feature learning. Neural Process. Lett. **52**(3), 1765–1791 (2020)
14. Liu, X., He, J., Liu, D., Lang, B.: Compact kernel hashing with multiple features. In: Proceedings of the 20th ACM International Conference on Multimedia, pp. 881–884 (2012)

15. Lu, X., Zhu, L., Cheng, Z., Li, J., Nie, X., Zhang, H.: Flexible online multi-modal hashing for large-scale multimedia retrieval. In: Proceedings of the 27th ACM International Conference on Multimedia, pp. 1129–1137 (2019)
16. Lu, X., Zhu, L., Liu, L., Nie, L., Zhang, H.: Graph convolutional multi-modal hashing for flexible multimedia retrieval. In: Proceedings of the 29th ACM International Conference on Multimedia, pp. 1414–1422 (2021)
17. Saul, L.K., Roweis, S.T.: Think globally, fit locally: unsupervised learning of low dimensional manifolds. J. Mach. Learn. Res. **4**(Jun), 119–155 (2003)
18. Shen, X., Shen, F., Liu, L., Yuan, Y.H., Liu, W., Sun, Q.S.: Multiview discrete hashing for scalable multimedia search. ACM Trans. Intell. Syst. Technol. (TIST) **9**(5), 1–21 (2018)
19. Shen, X., Shen, F., Sun, Q.S., Yuan, Y.H.: Multi-view latent hashing for efficient multimedia search. In: Proceedings of the 23rd ACM International Conference on Multimedia, pp. 831–834 (2015)
20. Song, J., Yang, Y., Huang, Z., Shen, H.T., Luo, J.: Effective multiple feature hashing for large-scale near-duplicate video retrieval. IEEE Trans. Multimedia **15**(8), 1997–2008 (2013)
21. Tan, W., Zhu, L., Guan, W., Li, J., Cheng, Z.: Bit-aware semantic transformer hashing for multi-modal retrieval. In: Proceedings of the 45th International ACM SIGIR Conference on Research and Development in Information Retrieval, pp. 982–991 (2022)
22. Welling, M., Kipf, T.N.: Semi-supervised classification with graph convolutional networks. J. Int. Conf. Learn. Represent. (ICLR 2017) (2016)
23. Xu, D., Chen, Y., Cui, N., Li, J.: Towards multi-dimensional knowledge-aware approach for effective community detection in LBSN. World Wide Web, pp. 1–24 (2022)
24. Yang, R., Shi, Y., Xu, X.S.: Discrete multi-view hashing for effective image retrieval. In: Proceedings of the 2017 ACM on International Conference on Multimedia Retrieval, pp. 175–183 (2017)
25. Yuan, L., et al.: Central similarity quantization for efficient image and video retrieval. In: Proceedings of the IEEE/CVF Conference on Computer Vision and Pattern Recognition, pp. 3083–3092 (2020)
26. Zheng, C., Zhu, L., Cheng, Z., Li, J., Liu, A.A.: Adaptive partial multi-view hashing for efficient social image retrieval. IEEE Trans. Multimedia **23**, 4079–4092 (2020)
27. Zhu, H., Long, M., Wang, J., Cao, Y.: Deep hashing network for efficient similarity retrieval. In: Proceedings of the AAAI Conference on Artificial Intelligence, vol. 30 (2016)
28. Zhu, J., Huang, Z., Ruan, X., Cui, Y., Cheng, Y., Zeng, L.: Deep metric multi-view hashing for multimedia retrieval. arXiv preprint arXiv:2304.06358 (2023)
29. Zhu, L., Lu, X., Cheng, Z., Li, J., Zhang, H.: Deep collaborative multi-view hashing for large-scale image search. IEEE Trans. Image Process. **29**, 4643–4655 (2020)
30. Zhu, L., Lu, X., Cheng, Z., Li, J., Zhang, H.: Flexible multi-modal hashing for scalable multimedia retrieval. ACM Trans. Intelli. Syst. Technol. (TIST) **11**(2), 1–20 (2020)

Entity Alignment Based on Multi-view Interaction Model in Vulnerability Knowledge Graphs

Jin Jiang and Mohan Li(✉)

Cyberspace Institute of Advance Technology, Guangzhou University, Guangzhou, China
csjiangjin@e.gzhu.edu.cn, limohan@gzhu.edu.cn

Abstract. Entity alignment (EA) aims to match the same entities in different Knowledge Graphs (KGs), which is a critical task in KG fusion. EA has recently attracted the attention of many researchers, but the performance of general methods on KGs in some professional fields is not satisfactory. Vulnerability KG is a kind of KG that stores vulnerability knowledge. The text and structure information is not the same as the general KG, so the EA task faces unique challenges. First, although some vulnerabilities have a unified CVE number, in reality, the CVE number attribute value of many vulnerability entities in KG is missing. Second, vulnerability KGs often contain a large number of 1−n and n−1 relations, and general entity embedding methods may generate similar vector representations for a large number of non-identical vulnerabilities. To address the above challenges, we propose a multi-view text-graph interaction model (TG-INT) for the EA task in vulnerability KG. We use cross-lingual BERT to learn text embeddings and an optimized model called QuatAE to embed two graphs into a unified vector space. After that, we employed a multi-view interactive modeling scheme for the EA task. On the vulnerability KGs built on the vulnerability database CNNVD and CNVD, we verified the effectiveness of TG-INT. The results show that our model is not only suitable for vulnerability KGs but also achieves promising results in general KGs.

Keywords: Vulnerability knowledge graph · Entity alignment · Graph embedding · Security database

1 Introduction

The quality of a knowledge graph (KG) is essential for most downstream tasks, including automatic question-answering [1] and anomaly detection [2]. However, ensuring high-quality information in a KG is challenging and can be costly. Therefore, fusing KGs of similar fields has become attractive and valuable. Entity alignment (EA) aims to identify pairs of entities that represent the same thing across different KGs, but this can be difficult due to the heterogeneity of KGs from different sources. EA is an even more challenging task for vulnerability KGs in cyberspace security, which consist of information about vulnerabilities and

X. Song et al. (Eds.): APWeb-WAIM 2023, LNCS 14332, pp. 501–516, 2024.
https://doi.org/10.1007/978-981-97-2390-4_34

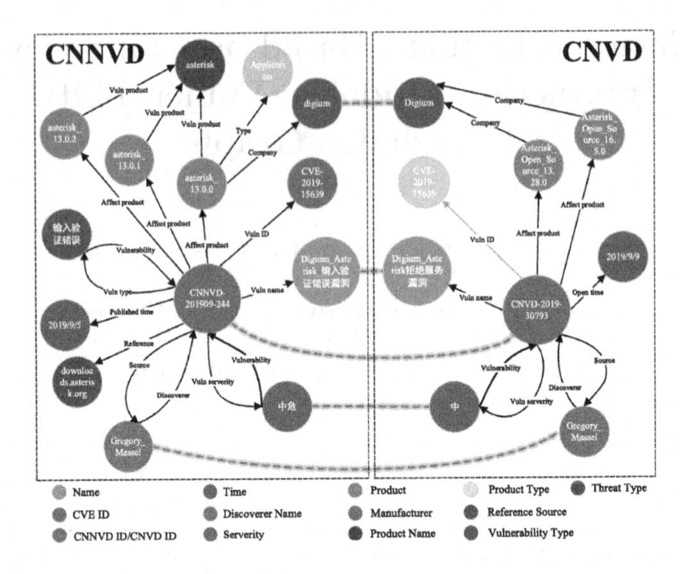

Fig. 1. An example of vulnerability KG alignment, where the alignment with missing CVE numbers is relatively difficult. (Dashed line connecting aligned entity pairs)

their associated levels of danger, products, manufacturers, etc. Unlike general KGs, vulnerability KGs contain a significant number of similar but not identical entities, such as different products from the same manufacturer and different vulnerabilities of the same type. As shown in Fig. 1, trained professionals are required to determine whether an entity pair is aligned when the CVE number is missing. Despite a large number of research on general KGs, there has been limited focus on entity alignment for vulnerability KGs.

Textual rules and logical reasoning [3,4] were the primary methods used for EA in general KGs, but these methods were not effective for cross-linguistic KGs. With the development of deep learning, geometric transformations [5,6] and graph neural networks [7–9] have been used to extract structural features for EA tasks. However, only calculating the similarity of graph structure features cannot achieve satisfactory accuracy. If text information such as entity names and descriptions are also considered, the accuracy will be greatly improved [10]. In recent studies, BERT-INT [11] constructs a BERT-based interaction model using only text information. Other models, such as UPLR [12], and SelfKG [13], combine text embedding into graph neural networks to build EA models.

In vulnerability KGs, there are many entities with similar names and some vulnerability databases mix different languages. For example, Chinese and English are both present in China National Vulnerability Database of Information Security (CNNVD) and China National Vulnerability Database (CNVD). Only relying on text similarity without considering the graph structure would not work well. However, relying only on graph embeddings is also insufficient. Due to the limited variety of relations in the vulnerability KG compared to a general KG, the semantic information contained in the graph structure is also limited. These challenges make the task of EA in vulnerability KGs difficult.

To jointly exploit text and graph structure information for EA in vulnerability KGs, we propose a multi-view **T**ext-**G**raph **INT**eraction model (TG-INT). First, we design QuatAE, a QuatE-based model that models symmetric and inverse relations and learns the semantic information of entities in the graph structure. Second, we employ a cross-language BERT to acquire text information about entities. Based on the acquired textual and structural features, we build a multi-view interaction model for EA in vulnerability KGs. Our model combines similarities in three types of views, i.e., text-graph view, neighbor view and attribute view. In the text-graph and neighbor views, we consider both the graph embedding and text embedding of entities, and only extracting text embeddings in the attribute view. Experimental results demonstrate that our approach is effective not only for vulnerability KGs but also improves general cross-linguistic KG alignment. To the best of our knowledge, this is the first study to address the vulnerability KG EA task. The contributions of this paper are as follows:

- We propose **QuatAE**, a graph embedding method for EA task of vulnerability KGs. QuatAE is an efficient approach for capturing the graph structure and effectively assists with EA task.
- We design **TG-INT**, a multi-view text-graph interaction model to improve the precision of EA. Our approach integrates information from the text-graph view, the neighbor view, and the attribute view.
- We experimentally verify that TG-INT not only achieves the best results on the EA task on vulnerability KGs, but also improves the EA accuracy on general KGs.[1]

2 Related Work

Knowledge Graph Embeddings. The current EA models mainly use GNN-based and geometric modeling-based methods for graph embedding. GNN [7–9] learns entity embeddings by aggregating information from their neighbors, while geometric modeling-based methods represent relations as geometric transformations between entities. TransE [5] models the relations as translational vectors in Euclidean space. QuatE [6] models the relation as a rotation in Quaternion space. These graph embedding methods capture important information about the graph structure that can be used for effective entity alignment.

EA Based on Graph Embeddings and Auxiliary Information. The initial EA models relied on graph embeddings, while MTransE [14] employs entity alignment seeds to train a transformation matrix. Later models such as IPTransE [15] and BootEA [16] have introduced iterative approaches to incorporate additional label data, while some models like NAEA [17] and AliNet [18] have integrated information about the entity's neighbors.

[1] The code is available at https://github.com/krypros/TG-INT.

Recent research has been focused on entity alignment using both graph structure information and side information. JAPE [19], GCN-Align [21] and MultiKE [20] assign structural and attribute features to entities. RDGCN [22] constructs a primal graph and dual relation graph to learn complex relational information for alignment. BERT-INT [11] learns text information using only BERT to model entity, neighbor, and attribute interactions. RPR-RHGT [23] aggregates relation embeddings and path embeddings using relation-aware Transform. UPLR [12] generates pseudo-labels using semantic information from the text and learns entity embedding using the GAT network. These methods incorporate side information from different perspectives to support EA. Although most of these methods incorporate text information into the graph neural network, the process of graph aggregation may weaken the textual features of the entities themselves. Therefore, we make all relevant text, graph, relation, and attribute information available for the EA task of vulnerability KGs.

3 Problem Formulation

Knowledge Graph. A knowledge graph can be defined as $\mathcal{G} = (\mathcal{E}, \mathcal{R}, \mathcal{A}, \mathcal{V}, \mathcal{T})$, which consists of relation triples and attribute triples. $\mathcal{E}$, $\mathcal{R}$, $\mathcal{A}$ and $\mathcal{V}$ denote the set of entities, relations, attributes and attribute values, respectively. $\mathcal{T} = \mathcal{T}_r \cup \mathcal{T}_a$ denotes the set of triples, where $\mathcal{T}_r$ is the set of relation triples and $\mathcal{T}_a$ is the set of attribute triples. A relation triple (h, r, t) consists of entities and relation. A attribute triple (h, a, v) consists of entity, attribute and attribute values. Entities tend to have multiple meanings, and attribute values are relatively more certain.

Entity Alignment. Given two graphs $\mathcal{G}_1 = (\mathcal{E}_1, \mathcal{R}_1, \mathcal{A}_1, \mathcal{V}_1, \mathcal{T}_1)$, $\mathcal{G}_2 = (\mathcal{E}_2, \mathcal{R}_2, \mathcal{A}_2, \mathcal{V}_2, \mathcal{T}_2)$ and few entity alignment seeds $\mathcal{S}_d = \{e, e' | e \equiv e', e \in \mathcal{E}_1, e' \in \mathcal{E}_2\}$. $\equiv$ indicates that the entity pairs are identical. Entity alignment requires that pairs of entities with the same actual meaning be found in $\mathcal{G}_1$ and $\mathcal{G}_2$.

The objective of vulnerability KG entity alignment is similar to that of general knowledge graphs, but it specifically addresses the challenge of aligning vulnerability entities that lack CVE numbers.

4 Methodology

In this chapter, we propose QuatAE, a KG embedding method for EA. And then combine these graph embeddings with a cross-linguistic alignment model based on BERT, utilizing a three-view interaction approach.

4.1 QuatAE

Given the large number of symmetric relations in the vulnerability KG, QuatE [6] can effectively model symmetric and inverse relations. Consequently, we have improved the QuatE model to better learn graph embeddings and enhance its performance in the EA task. First, we use the known label data $\mathcal{S}_d = \{(e, e') | e \in$

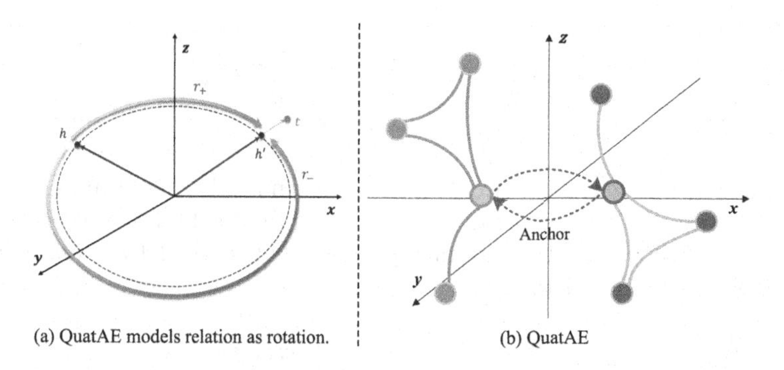

Fig. 2. QuatE and QuatAE.

$\mathcal{E}_1, e' \in \mathcal{E}_2, e \equiv e'\}$ to embed the two candidate knowledge graphs into a uniform vector space. For this purpose, we define a new relation 'anchor' r_a. By using $\mathcal{S}_d$ and r_a, we can generate a set of triples $\mathcal{T}_s = \{(e, r_a, e') \cup (e', r_a, e) | (e, e') \in \mathcal{S}_d\}$ to support alignment training. The training is supervised by these alignment triples so that the same entities in both graphs are pulled in by the 'anchor' during the training process. As depicted in Fig. 2 (b), the 'anchor' acts as a bridge between the two graphs. In contrast to the direct approach of sharing the same vector between known alignment pairs, the introduction of the 'anchor' [24] allows identical entities to be closer together in a unified vector space, while maintaining the original structural properties of the two graphs as much as possible.

QuatAE, just like QuatE, models relation as rotational transformation in quaternion space. Thus inverse and symmetric relations can be modeled, satisfying the properties of the 'anchor'. For a given triple (h, r, t), QuatAE embeds both entities and relations in the quaternion space. Quaternion embeddings $\boldsymbol{v_h}$, $\boldsymbol{v_r}$ and $\boldsymbol{v_t}$ are represented as:

$$\begin{aligned} \boldsymbol{v_h} &= \left(v_{h,w}, v_{h,i}, v_{h,j}, v_{h,k}\right), \\ \boldsymbol{v_r} &= \left(v_{r,w}, v_{r,i}, v_{r,j}, v_{r,k}\right), \\ \boldsymbol{v_t} &= \left(v_{t,w}, v_{t,i}, v_{t,j}, v_{t,k}\right), \end{aligned} \tag{1}$$

where each quaternion embedding consists of a real part and three imaginary parts. Each element in the embedding is a vector. To achieve a rotation in quaternion space, we need to normalize the relation $\boldsymbol{v_r}$ to the unit quaternion. This unitization likewise eliminates the effect of the embedding scale:

$$\boldsymbol{v}_r^{\triangleleft} = \frac{\boldsymbol{v_r}}{\|\boldsymbol{v_r}\|} = \left(\bar{v}_{r,w}, \bar{v}_{r,i}, \bar{v}_{r,j}, \bar{v}_{r,k}\right). \tag{2}$$

After obtaining the unit embedding. As shown in Fig. 2(a), QuatAE uses hamilton multiplication [6] to rotate the entity h to h' by isoclinic rotation r_+ or r_-, which makes the angle between h' and t to be zero. After the rotation, $\boldsymbol{v_h}$ transforms into $\boldsymbol{v}'_h$. That the inner product of the head and tail entity embedding is zero, from which the scoring function can be obtained:

$$f(h,r,t) = \boldsymbol{v}_h \otimes \boldsymbol{v}_r^{\triangleleft} \bullet \boldsymbol{v}_t = \boldsymbol{v}_h' \bullet \boldsymbol{v}_t, \tag{3}$$

where $\otimes$ denotes the hamilton product and $\bullet$ denotes the element-wise multiplication. Consider that we will measure the semantic similarity of two entity embeddings by comparing their cosine distances in a subsequent alignment task. So we constrain the cosine distances of the entity embeddings in the alignment seeds to be close during the training process, and learned by minimizing the following regularized logistic loss:

$$\begin{aligned} \mathcal{L}_1 = & \sum_{(h,r,t)\in\mathcal{T}\cup\mathcal{T}_-} \log(1+\exp(-l\cdot f(h,r,t))) \\ & + \lambda_1\left(\|\boldsymbol{v}_h\|_2^2 + \|\boldsymbol{v}_r\|_2^2 + \|\boldsymbol{v}_t\|_2^2\right) + \lambda_2 \sum_{(h,t)\in\mathcal{T}_s} \cos(\boldsymbol{v}_h, \boldsymbol{v}_t), \end{aligned} \tag{4}$$

where $l \in \{-1, 1\}$ denotes whether the triple (h,r,t) is a negative sample. $\mathcal{T}_-$ is the set of negative sample triples generated by sampling from $\mathcal{T}$. We utilized the same sampling strategies as QuatE [6]. To avoid overfitting, we adopt L2-regularization and set the parameter to λ_1. λ_2 is set to avoid the model being overly concerned with the aligned pairs. $\cos(\cdot)$ denotes the cosine distance between two entities.

QuatAE is capable of learning structure information on vulnerability KGs. By modeling entities in quaternion space, QuatAE can effectively address the widespread issue of 1−n and n−1 relations in the vulnerability KG. While alignment using QuatAE alone is superior to many graph embedding-based alignment models, it does not quite meet practical standards. To achieve optimal alignment, we combine the graph structure information learned by QuatAE with text information in the next section.

4.2 Text and Graph Interaction Model

Since the graph structure in the vulnerability KG provides limited semantic information, EA also requires considering the text-based semantic information such as entity names and descriptions. Fortunately, cross-language pre-training models like BERT can be used to learn text-based features, though the mixing of English and Chinese in CNNVD and CNVD can pose some challenges. For each entity, we use a BERT model to embed its name, description, and attribute values. We then concatenate the resulting text embeddings with the graph embeddings obtained from QuatAE and pass them through a MLP layer. Subsequently, we consider a combination of text-graph, neighbor, and attribute views, and use a unified dual aggregation function to compute matching scores between entity pairs based on information from the neighbor and attribute views. The overall framework of the model is shown in the Fig. 3.

Basic Model. Given that vulnerability data consists of both Chinese and English, pre-trained models in either language alone were weaker than cross-language

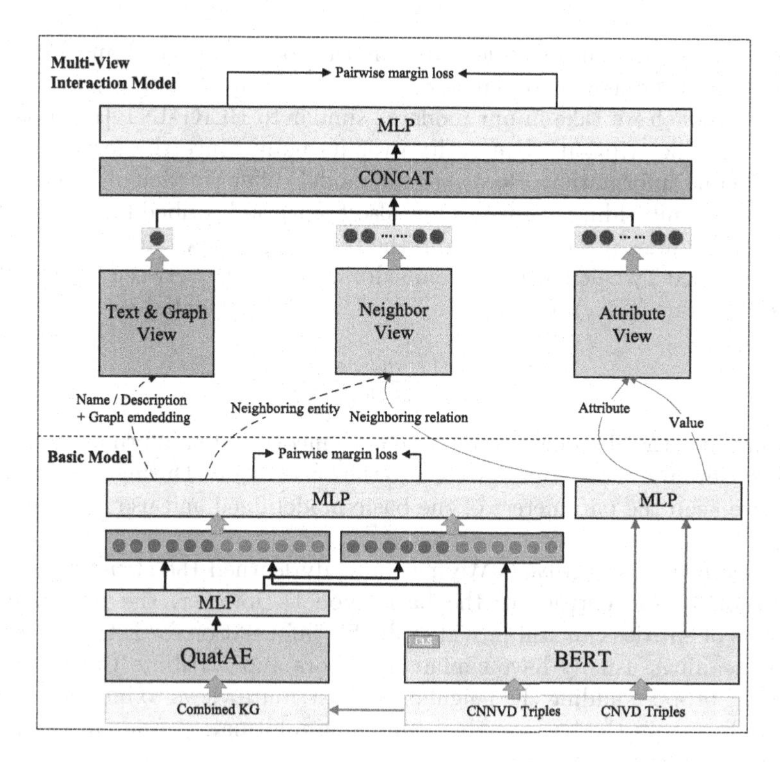

Fig. 3. The framework of TG-INT.

models. This characteristic makes our proposed solution suitable for cross linguistic KG alignment. In our scenario, we believe that relying solely on text information without taking into account the structure information would result in a loss of uniqueness of the dataset. Therefore, we designed a model that combines both text and graph information.

We first use the alignment seeds and randomly sample some negative samples $\mathcal{D} = \{(e, e_+, e_-)\}$ as training data for fine-tuning. $e_+ \in \mathcal{S}_d$ denotes the same entity as e. e_- denotes a negative sample that is not identical to e. The sampling of the negative sample is selected by the cosine similarity of the entity pairs. We take the name/description of each entity e in the training set as input to the BERT, concatenate the CLS embedding of the BERT with the QuatAE embedding, and then filter it by a MLP layer. Considering the discrepancy between the QuatAE embedding and the BERT embedding, we also pass the QuatAE embedding through a MLP layer for feature selection before concatenating:

$$G(e) = \mathrm{MLP}(\mathrm{QuatAE}(e)),\ TG(e) = \mathrm{MLP}\left(\mathrm{CLS}(e) \oplus G(e)\right), \tag{5}$$

where $\oplus$ denotes a concatenation between vectors. We use pairwise margin loss to fine tune the BERT:

$$\mathcal{L}_2 = \sum_{(e,e_+,e_-)\in\mathcal{D}} \max\{0, d(e,e_+) - d(e,e_-) + m\}, \tag{6}$$

where $d(\cdot)$ use l_1 distance to calculate similarity between $TG(e)$ and $TG(e_+)$(or $TG(e_-)$). m is the margin parameter.

The approach we take in our model is similar to BERT-INT [11] in that we prioritize the description of an entity over its name since the former provides more valuable information. However, our model differs in that it considers not only the text embedding of entities but also their graph embedding. In the basic model, embeddings are provided from the text-graph view, and aligned entities can be matched by calculating the embedding similarity between entities. For a given entity pair (e_i, e_j), its similarity score can be calculated as:

$$s_{e_i,e_j} = \frac{TG(e_i) \cdot TG(e_j)}{\|TG(e_i)\| \cdot \|TG(e_j)\|}. \tag{7}$$

To improve the alignment results, we have incorporated information from the neighbor view and the attribute view into the basic model. During the interaction model, we keep the parameters of the basic model fixed and use it as input.

Multi-view Interaction Model. We have already learned the structure and text information of the entities in the basic model. However, the neighbors and attributes of entities can still provide valuable information for EA. Entities with similar meanings usually have similar neighbors and attribute values. For this reason, we propose adding the neighbor and attribute views to interact with the text-graph view in the basic model, combining information from all views of an entity to complete the EA task.

We use a fixed number of candidate matching entity pairs from the basic model as input to the multi-view interaction model. The basic model filters out a significant number of mismatched entities and takes only a few pairs of entities for the interaction model to select. This not only saves a lot of computing time, but also makes better use of the basic model based on the text-graph view.

Our processing flow remains consistent for both the neighbor view and the attribute view. The only variation lies in the input data for each. The neighbor view requires a set of entities connected by the relation r to the target entity, while the attribute view needs a set of attribute values.

Take the neighbor view as an illustration, as shown in Fig. 4, we assist in matching entity pairs by comparing the similarity of neighboring entities. Given a pair of candidate (e_i, e_j), their respective embedding sets of neighboring entities $\{TG(e_i)\}_{i=1}^{|\mathcal{N}(e_i)|}$ and $\{TG(e_j)\}_{j=1}^{|\mathcal{N}(e_j)|}$ are obtained from the basic model. $|\mathcal{N}(e_i)|$ and $|\mathcal{N}(e_j)|$ denote the sets of neighboring entities for entity e_i and entity e_j, respectively. The embeddings of these entities are utilized to compute a similarity matrix $\mathbf{S}_r$ among the neighboring entities. Each element of the matrix is the similarity $s_{x,y}$ computed from Eq. (7) between the x-th neighbor of e_i and the y-th neighbor of e_j.

Meanwhile, we also consider the similarity of relations between entities and their neighbors as an influential factor in entity alignment. We use a neighboring relation similarity matrix as Mask to improve the reliability of the neighbor view. The similarity between relations is calculated similarly to the Eq. (7). It is

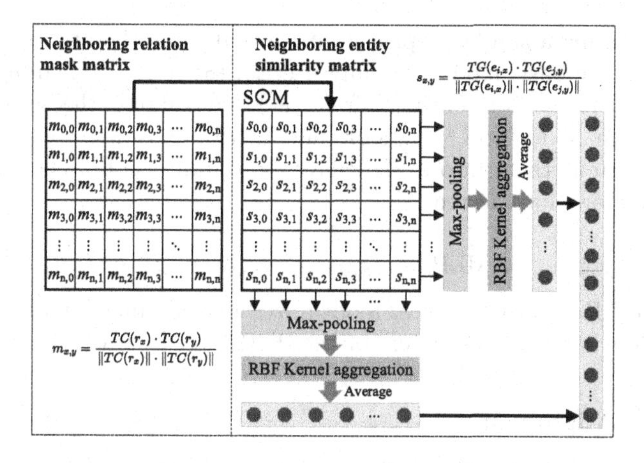

Fig. 4. Entity pair similarity matrix calculation from the neighbor view.

important to note that since the calculation of similarity between relations and attributes relies only on text similarity, as shown in Fig. 3, we only use the CLS embedding to calculate the similarity between relations/attributes:

$$TC(r) = \text{MLP}\left(\text{CLS}(r)\right) \ , m_{x,y} = \frac{TC(r_x) \cdot TC(r_y)}{\|TC(r_x)\| \cdot \|TC(r_y)\|}, \tag{8}$$

where $m_{x,y}$ denotes the similarity between the x-th relation of e_i and the y-th relation of e_j. These relation similarities can form a matrix of relation masks $\mathbf{M}$. The neighboring entity similarity matrix $\mathbf{S}_r$ and the corresponding relation mask matrix $\mathbf{M}$ are element-wise product to obtain the final similarity matrix $\mathbf{S}_r' = \mathbf{S}_r \odot \mathbf{M}$. We used two RBF kernel aggregation functions [25] to simultaneously extract similar features from the rows and columns of $\mathbf{S}_r'$. To make more accurate use of similar neighbor entity pairs, we take the maximum value of each row/column of $\mathbf{S}_r'$ as input to the RBF kernel function. That is, it only cares about how similar the most similar pairs of neighbor entities are. Thus, we can obtain the set of similarity scores for each row and column of entity e_i and entity e_j in the neighbor view $\mathbf{S}_r'$:

$$\begin{aligned}\varphi^{\mathcal{N}}(e_i, e_j) = &\left[\frac{1}{|\mathcal{N}(e_i)|} \sum_{x=1}^{|\mathcal{N}(e_i)|} \log\left(\text{RBF}_r(\mathbf{S}_{r,x}')\right)\right] \oplus \\ &\left[\frac{1}{|\mathcal{N}(e_j)|} \sum_{y=1}^{|\mathcal{N}(e_j)|} \log\left(\text{RBF}_c(\mathbf{S}_{r,y}')\right)\right],\end{aligned} \tag{9}$$

where $\text{RBF}_r(\cdot)$ aggregates the row elements of $\mathbf{S}_{r,x}'$ and $\text{RBF}_c(\cdot)$ aggregates the column elements of $\mathbf{S}_{r,y}'$. $\mathbf{S}_{r,x}'$ is the similarity matrix after row max-pooling. $\mathbf{S}_{r,y}'$ is the similarity matrix after column max-pooling.

Similarly, in the attribute view, the input triples are replaced with the attribute triples. We can obtain the set of similarity $\varphi^{\mathcal{A}}(e_i, e_j)$ for entity e_i

and entity e_j using a similar approach. Note that since the attribute values are not trained in the basic model, we only use Eq. (8) to get embeddings.

Up to now, for a given entity pair (e_i, e_j), we combine the set of similarity scores from the text-graph, neighbor, and attribute views, concatenate them together and input them into a MLP layer to obtain the final similarity scores for the three views interactions:

$$\varphi(e_i, e_j) = \text{MLP}\left([s_{e_i,e_j} \oplus \varphi^{\mathcal{N}}(e_i, e_j) \oplus \varphi^{\mathcal{A}}(e_i, e_j)]\right), \tag{10}$$

Note that the MLP layer here is still optimized by the same pairwise margin loss as in Eq. (6), while the parameters of the BERT are frozen.

4.3 Entity Alignment

For an entity e in the graph $\mathcal{G}_1$, We provide two options to find aligned entity in $\mathcal{G}_2$. The first is EA using only the text-graph view, i.e. using only the basic model. The similarity between e and $\mathcal{G}_2$ is determined using Eq. (7), and the entity with the highest similarity score is chosen as the matching entity. This approach has been able to achieve superior results compared to both the text-only and the graph-only approaches, as it leverages both the graph structure information and the text information.

If a more suitable match is required, we can select the second option. The top-K candidate matching entities are quickly found by the first scheme, and then the similarity of the three views is combined using the Eq. (10). This solution utilizes a combination of graph information, text information, neighbors, and attributes to substantially enhance EA.

5 Experiments

In this chapter, we experiment with a CNNVD-CNVD vulnerability KG and three widely used cross-linguistic knowledge graphs. Details of these datasets are shown in Table 1. The experimental results demonstrate that TG-INT achieves optimal results on all datasets. Our experiments were run on a GeForce GTX 2080Ti device with 12GB of RAM.

Table 1. Datasets.

Dataset	CVD19-6K		DBP15K_{ZH_EN}		DBP15K_{JA_EN}		DBP15K_{FR_EN}	
	CNNVD	CNVD	Chinese	English	Japanese	English	French	English
#Ent	28549	12310	19388	19572	19814	19780	19661	19993
#Rel	16	10	1701	1323	1299	1153	903	1208
#Attr	4	6	8113	7173	5882	6066	4547	6422
#Rel.Tri	172047	26584	70414	95142	77214	93484	105998	105722
#Attr.Tri	9941	16986	379684	567755	354619	497230	528665	576543
#Link	2286	2286	4500	4500	4500	4500	4500	4500
#Test Link	3895	3895	10500	10500	10500	10500	10500	10500

5.1 Datasets and Implementation

Vulnerability Database and DBP15K. We obtained public vulnerability data from the China National Vulnerability Database of Information Security (CNNVD)[2] and the China National Vulnerability Database (CNVD)[3], and defined their ontologies. For CNNVD, we defined 13 types of entities, 16 types of relations and 4 types of attributes. For CNVD, we defined 8 types of entities, 10 types of relations and 6 types of attributes. Notably, CNNVD's ontologies mostly include all of CNVD's ontologies, which allowed us to complement each other and obtain more comprehensive information.

Considering the limitation of GPU memory, we selected vulnerability data with shared CVE numbers between CNNVD and CNVD from September–December 2019 to create the dataset for the EA experiments. To simulate real-life situations where CVE numbers are missing, we randomly selected one-third of the data as the test set and removed the CVE number of these vulnerabilities in datasets. For vulnerabilities that have already been assigned a CVE number, we selected only the unambiguous entity pairs as seed for our EA experiments. We excluded vulnerabilities with retained CVE numbers from our test dataset since they are easier to align. Moreover, we did not include entities that can be aligned easily by text, such as time and threat types.

Despite having fewer relations and attributes than general cross-linguistic KGs, the vulnerability KG still has a sufficient number of relation triples because of the presence of similar entities. For instance, a vulnerability of a manufacturer may impact all of its products, which may share similar characteristics. Apart from the vulnerability dataset, we have conducted experiments on DBP15K datasets to test the versatility of our solution.

Parameter Settings. In our experiments, the embedding dimension of QuatAE is 800. Since QuatAE is a joint representation of an embedding by four vectors, we concatenate four 200-dimension vectors together. The hyperparameters λ_1 and λ_2 take values in $[0.1, 0.2, 0.5]$. The dimension of the BERT CLS embedding is 768. We use MLPs of 800 and 300 dimensions for $G(e)$ and $TG(e)$ respectively in Eq. (5). The 300-dimension MLP and 11 plus 1-dimension MLP are used in Eq. (8) and Eq. (10), respectively. In the Eq. (9), we use 20 semantic matching kernels [11]. We set both the maximum number of neighbors and the number of attributes to 50, and the number of candidates selected from the basic model of the text-graph view to 50. We found that the top-50 candidates contained 99% of the true alignment pairs in all datasets.

5.2 Experiments

Overall Performance. To validate the performance of the TG-INT model, 12 current models with high influence or good performance were selected as the

[2] https://www.cnnvd.org.cn.
[3] https://www.cnvd.org.cn.

Table 2. Overall entity alignment experimental results. (* indicates result from our experiments)

Model	CVD19-6K*			DBP15K$_{ZH_EN}$			DBP15K$_{JA_EN}$			DBP15K$_{FR_EN}$		
	Hit@1	H@10	MRR	Hit@1	H@10	MRR	Hit@1	H@10	MRR	Hit@1	H@10	MRR
Only use graph structures												
MTransE (2017) [14]	0.068	0.167	0.101	0.308	0.614	0.364	0.279	0.575	0.349	0.244	0.556	0.335
IPTransE (2017) [15]	0.053	0.138	0.082	0.406	0.735	0.516	0.367	0.693	0.474	0.333	0.685	0.451
BootEA (2018) [16]	0.112	0.328	0.182	0.629	0.848	0.703	0.622	0.854	0.701	0.653	0.874	0.731
NAEA (2019) [17]	–	–	–	0.650	0.867	0.720	0.641	0.873	0.718	0.673	0.894	0.752
AliNet (2020) [18]	0.085	0.181	0.120	0.539	0.826	0.628	0.549	0.831	0.645	0.552	0.852	0.657
Combine graph structures and side information.												
JAPE (2017) [19]	0.051	0.123	0.075	0.412	0.745	0.490	0.363	0.685	0.476	0.324	0.667	0.430
GCN-Align (2018) [21]	0.077	0.246	0.132	0.413	0.744	0.549	0.399	0.745	0.546	0.373	0.745	0.532
MultiKE (2019) [20]	0.144	0.303	0.199	0.509	0.576	0.532	0.393	0.489	0.426	0.639	0.712	0.665
RDGCN (2019) [22]	0.198	0.382	0.263	0.708	0.846	0.746	0.767	0.895	0.812	0.886	0.957	0.911
BERT-INT (2020) [11]*	<u>0.743</u>	<u>0.943</u>	<u>0.814</u>	<u>0.960</u>	<u>0.982</u>	<u>0.969</u>	<u>0.959</u>	<u>0.984</u>	<u>0.970</u>	<u>0.991</u>	<u>0.997</u>	<u>0.994</u>
RPR-RHGT(2022) [23]	–	–	–	0.693	–	0.754	0.886	–	0.912	0.889	–	0.919
UPLR (2022) [12]	0.514	0.676	0.570	0.902	0.970	0.927	0.912	0.978	0.937	0.967	0.994	0.974
TG-INT	**0.781**	**0.958**	**0.845**	**0.967**	**0.989**	**0.976**	**0.966**	**0.989**	**0.975**	**0.993**	**0.998**	**0.995**

baseline. These include MTransE [14], IPTransE [15], BootEA [16], NAEA [17] and AliNet [18] which use only graph embeddings. Also include JAPE [19], GCN-Align [21], MultiKE [20], BERT-INT [11], RPR-RHGT [23] and UPLR [12] which use various side information. To facilitate comparison experiments, we chose Hit@k (k=1,5,10), MRR (Mean Reciprocal Rank), and MR (Mean Rank) as the metrics. Hit@k indicates the percentage of correctly matched entity pairs ranked in the top k, and Hit@1 is the accuracy. Both Hit@k and MRR have higher values to indicate better performance, while MR is the opposite.

The results of the TG-INT experiments on the CVD19-6K and DBP15K datasets are shown in Table 2. The number in **bold** denotes the best results of all models and second best results are <u>underline</u>. We conducted our experiments on CVD19-6K using OpenEA [26] and also derived the results of the BERT-INT from our experiments, using the same parameters and environment for comparison. The rest is taken from the original papers. The experimental results indicate that the model based solely on graph embeddings performs poorly on the vulnerability KG due to limited interaction between entities and relations, resulting in less semantic information compared to a general KG. However, the alignment is greatly improved by models that incorporate side information, such as text. It should be noted that while alignment based solely on graph embeddings is inadequate, they can still provide a useful reference. Our TG-INT model outperforms all baseline models, and the inclusion of graph structure information performs better than BERT-INT, which relies solely on text information.

Comparative Experiments on CVD19-6K. The vulnerability KG contains a large number of entities that are very similar to one another, and the relations connecting these entities are not very diverse. This not only makes entity alignment methods based purely on graph structure ineffective, but also hinders traditional methods based on text matching.

We utilized Jaccard and edit distance algorithms to compute the similarity, and the results are shown in Table 3. As depicted, the character similarity-based technique outperforms the graph embedding-based approach. (char) indicates that all entities are compared based on characters, whereas name/des implies that for entities such as vulnerability numbers, their names or descriptions are employed instead of the numbers. Although the edit distance approach is simple yet effective, it is still not entirely satisfactory.

We also conducted EA experiments using graph embeddings, and evaluated them using cosine similarity. In addition, we employed the RA approach, which involves sharing relation embeddings between two graphs. Our proposed method, QuatAE, outperforms QuatE by 1.8% in terms of Hit@1. Furthermore, QuatAE achieves the best performance in CVD19-6K compared to the other alignment models in Table 2 that use only graph embeddings.

Table 3. Entity alignment based text-only and graph-only.

Model	CVD19-6K				
	Hit@1	H@5	H@10	MR	MRR
Only use text information.					
Jaccard (char)	0.260	0.349	0.383	871	0.304
Jaccard (char/name)	0.428	0.526	0.560	813	0.476
Jaccard (char/des)	0.532	0.654	0.694	450	0.590
Edit distance (char)	0.314	0.404	0.439	994	0.355
Edit distance (char/name)	0.524	0.663	0.710	**418**	0.587
Edit distance (char/des)	**0.583**	**0.720**	**0.764**	421	**0.644**
Only use graph embeddings.					
TransE (RA)	0.068	0.094	0.125	**1298**	0.090
QuatE (RA)	0.228	0.300	0.317	2491	0.263
QuatAE	**0.246**	**0.313**	**0.333**	1832	**0.280**

Ablation Study. We conducted ablation experiments on TG-INT and its basic model, and the results are presented in Table 4. The introduction of graph embeddings (GE) significantly improved the alignment of CVD19-6K and DBP15K datasets in the basic model, with an accuracy improvement of 0.7%–4.3%. After incorporating multi-view interaction, the effect of graph embeddings remained significant, resulting in a 3.8% accuracy improvement on CVD19-6K and outperforming the case without graph embeddings on the DBP15K datasets. The effect of introducing graph embeddings on CVD19-6K is illustrated in Fig. 5.

Table 4. Ablation study on CVD19-6K and DBP15K.

Model	CVD19-6K			DBP15K$_{ZH_EN}$			DBP15K$_{JA_EN}$			DBP15K$_{FR_EN}$		
	Hit@1	Hit@10	MRR	Hit@1	Hit@10	MRR	Hit@1	Hit@10	MRR	Hit@1	Hit@10	MRR
Basic Model (w/o GE)	0.692	0.929	0.775	0.774	0.951	0.841	0.799	0.954	0.846	0.955	0.993	0.971
Basic Model	**0.714**	**0.950**	**0.798**	**0.796**	**0.959**	**0.856**	**0.842**	**0.971**	**0.891**	**0.962**	**0.996**	**0.979**
TG-INT (w/o GE)	0.743	0.943	0.814	0.960	0.982	0.969	0.959	0.984	0.970	0.991	0.997	0.994
TG-INT	**0.781**	**0.958**	**0.845**	**0.967**	**0.989**	**0.976**	**0.966**	**0.989**	**0.975**	**0.993**	**0.998**	**0.995**

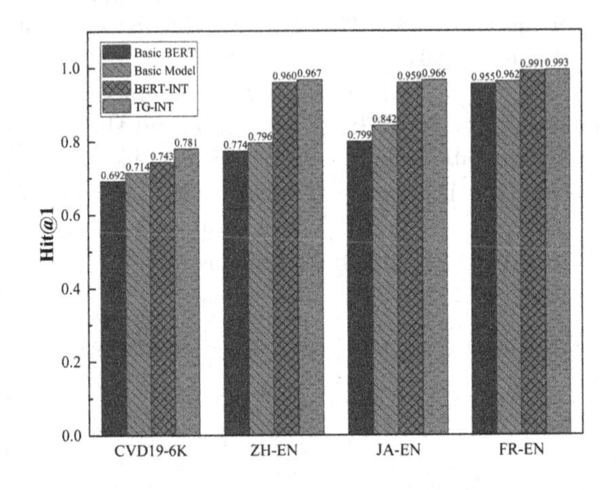

Fig. 5. Entity alignment with the addition of graph embeddings.

6 Conclusion

In this paper, we propose a model called TG-INT that uses a multi-view approach to align entities in vulnerability KGs by combining text-graph, neighbor, and attribute views. To achieve this, we use QuatAE, a graph embedding method, and a straightforward technique to combine graph and text embeddings. Our experiments show that adding graph embeddings to the entity alignment process provides valuable information without introducing significant noise. Our approach works not only for vulnerability KGs but also for general knowledge graphs, resulting in improved performance.

Acknowledgments. This work is funded by the National Key Research and Development Plan (Grant No. 2021YFB3101704), the National Natural Science Foundation of China (No. 62272119, 62072130, U20B2046), the Guangdong Basic and Applied Basic Research Foundation (No.2023A1515030142, 2020A15150104 50, 2021A1515012307), Guangdong Province Universities and Colleges Pearl River Scholar Funded Scheme (2019), and Guangdong Higher Education Innovation Group (No. 2020KCXTD007), Guangzhou Higher Education Innovation Group (No. 202032854), Consulting project of Chinese Academy of Engineering (2022-JB-04-05, 2021-HYZD-8-3), the Eleventh Key Project of Education Teaching Reform in Guangzhou Municipality.

References

1. Huang, X., Zhang, J., Li, D., et al.: Knowledge graph embedding based question answering. In: WSDM, pp. 105–113 (2019)
2. Dimitriadis, I., Poiitis, M., Faloutsos, C., et al.: TG-OUT: temporal outlier patterns detection in Twitter attribute induced graphs. World Wide Web **25**, 2429–2453 (2022)
3. Suchanek, F.M., Abiteboul, S., Senellart, P.: Paris: probabilistic alignment of relations, instances, and schema. In: VLDB, pp. 157–168 (2012)
4. Sassi, S., Tissaoui, A., Chbeir, R.: LEOnto+: a scalable ontology enrichment approach. World Wide Web **25**, 2347–2378 (2022)
5. Bordes, A., Usunier, N., Garcia-Duran, A., et al.: Translating embeddings for modeling multi-relational data. In: NIPS, pp. 2787–2795 (2013)
6. Zhang, S., Tay, Y., Yao, L., et al.: Quaternion knowledge graph embeddings. In: NIPS (2019)
7. Schlichtkrull, M., Kipf, T.N., Bloem, P., van den Berg, R., Titov, I., Welling, M.: Modeling relational data with graph convolutional networks. In: Gangemi, A., et al. (eds.) ESWC 2018. LNCS, vol. 10843, pp. 593–607. Springer, Cham (2018). https://doi.org/10.1007/978-3-319-93417-4_38
8. Li, R., Cao, Y., Zhu, Q., et al.: How does knowledge graph embedding extrapolate to unseen data: a semantic evidence view. In: AAAI, pp. 5781–5791 (2022)
9. Wang, H., Lian, D., Zhang, Y., et al.: Binarized graph neural network. World Wide Web **24**, 825–848 (2021)
10. Zhang, Z., Chen, J., Chen, X., et al.: An industry evaluation of embedding-based entity alignment. In: COLING, pp. 179–189 (2020)
11. Tang, X., Zhang, J., Chen, B., et al.: BERT-INT: a BERT-based interaction model for knowledge graph alignment. In: IJCAI (2020)
12. Li, J., Song, D.: Uncertainty-aware pseudo label refinery for entity alignment. In: Proceedings of the ACM Web Conference, pp. 829–837 (2022)
13. Liu, X., Hong, H., Wang, X., et al.: SelfKG: self-supervised entity alignment in knowledge graphs. In: Proceedings of the ACM Web Conference, pp. 860–870 (2022)
14. Chen, M., Tian, Y., Yang, M., et al.: Multilingual knowledge graph embeddings for cross-lingual knowledge alignment. In: IJCAI, pp. 1511–1517 (2017)
15. Zhu, H., Xie, R., Liu, Z., et al.: Iterative entity alignment via joint knowledge embeddings. In: IJCAI, pp. 4258–4264 (2017)
16. Sun, Z., Hu, W., Zhang, Q., et al.: Bootstrapping entity alignment with knowledge graph embedding. In: IJCAI, pp. 4396–4402 (2018)
17. Zhu, Q., Zhou, X., Wu, J., et al.: Neighborhood-aware attentional representation for multilingual knowledge graphs. In: IJCAI, pp. 3231–3237 (2019)
18. Sun, Z., Wang, C., Hu, W., et al.: Knowledge graph alignment network with gated multi-hop neighborhood aggregation. In: AAAI (2020)
19. Sun, Z., Hu, W., Li, C., et al.: Cross-lingual entity alignment via joint attribute preserving embedding. In: ISWC, pp. 628–644 (2017)
20. Zhang, Q., Sun, Z., Hu, W., et al.: Multi-view knowledge graph embedding for entity alignment. In: IJCAI, pp. 5429–5435 (2019)
21. Wang, Z., Lv, Q., Lan, X., et al.: Cross-lingual knowledge graph alignment via graph convolutional networks. In: EMNLP, pp. 349–357 (2018)
22. Wu, Y., Liu, X., Feng, Y., et al.: Relation-aware entity alignment for heterogeneous knowledge graphs. In: IJCAI, pp. 5278–5284 (2019)

23. Cai, W., Ma, W., Zhan, J., et al.: Entity alignment with reliable path reasoning and relation-aware heterogeneous graph transformer. In: IJCAI, pp. 1930–1937 (2022)
24. Huang, H., Li, C., Peng, X., et al.: Cross-knowledge-graph entity alignment via relation prediction. Knowl. Based Syst. **240**, 107813 (2022)
25. Xiong, C., Dai, Z., Callan, J., et al.: End-to-end neural ad-hoc ranking with kernel pooling. In: SIGIR, pp. 55–64 (2017)
26. Sun, Z., Zhang, Q., Hu, W., et al.: A benchmarking study of embedding-based entity alignment for knowledge graphs. In: Proceedings of the VLDB Endowment, pp. 2326–2340 (2020)

Author Index

X. Song et al. (Eds.): APWeb-WAIM 2023, LNCS 14332, pp. 517–518, 2024.
https://doi.org/10.1007/978-981-97-2390-4

Printed in the United States
by Baker & Taylor Publisher Services